Differential Geometry
of Manifolds

Second Edition

Textbooks in Mathematics

Series editors:

Al Boggess and Ken Rosen

CRYPTOGRAPHY: THEORY AND PRACTICE, FOURTH EDITION

Douglas R. Stinson and Maura B. Paterson

GRAPH THEORY AND ITS APPLICATIONS, THIRD EDITION

Jonathan L. Gross, Jay Yellen and Mark Anderson

COMPLEX VARIABLES: A PHYSICAL APPROACH WITH APPLICATIONS, SECOND EDITION

Steven G. Krantz

GAME THEORY: A MODELING APPROACH

Richard Alan Gillman and David Housman

FORMAL METHODS IN COMPUTER SCIENCE

Jiacun Wang and William Tepfenhart

AN ELEMENTARY TRANSITION TO ABSTRACT MATHEMATICS

Gove Effinger and Gary L. Mullen

ORDINARY DIFFERENTIAL EQUATIONS: AN INTRODUCTION TO THE FUNDAMENTALS, SECOND EDITION

Kenneth B. Howell

SPHERICAL GEOMETRY AND ITS APPLICATIONS

Marshall A. Whittlesey

COMPUTATIONAL PARTIAL DIFFERENTIAL PARTIAL EQUATIONS USING MATLAB®, SECOND EDITION

Jichun Li and Yi-Tung Chen

AN INTRODUCTION TO MATHEMATICAL PROOFS

Nicholas A. Loehr

DIFFERENTIAL GEOMETRY OF MANIFOLDS, SECOND EDITION

Stephen T. Lovett

MATHEMATICAL MODELING WITH EXCEL

Brian Albright and William P. Fox

THE SHAPE OF SPACE, THIRD EDITION

Jeffrey R. Weeks

CHROMATIC GRAPH THEORY, SECOND EDITION

Gary Chartrand and Ping Zhang

PARTIAL DIFFERENTIAL EQUATIONS: ANALYTICAL METHODS AND APPLICATIONS

Victor Henner, Tatyana Belozerova, and Alexander Nepomnyashchy

Advanced Problem Solving Using Maple: Applied Mathematics, Operation Research, Business Analytics, and Decision Analysis

William P. Fox and William C. Bauldry

DIFFERENTIAL EQUATIONS: A MODERN APPROACH WITH WAVELETS

Steven G. Krantz

https://www.crcpress.com/Textbooks-in-Mathematics/book-series/CANDHTEXBOOMTH

Differential Geometry
of Manifolds

Second Edition

Stephen Lovett

CRC Press
Taylor & Francis Group
Boca Raton London New York

CRC Press is an imprint of the
Taylor & Francis Group, an **informa** business

CRC Press
Taylor & Francis Group
6000 Broken Sound Parkway NW, Suite 300
Boca Raton, FL 33487-2742

First issued in paperback 2022

© 2020 by Taylor & Francis Group, LLC
CRC Press is an imprint of Taylor & Francis Group, an Informa business

No claim to original U.S. Government works

ISBN 13: 978-1-03-247490-8 (pbk)
ISBN 13: 978-0-367-18046-1 (hbk)

DOI: 10.1201/9780429059292

Visit the Taylor & Francis Web site at
http://www.taylorandfrancis.com

and the CRC Press Web site at
http://www.crcpress.com

Contents

Preface

Purpose of this Book

This book is the second in a pair of books which together are intended to bring the reader through classical differential geometry into the modern formulation of the differential geometry of manifolds. The first book in the pair, by Banchoff and Lovett, entitled *Differential Geometry of Curves and Surfaces* [6], introduces the classical theory of curves and surfaces, only assuming the calculus sequence and linear algebra. This book continues the development of differential geometry by studying manifolds – the natural generalization of regular curves and surfaces to higher dimensions. Though a background course in analysis is useful for this book, we have provided all the necessary analysis results in the text. Though [6] provides many examples of one- and two-dimensional manifolds that lend themselves well to visualization, this book does not rely on [6] and can be read independently.

Taken on its own, this book provides an introduction to differentiable manifolds, geared toward advanced undergraduate or beginning graduate readers in mathematics, retaining a view toward applications in physics. For readers primarily interested in physics, this book may fill a gap between the geometry typically offered in undergraduate programs and that expected in physics graduate programs. For example, some graduate programs in physics first introduce electromagnetism in the context of a manifold. The student who is unaccustomed to the formalism of manifolds may be lost in the notation at worst or, at best, be unaware of how to do explicit calculations on manifolds.

What is Differential Geometry?

Differential geometry studies properties of and analysis on curves, surfaces, and higher dimensional spaces using tools from calculus and linear algebra. Just as the introduction of calculus expands the descriptive and predictive abilities of nearly every scientific field, so the use of calculus in geometry brings about avenues of inquiry that extend far beyond classical geometry.

Though differential geometry does not possess the same restrictions as Euclidean geometry on what types of objects it studies, not every conceivable set of points falls within the purview of differential geometry. One of the underlying themes of this book is the development and description of the types of geometric sets on which it is possible to "do calculus." This leads to the definition of differentiable manifolds. A second, and somewhat obvious, theme is how to actually do calculus (measure rates of change of functions or interdependent variables) on manifolds. A third general theme is how to "do geometry" (measure distances, areas and angles) on such geometric objects. This theme leads us to the notion of a Riemannian manifold.

Applications of differential geometry outside of mathematics first arise in mechanics in the study of the dynamics of a moving particle or system of particles. The study of inertial frames is in common to both physics and differential geometry. Most importantly, however, differential geometry is necessary to study physical systems that involve functions on curved spaces. For example, just to make sense of directional derivatives of the surface temperature at a point on the earth (a sphere) requires analysis on manifolds. The study of mechanics and electromagnetism on a curved surface also requires analysis on a manifold. Finally, arguably the most revolutionary application of differential geometry to physics came from Einstein's theory of general relativity, in which spacetime becomes curved in the presence of mass/energy.

Organization of Topics

A typical calculus sequence analyzes one variable real functions ($\mathbb{R} \to \mathbb{R}$), parametric curves ($\mathbb{R} \to \mathbb{R}^n$), multivariable functions ($\mathbb{R}^n \to \mathbb{R}$) and vector fields ($\mathbb{R}^2 \to \mathbb{R}^2$ or $\mathbb{R}^3 \to \mathbb{R}^3$). This does not quite reach the full generality necessary for the definition of manifolds. Chapter 1 presents the analysis of functions $f : \mathbb{R}^n \to \mathbb{R}^m$ for any positive integers n and m.

Chapter 2 discusses the concept and calculus of variable frames. Variable frames arise naturally when using curvilinear coordinates, in the differential geometry of curves (see Chapters 1, 3, and 8 of [5]), and, in physics, in the mechanics of a moving particle. In special relativity, of critical importance are *momentarily comoving reference frames* (MCRFs), which are yet other examples of variable frames. Implicit in our treatment of variable frames is a view toward Lie algebras. However, to retain the chosen level of this book, we do not develop that theory here.

Chapter 3 defines the category of differentiable manifolds. Manifolds serve as the appropriate and most complete generalization to higher dimensions of regular curves and regular surfaces. The chapter also introduces the definition for the tangent space on a manifold and attempts to provide the underlying intuition behind the abstract definitions.

Before jumping into the analysis on manifolds, Chapter 4 introduces some necessary background in multilinear algebra. We focus on bilinear forms, dual spaces, automorphisms of nondegenerate bilinear forms, and tensor products of vector spaces.

Chapter 5 then develops the analysis on differentiable manifolds, including the differentials of functions between manifolds, vector fields, differential forms, and integration.

Chapter 6 introduces Riemannian geometry without any pretention of being comprehensive. One can easily take an entire course on Riemannian geometry, the proper context in which one can do both calculus and geometry on a curved space. The chapter introduces the notions of metrics, connections, geodesics, parallel transport and the curvature tensor.

Having developed the technical machinery of manifolds, in Chapter 7 we apply our the theory to a few areas in physics. We consider the Hamiltonian formulation of dynamics, with a view toward symplectic manifolds; the tensorial formulation of electromagnetism; a few geometric concepts involved in string theory, namely the properties of the world sheet which describes a string moving in a Minkowski space; and some fundamental concepts in general relativity.

In order to be rigorous and still only require the standard core in most undergraduate math programs, three appendices provide any necessary background from topology, calculus of variations, and a few additional results from multilinear algebra. The reader without any background in analysis would be served by consulting Appendix A on point set topology before Chapter 3.

A Comment on Using the Book

Because of the intended purpose of the book, it can serve well either as a textbook or for self-study. The conversational style attempts to introduce new concepts in an intuitive way, explaining why we formulate certain definitions as we do. As a mathematics text, this book provides proofs or references for all theorems. On the other hand, this book does not supply all the physical theory and discussion behind the all the application topics we broach.

Each section concludes with an ample collection of exercises. Problems marked with (*) indicate difficulty which may be related to technical ability, insight, or length.

As mentioned above, this book only assumes prior knowledge of multivariable calculus and linear algebra. A few key results presented in this textbook rely on theorems from the theory of differential equations but either the calculations are all spelled out or a reference to the appropriate theorem has been provided. Therefore, except in the case of exercises about geodesics, experience with differential equations is helpful though not necessary.

From the perspective of a faculty person using this as a course textbook, the author intends every section to correspond to one 60-minute lecture period. With the assumption of a 16-week semester, a course using this book should find the time to cover all main sections and the appendices on topology. If a faculty knows that his or her students have enough analysis or topology, Chapter 1 or Appendix A can be skipped.

Notation

It has been said jokingly that "differential geometry is the study of things that are invariant under a change of notation." A quick perusal of the literature on differential geometry shows that mathematicians and physicists usually present topics in this field in a variety of different ways. One could argue that notational differences have contributed to a communication gap between mathematicians and physicists. In addition, the classical and modern formulations of many differential geometric concepts vary significantly. Whenever different notations or modes of presentation exist for a topic (e.g. differentials, metric tensor, tensor fields), this book attempts to provide an explicit coordination between the notation variances.

As a comment on vector and tensor notation, this book consistently uses the following conventions. A vector or vector function in a Euclidean vector space is denoted by $\vec{v}$, $\vec{X}(t)$ or $\vec{X}(u,v)$. Vectors in an arbitrary vector space, curves on manifolds, tangent vectors to a manifold, vector fields or tensor fields have no over-right-arrow designation and are written, for example, as v, γ, X or T. A fair number of physics texts use a bold font like $\mathbf{g}$ or $\mathbf{A}$ to indicate tensors or tensor fields. Therefore, when discussing tensors taken from a physics context, we also use that notation.

Different texts also employ a variety of notations to express the coordinates of a vector with respect to a given basis. In this textbook, we regularly use the following notation. If V is a vector space with an ordered basis $\mathcal{B} = (e_1, e_2, \ldots, e_n)$, then the coordinates of a vector $v \in V$ with respect to $\mathcal{B}$ are denoted by $[v]_\mathcal{B}$. More precisely,

$$[v]_\mathcal{B} = \begin{pmatrix} v_1 \\ v_2 \\ \vdots \\ v_n \end{pmatrix} \quad \text{if and only if} \quad v = v_1 e_1 + v_2 e_2 + \cdots + v_n e_n.$$

As a point of precision, when discussing coordinates we must use an ordered basis since the order of vectors in the n-tuple matters for associating the correct coordinate.

Beginning in Chapter 2, we switch from this typical notation to writing the indices of coordinates in a superscript. So we will refer to the coordinates of $v \in V$ with respect to $\mathcal{B}$ as (v^i). This switch in notation from that developed in introductory linear algebra courses is standard in differential geometry and multilinear algebra. The reason for this switch is explained fully in Section 4.1. In this context, the superscript is not a power but an index. This modified notation is particularly useful to recognize the difference between a (contravariant) vector and a dual vector (also called covector) and then to use Einstein's summation convention. This new notation is standard in differential geometry, including applications in physics.

For linear transformations and their associated matrices, this book uses the following convention. Suppose also that W is a vector space with a basis $\mathcal{B}'$ and that T is a linear transformation $T : V \to W$. Then we denote by $[T]_\mathcal{B}^{\mathcal{B}'}$ the matrix representing T with respect to the basis $\mathcal{B}$ on V and the basis $\mathcal{B}'$ on W. We recall that this matrix is defined as the matrix such that

$$[T(v)]_{\mathcal{B}'} = [T]_{\mathcal{B}}^{\mathcal{B}'} [v]_{\mathcal{B}}$$

for all $v \in V$.

The authors of [6] chose the following notations for certain specific objects of interest in differential geometry of curves and surfaces. Often γ indicates a curve parametrized by $\vec{X}(t)$ while writing $\vec{X}(t) = \vec{X}(u(t), v(t))$ indicates a curve on a surface. The unit tangent and the binormal vectors of a curve in space are written in the standard notation $\vec{T}(t)$ and $\vec{B}(t)$ but the principal normal is written $\vec{P}(t)$, reserving $\vec{N}(t)$ to refer to the unit normal vector to a curve on a surface. For a plane curve, $\vec{U}(t)$ is the vector obtained by rotating $\vec{T}(t)$ by a positive angle of $\pi/2$. Furthermore, we denote by $\kappa_g(t)$ the curvature of a plane curve to identify it as the geodesic curvature of a curve on a surface. When these concepts occur in this text, we use the same conventions as [6].

Occasionally, there arise irreconcilable discrepancies in habits of notation, e.g., how to place the signs on a Minkowski metric, how one defines θ and ϕ in spherical coordinates, what units to use in electromagnetism, etc. In these instances the text makes a choice that best suits its purpose and philosophical leanings, and indicates commonly used alternatives.

Changes in the Second Edition

The second edition of this text arose from feedback from students and faculty using this book and the author seeing room for improvement of his personal experience teaching from it.

As a first major change to benefit faculty using this book, the second edition commits that each section should correspond to one 60-minute lecture period. Consequently, some of the sections in the first edition were split in two. Part of the reorganization required the creation of a few new sections to cover topics, which the author felt had been too compressed in the first edition, e.g., orientability of manifolds, the Lie derivative of vector fields, applications of integration.

The centrality of multilinear algebra in this text's approach encouraged us to take that content out of the appendices in the first edition to become Chapter 4 in the current edition. This may feel like an interlude between Chapter 3, which defines manifolds and differentiable maps between them, and Chapter 5, which studies the analysis on manifolds. Nonetheless, hopefully the location on this content makes sense since it first becomes necessary in Chapter 5. Having a regular chapter on multilinear algebra allows for a more natural introduction to tensors and the notation for tensor component notation.

Woven throughout, the second edition attempts to improve the presentation style and better foreshadow certain topics. For example, Equation (2.11) about how to decompose the partial derivatives in a frame of vector fields augurs the definition of a connection on a manifold.

Most of the exercises remained the same, though we improved the statements of some and modified the challenge level of the computations for others. In addition, we added a few new interesting problems.

Acknowledgments

I would first like to thank Thomas Banchoff my teacher, mentor and friend. After one class, he invited me to join his team of students on developing electronic books for differential geometry and multivariable calculus. Despite ultimately specializing in algebra, the exciting projects he led and his inspiring course in differential geometry instilled in me a passion for differential geometry. His ability to introduce differential geometry as a visually stimulating and mathematically interesting topic served as one of my personal motivations for writing this book.

I am grateful to the students and former colleagues at Eastern Nazarene College. In particular I would like to acknowledge the undergraduate students who served as a sounding board for the first draft of this manuscript: Luke Cochran, David Constantine, Joseph Cox, Stephen Mapes, and Chris Young. Special thanks are due to my colleagues Karl Giberson, Lee Hammerstrom and John Free. In addition, I am indebted to Ellie Waal who helped with editing and index creation.

The continued support from my colleagues at Wheaton College made writing the first edition of this text a gratifying project. In particular, I must thank Terry Perciante, Chair of the Department of Mathematics and Computer Science, for his enthusiasm and his interest. I am indebted to Dorothy Chapell, Dean of the Natural and Social Sciences, and to Stanton Jones, Provost of the College, for their encouragement and for a grant which freed up my time to finish writing. I am also grateful to Thomas VanDrunen and Darren Craig.

In preparation for the second edition, I need to thank Cole Adams, Dave Broaddus, Matthew McMillan, and Edwin Townsend for insightful feedback on the first edition. Their comments on exercises and content significantly improved on the first edition.

Finally, I am always grateful to my wife Carla Favreau Lovett and my daughter Anne. While I was absorbed in writing the first edition, they braved the significant time commitment and encouraged me at every step. They also continue to kindly put up with my occasional geometry musings such as how to see the Gaussian curvature in the reflection of the Cloud Gate in Chicago.

CHAPTER 1

Analysis of Multivariable Functions

Manifolds provide a generalization to the concept of a curve or a surface, objects introduced in the usual calculus sequence. Parametrized curves into $\mathbb{R}^n$ are continuous functions from an interval of $\mathbb{R}$ to $\mathbb{R}^n$; parametrized surfaces in $\mathbb{R}^3$ involve continuous functions from $\mathbb{R}^2$ to $\mathbb{R}^3$. In order to generalize the study of curves and surfaces to the theory of manifolds, we need a solid foundation in the analysis of multivariable functions $f : \mathbb{R}^n \to \mathbb{R}^m$.

1.1 Functions from $\mathbb{R}^n$ to $\mathbb{R}^m$

Let U be a subset of $\mathbb{R}^n$ and let $f : U \to \mathbb{R}^m$ be a function from U to $\mathbb{R}^m$. Writing the input variable as

$$\vec{x} = (x_1, x_2, \ldots, x_n),$$

we denote the output assigned to $\vec{x}$ by $f(\vec{x})$ or $f(x_1, \ldots, x_n)$. Since the codomain of f is $\mathbb{R}^m$, the images of f are m-tuples so we can write

$$\begin{aligned} f(\vec{x}) &= (f_1(\vec{x}), f_2(\vec{x}), \ldots, f_m(\vec{x})) \\ &= (f_1(x_1, x_2, \ldots, x_n), f_2(x_1, x_2, \ldots, x_n), \ldots, f_m(x_1, x_2, \ldots, x_n)) . \end{aligned}$$

The functions $f_i : U \to \mathbb{R}$, for $i = 1, 2, \ldots, m$, are called the *component functions* of f.

We sometimes use the notation $\vec{f}(\vec{x})$ to emphasize the fact that the codomain $\mathbb{R}^m$ is a vector space and that any operation on m-dimensional vectors is permitted on functions $\vec{f} : \mathbb{R}^n \to \mathbb{R}^m$. Therefore, some authors call such functions vector functions of a vector variable.

In any Euclidean space $\mathbb{R}^n$, the *standard basis* is the set of vectors written as

$\{\vec{e}_1, \vec{e}_2, \ldots, \vec{e}_n\}$, where

$$\vec{e}_i = \begin{pmatrix} 0 \\ \vdots \\ 1 \\ \vdots \\ 0 \end{pmatrix}$$

with the only nonzero entry 1 occurring in the ith coordinate. If no basis is explicitly specified for $\mathbb{R}^n$, then it is assumed that one uses the standard basis.

At this point, a remark is in order concerning the differences in notations between calculus and linear algebra. In calculus, one usually denotes an element of $\mathbb{R}^n$ as an n-tuple and writes this element on one line as $(x_1, x_2, \ldots, x_n)$. On the other hand, in order to reconcile vector notation with the usual manner we multiply a matrix by a vector, in linear algebra we denote an element of $\mathbb{R}^n$ as a column vector

$$\vec{x} = \begin{pmatrix} x_1 \\ x_2 \\ \vdots \\ x_n \end{pmatrix}.$$

At first pass, we might consider these differences of notation as an unfortunate result of history. However, the difference between column vectors and row vectors is not a mere variance of notation: one represents the coordinates of an element in a vector space V with respect to some basis, while the other represents the coordinates of an element in the dual vector space V^*, a concept which we develop later. In the rest of this book, we will write the components of a vector function on one line as per the n-tuple notation, but whenever a vector or vector function appears in a linear algebraic context, we write it as a column vector.

In the typical calculus sequence, we encounter vector functions or vector-valued functions in the following contexts.

Example 1.1.1 (Curves in $\mathbb{R}^n$). A parametrized curve into n-dimensional space is a continuous function $\vec{x} : I \to \mathbb{R}^n$, where I is some interval of $\mathbb{R}$. Parametrized curves are vector functions of a single variable. We can view the independent variable as coming from a one-dimensional real vector space.

Example 1.1.2 (Nonlinear Coordinate Changes). A general change of coordinates in $\mathbb{R}^2$ is a function $F : U \to \mathbb{R}^2$, where U is the subset of $\mathbb{R}^2$ in which the coordinates are defined. For example, the change from polar coordinates to Cartesian coordinates is given by the function $F : \mathbb{R}^2 \to \mathbb{R}^2$ defined by

$$F(r, \theta) = (r \cos \theta, r \sin \theta).$$

Example 1.1.3. In a multivariable calculus, we encounter functions $F : \mathbb{R}^n \to \mathbb{R}$, written as $F(x_1, x_2, \ldots, x_n)$. All such functions are just examples of vector functions of a vector variable with a codomain of $\mathbb{R}$.

Example 1.1.4. As an example of a function from $\mathbb{R}^2$ to $\mathbb{R}^3$, consider the function

$$F(u,v) = \left(\frac{2v(1-u^2)}{(1+u^2)(1+v^2)}, \frac{4uv}{(1+u^2)(1+v^2)}, \frac{1-v^2}{1+v^2} \right).$$

Notice that the component functions satisfy

$$
\begin{aligned}
F_1^2 + F_2^2 + F_3^2 &= \frac{4v^2(1-u^2)^2 + 16u^2v^2 + (1+u^2)^2(1-v^2)^2}{(1+u^2)^2(1+v^2)^2} \\
&= \frac{4v^2(1+u^2)^2 + (1+u^2)^2(1-v^2)^2}{(1+u^2)^2(1+v^2)^2} \\
&= \frac{(1+u^2)^2(1+v^2)^2}{(1+u^2)^2(1+v^2)^2} = 1.
\end{aligned}
$$

Thus, the image of F lies on the unit sphere $\mathbb{S}^2 = \{(x,y,z) \in \mathbb{R}^3 \mid x^2+y^2+z^2 = 1\}$.

Note that F does not surject onto $\mathbb{S}^2$. Assuming $x^2 + y^2 + z^2 = 1$, if $F(u,v) = (x,y,z)$, then in particular

$$z = \frac{1-v^2}{1+v^2} \iff v = \sqrt{\frac{1-z}{1+z}}$$

which implies that $-1 < z \leq 1$, and hence, the point $(0,0,-1)$ is not in the range of F. Furthermore, since

$$z^2 + \left(\frac{2v}{1+v^2} \right)^2 = 1, \quad \text{and thus} \quad \frac{2v}{1+v^2} = \sqrt{1-z^2},$$

for any fixed z, we have

$$x = \frac{1-u^2}{1+u^2}\sqrt{1-z^2} \quad \text{and} \quad y = \frac{2u}{1+u^2}\sqrt{1-z^2}.$$

But then, if $y = 0$, it is impossible to obtain $x = -\sqrt{1-z^2}$. Consequently, the image of F is

$$F(\mathbb{R}^2) = \mathbb{S}^2 - \{(x,y,z) \in \mathbb{S}^2 \mid x = -\sqrt{1-z^2} \text{ with } z < 1\}.$$

Figure 1.1 shows the image of F over the rectangle $(x_1, x_2) \in [-2,5] \times [0.5,5]$.

There are a few different ways to visualize functions, particularly when n and m are less than or equal to 3. Recall that the *graph* of a function $f : \mathbb{R}^n \to \mathbb{R}^m$ is the subset of $\mathbb{R}^n \times \mathbb{R}^m = \mathbb{R}^{n+m}$ defined by

$$\{(x_1, \ldots, x_n, y_1, \ldots, y_m) \in \mathbb{R}^{n+m} \mid (y_1, \ldots, y_m) = f(x_1, \ldots, x_n)\}.$$

We can visualize this explicitly when $m + n \leq 3$ with a three dimensional graphic. When $m = 1$, we recover the usual method to depict functions $f : \mathbb{R} \to \mathbb{R}$ and $f : \mathbb{R}^2 \to \mathbb{R}$.

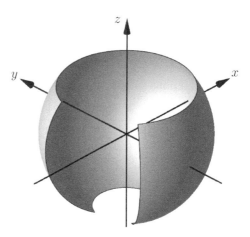

Figure 1.1: Portion of the image for Example 1.1.4.

For functions $F : \mathbb{R}^2 \to \mathbb{R}$ (respectively $F : \mathbb{R}^3 \to \mathbb{R}$), another way to attempt to visualize F is by plotting together (or in succession if one has dynamical graphing capabilities) a collection of level curves (respectively surfaces) defined by $F(x, y) = c_i$ (respectively $F(x, y, z) = c_i$) for a discrete set of values c_i. This is typically called a *contour diagram* of F. Figure 1.2 depicts a contour diagram of $2y/(x^2 + y^2 + 1)$ with $c = 0, \pm 0.2, \pm 0.4, \pm 0.6, \pm 0.8$.

In multivariable calculus or in a basic differential geometry course ([5]), one typically uses yet another technique to visualize functions of the form $\vec{f} : \mathbb{R} \to \mathbb{R}^m$, for $m = 2$ or 3. By plotting the points that consist of the image of $\vec{f}$ we see a plane or space curve. In doing so, we lose visual information about how fast one travels along the curve. Figure 1.3 shows the image of the so-called *space cardioid*, given by the function

$$\vec{f}(t) = \big((1 - \cos t) \cos t, (1 - \cos t) \sin t, \sin t\big).$$

Similarly, in the study of surfaces, it is common to depict a function $\vec{F} : \mathbb{R}^2 \to \mathbb{R}^3$ by plotting its image in $\mathbb{R}^3$. (The graph of a function of the form $\mathbb{R}^2 \to \mathbb{R}^3$ is a subset of $\mathbb{R}^5$, which is quite difficult to visualize no matter what computer tools one has at one's disposal!)

We define the usual operations on functions as expected.

Definition 1.1.5. Let $\vec{f}$ and $\vec{g}$ be two functions defined over a subset U of $\mathbb{R}^n$ with codomain $\mathbb{R}^m$. Then we define the following functions:

1. $(\vec{f} + \vec{g}) : U \to \mathbb{R}^m$, where $(\vec{f} + \vec{g})(\vec{x}) = \vec{f}(\vec{x}) + \vec{g}(\vec{x})$.

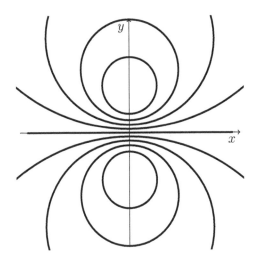

Figure 1.2: A contour diagram.

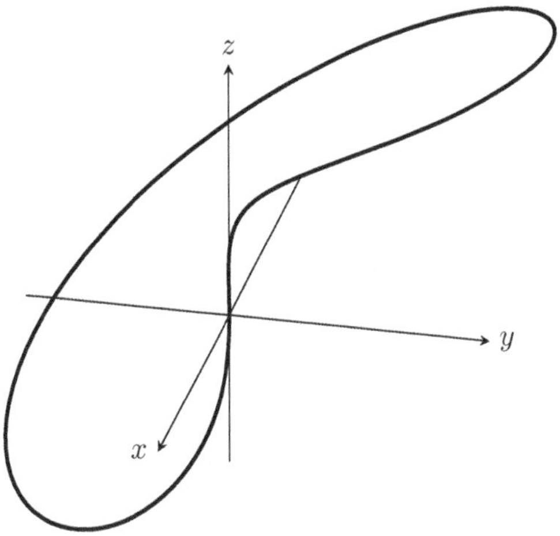

Figure 1.3: A space curve.

2. $(\vec{f} \cdot \vec{g}) : U \to \mathbb{R}$, where $(\vec{f} \cdot \vec{g})(\vec{x}) = \vec{f}(\vec{x}) \cdot \vec{g}(\vec{x})$.

3. If $m = 3$, $(\vec{f} \times \vec{g}) : U \to \mathbb{R}^3$, where $(\vec{f} \times \vec{g})(\vec{x}) = \vec{f}(\vec{x}) \times \vec{g}(\vec{x})$.

Definition 1.1.6. Let $\vec{f}$ be a function from a subset $U \subset \mathbb{R}^n$ to $\mathbb{R}^m$, and let $\vec{g}$ be a function from $V \subset \mathbb{R}^m$ to $\mathbb{R}^s$. If the image of $\vec{f}$ is a subset of V, then the composition function $\vec{g} \circ \vec{f}$ is the function $U \to \mathbb{R}^s$ defined by

$$(\vec{g} \circ \vec{f})(\vec{x}) = \vec{g}(\vec{f}(\vec{x})).$$

Out of the vast variety of possible functions one could study, the class of linear functions serves a fundamental role in the analysis of multivariable functions. We remind the reader of various properties of linear functions.

Definition 1.1.7. A function $F : \mathbb{R}^n \to \mathbb{R}^m$ is called a *linear function* if

$$F(\vec{x} + \vec{y}) = F(\vec{x}) + F(\vec{y}) \qquad \text{for all } \vec{x}, \vec{y} \in \mathbb{R}^n,$$
$$F(k\vec{x}) = kF(\vec{x}) \qquad \text{for all } k \in \mathbb{R} \text{ and all } \vec{x} \in \mathbb{R}^n.$$

If a function $F : \mathbb{R}^n \to \mathbb{R}^m$ is linear, then

$$F(\vec{0}) = F(\vec{0} - \vec{0}) = F(\vec{0}) - F(\vec{0}) = \vec{0},$$

and hence F maps the origin of $\mathbb{R}^n$ to the origin of $\mathbb{R}^m$.

If $\mathcal{B} = \{\vec{f}_1, \vec{f}_2, \ldots, \vec{f}_n\}$ is a basis of $\mathbb{R}^n$, then any vector $\vec{v} \in \mathbb{R}^n$ can be written uniquely as a linear combination of vectors in $\mathcal{B}$ as

$$\vec{v} = c_1 \vec{f}_1 + c_2 \vec{f}_2 + \cdots + c_n \vec{f}_n.$$

One often writes the coefficients in linear algebra as the column vector

$$[\vec{v}]_{\mathcal{B}} = \begin{pmatrix} c_1 \\ c_2 \\ \vdots \\ c_n \end{pmatrix}.$$

If the basis $\mathcal{B}$ is not specified, one assumes that the coefficients are given in terms of the standard basis. If F is a linear function, then

$$F(\vec{v}) = c_1 F(\vec{f_1}) + \cdots + c_n F(\vec{f_n}),$$

hence, to know all outputs of F one needs to know the coefficients of $[\vec{v}]_{\mathcal{B}}$ and the output of the basis vectors of $\mathcal{B}$. Suppose also that $\mathcal{B}' = \{\vec{w}_1, \vec{w}_2, \ldots, \vec{w}_m\}$ is a basis of $\mathbb{R}^m$. If the $\mathcal{B}'$-coordinates of the outputs of the vectors in $\mathcal{B}$ are

$$[F(\vec{f_1})]_{\mathcal{B}'} = \begin{pmatrix} a_{11} \\ a_{21} \\ \vdots \\ a_{m1} \end{pmatrix}, \quad [F(\vec{f_2})]_{\mathcal{B}'} = \begin{pmatrix} a_{12} \\ a_{22} \\ \vdots \\ a_{m2} \end{pmatrix}, \quad \cdots, \quad [F(\vec{f_n})]_{\mathcal{B}'} = \begin{pmatrix} a_{1n} \\ a_{2n} \\ \vdots \\ a_{mn} \end{pmatrix},$$

then the image of the vector $\vec{v} \in \mathbb{R}^n$ is given by

$$[F(\vec{v})]_{\mathcal{B}'} = c_1 \begin{pmatrix} a_{11} \\ a_{21} \\ \vdots \\ a_{m1} \end{pmatrix} + c_2 \begin{pmatrix} a_{12} \\ a_{22} \\ \vdots \\ a_{m2} \end{pmatrix} + \cdots + c_n \begin{pmatrix} a_{1n} \\ a_{2n} \\ \vdots \\ a_{mn} \end{pmatrix} = \begin{pmatrix} a_{11} & a_{12} & \cdots & a_{1n} \\ a_{21} & a_{22} & \cdots & a_{2n} \\ \vdots & \vdots & \ddots & \vdots \\ a_{m1} & a_{m2} & \cdots & a_{mn} \end{pmatrix} \begin{pmatrix} c_1 \\ c_2 \\ \vdots \\ c_n \end{pmatrix}.$$

The matrix

$$A = \begin{pmatrix} a_{11} & a_{12} & \cdots & a_{1n} \\ a_{21} & a_{22} & \cdots & a_{2n} \\ \vdots & \vdots & \ddots & \vdots \\ a_{m1} & a_{m2} & \cdots & a_{mn} \end{pmatrix}$$

is called the $\mathcal{B}, \mathcal{B}'$-matrix representing the linear function F and is denoted by $[F]_{\mathcal{B}'}^{\mathcal{B}}$. Therefore,

$$[F(\vec{v})]_{\mathcal{B}'} = [F]_{\mathcal{B}'}^{\mathcal{B}} [\vec{v}]_{\mathcal{B}}$$

for all $\vec{v} \in \mathbb{R}^n$.

Given a linear function $F : \mathbb{R}^n \to \mathbb{R}^m$, one calls the *image* of F the set $\operatorname{Im} F = F(\mathbb{R}^n)$, also called the range. The *kernel* of F is the zero set

$$\ker F = \{\vec{u} \in \mathbb{R}^n \,|\, F(\vec{u}) = \vec{0}\}.$$

The image $\operatorname{Im} F$ is a vector subspace of the codomain $\mathbb{R}^m$ and the kernel is a subspace of the domain $\mathbb{R}^n$. The *rank* of F is the dimension $\dim(\operatorname{Im} F)$ and can be shown to be equal to the size of the largest nonvanishing minor of any matrix representing F, which is independent of the bases. The image of F cannot have a greater dimension than either the domain or the codomain, so

$$\operatorname{rank} F \leq \min\{m, n\},$$

and one says that F has *maximal rank* if $\operatorname{rank} F = \min\{m, n\}$. It is not hard to show that a linear function $F : \mathbb{R}^n \to \mathbb{R}^m$ is surjective if and only if $\operatorname{rank} F = m$ and F is injective if and only if $\operatorname{rank} F = n$.

The rank is also useful in determining the linear dependence between a set of vectors. If $\{\vec{u}_1, \vec{u}_2, \ldots, \vec{u}_n\}$ is a set of vectors in $\mathbb{R}^m$, then the matrix

$$A = \begin{pmatrix} \vec{u}_1 & \vec{u}_2 & \cdots & \vec{u}_n \end{pmatrix},$$

where the $\vec{u}_i$ are viewed as column vectors, represents a linear function $F : \mathbb{R}^n \to \mathbb{R}^m$, with

$$\operatorname{Im} F = \operatorname{Span}\{\vec{u}_1, \vec{u}_2, \ldots, \vec{u}_n\}.$$

Thus, the set of vectors $\{\vec{u}_1, \vec{u}_2, \ldots, \vec{u}_n\}$ is linearly independent if and only if $\operatorname{rank} F = n$.

In the case of $n = m$, the determinant provides an alternative characterization to linear independence. If F is a linear function from $\mathbb{R}^n$ to itself with associated matrix A, then $|\det A|$ is the n-volume of the image under F of the unit n-cube. Consequently, if the columns of A are not linearly independent, the n-volume of this parallelopiped will be 0. This leads one to a fundamental summary theorem in linear algebra.

Theorem 1.1.8. *For a linear function $F : \mathbb{R}^n \to \mathbb{R}^n$ with associated square matrix A, the following statements are equivalent:*

1. $\operatorname{rank} F = n$.

2. $\det A \neq 0$.

3. $\operatorname{Im} F = \mathbb{R}^n$.

4. $\ker F = \{\vec{0}\}$.

5. The column vectors of A are linearly independent.

6. The column vectors of A form a basis of $\mathbb{R}^n$.

7. The column vectors of A span $\mathbb{R}^n$.

8. F has an inverse function.

We remind the reader that matrix multiplication is defined in such a way so that if A is the matrix for a linear function $F : \mathbb{R}^n \to \mathbb{R}^m$ and B is the matrix for a linear function $G : \mathbb{R}^p \to \mathbb{R}^n$, then the product AB is the matrix representing the composition $F \circ G : \mathbb{R}^p \to \mathbb{R}^m$. In other words,

$$[F \circ G]_{\mathcal{C}}^{\mathcal{A}} = [F]_{\mathcal{C}}^{\mathcal{B}} [G]_{\mathcal{B}}^{\mathcal{A}},$$

where $\mathcal{A}$, $\mathcal{B}$ and $\mathcal{C}$ are bases on $\mathbb{R}^p$, $\mathcal{R}^n$, and $\mathcal{R}^m$ respectively. Furthermore, if $m = n$ and rank $F = n$, then the matrix A^{-1} is the matrix that represents the inverse function of F.

A particularly important case of matrices representing linear transformations is the *change of basis matrix*. Let $\mathcal{B}$ and $\mathcal{B}'$ be two bases on $\mathbb{R}^n$. The change of basis matrix from $\mathcal{B}$ to $\mathcal{B}'$ coordinates is $M = [\mathrm{id}]_{\mathcal{B}'}^{\mathcal{B}}$, where $\mathrm{id} : \mathbb{R}^n \to \mathbb{R}^n$ is the identity transformation. In other words, for all $\vec{v} \in \mathbb{R}^n$,

$$[\vec{v}]_{\mathcal{B}'} = M[\vec{v}]_{\mathcal{B}}.$$

If $\mathcal{B} = \{\vec{f_1}, \vec{f_2}, \dots, \vec{f_n}\}$, then $M = \left([\vec{f_1}]_{\mathcal{B}'} \quad [\vec{f_2}]_{\mathcal{B}'} \quad \cdots \quad [\vec{f_n}]_{\mathcal{B}'} \right)$.

PROBLEMS

1.1.1. Consider the function F in Example 1.1.4. Prove algebraically that if the domain is restricted to $\mathbb{R} \times (0, +\infty)$, it is injective. What is the image of F in this case?

1.1.2. Let $F : \mathbb{R}^2 \to \mathbb{R}^2$ be the function defined by $F(s, t) = (s^2 - t^2, 2st)$, and let $G : \mathbb{R}^2 \to \mathbb{R}^2$ be the function defined by $G(u, v) = (2u^2 - 3v, uv + v^3)$. Calculate the component functions of $F \circ G$ and of $G \circ F$.

1.1.3. Show that the function $\vec{X} : [0, 2\pi] \times [0, \pi] \to \mathbb{R}^3$, with

$$\vec{X}(x_1, x_2) = (\cos x_1 \sin x_2, \sin x_1 \sin x_2, \cos x_2),$$

defines a mapping *onto* the unit sphere in $\mathbb{R}^3$. Which points on the unit sphere have more than one preimage?

1.1.4. Consider the function F from $\mathbb{R}^3$ to itself defined by

$$F(x_1, x_2, x_3) = (x_1 + 2x_2 + 3x_3, 4x_1 + 5x_2 + 6x_3, 7x_1 + 8x_2 + 9x_3).$$

Prove that this is a linear function. Find the matrix associated to F (with respect to the standard basis). Find the rank of F, and if the rank is less than 3, find equations for the image of F.

1.1.5. Consider a line L in $\mathbb{R}^n$ traced out by the parametric equation $\vec{x}(t) = t\vec{a} + \vec{b}$. Prove that for any linear function $F : \mathbb{R}^n \to \mathbb{R}^m$, the image $F(L)$ is either a line or a point.

1.1.6. Let $F : \mathbb{R}^n \to \mathbb{R}^m$ be a linear function, and let L_1 and L_2 be parallel lines in $\mathbb{R}^n$. Prove that $F(L_1)$ and $F(L_2)$ are either both points or both lines in $\mathbb{R}^m$. If $F(L_1)$ and $F(L_2)$ are both lines, prove that they are parallel.

1.1.7. Let $F : \mathbb{R}^n \to \mathbb{R}^m$ be a linear function represented by a matrix A with respect to a basis $\mathcal{B}$ on $\mathbb{R}^n$ and a basis $\mathcal{B}'$ on $\mathbb{R}^m$. Prove that F maps every pair of perpendicular lines in $\mathbb{R}^n$ to another pair of perpendicular lines in $\mathbb{R}^m$ if and only if $A^T A = \lambda I_n$ for some nonzero real number λ.

1.1.8. Let $\vec{\omega}$ be a nonzero vector in $\mathbb{R}^n$. Define the function $F : \mathbb{R}^n \to \mathbb{R}$ as

$$F(\vec{x}) = \vec{\omega} \cdot \vec{x}.$$

Prove that F is a linear function. Find the matrix associated to F (with respect to the standard basis).

1.1.9. Let $\vec{\omega}$ be a nonzero vector in $\mathbb{R}^3$. Define the function $F : \mathbb{R}^3 \to \mathbb{R}^3$ as

$$F(\vec{x}) = \vec{\omega} \times \vec{x}.$$

Prove that F is a linear function. Find the matrix associated to F (with respect to the standard basis). Prove that rank $F = 2$.

1.2   Continuity, Limits, and Differentiability

Intuitively, a function is called *continuous* if it preserves "nearness." A rigorous mathematical definition for continuity for functions from $\mathbb{R}^n$ to $\mathbb{R}^m$ is hardly any different for functions from $\mathbb{R} \to \mathbb{R}$.

In calculus of a real variable, one does not study functions defined over a discrete set of real values because the notions behind continuity and differentiability do not make sense over such sets. Instead, one often assumes the function is defined over some interval. Similarly, for the analysis of functions $\mathbb{R}^n$ to $\mathbb{R}^m$, one does not study functions defined from any subset of $\mathbb{R}^n$ into $\mathbb{R}^m$. One typically considers functions defined over what is called an *open set* in $\mathbb{R}^n$, a notion we define now.

Definition 1.2.1. The *open ball* around $\vec{x}_0$ of radius r is the set

$$B_r(\vec{x}_0) = \left\{ \vec{x} \in \mathbb{R}^n \ : \ \|\vec{x} - \vec{x}_0\| < r \right\}.$$

A subset $U \subset \mathbb{R}^n$ is called *open* if for all $\vec{x} \in U$ there exists an $r > 0$ such that $B_r(\vec{x}) \subset U$.

Intuitively speaking, the definition of an open set U in $\mathbb{R}^n$ implies that at every point $p \in U$ it is possible to "move" in any direction by at least a little amount ϵ and still remain in U. This means that in some sense U captures the full dimensionality of the ambient space $\mathbb{R}^n$. This is why, when studying the analysis of functions from $\mathbb{R}^n$ to $\mathbb{R}^m$, we narrow our attention to functions $F : U \to \mathbb{R}^m$, where U is an open subset of $\mathbb{R}^n$.

The reader is encouraged to consult Subsection A.1.2 in Appendix A for more background on open and closed sets. The situation in which we need to consider an open set U and a point $\vec{x}_0$ in U is so common that another terminology exists for U in this case.

Definition 1.2.2. Let $\vec{x}_0 \in \mathbb{R}^n$. Any open set U in $\mathbb{R}^n$ such that $\vec{x}_0 \in U$ is called an *open neighborhood*, or more simply, a neighborhood, of $\vec{x}_0$.

We are now in a position to formally define continuity.

Definition 1.2.3. Let U be an open subset of $\mathbb{R}^n$, and let F be a function from U into $\mathbb{R}^m$. The function F is called continuous at the point $\vec{x}_0 \in U$ if $F(\vec{x}_0)$ exists and if, for all $\varepsilon > 0$, there exists a $\delta > 0$ such that for all $\vec{x} \in \mathbb{R}$,

$$\|\vec{x} - \vec{x}_0\| < \delta \Longrightarrow \|F(\vec{x}) - F(\vec{x}_0)\| < \epsilon.$$

The function F is called continuous on U if it is continuous at every point of U.

With the language of open balls, one can rephrase the definition of continuity as follows. Let U be an open subset of $\mathbb{R}^n$. A function $F : U \to \mathbb{R}^m$ is continuous at a point $\vec{x}_0$ if for all $\varepsilon > 0$ there exists a $\delta > 0$ such that

$$F(B_\delta(\vec{x}_0)) \subset B_\varepsilon(F(\vec{x}_0)). \tag{1.1}$$

Sections A.1.2 and A.1.4 in Appendix A provide a comprehensive discussion about open sets in a metric space, a generalization of $\mathbb{R}^n$, and continuity of functions between metric spaces. We point out that using the language of open sets, Definition 1.2.3 can be rephrased once more in a manner that lines up with the definition of continuity of functions between topological spaces.

Proposition 1.2.4. *Let $F : U \to \mathbb{R}^m$ be a function, where $U \subset \mathbb{R}^n$ is open. The function F is continuous if and only if $F^{-1}(V)$ is open for all open sets $V \in \mathbb{R}^m$.*

Proof. Suppose the function F is continuous. Let V be open in $\mathbb{R}^m$ and let $\vec{x}_0 \in F^{-1}(V)$, which means that $F(\vec{x}_0) \in V$. Since V is open, there exists $\varepsilon > 0$ such that $B_\varepsilon(F(\vec{x}_0)) \subset V$. By (1.1), there exists $\delta > 0$ such that $F(B_\delta(\vec{x}_0)) \subset B_\varepsilon(F(\vec{x}_0))$. This means that $B_\delta(\vec{x}_0) \subseteq F^{-1}(V)$, which, since $\vec{x}_0$ was arbitrary in V, implies that $F^{-1}(V)$ is open.

Conversely, suppose that $F^{-1}(V)$ is open for all open sets $V \subset \mathbb{R}^m$. Let $\vec{x}_0 \in U$ be any point and let $\varepsilon > 0$ be a real number. Consider the open ball $B_\varepsilon(F(\vec{x}_0))$. By hypothesis, $F^{-1}(B_\varepsilon(F(\vec{x}_0)))$ is open in $\mathbb{R}^n$. Since $\vec{x}_0 \in F^{-1}(B_\varepsilon(F(\vec{x}_0)))$, we deduce that there exists $\delta > 0$ such that $B_\delta(\vec{x}_0) \subset F^{-1}(B_\varepsilon(F(\vec{x}_0)))$. This is equivalent to (1.1), so the proposition follows. $\qquad\square$

Example 1.2.5. Consider the function $F : \mathbb{R}^n \to \mathbb{R}^n$ defined by

$$F(\vec{x}) = \begin{cases} \vec{x}/\|\vec{x}\|, & \text{if } \vec{x} \neq \vec{0}, \\ \vec{0}, & \text{if } \vec{x} = \vec{0}. \end{cases}$$

This function leaves $\vec{0}$ fixed and projects the rest of $\mathbb{R}^n$ onto the unit sphere. If $\vec{x} \neq \vec{0}$, then

$$\|F(\vec{x}) - F(\vec{x}_0)\| = \left\| \frac{\vec{x}}{\|\vec{x}\|} - \frac{\vec{x}_0}{\|\vec{x}_0\|} \right\| \leq \left\| \frac{\vec{x}}{\|\vec{x}\|} - \frac{\vec{x}}{\|\vec{x}_0\|} \right\| + \left\| \frac{\vec{x}}{\|\vec{x}_0\|} - \frac{\vec{x}_0}{\|\vec{x}_0\|} \right\|.$$

However,

$$\left\| \frac{\vec{x}}{\|\vec{x}\|} - \frac{\vec{x}}{\|\vec{x}_0\|} \right\| = \left| \frac{1}{\|\vec{x}\|} - \frac{1}{\|\vec{x}_0\|} \right| \|\vec{x}\| = \|\vec{x}\| \frac{|\,\|\vec{x}_0\| - \|\vec{x}\|\,|}{\|\vec{x}\| \, \|\vec{x}_0\|} \leq \frac{1}{\|\vec{x}_0\|} \|\vec{x} - \vec{x}_0\|,$$

and thus,

$$\|F(\vec{x}) - F(\vec{x}_0)\| \leq \frac{2}{\|\vec{x}_0\|} \|\vec{x} - \vec{x}_0\|.$$

Consequently, given any $\varepsilon > 0$ and setting

$$\delta = \min \left(\|\vec{x}_0\|, \frac{1}{2}\varepsilon\|\vec{x}_0\| \right),$$

we know that $\vec{x} \neq \vec{0}$ and also that $\|F(\vec{x}) - F(\vec{x}_0)\| < \varepsilon$. Hence, F is continuous at all $\vec{x}_0 \neq \vec{0}$.

On the other hand, if $\vec{x}_0 = \vec{0}$, for all $\vec{x} \neq \vec{x}_0$,

$$\|F(\vec{x}) - F(\vec{0})\| = \|F(\vec{x}) - \vec{0}\| = \|F(\vec{x})\| = 1,$$

which can never be less than ε if $\varepsilon \leq 1$.

Example 1.2.6. As a contrast to Example 1.2.5, consider the function

$$F(\vec{x}) = \begin{cases} \vec{x}, & \text{if all components of } \vec{x} \text{ are rational,} \\ \vec{0}, & \text{otherwise.} \end{cases}$$

The function F is obviously continuous at $\vec{0}$, with $\delta = \varepsilon$ satisfying the requirements of Definition 1.2.3. On the other hand, if $\vec{x}_0 \neq \vec{0}$, then in $B_\delta(\vec{x}_0)$, for any $\delta > 0$, one can always find an $\vec{x}$ that has either all rational components or has at least one irrational component. Thus, if $\varepsilon < \|\vec{x}_0\|$, for all $\delta > 0$, we have

$$F(B_\delta(\vec{x}_0)) \not\subset B_\varepsilon(F(\vec{x}_0)).$$

Thus, F is discontinuous everywhere except at $\vec{0}$.

The following theorem implies many other corollaries concerning continuity of multivariable functions.

Theorem 1.2.7. *Let U be an open subset of $\mathbb{R}^n$, let $F : U \to \mathbb{R}^m$ be a function, and let F_i, with $i = 1, \ldots, m$, be the component functions. The function F is continuous at the point $\vec{a} \in U$ if and only if, for all $i = 1, \ldots, m$, the component function $F_i : U \to \mathbb{R}$ is continuous at $\vec{a}$.*

Proof. Suppose that F is continuous at $\vec{a}$. Thus, for all $\varepsilon > 0$, there exists a $\delta > 0$ such that $\|\vec{x} - \vec{a}\| < \delta$ implies $\|F(\vec{x}) - F(\vec{a})\| < \varepsilon$. Since

$$\|F(\vec{x}) - F(\vec{a})\| = \sqrt{(F_1(\vec{x}) - F_1(\vec{a})_1)^2 + \cdots + (F_m(\vec{x}) - F_m(\vec{a}))^2}$$
$$\geq |F(\vec{x})_i - F(\vec{a})_i|,$$

then for all $\varepsilon > 0$, having $\|\vec{x} - \vec{a}\| < \delta$ implies that

$$|F_i(\vec{x}) - F_i(\vec{a})| \leq \|F(\vec{x}) - F(\vec{a})\| < \varepsilon$$

for any i. Hence, each function $F_i : U \to \mathbb{R}$ is continuous at $\vec{a}$.

Conversely, suppose that all the functions F_i are continuous at $\vec{a}$. Thus, for any ε and for all i, there exist $\delta_i > 0$ such that $\|\vec{x}-\vec{a}\| < \delta_i$ implies $|F_i(\vec{x})-F_i(\vec{a})| < \varepsilon/\sqrt{m}$. Then taking $\delta = \min(\delta_1,\ldots,\delta_m)$, if $\|\vec{x}-\vec{a}\| < \delta$, then

$$\|F(\vec{x}) - F(\vec{a})\| = \sqrt{|F_1(\vec{x}) - F_1(\vec{a})|^2 + \cdots + |F_m(\vec{x}) - F_m(\vec{a})|^2}$$

$$\leq \sqrt{\frac{\varepsilon^2}{m} + \cdots + \frac{\varepsilon^2}{m}} = \varepsilon.$$

Thus F is continuous. $\square$

If U is an open set containing a point $\vec{a}$, then the set $U - \{\vec{a}\}$ is called a *deleted neighborhood* of $\vec{a}$. If a function F is a function into $\mathbb{R}^m$ defined on a deleted neighborhood of a point $\vec{a} \in \mathbb{R}^n$, it is possible to define the limit of F at $\vec{a}$. The *limit* of F at $\vec{a}$ is the value $\vec{L}$ such that if $F(\vec{a})$ were $\vec{L}$, then $F(\vec{a})$ would be continuous at $\vec{a}$. We make this more precise as follows.

Definition 1.2.8. Let $\vec{a} \in \mathbb{R}^n$. Let F be a function from an open subset $U - \{\vec{a}\} \subset \mathbb{R}^n$ into $\mathbb{R}^m$. The *limit* of F at $\vec{a}$ is defined as the point $\vec{L}$, and we write

$$\lim_{\vec{x}\to\vec{a}} F(\vec{x}) = \vec{L},$$

if for all ε there exists a δ such that

$$F(B_\delta(\vec{a}) - \{\vec{a}\}) \subset B_\varepsilon(\vec{L}).$$

We point out right away that a function $F : U \to \mathbb{R}^m$, where U is open in $\mathbb{R}^n$ is continuous at $\vec{a} \in U$ if and only if

$$\lim_{\vec{x}\to\vec{a}} F(\vec{x}) = F(\vec{a}).$$

Key results in calculus and analysis are the limit laws along with their implications for continuity.

Theorem 1.2.9. *Let U be an open set in $\mathbb{R}^n$, let $\vec{a} \in U$, and let F and G be functions from $U - \{\vec{a}\}$ to $\mathbb{R}^m$ and $w : U - \{\vec{a}\} \to \mathbb{R}$. Suppose that the limits of F, G, and w at $\vec{a}$ exist. Then*

$$\lim_{\vec{x}\to\vec{a}} (F(\vec{x}) + G(\vec{x})) = \left(\lim_{\vec{x}\to\vec{a}} F(\vec{x})\right) + \left(\lim_{\vec{x}\to\vec{a}} G(\vec{x})\right)$$

$$\lim_{\vec{x}\to\vec{a}} (w(\vec{x})F(\vec{x})) = \left(\lim_{\vec{x}\to\vec{a}} w(\vec{x})\right)\left(\lim_{\vec{x}\to\vec{a}} F(\vec{x})\right)$$

$$\lim_{\vec{x}\to\vec{a}} (F(\vec{x}) \cdot G(\vec{x})) = \left(\lim_{\vec{x}\to\vec{a}} F(\vec{x})\right) \cdot \left(\lim_{\vec{x}\to\vec{a}} G(\vec{x})\right) \qquad (dot\ product)$$

$$\lim_{\vec{x}\to\vec{a}} \|F(\vec{x})\| = \left\|\lim_{\vec{x}\to\vec{a}} F(\vec{x})\right\|.$$

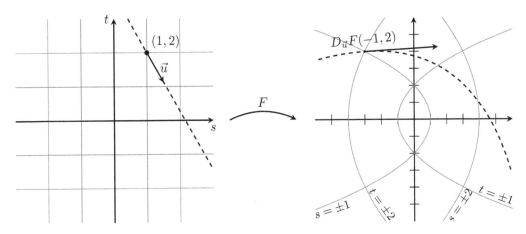

Figure 1.4: Example 1.2.12.

Proof. (Left as an exercise for the reader.) $\qquad\square$

Theorem 1.2.10. *Let U be an open set in $\mathbb{R}^n$, let F and G be functions from U to $\mathbb{R}^m$, let $w : U \to \mathbb{R}$, and suppose that F, G, and w are all continuous at $\vec{a} \in U$. Then the functions $\|F\|$, $F + G$, wF, and $F \cdot G$ are also continuous at $\vec{a}$. If $m = 3$, then the vector function $F \times G$ is also continuous at $\vec{a}$.*

Proof. (Left as an exercise for the reader.) $\qquad\square$

Similar to most multivariable calculus courses, before addressing partial derivatives, we introduce the notion of a directional derivative, which measures the rate of change of a function in a given direction.

Definition 1.2.11. Let F be a function from an open subset $U \subset \mathbb{R}^n$ into $\mathbb{R}^m$, let $\vec{x}_0 \in U$ be a point, and let $\vec{u}$ be a unit vector. The directional derivative of F in the direction $\vec{u}$ at the point $\vec{x}_0$ is

$$D_{\vec{u}} F(\vec{x}_0) = \lim_{h \to 0} \frac{F(\vec{x}_0 + h\vec{u}) - F(\vec{x}_0)}{h}$$

whenever the limit exists.

Another way to understand this definition is to consider the curve $\vec{\gamma} : (-\varepsilon, \varepsilon) \to \mathbb{R}^m$, for some $\varepsilon > 0$, defined by $\vec{\gamma}(t) = F(\vec{x}_0 + t\vec{u})$. Then $D_{\vec{u}} F(\vec{x}_0)$ is equal to the derivative $\vec{\gamma}\,'(0)$.

We note that though F is a multivariable function, the definition of $D_{\vec{u}} F(\vec{x}_0)$ reduces to a single variable, vector-valued function before taking a derivative.

Example 1.2.12. Consider the function $F(s,t) = (s^2 - t^2, 2st)$ from $\mathbb{R}^2$ to itself. We will calculate the directional derivative of F at $\vec{x}_0 = (1,2)$ in the direction of $\vec{u} = (1/2, -\sqrt{3}/2)$.

We can picture this kind of function by plotting a discrete set of coordinate lines mapped under F (see Figure 1.4). However, for functions like F that are not injective, even this method of picturing F can be misleading since every point in the codomain can have multiple preimages.

Now,

$$F(\vec{x}_0 + t\vec{u}) = \left((1 + \frac{1}{2}t)^2 - (2 - \frac{\sqrt{3}}{2}t)^2, 2(1 + \frac{1}{2}t)(2 - \frac{\sqrt{3}}{2}t)\right)$$
$$= \left(-3 + (2 + 4\sqrt{3})t - 2t^2, 4 + (4 - 2\sqrt{3})t - 2\sqrt{3}t^2\right),$$

so

$$D_{\vec{u}}F(\vec{x}_0) = \left((2 + 4\sqrt{3}) - 4t, (4 - 2\sqrt{3}) - 4\sqrt{3}t\right)\Big|_{t=0} = (2 + 4\sqrt{3}, 4 - 2\sqrt{3}).$$

Figure 1.4 shows the curve $F(\vec{x}_0 + t\vec{u})$ and illustrates the directional derivative as being the derivative of $F(\vec{x}_0 + t\vec{u})$ at $t = 0$. The figure shows that though $\vec{u}$ must be a unit vector, the directional derivative is usually not.

Let F be a function from an open set $U \subset \mathbb{R}^n$ to $\mathbb{R}^m$. For any point $\vec{x}_0 \in U$, the directional derivative of F in the direction $\vec{u}_k$ at $\vec{x}_0$ is called the *kth partial derivative* of F at $\vec{x}_0$. The kth partial derivative of F is itself a vector function possibly defined on a smaller set than U. Writing

$$F(\vec{x}) = \left(F_1(x_1, \ldots, x_n), \ldots, F_m(x_1, \ldots, x_n)\right),$$

some common notations for the kth partial derivative $D_{\vec{u}_k}F$ are

$$F_{x_k}, \quad \frac{\partial F}{\partial x_k}, \quad D_k F, \quad F_{,k}.$$

In the last notation, the comma distinguishes the derivative operation from an index. It is not hard to show that

$$\frac{\partial F}{\partial x_k}(\vec{x}) = \left(\frac{\partial F_1}{\partial x_k}(x_1, \ldots, x_n), \ldots, \frac{\partial F_m}{\partial x_k}(x_1, \ldots, x_n)\right).$$

Example 1.2.13. Consider the real-valued function $f(x_1, x_2)$ defined by

$$f(x_1, x_2) = \begin{cases} \dfrac{x_1 x_2^2}{x_1^2 + x_2^4}, & \text{if } (x_1, x_2) \neq (0,0), \\ 0, & \text{otherwise.} \end{cases}$$

See Figure 1.5. We study the behavior of f near $\vec{x} = \vec{0}$.

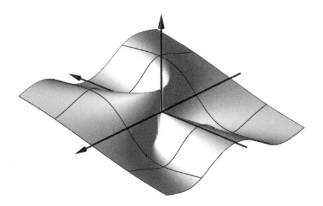

Figure 1.5: Graph of the function in Example 1.2.13.

Let $\vec{u} = (u_1, u_2)$ be a unit vector, with $u_1 \neq 0$. Then

$$D_{\vec{u}}f(\vec{0}) = \lim_{h \to 0} \frac{f(\vec{0} + h\vec{u}) - f(\vec{0})}{h} = \lim_{h \to 0} \frac{h^3 u_1 u_2^2}{h(h^2 u_1^2 + h^4 u_2^4)}$$

$$= \lim_{h \to 0} \frac{u_1 u_2^2}{(u_1^2 + h^2 u_2^4)} = \frac{u_2^2}{u_1}.$$

If $u_1 = 0$, then $f(\vec{0} + h\vec{u}) = 0$ for all h, so $D_{\vec{u}}f(\vec{0}) = 0$. Thus, the directional derivative $D_{\vec{u}}f(\vec{0})$ is defined for all unit vectors $\vec{u}$.

On the other hand, consider the curve $\vec{x}(t) = (t^2, t)$. Along this curve, if $t \neq 0$,

$$f(\vec{x}(t)) = \frac{t^4}{t^4 + t^4} = \frac{1}{2}.$$

Thus,

$$f(\vec{x}(t)) = \begin{cases} \frac{1}{2}, & \text{if } t \neq 0, \\ 0, & \text{if } t = 0, \end{cases}$$

which is not continuous. Notice that this implies that f as a function from $\mathbb{R}^2$ to $\mathbb{R}$ is not continuous at $\vec{0}$ since taking $\varepsilon = \frac{1}{4}$, for all $\delta > 0$, there exist points $\vec{x}$ (in this case, points of the form $\vec{x} = (t^2, t)$) such that $\|\vec{x}\| < \delta$ have $|f(\vec{x})| > \varepsilon$.

Therefore, the function f is defined at $\vec{0}$, has directional derivatives in every direction at $\vec{0}$, but is not continuous at $\vec{0}$.

Example 1.2.13 shows that it is possible for a vector function to have directional derivatives in every direction at some point $\vec{a}$ but, at the same time, fail to be continuous at $\vec{a}$. The reason for this is that the directional derivative depends only upon the behavior of a function along a line through $\vec{a}$, while approaching $\vec{a}$ along other families of curves may exhibit a different behavior of the function.

Example 1.2.13 also illustrates that even if all the partial derivatives of a function F exist at a point, we should not yet consider it as differentiable there. A better approach is to call a function differentiable at some point if it can be approximated by a linear function.

Definition 1.2.14. Let F be a function from an open set $U \subset \mathbb{R}^n$ to $\mathbb{R}^m$ and let $\vec{a} \in U$. We call F *differentiable* at $\vec{a}$ if there exist a linear transformation $L : \mathbb{R}^n \to \mathbb{R}^m$ and a function R defined in a neighborhood V of $\vec{0}$ such that for all $\vec{h} \in V$,

$$F(\vec{a} + \vec{h}) = F(\vec{a}) + L(\vec{h}) + R(\vec{h}),$$

with

$$\lim_{\vec{h} \to \vec{0}} \frac{R(\vec{h})}{\|\vec{h}\|} = \vec{0}.$$

If F is differentiable at $\vec{a}$, the linear transformation L is denoted by $dF_{\vec{a}}$ and is called the *differential* of F at $\vec{a}$.

Notations for the differential vary widely. Though we will consistently use $dF_{\vec{a}}$ for the differential of F at $\vec{a}$, some authors write $dF(\vec{a})$ instead. The notation in this text attempts to use the most common notation in differential geometry texts and to incorporate some notation that is standard among modern linear algebra texts.

If bases $\mathcal{B}$ and $\mathcal{B}'$ are given for $\mathbb{R}^n$ and $\mathbb{R}^m$, then we denote the matrix for $dF_{\vec{a}}$ by

$$\left[dF_{\vec{a}}\right]_{\mathcal{B}}^{\mathcal{B}'}.$$

Assuming we use the standard bases for $\mathbb{R}^n$ and $\mathbb{R}^m$, we write the matrix for $dF_{\vec{a}}$ as $\left[dF_{\vec{a}}\right]$.

If F is differentiable over an open set $U \subset \mathbb{R}^n$, the differential dF (not evaluated at any point) is a function from U to $\mathrm{Hom}(\mathbb{R}^n, \mathbb{R}^m)$, the set of linear transformations from $\mathbb{R}^n$ to $\mathbb{R}^m$. Its associated matrix $\left[dF\right]$ is a matrix of functions, each defined over U, and we call $\left[dF\right]$ the *Jacobian matrix* of F.

If $m = n$, the determinant of the Jacobian matrix is simply called the *Jacobian* of F. The Jacobian of F is a function $U \to \mathbb{R}$ and some common notations include

$$J(F), \quad \frac{\partial(F_1, \ldots, F_n)}{\partial(x_1, \ldots, x_n)}, \quad \det\left(\frac{\partial F_i}{\partial x_j}\right), \quad \text{and} \quad \det(dF).$$

Differentiability at a point is a strong condition that implies both continuity and the existence of directional derivatives. In the propositions in the rest of the section, F is a function from an open set $U \subset \mathbb{R}^n$ to $\mathbb{R}^m$ and $\vec{a}$ is any point in U.

Proposition 1.2.15. *If F is differentiable at $\vec{a}$, then F is continuous at $\vec{a}$.*

Proof. Suppose we have the condition of Definition 1.2.14. From Theorem 1.2.7, since each component function of a linear transformation is a polynomial in the input variables, we deduce that the linear transformation L is continuous everywhere. Hence,

$$\lim_{\vec{h}\to\vec{0}} L(\vec{h}) = \vec{0}.$$

The condition on R also implies that $\lim_{\vec{h}\to\vec{0}} R(\vec{h}) = \vec{0}$. Hence,

$$\lim_{\vec{x}\to\vec{a}} F(\vec{x}) = \lim_{\vec{h}\to\vec{0}} F(\vec{a}+\vec{h}) = F(\vec{a}) + \lim_{\vec{h}\to\vec{0}}(L(\vec{h}) + R(\vec{h})) = F(\vec{a}).$$

Hence, F is continuous at $\vec{a}$.   □

Proposition 1.2.16. *If F is differentiable at $\vec{a}$, then it has a directional derivative in every direction at $\vec{a}$. Furthermore, $D_{\vec{u}}F(\vec{a}) = dF_{\vec{a}}(\vec{u})$.*

Proof. (Left as an exercise for the reader.)   □

Since the differential $dF_{\vec{a}}$ is a linear function from $\mathbb{R}^n$ to $\mathbb{R}^m$, for a vector $\vec{v} = (v_1, v_2, \ldots, v_n)$ with coordinates given with respect to the standard basis, at any point $\vec{a}$ we have

$$dF_{\vec{a}}(v_1\vec{u}_1 + \cdots + v_n\vec{u}_n) = v_1 dF_{\vec{a}}(\vec{u}_1) + \cdots + v_n dF_{\vec{a}}(\vec{u}_n)$$

$$= v_1 \frac{\partial F}{\partial x_1}\Big|_{\vec{a}} + \cdots + v_n \frac{\partial F}{\partial x_n}\Big|_{\vec{a}},$$

where the second line follows from the last part of Proposition 1.2.16. Finally, viewing each partial derivative $\frac{\partial F}{\partial x_i}(\vec{a})$ as a column vector, we have

$$dF_{\vec{a}}(\vec{v}) = \left(\frac{\partial F}{\partial x_1}(\vec{a}) \quad \frac{\partial F}{\partial x_2}(\vec{a}) \quad \cdots \quad \frac{\partial F}{\partial x_n}(\vec{a}) \right) \vec{v}.$$

This proves the following proposition.

Proposition 1.2.17. *Writing $F = (F_1, F_2, \ldots, F_m)$ in component functions, at any point where F is differentiable, the Jacobian matrix of F is*

$$[dF] = \left(\frac{\partial F}{\partial x_1} \quad \frac{\partial F}{\partial x_2} \quad \cdots \quad \frac{\partial F}{\partial x_n} \right) = \begin{pmatrix} \frac{\partial F_1}{\partial x_1} & \frac{\partial F_1}{\partial x_2} & \cdots & \frac{\partial F_1}{\partial x_n} \\ \frac{\partial F_2}{\partial x_1} & \frac{\partial F_2}{\partial x_2} & \cdots & \frac{\partial F_2}{\partial x_n} \\ \vdots & \vdots & \ddots & \vdots \\ \frac{\partial F_m}{\partial x_1} & \frac{\partial F_m}{\partial x_2} & \cdots & \frac{\partial F_m}{\partial x_n} \end{pmatrix}, \tag{1.2}$$

where in the middle expression we view $\partial F/\partial x_i$ as a column vector.

Example 1.2.13 shows that the implication statement in Proposition 1.2.16 cannot be replaced with an equivalence statement. Therefore, one should remember the caveat that the Jacobian matrix may exist at a point $\vec{a}$ without F being differentiable at $\vec{a}$, but in that case, $dF_{\vec{a}}$ does not even exist.

Example 1.2.18. Consider a function f from an open set $U \subset \mathbb{R}^n$ to $\mathbb{R}$. The differential df has the matrix

$$[df] = \left(\frac{\partial f}{\partial x_1} \quad \frac{\partial f}{\partial x_2} \quad \cdots \quad \frac{\partial f}{\partial x_n} \right),$$

which is in fact the gradient of f, though viewed as a row vector.

Example 1.2.19. As a simple example of calculating the Jacobian matrix, consider the function

$$F(x_1, x_2) = \left(3x_1 + x_2^2, x_1 \cos x_2, e^{x_1 - 2x_2} + 2x_2 \right).$$

It is defined over all $\mathbb{R}^2$. The Jacobian matrix is

$$\begin{pmatrix} \frac{\partial F_1}{\partial x_1} & \frac{\partial F_1}{\partial x_2} \\ \frac{\partial F_2}{\partial x_1} & \frac{\partial F_2}{\partial x_2} \\ \frac{\partial F_3}{\partial x_1} & \frac{\partial F_2}{\partial x_2} \end{pmatrix} = \begin{pmatrix} 3 & 2x_2 \\ \cos x_2 & -x_1 \sin x_2 \\ e^{x_1 - 2x_2} & -2e^{x_1 - 2x_2} + 2 \end{pmatrix}.$$

If, for example, $\vec{a} = (2, \pi/2)$, then the matrix for $dF_{\vec{a}}$ is

$$[dF_{\vec{a}}] = \begin{pmatrix} 3 & \pi \\ 0 & -2 \\ e^{2-\pi} & -2e^{2-\pi} + 2 \end{pmatrix}.$$

If, in addition, $\vec{v} = (3, -4)$ with coordinates given in the standard basis, then

$$dF_{\vec{a}}(\vec{v}) = \begin{pmatrix} 3 & \pi \\ 0 & -2 \\ e^{2-\pi} & -2e^{2-\pi} + 2 \end{pmatrix} \begin{pmatrix} 3 \\ -4 \end{pmatrix} = \begin{pmatrix} 9 - 4\pi \\ 8 \\ 11e^{2-\pi} - 8 \end{pmatrix}.$$

To calculate the directional derivative in the direction of $\vec{v}$, we must use the unit vector $\vec{u} = \vec{v}/\|\vec{v}\| = (0.6, -0.8)$ and

$$D_{\vec{u}}F(\vec{a}) = dF_{\vec{a}}(\vec{u}) = \begin{pmatrix} 1.8 - 0.8\pi \\ 1.6 \\ 2.2e^{2-\pi} - 1.6 \end{pmatrix}.$$

PROBLEMS

1.2.1. Let $F(x, y) = (3x - 2y + 4xy, x^4 - 3x^3y^2 + 3xy + 1)$. Determine the domain of the function, explain why or why not the function is continuous over its domain, and find all its (first) partial derivatives.

1.2.2. Repeat Problem 1 with $F(x, y) = \left(\frac{x - y}{x + y}, y \ln x \right)$.

1.2.3. Repeat Problem 1 with $F(x, y, z) = (\tan(x/y), x^3 e^{y+3z}, \sqrt{x^2 + y^2 + z^2})$.

1.2.4. Let $F(x, y, z) = \left(\cos(4x + 3yz), xz/(1 + x^2 + y^2) \right)$. Calculate F_{xx}, F_{yz} and F_{xyz}.

1.2.5. Let $F : \mathbb{R}^n \to \mathbb{R}$ be a function defined by $F(\vec{x}) = e^{\vec{u} \cdot \vec{x}}$, where $\vec{u}$ is a unit vector in $\mathbb{R}^n$. Prove that
$$\frac{\partial^2 F}{\partial x_1^2} + \frac{\partial^2 F}{\partial x_2^2} + \cdots + \frac{\partial^2 F}{\partial x_n^2} = F.$$

1.2.6. If F is a linear function, show that F is continuous.

1.2.7. Show that the following function is continuous everywhere
$$F(u, v) = \begin{cases} x_1 \sin\left(\frac{1}{x_2}\right) + x_2 \sin\left(\frac{1}{x_1}\right), & \text{if } x_1 x_2 \neq 0, \\ 0, & \text{if } x_1 x_2 = 0. \end{cases}$$

1.2.8. Find the directional derivative of $F(s, t) = (s^3 - 3st^2, 3s^2 t - t^3)$ at $(2, 3)$ in the direction $\vec{u} = (1/\sqrt{2}, 1/\sqrt{2})$.

1.2.9. Find the directional derivative of $F(x_1, x_2, x_3) = (x_1 + x_2 + x_3, x_1 x_2 + x_2 x_3 + x_1 x_3, x_1 x_2 x_3)$ at $(1, 2, 3)$ in the direction of $\vec{u} = (1/\sqrt{2}, 1/\sqrt{3}, 1/\sqrt{6})$.

1.2.10. Let $F : \mathbb{R}^2 \to \mathbb{R}^2$ be defined by $F(u, v) = (u^2 - v^2, 2uv)$. Calculate the Jacobian matrix of F. Find all points in $\mathbb{R}^2$ where $J(F) = 0$.

1.2.11. Define F over $\mathbb{R}^2$ by $F(x, y) = (e^x \cos y, e^x \sin y)$. Calculate the partial derivatives F_x and F_y. Show that the Jacobian $J(F)$ is never 0. Conclude that F_x and F_y are never collinear.

1.2.12. Let $F(u, v) = (\cos u \sin v, \sin u \sin v, \cos v)$ be defined over $[0, 2\pi] \times [0, \pi]$. Show that the image of F lies on the unit sphere in $\mathbb{R}^3$. Calculate $dF_{(u,v)}$ for all (u, v) in the domain.

1.2.13. Define $F : \mathbb{R}^3 \to \mathbb{R}^3$ by
$$F(u, v, w) = \left((u^3 + uv) \cos w, (u^3 + uv) \sin w, u^2\right).$$
Calculate the partial derivatives F_u, F_v, and F_w. Calculate the Jacobian $J(F)$. Determine where F does not have maximal rank.

1.2.14. Define F over the open set $\{(x, y, z) \in \mathbb{R}^3 \mid x > 0, y > 0, z > 0\}$ by $F(x, y, z) = (x \cdot y^z, y \cdot z^x, z \cdot x^y)$. Calculate the partial derivatives F_x, F_y, and F_z. Calculate the Jacobian $J(F)$.

1.2.15. Let $F : \mathbb{R}^n \to \mathbb{R}^m$ be a linear function, with $F(\vec{v}) = A\vec{v}$ for some $m \times n$ matrix A. Prove that the Jacobian matrix is the constant matrix A and that for all $\vec{a}$, $dF_{\vec{a}} = F$.

1.2.16. Let $F(u, v) = (u \cos v, u \sin v, u)$ defined over $\mathbb{R}^2$. Show that the image of F is a cone $x^2 + y^2 - z^2 = 0$. Calculate the differential, and determine where the differential does not have maximal rank.

1.2.17. Prove the limit laws listed in Theorem 1.2.9.

1.2.18. Prove Theorem 1.2.10.

1.2.19. Prove Proposition 1.2.16. [Hint: Using Definition 1.2.14, set $\vec{v} = h\vec{u}$, where $\vec{u}$ is a unit vector.]

1.2.20. Prove that if a function F is differentiable at $\vec{a}$, then F is continuous at $\vec{a}$.

1.2.21. *Mean Value Theorem.* Let F be a real-valued function defined over an open set $U \in \mathbb{R}^n$ and differentiable at every point of U. If the segment $[\vec{a}, \vec{b}] \subset U$, then there exists a point $\vec{c}$ in the segment $[\vec{a}, \vec{b}]$ such that

$$F(\vec{b}) - F(\vec{a}) = dF_{\vec{c}}(\vec{b} - \vec{a}).$$

1.2.22. (*) Let $n \leq m$, and consider a function $F : U \to \mathbb{R}^m$ of class C^1, where U is an open set in $\mathbb{R}^n$. Let $p \in U$, and suppose that dF_p is injective.

   (a) Prove there exists a positive number A_p such that $\|dF_p(\vec{v})\| \geq A_p\|\vec{v}\|$ for $\vec{v} \in \mathbb{R}^n$.

   (b) Use part (a) and the Mean Value Theorem to show that F is *locally injective* near p, i.e., there exists an open neighborhood U' of p such that $F : U' \to F(U')$ is injective.

1.3   Differentiation Rules; Functions of Class C^r

In a single-variable calculus course, one learns a number of differentiation rules. With functions F from $\mathbb{R}^n$ to $\mathbb{R}^m$, one must use some caution since the matrix $[dF]$ of the differential dF is not a vector function but a matrix of functions. (Again, we remind the reader that our notation for evaluating the matrix of functions $[dF]$ at a point $\vec{a}$ is $[dF_{\vec{a}}]$.)

Theorem 1.3.1. *Let U be an open set in $\mathbb{R}^n$. Let F and G be functions from U to $\mathbb{R}^m$, and let $w : U \to \mathbb{R}$ be a scalar function. If F, G, and w are differentiable at $\vec{a}$, then $F + G$ and wF are differentiable at $\vec{a}$ and*

   1. $d(F + G)_{\vec{a}} = dF_{\vec{a}} + dG_{\vec{a}}$;

   2. $\left[d(wF)_{\vec{a}}\right] = w(\vec{a})\left[dF_{\vec{a}}\right] + \left[F(\vec{a})\right]\left[dw_{\vec{a}}\right]$.

Proof. The proof for both parts follows from Proposition 1.2.17. Explicitly for the second part, the ij-entry of $\left[d(wF)_{\vec{a}}\right]$ is

$$\frac{\partial(wF_i)}{\partial x_j} = w(\vec{a})\frac{\partial F_i}{\partial x_j}(\vec{a}) + \frac{\partial w}{\partial x_j}(\vec{a})F_i(\vec{a}).$$

The first term on the right side is the ij-entry of $w(\vec{a})\left[dF_{\vec{a}}\right]$ while the second term is the ij-entry of $\left[F(\vec{a})\right]\left[dw_{\vec{a}}\right]$, which is the product of a columns by a row vector. The result follows.   □

Note that in Theorem 1.3.1(2), $[F(\vec{a})]$ is a column vector of dimension m, while $[dw_{\vec{a}}]$ is a row vector of dimension n. Hence $[F(\vec{a})][dw_{\vec{a}}]$ is an $m \times n$ matrix of rank 1.

Example 1.3.2. Let $F(u, v) = (u^2 - v, v^3, u + 2v + 1)$, and let $w(u, v) = u^3 + uv - 2$. The differentials of F and w are

$$[dF] = \begin{pmatrix} 2u & -1 \\ 0 & 3v^2 \\ 1 & 2 \end{pmatrix} \qquad \text{and} \qquad [dw] = \begin{pmatrix} 3u^2 + v & u \end{pmatrix}.$$

According to Theorem 1.3.1, the Jacobian matrix of wF is

$$[d(wF)] = w[dF] + [F][dw]$$

$$= (u^3 + uv - 2) \begin{pmatrix} 2u & -1 \\ 0 & 3v^2 \\ 1 & 2 \end{pmatrix} + \begin{pmatrix} u^2 - v \\ v^3 \\ u + 2v + 1 \end{pmatrix} \begin{pmatrix} 3u^2 + v & u \end{pmatrix}$$

$$= \begin{pmatrix} 2u(u^3 + uv - 2) & -(u^3 + uv - 2) \\ 0 & 3v^2(u^3 + uv - 2) \\ (u^3 + uv - 2) & 2(u^3 + uv - 2) \end{pmatrix} + \begin{pmatrix} (u^2 - v)(3u^2 + v) & u(u^2 - v) \\ v^3(3u^2 + v) & uv^3 \\ (u + 2v + 1)(3u^2 + v) & u(u + 2v + 1) \end{pmatrix}$$

$$= \begin{pmatrix} 5u^4 - v^2 - 4u & -2uv + 2 \\ 3u^2v^4 + v^4 & 3u^3v^2 + 4uv^3 - 6v^2 \\ 4u^3 + 6u^2v + 3u^2 + 2uv + 2v^2 + v - 2 & 2u^3 + u^2 + 4uv + u - 4 \end{pmatrix}.$$

If we had to find $[d(wF)_{(1,2)}]$, we could simplify the work and do

$$[d(wF)_{(1,2)}] = w(1, 2)[dF_{(1,2)}] + [F(1, 2)][dw_{(1,2)}]$$

$$= 1 \begin{pmatrix} 2 & -1 \\ 0 & 12 \\ 1 & 2 \end{pmatrix} + \begin{pmatrix} -1 \\ 8 \\ 6 \end{pmatrix} \begin{pmatrix} 5 & 1 \end{pmatrix}$$

$$= \begin{pmatrix} 2 & -1 \\ 0 & 12 \\ 1 & 2 \end{pmatrix} + \begin{pmatrix} -5 & -1 \\ 40 & 8 \\ 30 & 6 \end{pmatrix} = \begin{pmatrix} -3 & -2 \\ 40 & 20 \\ 31 & 8 \end{pmatrix}.$$

We now consider the *composition* of two multivariable functions. Let F be a function from a set $U \subset \mathbb{R}^n$ to $\mathbb{R}^m$, and let G be a function from a set $V \subset \mathbb{R}^p$ to $\mathbb{R}^n$ such that $G(V) \subset U$. The composite function $F \circ G : V \to \mathbb{R}^m$, depicted by the diagram

$$V(\subset \mathbb{R}^p) \xrightarrow{\quad G \quad} U(\subset \mathbb{R}^n) \xrightarrow{\quad F \quad} \mathbb{R}^m$$

is the function such that, for each $\vec{a} \in V$,

$$(F \circ G)(\vec{a}) = F\big(G(\vec{a})\big).$$

As a consequence of a general theorem in topology (see Proposition A.1.28), we know that the composition of two continuous functions is continuous. The same is true for differentiable functions, and the chain rule tells us how to compute the differential of the composition of two functions.

Theorem 1.3.3 (The Chain Rule). *Let F be a function from an open set $U \subset \mathbb{R}^n$ to $\mathbb{R}^m$, and let G be a function from an open set $V \subset \mathbb{R}^p$ to $\mathbb{R}^n$ such that $G(V) \subset U$. Let $\vec{a} \in V$. If G is differentiable at $\vec{a}$ and G is differentiable at $G(\vec{a})$, then $F \circ G$ is differentiable at $\vec{a}$ and*

$$d(F \circ G)_{\vec{a}} = dF_{G(\vec{a})} \circ dG_{\vec{a}}. \tag{1.3}$$

The Jacobian matrices satisfy the matrix product

$$\left[d(F \circ G)_{\vec{a}} \right] = \left[dF_{G(\vec{a})} \right] \left[dG_{\vec{a}} \right]. \tag{1.4}$$

Before proving this theorem, we establish a lemma.

Lemma 1.3.4. *Let A be an $m \times n$ matrix. For all $\vec{v} \in \mathbb{R}^n$, with $\|\vec{v}\| = 1$, the length $\|A\vec{v}\|$ is less than or equal to the square root of the largest eigenvalue λ_1 of $A^\top A$. Furthermore, if $\vec{u}_1$ is a unit eigenvector of $A^\top A$ corresponding to λ_1, then $\|A\vec{u}_1\| = \sqrt{\lambda_1}$.*

Proof. Assuming that we use standard bases in $\mathbb{R}^n$ and $\mathbb{R}^m$, then

$$\|A\vec{v}\|^2 = (A\vec{v}) \cdot (A\vec{v}) = (A\vec{v})^\top (A\vec{v}) = \vec{v}^\top A^\top A \vec{v}.$$

By the Spectral Theorem from linear algebra, since $A^\top A$ is a symmetric matrix, it is diagonalizable, has an orthonormal eigenbasis (i.e., a basis of eigenvectors of $A^\top A$) $\{\vec{u}_1, \ldots, \vec{u}_n\}$, and all the eigenvalues are real. Assume $\vec{u}_i$ has eigenvalue λ_i. Note that

$$\lambda_i = \vec{u}_i \cdot (\lambda_i \vec{u}_i) = \vec{u}_i \cdot (A^\top A \vec{u}_i) = \vec{u}_i^\top A^\top A \vec{u}_i = \|A\vec{u}_i\|^2,$$

so $\lambda_i \geq 0$. We also suppose that the eigenvalues are labeled so that $\lambda_1 \geq \lambda_2 \geq \cdots \geq \lambda_n$.

If $\vec{v}$ has unit length, we can write $\vec{v} = x_1 \vec{u}_1 + \cdots + x_n \vec{u}_n$, with $x_1^2 + \cdots + x_n^2 = 1$. Then

$$\|A\vec{v}\|^2 = \vec{v}^\top A^\top A \vec{v} = \lambda_1 x_1^2 + \cdots + \lambda_n x_n^2.$$

A simple calculation using Lagrange multipliers shows that $\|A\vec{v}\|^2$, subject to the constraint $\|\vec{v}\| = 1$, is maximized when $\lambda = \lambda_1$ and $(x_1, \ldots, x_n) = (1, 0, \ldots, 0)$. The lemma follows. $\qquad\qquad\Box$

We call $\sqrt{\lambda_1}$ in the above lemma the *matrix norm* of A and denote it by $|A|$. Note that for all $\vec{v} \in \mathbb{R}^n$, $\|A\vec{v}\| \leq |A| \|\vec{v}\|$.

of Theorem 1.3.3. Let F and G be functions as defined in the hypotheses of the theorem. Then there exist an $m \times n$ matrix A and an $n \times p$ matrix B such that

$$G(\vec{a} + \vec{h}) = G(\vec{a}) + B\vec{h} + R_1(\vec{h}),$$
$$F(G(\vec{a}) + \vec{k}) = F(\vec{g}(\vec{a})) + A\vec{k} + R_2(\vec{k}), \tag{1.5}$$

with

$$\lim_{\vec{h}\to\vec{0}} \frac{R_1(\vec{h})}{\|\vec{h}\|} = \vec{0} \quad \text{and} \quad \lim_{\vec{k}\to\vec{0}} \frac{R_2(\vec{k})}{\|\vec{k}\|} = \vec{0}. \tag{1.6}$$

Then for the composition $(F \circ G)(\vec{a} + \vec{h})$, we have

$$(F \circ G)(\vec{a} + \vec{h}) = F(G(\vec{a})) + AB\vec{h} + AR_1(\vec{h}) + R_2(B\vec{h} + R_1(\vec{h})). \tag{1.7}$$

Note that $\|AR_1(\vec{h})\| \le |A| \, \|R_1(\vec{h})\|$, so

$$0 \le \lim_{\vec{h}\to\vec{0}} \frac{\|AR_1(\vec{h})\|}{\|\vec{h}\|} \le \lim_{\vec{h}\to\vec{0}} |A| \frac{\|R_1(\vec{h})\|}{\|\vec{h}\|}.$$

By the Squeeze Theorem, since $\lim_{\vec{h}\to\vec{0}} \|R_1(\vec{h})\|/\|\vec{h}\| = 0$, we deduce that

$$\lim_{\vec{h}\to\vec{0}} \frac{AR_1(\vec{h})}{\|\vec{h}\|} = \vec{0}.$$

Also because $\lim_{\vec{h}\to\vec{0}} \|R_1(\vec{h})\|/\|\vec{h}\| = 0$, for any $\varepsilon > 0$, there exists a $\delta > 0$ such that if $\vec{h} \in \mathbb{R}^n$, with $\|\vec{h}\| < \delta$, then $\|R_1(\vec{h})\| < \varepsilon\|\vec{h}\|$. In particular, pick $\varepsilon = 1$ and let δ_0 be the corresponding value of δ. Then if $\|\vec{h}\| < \delta_0$, we have

$$\|B\vec{h} + R_1(\vec{h})\| \le \|B\vec{h}\| + \|R_1(\vec{h})\| \le (|B| + 1)\|\vec{h}\|.$$

This leads to

$$0 \le \frac{\|R_2(B\vec{h} + R_1(\vec{h}))\|}{\|\vec{h}\|} \le (|B| + 1)\frac{\|R_2(B\vec{h} + R_1(\vec{h}))\|}{\|B\vec{h} + R_1(\vec{h})\|}. \tag{1.8}$$

However, by Equation (1.6), one concludes that

$$\lim_{\vec{h}\to\vec{0}} \frac{\|R_2(B\vec{h} + R_1(\vec{h}))\|}{\|B\vec{h} + R_1(\vec{h})\|} = 0$$

and consequently, by Equation (1.8),

$$\lim_{\vec{h}\to\vec{0}} \frac{\|R_2(B\vec{h} + R_1(\vec{h}))\|}{\|\vec{h}\|} = 0.$$

Setting $R_3(\vec{h}) = AR_1(\vec{h}) + R_2(B\vec{h} + R_1(\vec{h}))$, we have $\lim_{\vec{h}\to\vec{0}} R_3(\vec{h})/\|\vec{h}\| = \vec{0}$, so from Equation (1.7), we see that all parts of the theorem hold. $\square$

Example 1.3.5. Consider the functions $F(r, \theta) = (r\cos\theta, r\sin\theta)$ and $G(s, t) = (s^2 - t^2, 2st)$. Calculating the composition function directly, we have

$$(G \circ F)(r, \theta) = (r^2 \cos^2\theta - r^2\sin^2\theta, 2r^2\cos\theta\sin\theta) = (r^2\cos 2\theta, r^2\sin 2\theta).$$

Thus,

$$[d(G \circ F)_{(r,\theta)}] = \begin{pmatrix} 2r\cos 2\theta & -2r^2\sin 2\theta \\ 2r\sin 2\theta & 2r^2\cos 2\theta \end{pmatrix}.$$

On the other hand, we have

$$[dG_{(s,t)}] = \begin{pmatrix} 2s & -2t \\ 2t & 2s \end{pmatrix} \qquad \text{and} \qquad [dF_{(r,\theta)}] = \begin{pmatrix} \cos\theta & -r\sin\theta \\ \sin\theta & r\cos\theta \end{pmatrix}.$$

Using the right-hand side of the chain rule, we calculate

$$\begin{aligned}
[dG_{F(r,\theta)}]\,[dF_{(r,\theta)}] &= [dG_{(r\cos\theta, r\sin\theta)}]\,[dF_{(r,\theta)}] \\
&= \begin{pmatrix} 2r\cos\theta & -2r\sin\theta \\ 2r\sin\theta & 2r\cos\theta \end{pmatrix}\begin{pmatrix} \cos\theta & -r\sin\theta \\ \sin\theta & r\cos\theta \end{pmatrix} \\
&= \begin{pmatrix} 2r\cos 2\theta & -2r^2\sin 2\theta \\ 2r\sin 2\theta & 2r^2\cos 2\theta \end{pmatrix} = [d(G \circ F)_{(r,\theta)}]
\end{aligned}$$

as expected.

The style of presentation of the chain rule in Theorem 1.3.3 is often attributed to Newton's notation. Possible historical inaccuracies aside, Equation (1.3) is commonly used by mathematicians. In contrast, physicists tend to use Leibniz's notation, which we present now.

Suppose that the vector variable $\vec{y} = (y_1, \ldots, y_n)$ is given as a function of a variable $\vec{x} = (x_1, \ldots, x_p)$ (this function corresponds to $\vec{g}$ in Equation (1.3)) and suppose that the vector variable $\vec{z} = (z_1, \ldots, z_m)$ is given as a function of the variable $\vec{y}$ (this function corresponds to $\vec{f}$). With Leibniz's notation, one writes the chain rule as

$$\frac{\partial z_i}{\partial x_j} = \sum_{k=1}^{n} \frac{\partial z_i}{\partial y_k} \frac{\partial y_k}{\partial x_j} \qquad \text{for all } i = 1, \ldots, m \text{ and } j = 1, \ldots, p. \qquad (1.9)$$

When evaluating $\partial z_i / \partial x_j$ at a point $\vec{a} \in \mathbb{R}^p$, one should understand the chain rule in Equation (1.9) explicitly as

$$\left.\frac{\partial z_i}{\partial x_j}\right|_{\vec{a}} = \sum_{k=1}^{n} \left.\frac{\partial z_i}{\partial y_k}\right|_{\vec{y}(\vec{a})} \left.\frac{\partial y_k}{\partial x_j}\right|_{\vec{a}}.$$

Suppose a function F is differentiable over an open set $U \subset \mathbb{R}^n$ to $\mathbb{R}^m$. Then for any unit vector $\vec{u} \subset \mathbb{R}^n$, the directional derivative $D_{\vec{u}}F$ is itself a vector function from U to $\mathbb{R}^m$, and we can consider the directional derivative $D_{\vec{v}}(D_{\vec{u}}F)$ along some unit vector $\vec{v}$. This *second-order directional derivative* is denoted by $D^2_{\vec{v}\vec{u}}F$. Higher-order directional derivatives are defined in the same way.

If $\mathbb{R}^n$ is given a basis, then one can take higher-order partial derivatives with respect to this basis. Some common notations for the second partial derivative $\frac{\partial}{\partial x_j}\left(\frac{\partial F}{\partial x_i}\right)$ are

$$\frac{\partial^2 F}{\partial x_j \partial x_i}\ ,\qquad \partial_j \partial_i F\ ,\qquad D_j D_i F\ ,\qquad F_{x_i x_j}\ ,\qquad F_{,i,j}$$

and notations for third partial derivatives are

$$\frac{\partial^3 F}{\partial x_k \partial x_j \partial x_i}\ ,\qquad \partial_k \partial_j \partial_i F\ ,\qquad D_k D_j D_i F\ ,\qquad F_{x_i x_j x_k}\ ,\qquad F_{,i,j,k}\ .$$

Most advanced physics texts use the notation $\partial_i F$ for the partial derivative $\partial F/\partial x_i$. In that case, the second and third partial derivatives are $\partial_j \partial_i \vec{f}$ and $\partial_k \partial_j \partial_i F$, as indicated above.

Note that the order of the indices or subscripts is important since it is possible that

$$\frac{\partial^2 F}{\partial x_1 \partial x_2} \neq \frac{\partial^2 F}{\partial x_2 \partial x_1},$$

though we will see momentarily a condition that implies their equality.

We conclude this section with two theorems from analysis and a comment on the C^r notation.

Theorem 1.3.6. *Let U be an open set in $\mathbb{R}^n$, let $F : U \to \mathbb{R}^m$ be a function, and let $\vec{a} \in U$. Suppose that for each $i = 1, 2, \ldots, n$, the partial derivative $\partial F/\partial x_i$ exists in a neighborhood of $\vec{a}$ and is continuous at $\vec{a}$. Then F is differentiable at $\vec{a}$.*

Proof. (See Theorem 8.23 in [15].) □

Theorem 1.3.7 (Clairaut's Theorem). *Let U be an open set in $\mathbb{R}^n$, let $F : U \to \mathbb{R}^m$ be a function, and let $\vec{a} \in U$. Suppose that*

$$\frac{\partial F}{\partial x_i},\ \frac{\partial F}{\partial x_j},\ \text{and}\ \frac{\partial^2 F}{\partial x_j \partial x_i}$$

exists in a neighborhood of $\vec{a}$ and that $\partial^2 F/\partial x_j \partial x_i$ is continuous at $\vec{a}$. Then $\partial^2 F/\partial x_i \partial x_j(\vec{a})$ exists and

$$\frac{\partial^2 F}{\partial x_i \partial x_j}(\vec{a}) = \frac{\partial^2 F}{\partial x_j \partial x_i}(\vec{a}).$$

Proof. (See Theorem 8.24 in [15].) □

Theorems 1.3.6 and 1.3.7 illustrate that certain nice properties occur when we not only assume that partial derivatives exist but that they are continuous at a particular point. For this reason, if U is an open set in $\mathbb{R}^n$, we say that a function $F : U \to \mathbb{R}^m$ is of *class* C^1 if all of its partial derivatives exist and are continuous. By Theorem 1.3.6, a function of class C^1 is differentiable. We denote by $C^1(U, \mathbb{R}^m)$ the set of such functions.

More generally, we say that the function F is of *class* C^r, or write $F \in C^r(U, \mathbb{R}^m)$, if all of its first rth partial derivatives exist and are continuous. By Clairaut's Theorem, we see that for a function of class C^r all the mixed partial derivatives up to order r involving the same number of the same index of variable are equal. To be consistent with this notation, we say that F is of class C^0 if it is continuous and we say that it is of class C^∞ if all of its higher partial derivatives exist and are continuous. Functions of class C^∞ are called *smooth*.

Finally, we say that a function $F : U \to \mathbb{R}^m$ is *analytic* if for all $\vec{a} \in U$, there exists an open ball $B_\delta(\vec{a}) \subseteq U$ such that over $B_\delta(\vec{a})$ the Taylor series of F centered at $\vec{a}$ converges to $F(\vec{x})$ in $B_\delta(\vec{a})$. If $F : U \to \mathbb{R}^m$ is analytic, we write $F \in C^\omega(U, \mathbb{R}^m)$. This is a stronger condition than smooth since in order for a function to be analytic at $\vec{a}$, all of its partial derivatives must exist at $\vec{a}$.

There is a natural chain of containment among these classes of functions

$$C^\omega(U, \mathbb{R}^m) \subset C^\infty(U, \mathbb{R}^m) \subset \cdots \subset C^r(U, \mathbb{R}^m) \subset \cdots \subset C^0(U, \mathbb{R}^m).$$

Theorem 1.3.8 (First-Order Taylor Series). *Let $\vec{a} \in \mathbb{R}^n$ and let $U = B_r(\vec{a})$ be the open ball of radius r and center $\vec{a}$. Suppose that $f \in C^k(U, \mathbb{R})$ for $k \geq 1$. Then*

$$f(\vec{x}) = f(\vec{a}) + \sum_{i=1}^{n} \frac{\partial f}{\partial x_i}(\vec{a})(x_i - a_i) + \sum_{i=1}^{n} g_i(\vec{x})(x_i - a_i) \qquad (1.10)$$

for some functions $g_1, g_2, \ldots, g_n \in C^{k-1}(U, \mathbb{R})$ such that $g_i(\vec{a}) = 0$.

Proof. Let $\vec{x}$ be any element in the ball U. The Fundamental Theorem of Calculus gives

$$f(\vec{x}) - f(\vec{a}) = \int_0^1 \frac{d}{dt} f(\vec{a} + t(\vec{x} - \vec{a})) \, dt.$$

By the chain rule

$$f(\vec{x}) - f(\vec{a}) = \int_0^1 \sum_{i=1}^{n} \frac{\partial f}{\partial x_i}(\vec{a} + t(\vec{x} - \vec{a}))(x_i - a_i) \, dt$$

and since $\partial f / \partial x_i(\vec{a})(x_i - a_i)$ is constant with respect to t, we have

$$f(\vec{x}) - f(\vec{a}) = \sum_{i=1}^{n} \frac{\partial f}{\partial x_i}(\vec{a})(x_i - a_i) + \sum_{i=1}^{n}(x_i - a_i) \int_0^1 \left(\frac{\partial f}{\partial x_i}(\vec{a} + t(\vec{x} - \vec{a})) - \frac{\partial f}{\partial x_i}(\vec{a}) \right) dt.$$

Setting

$$g_i(\vec{x}) = \int_0^1 \left(\frac{\partial f}{\partial x_i}(\vec{a} + t(\vec{x} - \vec{a})) - \frac{\partial f}{\partial x_i}(\vec{a}) \right) dt,$$

we obtain (1.10). We note that $g_i(\vec{a}) = 0$. Furthermore, it is possible to differentiate each $g_i(x)$ by passing the differentiation with respect to any x_i variable underneath the integral with respect to t. Hence, we see that $g_i \in C^{k-1}(U, \mathbb{R})$ for all $i = 1, 2, \ldots, n$. $\qquad\square$

PROBLEMS

1.3.1. Prove Theorem 1.3.1.

1.3.2. Suppose that $\vec{f}$ and $\vec{g}$ are differentiable at $\vec{a} \in \mathbb{R}^n$. Prove that the function $\vec{f} \cdot \vec{g}$ is differentiable at $\vec{a}$ and that

$$d(F \cdot G)_{\vec{a}} = F(\vec{a}) \cdot dG_{\vec{a}} + G(\vec{a}) \cdot dF_{\vec{a}}$$

are linear functions.

1.3.3. Let $F(r, \theta, \phi) = (r \cos \theta \sin \phi, r \sin \theta \sin \phi, r \cos \phi)$. Calculate the Jacobian matrix. Prove that the Jacobian is the function $r^2 \sin \phi$.

1.3.4. Let

$$\begin{cases} z_1 &= 2y_1 + 3y_2, \\ z_2 &= y_1 y_2^2, \end{cases} \quad \text{and} \quad \begin{cases} y_1 &= e^{x_1} + x_2 + x_3, \\ y_2 &= e^{x_2 - x_3} + x_1. \end{cases}$$

Use the chain rule to calculate the partial derivatives $\dfrac{\partial z_i}{\partial x_j}$ for $i = 1, 2$ and $j = 1, 2, 3$.

1.3.5. Let F be a differentiable function from an open set $U \subset \mathbb{R}^n$ to $\mathbb{R}^n$, and let G be a differentiable function from an open set $V \subset \mathbb{R}^n$ to U. Prove that $J(F \circ G) = J(F)J(G)$.

1.3.6. Suppose that U and V are open sets in $\mathbb{R}^n$ and that F is bijective from U to V. Suppose in addition that F is differentiable on U and F^{-1} is differentiable on V. Prove that for all $\vec{a} \in U$, the linear function $dF_{\vec{a}}$ is invertible and that

$$(dF_{\vec{a}})^{-1} = dF_{F(\vec{a})}^{-1}.$$

Conclude that $J(F^{-1}) = 1/J(F)$.

1.3.7. Let F be a function from $U \subset \mathbb{R}^2$ to $\mathbb{R}^3$ such that $dF_{\vec{x}}$ has rank 2 for all $\vec{x} \in U$. Let $\vec{\alpha}$ be a regular curve from an interval I to U. Show that

(a) the function $\vec{\beta}(t) = F(\vec{\alpha}(t))$ is a regular curve in $\mathbb{R}^3$;

(b) the speed of $\vec{\beta}$ satisfies

$$\left\| \frac{\partial \vec{\beta}}{\partial t} \right\|^2 = \left\| \frac{\partial F}{\partial x_1} \right\|^2 \left(\frac{d\alpha_1}{dt} \right)^2 + 2 \left(\frac{\partial F}{\partial x_1} \cdot \frac{\partial F}{\partial x_2} \right) \frac{d\alpha_1}{dt} \frac{d\alpha_2}{dt} + \left\| \frac{\partial F}{\partial x_2} \right\|^2 \left(\frac{d\alpha_2}{dt} \right)^2.$$

1.3.8. Repeat part (b) of Problem 1.3.7, but prove that

$$\|\vec{\beta}'(t)\|^2 = (\vec{\alpha}'(t))^\top [dF]^\top [dF] \vec{\alpha}'(t).$$

[Hint: Recall that we view the vectors $\vec{a}, \vec{b} \in \mathbb{R}^n$ as column vectors and $\vec{a} \cdot \vec{b} = \vec{a}^\top \vec{b}$ as a matrix product.]

1.3.9. Let $F(s, t) = (s^2 t + t^3, t e^s + s e^t)$, and let $\vec{u}$ be the unit vector in the direction $(1, 1)$ and $\vec{v}$ be the unit vector in the direction $(2, 3)$. Calculate the second directional derivative function $D^2_{\vec{v}\vec{u}} F$. [Hint: This is a function of (s, t).]

1.3.10. Let F be a function from an open set $U \subset \mathbb{R}^n$ to $\mathbb{R}^m$. Let $\vec{v}$ and $\vec{u}$ be two unit vectors in $\mathbb{R}^n$. Prove that

$$D^2_{\vec{v}\vec{u}} F = \sum_{i,j=1}^{n} \frac{\partial^2 F}{\partial x_i \partial x_j} v_i u_j.$$

1.3.11. Let $F(r, \theta) = (r \cos \theta, r \sin \theta)$. Calculate all the second partial derivatives of F. Prove that F is of class C^∞ over all of $\mathbb{R}^2$.

1.3.12. Let $F(u, v) = (u^2 + v e^{2u}, v + \tan^{-1}(u + 3), \sin v)$. Find the domain of F. Calculate all of its second partial derivatives. Calculate the following third partial derivatives: F_{vvu}, F_{vuv}, and F_{uuv}.

1.3.13. If $(w_1, w_2) = \left(e^{-x_1 + x_2^2}, \cos(x_2 + x_3) \right)$, calculate

$$\frac{\partial^2 w_1}{\partial x_1 \partial x_3}, \frac{\partial^2 w_1}{\partial x_3 \partial x_2}, \text{ and } \frac{\partial^3 w_2}{\partial x_1 \partial x_2 \partial x_3}.$$

1.3.14. Let the function $f : \mathbb{R}^2 \to \mathbb{R}$ be defined by

$$f(s, t) = \begin{cases} \dfrac{2st(s^2 - t^2)}{s^2 + t^2}, & \text{if } (s, t) \neq (0, 0), \\ 0, & \text{if } (s, t) = (0, 0). \end{cases}$$

Show that f is of class C^1. Show that the mixed second partial derivatives f_{st} and f_{ts} exist at every point of $\mathbb{R}^2$. Show that $f_{st}(0, 0) \neq f_{ts}(0, 0)$.

1.4   Inverse and Implicit Function Theorems

In single- and multivariable calculus of a function $F : \mathbb{R}^n \to \mathbb{R}$, one defines a critical point as a point $\vec{a} = (a_1, \ldots, a_n)$ such that the gradient of F at $\vec{a}$ is $\vec{0}$, i.e.,

$$\nabla F(\vec{a}) = \left(\frac{\partial F}{\partial x_1}(\vec{a}), \ldots, \frac{\partial F}{\partial x_n}(\vec{a}) \right) = \vec{0}.$$

At such a point, F is said to have a flat tangent line or tangent plane, and, according to standard theorems in calculus, $F(\vec{a})$ is either a local minimum, local maximum, or a "saddle point." This notion is a special case of the following general definition.

Definition 1.4.1. Let U be an open subset of $\mathbb{R}^n$ and $F : U \to \mathbb{R}^m$ a differentiable function. We call $q \in U$ a *critical point* of F if F is not differentiable at q or if $dF_q : \mathbb{R}^n \to \mathbb{R}^m$ is not of maximum rank, i.e., if $\text{rank}(dF_q) < \min(m, n)$. If q is a critical point of F, we call $F(q)$ a *critical value*. If $p \in \mathbb{R}^m$ is not a critical value of F (even if p is not in the image of F), then we call p a *regular value* of F.

We point out that this definition simultaneously generalizes the notion of a critical point for functions $F : U \to \mathbb{R}$, with U an open subset of $\mathbb{R}^n$, and the definition for a critical point of a parametric curve in $\mathbb{R}^n$ (Definition 3.2.1 in [5]). If $m = n$, the notion of a critical point has a few alternate equivalent criteria.

Proposition 1.4.2. *Let U be an open subset of $\mathbb{R}^n$, $F : U \to \mathbb{R}^n$ a differentiable function, and q a point in U such that F is differentiable at q. The following are equivalent:*

1. *q is a critical point of U.*

2. *$J(F)(q) = 0$.*

3. *The set of partial derivatives $\{\partial F / \partial x_1(q), \ldots, \partial F / \partial x_n(q)\}$ is a linearly dependent set of vectors.*

4. *The differential dF_q is not invertible.*

Proof. These all follow from Theorem 1.1.8. $\qquad\qquad\qquad\qquad\qquad$ $\square$

More generally, when n is not necessarily equal to m, linear algebra gives the following equivalent statements for when q is a critical point.

Determining for what values of q in the domain U the differential dF_q does not have maximal rank is not easy if done simply by looking at the matrix of functions $[dF_q]$. The following proposition provides a concise criterion.

Proposition 1.4.3. *Let $F : U \to \mathbb{R}^m$ be a function where U is an open subset of $\mathbb{R}^n$. Let $q \in U$ such that F is differentiable at q. Then the following are equivalent:*

1. *q is a critical point of F.*

2. *The determinants of all the maximal square submatrices of $[dF_q]$ are 0.*

3. *The sum of the squares of the determinants of all the maximal square submatrices of $[dF_q]$ is 0.*

Furthermore, if $n \geq m$ and $A = [dF_q]$, then q is a critical point of F if and only if $\det(AA^\top) \neq 0$.

Proof. To prove $1 \Leftrightarrow 2$, note that by definition, q is a critical point if dF_q does not have maximal rank, which means that the set of column vectors or the set of row vectors of $[dF_q]$ is linearly dependent. This is equivalent to the determinants of all maximal submatrices of A (sometimes referred to as the *maximal minors* of A) being 0 since, if one such determinant were not 0, then no nontrivial linear combination of the columns of $[dF_q]$ or of the rows of $[dF_q]$ would be 0, and hence, this set would be linearly independent.

The equivalence $2 \Leftrightarrow 3$ is trivial.

To prove the last part of the proposition, assuming that $n \geq m$, recall that if $\{\vec{v}_1, \ldots, \vec{v}_m\}$ are vectors in $\mathbb{R}^n$, the m-volume of the parallelepiped formed by $\{\vec{v}_1, \ldots, \vec{v}_m\}$ is

$$\sqrt{\det(B^\top B)},$$

where B is the $n \times m$ matrix, with the $\vec{v}_i$ as columns (see [14, Fact 6.3.7]). Now the m-volume of this parallelepiped is 0 if and only if $\{\vec{v}_1, \ldots, \vec{v}_m\}$ are linearly dependent. Thus, taking $B = A^\top$ and taking the $\vec{v}_i$ as the columns of $A^\top$ establishes the result. $\qquad\square$

By referring to some advanced linear algebra, it is possible to prove directly that, if $n > m$, Condition 3 in the above proposition implies that $\det(AA^\top) \neq 0$. In fact, even more can be said. If A is an $m \times n$ matrix with $n > m$, then $\det(AA^\top)$ is equal to the sum of the squares of the maximal minors of A. (See Proposition C.1.2 in Appendix C.)

Example 1.4.4. For example, consider the function $F : \mathbb{R}^3 \to \mathbb{R}^2$ defined by $F(x, y, z) = (x^2 + 3y + z^3, xy + z^2 + 1)$. The Jacobian matrix for this function is

$$[dF] = \begin{pmatrix} 2x & 3 & 3z^2 \\ y & x & 2z \end{pmatrix}.$$

In this case, the easiest way to find the critical points of this function is to use the second equivalence statement in Proposition 1.4.3. The maximal 2×2 submatrices are

$$\begin{pmatrix} 2x & 3 \\ y & x \end{pmatrix}, \quad \begin{pmatrix} 2x & 3z^2 \\ y & 2z \end{pmatrix}, \quad \begin{pmatrix} 3 & 3z^2 \\ x & 2z \end{pmatrix},$$

so since critical points occur where all of these have determinant 0, the critical points satisfy the system of equations

$$\begin{cases} 2x^2 - 3y = 0, \\ 4xz - 3yz^2 = 0, \\ 6z - 3xz^2 = 0. \end{cases}$$

This is equivalent to

$$\begin{cases} y = \frac{2}{3}x^2, \\ 4xz - 2x^2z^2 = 0, \\ z(2 - xz) = 0, \end{cases} \iff \begin{cases} y = \frac{2}{3}x^2, \\ xz(2 - xz) = 0, \\ z(2 - xz) = 0. \end{cases}$$

Thus, the set of critical points of F is

$$\left\{ \left(x, \frac{2}{3}x^2, \frac{2}{x} \right) \in \mathbb{R}^3 \,\Big|\, x \in \mathbb{R} - \{0\} \right\} \cup \left\{ \left(x, \frac{2}{3}x^2, 0 \right) \in \mathbb{R}^3 \,\Big|\, x \in \mathbb{R} \right\}.$$

The set of critical values is then

$$\left\{ \left(3x^2 + \frac{8}{x^3}, \frac{2}{3}x^3 + \frac{4}{x^2} + 1 \right) \in \mathbb{R}^2 \,\Big|\, x \in \mathbb{R} - \{0\} \right\} \cup \left\{ \left(3x^2, \frac{2}{3}x^3 + 1 \right) \in \mathbb{R}^2 \,\Big|\, x \in \mathbb{R} \right\}.$$

One important aspect of critical points already arises with real functions. With a real differentiable function $f : [a, b] \to \mathbb{R}$, if $f'(x_0) = 0$, one can show that f does not have an inverse function that is differentiable over a neighborhood of x_0. Conversely, if $f'(x_0) \neq 0$, the function f has a differentiable inverse in a neighborhood of x_0, with

$$(f^{-1})'(y_0) = \frac{1}{f'(f^{-1}(y_0))}.$$

A similar fact holds for multivariable functions and is called the Inverse Function Theorem.

The proof of the Inverse Function Theorem and the following Implicit Function Theorem are quite long and not necessary for the purposes of this book, so we refer the reader to a book on analysis for a proof (see, for example, Section 8.5 in [15]). Instead, we simply state the theorems and present a few examples.

Theorem 1.4.5 (Inverse Function Theorem). *Let F be a function from an open set $U \subset \mathbb{R}^n$ to $\mathbb{R}^n$, and suppose that F is of class C^r, with $r \geq 1$. If $q \in U$ is not a critical point of F, then dF_q is invertible and there exists a neighborhood V of q such that F is one-to-one on V, $F(V)$ is open, and the inverse function $F^{-1} : F(V) \to V$ is of class C^r. Furthermore, for all $p \in F(V)$, with $p = F(q)$,*

$$d(F^{-1})_p = (dF_q)^{-1}.$$

In many situations, it is impossible to explicitly calculate the inverse function F^{-1}. The following example illustrates the Implicit Function Theorem in a situation in which we can calculate the inverse function.

Example 1.4.6. Consider the function $F(s, t) = (s^2 - t^2, 2st)$ and $q = (2, 3)$. Note that F is defined on all $U = \mathbb{R}^2$. The Jacobian matrix is

$$\begin{pmatrix} 2s & -2t \\ 2t & 2s \end{pmatrix},$$

so the Jacobian is the function $J(F)(s, t) = 4(s^2 + t^2)$. By Proposition 1.4.2, the only critical point of F is $(0, 0)$, so F satisfies the conditions of the Inverse Function Theorem at q.

Now with $q = (2, 3)$, by the Inverse Function Theorem, since $p = F(q) = (-5, 12)$, we have

$$[dF_q] = \begin{pmatrix} 4 & -6 \\ 6 & 4 \end{pmatrix} \quad \text{and} \quad [d(F^{-1})_p] = [dF_q]^{-1} = \frac{1}{26} \begin{pmatrix} 2 & 3 \\ -3 & 2 \end{pmatrix}.$$

For simplicity, let us assume $V = \{(s, t) \in \mathbb{R}^2 \mid s > 0, t > 0\}$ and note that $q \in V$. Setting $(x, y) = F(s, t)$ and solving for (s, t), we find that $F(V) = \{(x, y) \in \mathbb{R}^2 \mid y > 0\}$ and that the inverse of F is given by

$$s = \sqrt{\frac{\sqrt{x^2 + y^2} + x}{2}} \quad \text{and} \quad t = \sqrt{\frac{\sqrt{x^2 + y^2} - x}{2}}.$$

Calculating the partial derivative $\partial s/\partial x$, we have

$$\frac{\partial s}{\partial x} = \frac{1}{2\sqrt{2}\sqrt{\sqrt{x^2+y^2}+x}}\left(\frac{x}{\sqrt{x^2+y^2}}+1\right)$$

$$= \frac{1}{2\sqrt{x^2+y^2}}\sqrt{\frac{\sqrt{x^2+y^2}+x}{2}},$$

and similarly, the Jacobian matrix of F^{-1} is

$$\left[dF^{-1}\right] = \frac{1}{2\sqrt{x^2+y^2}}\begin{pmatrix} \sqrt{\frac{\sqrt{x^2+y^2}+x}{2}} & \sqrt{\frac{\sqrt{x^2+y^2}-x}{2}} \\ -\sqrt{\frac{\sqrt{x^2+y^2}-x}{2}} & \sqrt{\frac{\sqrt{x^2+y^2}+x}{2}} \end{pmatrix}.$$

Plugging in $p = (-5, 12) = F(q)$, we calculate directly that

$$\left[dF^{-1}_{F(q)}\right] = \frac{1}{2\sqrt{(-5)^2+12^2}}\begin{pmatrix} \sqrt{\frac{\sqrt{(-5)^2+12^2}-5}{2}} & \sqrt{\frac{\sqrt{(-5)^2+12^2}+5}{2}} \\ -\sqrt{\frac{\sqrt{(-5)^2+12^2}+5}{2}} & \sqrt{\frac{\sqrt{(-5)^2+12^2}-5}{2}} \end{pmatrix}$$

$$= \frac{1}{26}\begin{pmatrix} 2 & 3 \\ -3 & 2 \end{pmatrix} = [dF_q]^{-1},$$

thereby illustrating the Inverse Function Theorem.

Another important theorem about functions in the neighborhood of a point p that is not critical, is the fact that the level set through p, can be parametrized by (is the image of) an appropriate function. This is the Implicit Function Theorem.

Theorem 1.4.7 (Implicit Function Theorem). *Let F be a function from an open set $U \subset \mathbb{R}^n$ to $\mathbb{R}^m$, with $n > m$, and suppose that F is of class C^r, with $r \geq 1$. Let $q \in U$, and let Σ be the level set of F through q, defined as*

$$\Sigma = \{\vec{x} \in \mathbb{R}^n \mid F(\vec{x}) = F(q)\}.$$

If $q \in U$ is not a critical point, then the coordinates of $\mathbb{R}^n$ can be relabeled so that

$$dF_q = \begin{pmatrix} \overset{n-m}{S} & \Big| & \overset{m}{T} \end{pmatrix} m,$$

with T an $m \times m$ invertible matrix. Then there exist an open neighborhood V of q in $\mathbb{R}^n$, an open neighborhood W of $a = (q_1, \ldots, q_{n-m})$ in $\mathbb{R}^{n-m}$, and a function $g: W \to \mathbb{R}^m$ that is of class C^r such that $\Sigma \cap V$ is the graph of g, i.e.,

$$\Sigma \cap V = \{(\vec{s}, g(\vec{s})) \mid \vec{s} \in W\}.$$

Furthermore, the Jacobian matrix of g at a is

$$[dg_a] = -T^{-1}S.$$

Example 1.4.8. Let $F : \mathbb{R}^3 \to \mathbb{R}^1$ be a function. If c is some constant, we expect that the solution set Σ to the equation $F(x, y, z) = c$ is a surface in $\mathbb{R}^3$. Suppose that around some point p that satisfies the equation, we could consider Σ as the graph of a function $z = f(x, y)$. If it is not tractable to exactly solve for z and get $f(x, y)$ implicitly, then f is called an implicit function. By chain rule we have

$$\frac{\partial F}{\partial x}\frac{\partial x}{\partial x} + \frac{\partial F}{\partial y}\frac{\partial y}{\partial x} + \frac{\partial F}{\partial z}\frac{\partial z}{\partial x} = 0 \quad \text{and} \quad \frac{\partial F}{\partial x}\frac{\partial x}{\partial y} + \frac{\partial F}{\partial y}\frac{\partial y}{\partial y} + \frac{\partial F}{\partial z}\frac{\partial z}{\partial y} = 0.$$

Hence,

$$\frac{\partial F}{\partial x} + \frac{\partial F}{\partial z}\frac{\partial z}{\partial x} = 0 \quad \text{and} \quad \frac{\partial F}{\partial y} + \frac{\partial F}{\partial z}\frac{\partial z}{\partial y} = 0,$$

so

$$\frac{\partial z}{\partial x} = -\frac{\partial F}{\partial x}\bigg/\frac{\partial F}{\partial z} \quad \text{and} \quad \frac{\partial z}{\partial y} = -\frac{\partial F}{\partial y}\bigg/\frac{\partial F}{\partial z}.$$

This work is called implicit differentiation. Organizing this last line into a matrix of a differential, we have

$$[df] = -\left(\frac{\partial F}{\partial z}\right)^{-1}\left(\frac{\partial F}{\partial x} \quad \frac{\partial F}{\partial y}\right) = -T^{-1}S,$$

where $[dF] = (\ S\ |\ T\ )$ as in the Implicit Function Theorem. In this example, we began by assuming that a neighborhood of $p \in \Sigma$ could be viewed as the graph of $z = f(x, y)$ and proceeded from there without knowing that we were allowed to do so. The Implicit Function Theorem gives a condition in which we are allowed to proceed as we did. In this specific case, the theorem states that we can make this assumption when $\partial F/\partial z \neq 0$, which is precisely what is required for our calculations to have meaning.

Example 1.4.9. We use the Implicit Function Theorem to tell us something about the set

$$\Sigma = \{(x, y, z) \in \mathbb{R}^3 \,|\, x^2 + y^2 + z^2 = 1 \text{ and } x + y + z = 1\}.$$

This is the intersection between a sphere and a plane, which is a circle lying in $\mathbb{R}^3$. (In Figure 1.6, Σ is the circle shown as the intersection of the sphere and the plane.) Consider the point $q = \left(\frac{4}{13}, -\frac{3}{13}, \frac{12}{13}\right) \in \Sigma$. To study Σ near q, consider the function $F : \mathbb{R}^3 \to \mathbb{R}^2$ defined by $F(x, y, z) = (x^2 + y^2 + z^2, x + y + z)$. The Jacobian matrix of F is

$$[dF] = \begin{pmatrix} 2x & 2y & 2z \\ 1 & 1 & 1 \end{pmatrix},$$

and so the critical points of F are points $(x, y, z) \in \mathbb{R}^3$ such that $x = y = z$. Thus, q is not a critical point and

$$[dF_q] = \begin{pmatrix} \frac{8}{13} & -\frac{6}{13} & \frac{24}{13} \\ 1 & 1 & 1 \end{pmatrix}.$$

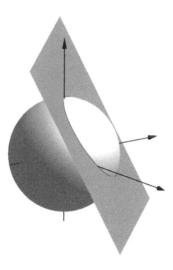

Figure 1.6: Example 1.4.9.

Writing

$$S = \begin{pmatrix} \frac{8}{13} \\ 1 \end{pmatrix} \qquad \text{and} \qquad T = \begin{pmatrix} -\frac{6}{13} & \frac{24}{13} \\ 1 & 1 \end{pmatrix},$$

since T is invertible, F and q satisfy the criteria of the Implicit Function Theorem. Thus, there exist an open neighborhood V of q in $\mathbb{R}^3$, an open interval W around $a = \frac{4}{13}$ in $\mathbb{R}$, and a function $g : W \to \mathbb{R}^2$ such that the portion of the circle $\Sigma \cap V$ is the graph of g. Also, the Jacobian matrix of g at a (the gradient of g at a) is

$$dg_a = \vec{\nabla} g\left(\frac{4}{13}\right) = -T^{-1}S = -\begin{pmatrix} -\frac{13}{30} & \frac{24}{30} \\ \frac{13}{30} & \frac{6}{30} \end{pmatrix}\begin{pmatrix} \frac{8}{13} \\ 1 \end{pmatrix} = \begin{pmatrix} -\frac{8}{15} \\ -\frac{7}{15} \end{pmatrix}. \tag{1.11}$$

One can find Σ by first noting that the subspace $x + y + z = 0$ has $\{(0, -1, 1), (-2, 1, 1)\}$ as an orthogonal basis. Thus, the plane $x + y + z = 1$ can be parametrized by

$$\vec{X}(u, v) = \left(\frac{1}{3}, \frac{1}{3}, \frac{1}{3}\right) + u(0, -1, 1) + v(-2, 1, 1),$$

and all vectors in this expression are orthogonal to each other. The additional condition that $x^2 + y^2 + z^2 = 1$ be equivalent to $\vec{X} \cdot \vec{X} = 1$ leads to $2u^2 + 6v^2 = \frac{2}{3}$. This shows that the set Σ can be parametrized by

$$\begin{cases} x &= \frac{1}{3} - \frac{2}{3}\sin t, \\ y &= \frac{1}{3} - \frac{1}{\sqrt{3}}\cos t + \frac{1}{3}\sin t, \\ z &= \frac{1}{3} + \frac{1}{\sqrt{3}}\cos t + \frac{1}{3}\sin t. \end{cases}$$

However, this parametrization is not the one described by the Implicit Function Theorem. But by using it, one can find that in a neighborhood of $q = (4/13, -3/13, 12/13)$,
Σ is parametrized by

$$\left(x, \frac{1-x}{2} - \frac{1}{2}\sqrt{1+2x-3x^2}, \frac{1-x}{2} + \frac{1}{2}\sqrt{1+2x-3x^2} \right),$$

and thus the implicit function g in Theorem 1.4.7 is

$$g(x) = \left(\frac{1-x}{2} - \frac{1}{2}\sqrt{1+2x-3x^2}, \frac{1-x}{2} + \frac{1}{2}\sqrt{1+2x-3x^2} \right).$$

From here it is not difficult to verify Equation (1.11) directly.

Example 1.4.9 illustrates the use of the Implicit Function Theorem. However,
though the theorem establishes the existence of the implicit function g and provides
a method to calculate $[dg_a]$, the theorem provides no method to calculate the
function g. In fact, unlike in Example 1.4.9, in most cases, one cannot calculate g
with elementary functions.

PROBLEMS

1.4.1. Find the critical points of the following $\mathbb{R} \to \mathbb{R}$ functions: (a) $f(x) = x^3$, (b)
$g(x) = \sin x$, and (c) $h(x) = x^3 - 3x^2 + x + 1$.

1.4.2. Find all the critical points of the function $F(x, y) = (x^3 - xy + y^2, x^2 - y)$ defined
over all $\mathbb{R}^2$.

1.4.3. Let $F : \mathbb{R}^3 \to \mathbb{R}^3$ be defined by $F(x, y, z) = (z^2 - xy, x^3 - 3xyz, x^2 + y^2 + z^2)$.

(a) Find an equation describing the critical points of this function. (If you have
access to a computer algebra system, plot it.)

(b) Prove that if (x_0, y_0, z_0) is a critical point of F, then any point $(\lambda x_0, \lambda y_0, \lambda z_0)$,
with $\lambda \in \mathbb{R}$, is also a critical point. (That is, if (x_0, y_0, z_0) is a critical point,
then any point on the line through $(0, 0, 0)$ and (x_0, y_0, z_0) is also critical.
We say that the equation for the critical points is a homogeneous equation.)

1.4.4. Let $F : \mathbb{R}^3 \to \mathbb{R}^2$ be defined by $F(x, y, z) = (e^{xy}, z \cos x)$. Find all the critical
points of F.

1.4.5. Consider the function $f : \mathbb{R}^3 \to \mathbb{R}^3$ defined by

$$f(x_1, x_2, x_3) = (x_1 \cos x_2 \sin x_3, x_1 \sin x_2 \sin x_3, x_1 \cos x_3).$$

Find the critical points and the critical values of f.

1.4.6. Let $F : \mathbb{R}^2 \to \mathbb{R}^2$ be the function defined by $F(s, t) = (s^3 - 3st^2, 3s^2t - t^3)$, and let
$q = (2, 3)$. Find the critical points of F. Prove that there exists a neighborhood
V of q such that F is one-to-one on V so that $F^{-1} : F(V) \to V$ exists. Let
$p = F(q) = (-46, 9)$. Find $d(F^{-1})_{(-46,9)}$.

1.4.7. Let $F : \mathbb{R}^2 \to \mathbb{R}^2$ be defined by $F(x,y) = (y^2 \sin x + 1, (x + 2y) \cos y)$. Show that $(0, \pi/2)$ is not a critical point of F. Show that F is a bijection from a neighborhood U of $(0, \pi/2)$ to a neighborhood V of $(1, 0)$. If $G : V \to U$ is the inverse function $G = F^{-1}$, then find the matrix of $dG_{(1,0)}$.

1.4.8. Consider the function $F : (1, +\infty)^2 \to (1, +\infty)^2$ defined by $F(x, y) = (x^y, y^x)$.

    (a) Find the set of critical points of F.

    (b) Show that on a neighborhood V of $q = (2, 3)$ the function F is one-to-one.

    (c) Calling $F :^{-1} : F(V) \to V$ the inverse function near q, use the Inverse Function Theorem to determine $d(F^{-1})_{(8,9)}$.

1.4.9. Consider the function

$$f(x_1, x_2, x_3) = \left(\frac{x_2 + x_3}{1 + x_1 + x_2 + x_3}, \frac{x_1 + x_3}{1 + x_1 + x_2 + x_3}, \frac{x_1 + x_2}{1 + x_1 + x_2 + x_3} \right)$$

defined over the domain $U = \mathbb{R}^3 - \{(x_1, x_2, x_3) \,|\, 1 + x_1 + x_2 + x_3 = 0\}$.

    (a) Show that no point in the domain of f is a critical point. [Hint: Prove that $J(f) = 2/(1 + x_1 + x_2 + x_3)^4$.]

    (b) Prove that f is injective.

    (c) Find $\left[df^{-1} \right]$ in terms of (x_1, x_2, x_3) at every point using the Inverse Function Theorem.

    (d) Show that the inverse function is

$$f^{-1}(y_1, y_2, y_3) = \left(\frac{-y_1 + y_2 + y_3}{2 - y_1 - y_2 - y_3}, \frac{y_1 - y_2 + y_3}{2 - y_1 - y_2 - y_3}, \frac{y_1 + y_2 - y_3}{2 - y_1 - y_2 - y_3} \right).$$

    (e) Prove that f is a bijection between $U = \mathbb{R}^3 - \{(x_1, x_2, x_3) \,|\, 1 + x_1 + x_2 + x_3 = 0\}$ and $V = \mathbb{R}^3 - \{(x_1, x_2, x_3) \,|\, 2 - x_1 - x_2 - x_3 = 0\}$.

1.4.10. Verify all the calculations of Example 1.4.9.

1.4.11. Let Σ be the curve in $\mathbb{R}^3$ defined by

$$\begin{cases} 4x^2 + 5y^2 + z^2 = 33 \\ x^2 + 4y^2 + 2z^2 = 35. \end{cases}$$

Using the Implicit Function Theorem, show that near the point $q = (1, 2, 3)$, Σ can be parametrized by $(x, g_1(x), g_2(x))$. Find $[dg_1]$ and use this to give a parametrization of the tangent line to Σ at q.

1.4.12. Let Σ be the level set in $\mathbb{R}^4$ defined by

$$\begin{cases} x^2 + 2y^2 + 3z^2 + 4w^2 = 24 \\ x^3 w - 2y^2 z^2 + w^3 = 20. \end{cases}$$

Let $F(x, y, z, w) = (x^2 + 2y^2 + 3z^2 + 4w^2, x^3 w - 2y^2 z^2 + w^3)$.

    (a) Prove that $q = (3, 2, 1, 1)$ is not a critical point of F and observe that $q \in \Sigma$.

(b) Using the Implicit Function Theorem, show that there is an open neighborhood W of $a = (3, 2)$ in $\mathbb{R}^2$ and a function $g : W \to \mathbb{R}^2$ such that a neighborhood of q in Σ is the graph of g.

(c) Calculate $[dg]$ over W.

(d) Use this to provide a parametrization of the tangent plane to Σ at q.

CHAPTER 2

Variable Frames

The strategy of choosing a particular coordinate system or frame to perform a calculation or to present a concept is ubiquitous in both mathematics and physics. For example, Newton's equations of planetary motion are much easier to solve in polar coordinates than in Cartesian coordinates. In the differential geometry of curves, calculations of local properties are often simpler when carried out in the Frenet frame associated to the curve at a point. (See [5, Chapter 3].) This chapter introduces general coordinate systems on $\mathbb{R}^n$ and the concept of variable frames in a consistent and general manner.

2.1 Frames Associated to Coordinate Systems

Many problems in introductory mechanics involve finding the trajectory of a particle under the influence of various forces and/or subject to certain constraints. The first approach uses the coordinate functions and describes the trajectory as

$$\vec{r}(t) = (x(t), y(t), z(t)) = x(t)\vec{\imath} + y(t)\vec{\jmath} + z(t)\vec{k}.$$

Newton's equations of motion then lead to differential equations in the three coordinate functions $x(t)$, $y(t)$, and $z(t)$. The velocity function is the derivative, namely

$$
\begin{aligned}
\vec{r}\,'(t) &= \frac{d}{dt}(x(t)\vec{\imath}) + \frac{d}{dt}(y(t)\vec{\jmath}) + \frac{d}{dt}(z(t)\vec{k}) \\
&= x'(t)\vec{\imath} + x(t)\frac{d}{dt}(\vec{\imath}) + y'(t)\vec{\jmath} + y(t)\frac{d}{dt}(\vec{\jmath}) + z'(t)\vec{k} + z(t)\frac{d}{dt}(\vec{k}) \\
&= x'(t)\vec{\imath} + y'(t)\vec{\jmath} + z'(t)\vec{k},
\end{aligned}
$$

because $\frac{d}{dt}\vec{\imath} = 0$, $\frac{d}{dt}\vec{\jmath} = 0$, and $\frac{d}{dt}\vec{k} = 0$. This last remark shows that the frame $(\vec{\imath}, \vec{\jmath}, \vec{k})$ associated to the Cartesian coordinate systems is a constant frame.

As we discuss variable frames, we introduce a nice way to describe the rate of change of a variable frame. Suppose that $\{\vec{u}_1, \vec{u}_2, \vec{u}_3\}$ is a basis of $\mathbb{R}^3$ and let $\vec{a}$ and $\vec{b}$ be two other vectors with components $\vec{a} = a_1\vec{u}_1 + a_2\vec{u}_2 + a_3\vec{u}_3$ and $\vec{b} =$

$b_1 \vec{u}_1 + b_2 \vec{u}_2 + b_3 \vec{u}_3$. Assuming that all vectors are column vectors, we can write these component definitions of $\vec{a}$ and $\vec{b}$ in the matrix expression

$$\begin{pmatrix} \vec{a} & \vec{b} \end{pmatrix} = \begin{pmatrix} \vec{u}_1 & \vec{u}_2 & \vec{u}_3 \end{pmatrix} \begin{pmatrix} a_1 & b_1 \\ a_2 & b_2 \\ a_3 & b_3 \end{pmatrix}.$$

Using this notation, we can express the relationships $\frac{d}{dt}\vec{i} = 0$, $\frac{d}{dt}\vec{j} = 0$, and $\frac{d}{dt}\vec{k} = 0$ by

$$\frac{d}{dt}\begin{pmatrix} \vec{i} & \vec{j} & \vec{k} \end{pmatrix} = \begin{pmatrix} \vec{i} & \vec{j} & \vec{k} \end{pmatrix} \begin{pmatrix} 0 & 0 & 0 \\ 0 & 0 & 0 \\ 0 & 0 & 0 \end{pmatrix}. \tag{2.1}$$

This notation appears trivial but it will become important as we study the behavior of frames associated to other natural coordinate systems.

Using cylindrical coordinates, we locate a point in $\mathbb{R}^3$ using the distance r between the origin and the projection of the point onto the xy-plane, the angle from the positive x-axis θ, and the height z above the xy-plane. See Figure 2.1. We have the following relationship between Cartesian coordinates and cylindrical coordinates:

$$\begin{cases} x = r\cos\theta, \\ y = r\sin\theta, \\ z = z, \end{cases} \longleftrightarrow \begin{cases} r = \sqrt{x^2 + y^2}, \\ \theta = \tan^{-1}\left(\frac{y}{x}\right), \\ z = z. \end{cases} \tag{2.2}$$

Of course, by the expression $\tan^{-1}\left(\frac{y}{x}\right)$, one must understand that we assume that $x > 0$. For $x \le 0$, one must adjust the formula to obtain the appropriate corresponding angle. Using cylindrical coordinates, one would locate a point in space by

$$\vec{r} = (r\cos\theta, r\sin\theta, z).$$

We define the natural frame with respect to this coordinate system as follows. To each independent variable in the coordinate system, one associates the unit vector that corresponds to the directions of change with respect to that variable. For example, with cylindrical coordinates, we have the following three unit vectors:

$$\vec{e}_r = \frac{\partial \vec{r}}{\partial r} \Big/ \left\| \frac{\partial \vec{r}}{\partial r} \right\|, \quad \vec{e}_\theta = \frac{\partial \vec{r}}{\partial \theta} \Big/ \left\| \frac{\partial \vec{r}}{\partial \theta} \right\|, \quad \text{and} \quad \vec{e}_z = \frac{\partial \vec{r}}{\partial z} \Big/ \left\| \frac{\partial \vec{r}}{\partial z} \right\|. \tag{2.3}$$

These formulas give us explicitly

$$\begin{aligned} \vec{e}_r &= (\cos\theta, \sin\theta, 0) = \cos\theta\,\vec{i} + \sin\theta\,\vec{j}, \\ \vec{e}_\theta &= (-\sin\theta, \cos\theta, 0) = -\sin\theta\,\vec{i} + \cos\theta\,\vec{j}, \\ \vec{e}_z &= (0, 0, 1) = \vec{k}. \end{aligned} \tag{2.4}$$

Using this new frame, the position vector of a point with cylindrical coordinates (r, θ, z) is

$$\vec{r} = r\vec{e}_r + z\vec{e}_z, \tag{2.5}$$

As opposed to the fixed frame $(\vec{i}, \vec{j}, \vec{k})$, many frames associated to non-Cartesian coordinates often depend on the position of the base point p of the frame. In this case, the frame $(\vec{e}_r, \vec{e}_\theta, \vec{e}_z)$ associated to cylindrical coordinates depends explicitly on the coordinates (r, θ, z) (in this case, only on θ) of the frame's origin point.

To see how this frame varies with respect to any parameter, consider a space curve parametrized by $\vec{r} : I \to \mathbb{R}^3$, where I is an interval of $\mathbb{R}$. We can attach the frame $(\vec{e}_r, \vec{e}_\theta, \vec{e}_z)$ to each point $\vec{r}(t)$ of the curve, but, unlike with the fixed Cartesian frame, the frame $(\vec{e}_r, \vec{e}_\theta, \vec{e}_z)$ is not constant. As we study motion in the new coordinate system, we are led to take higher derivatives of $\vec{r}(t)$ and express them with components in the frame associated to the particular coordinate system.

If $\vec{r}(t)$ is a space curve, then r, θ, and z are functions of t. Therefore, taking the derivative with respect to t, we get

$$\vec{r}' = \frac{d}{dt}(r\vec{e}_r) + \frac{d}{dt}(z\vec{e}_z) = r'\vec{e}_r + r\frac{d}{dt}\vec{e}_r + z'\vec{e}_z + z\frac{d}{dt}\vec{e}_z.$$

Thus, in order to write equations of motion in cylindrical coordinates, we must determine $\frac{d}{dt}\vec{e}_r$, $\frac{d}{dt}\vec{e}_\theta$, and $\frac{d}{dt}\vec{e}_z$. We obtain

$$\vec{e}_r' = \frac{d}{dt}(\cos\theta, \sin\theta, 0) = (-\theta'\sin\theta, \theta'\cos\theta, 0) = \theta'\vec{e}_\theta,$$

$$\vec{e}_\theta' = \frac{d}{dt}(-\sin\theta, \cos\theta, 0) = (-\theta'\cos\theta, -\theta'\sin\theta, 0) = -\theta'\vec{e}_r,$$

$$\vec{e}_z' = \frac{d}{dt}(0, 0, 1) = \vec{0}.$$

Following the same method of presentation as in (2.1) the change of the cylindrical coordinates frame can be expressed as

$$\frac{d}{dt}\begin{pmatrix} \vec{e}_r & \vec{e}_\theta & \vec{e}_z \end{pmatrix} = \begin{pmatrix} \vec{e}_r & \vec{e}_\theta & \vec{e}_z \end{pmatrix} \begin{pmatrix} 0 & -\theta' & 0 \\ \theta' & 0 & 0 \\ 0 & 0 & 0 \end{pmatrix}. \tag{2.6}$$

An application of the cylindrical frame and its rate of change arises when describing the velocity vector and acceleration vector:

$$\vec{r}' = r'\vec{e}_r + r\theta'\vec{e}_\theta + z'\vec{e}_z,$$
$$\vec{r}'' = r''\vec{e}_r + r'\theta'\vec{e}_\theta + r'\theta'\vec{e}_\theta + r\theta''\vec{e}_\theta + r(\theta')^2(-\vec{e}_r) + z''\vec{e}_z$$
$$= \left(r'' - r(\theta')^2\right)\vec{e}_r + (2r'\theta' + r\theta'')\vec{e}_\theta + z''\vec{e}_z.$$

If we restrict ourselves to polar coordinates, the above formula would still hold but with no z-component. In the study of trajectories in the plane, the first four terms

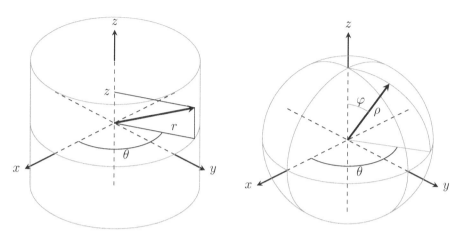

Figure 2.1: Cylindrical coordinates.   Figure 2.2: Spherical coordinates.

in the last expression have particular names (see [22, Section 5.2]). We call

$$r'' \vec{e}_r \quad \text{the radial acceleration,}$$
$$-r(\theta')^2 \vec{e}_r \quad \text{the centripetal acceleration,}$$
$$2r' \theta' \vec{e}_\theta \quad \text{the Coriolis acceleration, and}$$
$$r\theta'' \vec{e}_\theta \quad \text{the component due to angular acceleration.}$$

Example 2.1.1. Using spherical coordinates, we locate a point P in $\mathbb{R}^3$ as follows. Let P' be the projection of P onto the xy-plane. Use the distance from the origin $\rho = OP$, longitude θ (i.e., the angle from the positive x-axis to the ray $[OP')$), and the angle φ, which is the angle between the positive z-axis and the ray $[OP)$. See Figure 2.2. Elementary geometry gives the relationship between Cartesian coordinates and spherical coordinates:

$$\begin{cases} x = \rho \cos\theta \sin\varphi, \\ y = \rho \sin\theta \sin\varphi, \\ z = \rho \cos\varphi, \end{cases} \longleftrightarrow \begin{cases} \rho = \sqrt{x^2 + y^2 + z^2}, \\ \theta = \tan^{-1}\left(\frac{y}{x}\right), \\ \varphi = \cos^{-1}\left(\frac{z}{\sqrt{x^2+y^2+z^2}}\right), \end{cases} \tag{2.7}$$

with the same caveat for θ as discussed with cylindrical coordinates. With spherical coordinates, we usually assume that $\rho \geq 0$, $0 \leq \theta < 2\pi$, and $0 \leq \varphi \leq \pi$.

We leave it as an exercise for the reader to determine the frame associated to a spherical coordinate system as

$$\vec{e}_\rho = \frac{\partial \vec{r}}{\partial \rho} \Big/ \left\| \frac{\partial \vec{r}}{\partial \rho} \right\| = (\cos\theta \sin\varphi, \sin\theta \sin\varphi, \cos\varphi),$$

$$\vec{e}_\theta = \frac{\partial \vec{r}}{\partial \theta} \Big/ \left\| \frac{\partial \vec{r}}{\partial \theta} \right\| = (-\sin\theta, \cos\theta, 0),$$

$$\vec{e}_\varphi = \frac{\partial \vec{r}}{\partial \varphi} \Big/ \left\| \frac{\partial \vec{r}}{\partial \varphi} \right\| = (\cos\theta \cos\varphi, \sin\theta \cos\varphi, -\sin\varphi).$$

In contrast to Cartesian coordinates, where the position vector of a point is $\vec{r} = x\vec{i} + y\vec{j} + z\vec{k}$, and in contrast to cylindrical coordinates where the position vector is given by (2.5), in spherical coordinates, the position vector is simply

$$\vec{r} = \rho \vec{e}_\rho.$$

To discuss how the frame associated to spherical coordinates changes, consider a parametric curve $\vec{r} : I \to \mathbb{R}^3$ and calculate how $\vec{e}_\rho$, $\vec{e}_\theta$ and $\vec{e}_\varphi$ change as t changes. Again, we leave it as an exercise for the reader to show that

$$\frac{d}{dt} \begin{pmatrix} \vec{e}_\rho & \vec{e}_\theta & \vec{e}_\varphi \end{pmatrix} = \begin{pmatrix} \vec{e}_\rho & \vec{e}_\theta & \vec{e}_\varphi \end{pmatrix} \begin{pmatrix} 0 & -\theta' \sin\varphi & -\varphi' \\ \theta' \sin\varphi & 0 & \theta' \cos\varphi \\ \varphi' & -\theta' \cos\varphi & 0 \end{pmatrix}. \tag{2.8}$$

All the coordinate systems we have considered thus far, though curvilinear, are examples of *orthogonal* coordinate systems; the basis vectors associated to the coordinate system form an orthogonal basis of $\mathbb{R}^n$. In general, this is not the case. We point out that, as shown in (2.4), both mathematicians and physicists make the traditional choice when they impose that the frames associated to the cylindrical and spherical coordinate systems be composed of unit vectors. As useful as this is for calculations involving distances or angles, this choice has some drawbacks.

We now consider general coordinate systems in $\mathbb{R}^n$. Already in polar, cylindrical, and spherical coordinates, we encounter some challenges in bringing together practical application and precision. For example, polar coordinates (r, θ) do locate points uniquely in the plane and for every point p in the plane, there do exist some (r_0, θ_0) that correspond to p. However, the assignment $p = f(r, \theta)$ is not injective.

Let S be an open set in $\mathbb{R}^n$. A continuous surjective function $f : U \to S$, where U is an open set in $\mathbb{R}^n$, defines a coordinate system on S by associating to every point $P \in S$ an n-tuple $x(P) = (x^1(P), x^2(P), \cdots, x^n(P))$ such that $f(x(P)) = P$. In this notation, the superscripts do not indicate powers of a variable x but the ith coordinate for that point in the given coordinate system. Though a possible source of confusion at first, differential geometry literature uses superscripts instead of the usual subscripts in order to mesh properly with subsequent tensor notation. As with polar coordinates where (r_0, θ_0) and $(r_0, \theta_0 + 2\pi)$ correspond to the same point in the plane, in practice the n-tuple need not be uniquely associated to the point P. However, it is not uncommon for the sake of proofs to restrict f to a smaller domain $V \subset U$ so that $f|_V$ is a bijection with the corresponding $x : f(V) \to V$ as the inverse. Note that in this latter case, we call $x = (x^1, x^2, \ldots, x^n)$ the coordinate functions, or the coordinate system

Let $(x^1, x^2, \ldots, x^n)$ be a coordinate system in $\mathbb{R}^n$. Since $\mathbb{R}^n$ is a vector space, we can talk about position vectors of points in $\mathbb{R}^n$. To say that the n-tuple $(x^1, x^2, \ldots, x^n)$ gives coordinates of a point p means that p has a position vector $\vec{r}$ that is a function in the n variables $(x^1, x^2, \ldots, x^n)$. In our present formulation, the position function $\vec{r}(x^1, x^2, \ldots, x^n)$ is precisely the function f.

Definition 2.1.2. Let $x : S \to U$ be a coordinate system on an open subset $S \subset \mathbb{R}^n$. If p is not a critical point of x, then the frame (or basis) associated to this coordinate system at p is the set of vectors

$$\left\{ \left.\frac{\partial \vec{r}}{\partial x^1}\right|_p, \left.\frac{\partial \vec{r}}{\partial x^2}\right|_p, \ldots, \left.\frac{\partial \vec{r}}{\partial x^n}\right|_p \right\}. \tag{2.9}$$

If there is no cause for confusion, we drop the $|_p$ but understands from context that derivatives are evaluated at a point p. We say that the components of a vector $\vec{A}$ at p in this system of coordinates are $(A^1, A^2, \ldots, A^n)$ if we can write

$$\vec{A} = \sum_{i=1}^{n} A_i \frac{\partial \vec{r}}{\partial x^i}. \tag{2.10}$$

Note that since p is not a critical point of x, then dx_p is invertible with inverse $(dx_p)^{-1} = df_{x(p)}$. Hence, the columns of $[df_{x(p)}]$, which are precisely these vectors $\partial \vec{r}/\partial x^i|_{x(p)}$, form a linearly independent set. In general, this condition of linear independence is all we can assume from a frame associated to a general coordinate system at p, namely, it need not be an orthogonal set of vectors or consist of unit vectors. If the set of vectors (2.9) is an orthogonal set, then the system of coordinates is called an *orthogonal* coordinate system.

Definition 2.1.3. Let $x : S \to U$ be an orthogonal coordinate system on an open subset $S \subset \mathbb{R}^n$. The *scale factors* of this coordinate system at point p that is not a critical point are $h_{x^1}, h_{x^2}, \ldots, h_{x^n}$, where

$$h_{x^i} = \left\| \frac{\partial \vec{r}}{\partial x^i} \right\|.$$

When a coordinate system $(x^1, x^2, \ldots, x^n)$ is orthogonal, it is common to divide the basis vectors $\partial \vec{r}/\partial x^i$ by the scale factors to obtain an orthonormal basis associated to the coordinate system. This is precisely what we did with both the cylindrical and spherical coordinate systems.

Another interesting aspect to using frames associated to coordinate systems involves how to consider rates of change of a vector field when expressed with respect to a variable frame. Let U be an open subset of $\mathbb{R}^n$. Let $\{\vec{u}_1, \vec{u}_2, \ldots, \vec{u}_n\}$ be a variable frame defined over U, i.e., each vector $\vec{u}_i$ is a vector function $\vec{u}_i(x^1, x^2, \ldots, x^n)$

that is differentiable on U and for each $(x^1, x^2, \ldots, x^n)$ the collection of vectors is linearly independent. Let $\vec{V}$ be a vector field defined on U. At each point $p \in U$, the vector $\vec{V}(p)$ can be decomposed into components $V^i(p)$ as

$$\vec{V}(p) = V^1(p)\vec{u}_1(p) + V^2(p)\vec{u}_2(p) + \cdots + V^n(p)\vec{u}_n(p).$$

More concisely, $\vec{V} = V^1\vec{u}_1 + V^2\vec{u}_2 + \cdots + V^n\vec{u}_n$, where we understand each V^j to be a function on U. (Again, the superscripts are indices and not powers. We explain this convention in more detail when we discuss multilinear algebra in Chapter 4.)

When we take partial derivatives of the vector field $\vec{V}$, we can express these derivatives in terms of the local frame $\{\vec{u}_1, \vec{u}_2, \ldots, \vec{u}_n\}$. We have

$$\frac{\partial \vec{V}}{\partial x^i} = \frac{\partial}{\partial x^i}\left(\sum_{j=1}^n V^j \vec{u}_j\right) = \sum_{j=1}^n \frac{\partial V^j}{\partial x^i}\vec{u}_j + \sum_{j=1}^n V^j \frac{\partial \vec{u}_j}{\partial x^i}.$$

In order to proceed and find the component functions of $\partial\vec{V}/\partial x^i$, we need to decompose $\partial\vec{u}_j/\partial x^i$ into its components with respect to $\{\vec{u}_1, \vec{u}_2, \ldots, \vec{u}_n\}$. This leads to the collection of n^3 functions Γ_{ij}^k defined as

$$\frac{\partial \vec{u}_j}{\partial x^i} = \sum_{k=1}^n \Gamma_{ij}^k \vec{u}_k.$$

Then

$$\frac{\partial \vec{V}}{\partial x^i} = \sum_{j=1}^n \frac{\partial V^j}{\partial x^i}\vec{u}_j + \sum_{j=1}^n V^j \frac{\partial \vec{u}_j}{\partial x^i} = \sum_{j=1}^n \frac{\partial V^j}{\partial x^i}\vec{u}_j + \sum_{j=1}^n V^j\left(\sum_{k=1}^n \Gamma_{ij}^k \vec{u}_k\right)$$

$$= \sum_{k=1}^n \frac{\partial V^k}{\partial x^i}\vec{u}_k + \sum_{k=1}^n\sum_{j=1}^n \Gamma_{ij}^k V^j \vec{u}_k = \sum_{k=1}^n\left(\frac{\partial V^k}{\partial x_i} + \sum_{j=1}^n \Gamma_{ij}^k V^j\right)\vec{u}_k.$$

Hence, because we work in variable frames, the kth component of the vector field $\partial\vec{V}/\partial x^i$ is not just $\partial V^k/\partial x^i$, but rather

$$\left(\frac{\partial \vec{V}}{\partial x^i}\right)^k = \frac{\partial V^k}{\partial x^i} + \sum_{j=1}^n \Gamma_{ij}^k V^j. \tag{2.11}$$

Equation (2.11) will reappear in a more general context in the analysis on manifolds. In that context, the collection of functions Γ_{jk}^i are called the components of a *connection*.

Example 2.1.4 (Spherical Coordinates, 1). We illustrate how to calculate the Γ_{jk}^i functions for the normalized spherical coordinate frame. Consider the variable frame $\{\vec{e}_\rho, \vec{e}_\theta, \vec{e}_\varphi\}$ and use (2.8) where instead of t, we use ρ, θ, and φ successively for the derivatives, i.e., for $i = 1, 2, 3$. Thus, with k representing the row and j representing the column, we have

$$\Gamma^k_{1j} = \begin{pmatrix} 0 & 0 & 0 \\ 0 & 0 & 0 \\ 0 & 0 & 0 \end{pmatrix}, \ \Gamma^k_{2j} = \begin{pmatrix} 0 & -\sin\varphi & 0 \\ \sin\varphi & 0 & \cos\varphi \\ 0 & -\cos\varphi & 0 \end{pmatrix}, \ \Gamma^k_{3j} = \begin{pmatrix} 0 & 0 & -1 \\ 0 & 0 & 0 \\ 1 & 0 & 0 \end{pmatrix}.$$

Example 2.1.5 (Spherical Coordinates, 2). The previous example used the normalized frame for spherical coordinates. We could also use the basis described by (2.9), which consists of the three vectors

$$\vec{u}_1 = \frac{\partial \vec{r}}{\partial \rho} = (\cos\theta \sin\varphi, \sin\theta \sin\varphi, \cos\varphi)$$

$$\vec{u}_2 = \frac{\partial \vec{r}}{\partial \theta} = (-\rho \sin\theta \sin\varphi, \rho \cos\theta \sin\varphi, 0)$$

$$\vec{u}_3 = \frac{\partial \vec{r}}{\partial \varphi} = (\rho \cos\theta \sin\varphi, \rho \sin\theta \cos\varphi, -\rho \sin\varphi).$$

Calculating the Γ^k_{ij} components requires us to take derivatives of each of the above vector functions with respect to each of the coordinates and then decompose back into the basis $\{\vec{u}_1, \vec{u}_2, \vec{u}_3\}$. Because these three vectors are orthogonal, though not unit vectors we find the components of a vector in this frame by

$$\vec{v} = \frac{\vec{v} \cdot \vec{u}_1}{\vec{u}_1 \cdot \vec{u}_1} \vec{u}_1 + \frac{\vec{v} \cdot \vec{u}_2}{\vec{u}_2 \cdot \vec{u}_2} \vec{u}_2 + \frac{\vec{v} \cdot \vec{u}_3}{\vec{u}_3 \cdot \vec{u}_3} \vec{u}_3.$$

The calculations are straightforward and we leave it as an exercise to prove that

$$\Gamma^k_{1j} = \begin{pmatrix} 0 & 0 & 0 \\ 0 & \frac{1}{\rho} & 0 \\ 0 & 0 & \frac{1}{\rho} \end{pmatrix} \ \Gamma^k_{2j} = \begin{pmatrix} 0 & -\rho \sin^2\varphi & 0 \\ \frac{1}{\rho} & 0 & \cot\varphi \\ 0 & -\sin\varphi \cos\varphi & 0 \end{pmatrix} \ \Gamma^k_{3j} = \begin{pmatrix} 0 & 0 & -\rho \\ 0 & \cot\varphi & 0 \\ \frac{1}{\rho} & 0 & 0 \end{pmatrix}.$$

We have chosen to list the functions with fixed i since this is the variable with respect to which we take the derivative. However, it is interesting to organize the functions into three matrices, each corresponding to a fixed k. We get

$$\Gamma^1_{ij} = \begin{pmatrix} 0 & 0 & 0 \\ 0 & -\rho \sin^2\varphi & 0 \\ 0 & 0 & -\rho \end{pmatrix} \ \Gamma^2_{ij} = \begin{pmatrix} 0 & \frac{1}{\rho} & 0 \\ \frac{1}{\rho} & 0 & \cot\varphi \\ 0 & \cot\varphi & 0 \end{pmatrix} \ \Gamma^3_{ij} = \begin{pmatrix} 0 & 0 & \frac{1}{\rho} \\ 0 & -\sin\varphi \cos\varphi & 0 \\ \frac{1}{\rho} & 0 & 0 \end{pmatrix},$$

each of which is a symmetric matrix.

PROBLEMS

2.1.1. Prove Equation (2.8) for the rate of change of the spherical coordinates frame.

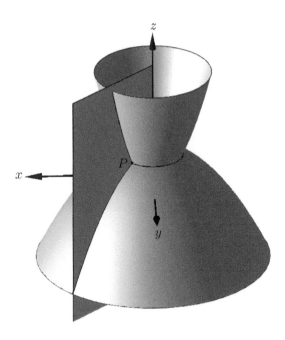

Figure 2.3: Coordinate planes for parabolic coordinates.

2.1.2. Let $\vec{r} : I \to \mathbb{R}^3$ be a smooth curve in space. Express $\vec{r}\,'$ and $\vec{r}\,''$ in terms of functions of spherical coordinates $\rho(t)$, $\theta(t)$, $\varphi(t)$, and the local frame $\{\vec{e}_\rho, \vec{e}_\theta, \vec{e}_\varphi\}$.

2.1.3. Calculate the Γ_{ij}^k functions for the spherical coordinate frame as decribed in Example 2.1.5.

2.1.4. Fix a positive real number a. Elliptic coordinates on $\mathbb{R}^2$ consists of the pair (μ, ν) with $\mu \geq 0$ and $0 \leq \nu < 2\pi$, connected to Cartesian coordinates by

$$\begin{cases} x = a \cosh \mu \cos \nu \\ y = a \sinh \mu \sin \nu. \end{cases}$$

(a) Prove that the curves of constant μ form ellipses; and that curves of constant ν form hyperbolas.

(b) Calculate $\partial \vec{r}/\partial \mu$ and $\partial \vec{r}/\partial \nu$ and observe that the elliptic coordinate system is an orthogonal system.

(c) Show that the scale factors are $h_\mu = h_\nu = a\sqrt{\cosh^2 \mu - \cos^2 \nu}$ and calculate $\vec{e}_\mu$ and $\vec{e}_\nu$.

(d) Calculate the eight connection functions Γ_{jk}^i for $i, j, k = 1, 2$ associated to the frame $\{\vec{e}_\mu, \vec{e}_\nu\}$.

2.1.5. The *parabolic coordinates* system of $\mathbb{R}^3$ consists of the triple (u, v, θ), with $u \in [0, +\infty)$, $v \in [0, +\infty)$, and $\theta \in [0, 2\pi)$ with equations

$$\begin{cases} x = uv \cos \theta, \\ y = uv \sin \theta, \\ z = \frac{1}{2}(u^2 - v^2). \end{cases}$$

These equations are also called the transition functions from parabolic to Cartesian coordinate. (Figure 2.3 shows the three coordinate "planes" for parabolic coordinates in $\mathbb{R}^3$ passing through the point $P \in \mathbb{R}^3$ with coordinates $(u, v, \theta) = (1, 1/2, \pi/4)$.)

(a) Find the basis vectors for the associated frame according to (2.9) and show that the parabolic coordinate system is an orthogonal coordinate system.

(b) Consider also the basis $\{\vec{e}_u, \vec{e}_v, \vec{e}_\theta\}$ given by

$$\vec{e}_u = \frac{\partial \vec{r}}{\partial u} \Big/ \left\| \frac{\partial \vec{r}}{\partial u} \right\|, \quad \vec{e}_v = \frac{\partial \vec{r}}{\partial v} \Big/ \left\| \frac{\partial \vec{r}}{\partial v} \right\|, \quad \vec{e}_\theta = \frac{\partial \vec{r}}{\partial \theta} \Big/ \left\| \frac{\partial \vec{r}}{\partial \theta} \right\|.$$

Calculate the rate of change matrix for this frame similar to (2.8) as done for spherical coordinates.

(c) Calculate the Γ^i_{jk} connection functions associated to the $\{\vec{e}_u, \vec{e}_v, \vec{e}_\theta\}$ frame of parabolic coordinates.

2.1.6. Toroidal coordinates in $\mathbb{R}^3$ are denoted by the triple $(\sigma \tau, \phi)$ and transform into Cartesian coordinates via

$$\begin{cases} x = \dfrac{\sinh \tau}{\cosh \tau - \cos \sigma} \cos \phi \\ y = \dfrac{\sinh \tau}{\cosh \tau - \cos \sigma} \sin \phi \\ z = \dfrac{\sin \sigma}{\cosh \tau - \cos \sigma} \end{cases}$$

typically used with $-\pi < \sigma \leq \pi$, $0 \leq \tau$, and $0 \leq \phi < 2\pi$.

(a) Show that surfaces of constant σ are spheres of center $(0, 0, \cot \sigma)$ and radius $\csc \sigma$; that surfaces of constant τ are tori with the z-axis as the axis of rotation; and that surfaces of constant ϕ are planes through the z-axis.

(b) Find the frame associated to this coordinate system and show that this coordinate system is an orthogonal system.

(c) Show that scale factors are $h_\sigma = h_\tau = 1/(\cosh \tau - \cos \sigma)$, and $h_\phi = \sinh \tau / (\cosh \tau - \cos \sigma)$; and calculate the associated orthonormal frame $\{\vec{e}_\sigma, \vec{e}_\tau, \vec{e}_\phi\}$.

(d) Calculate the rate of change matrix for $\{\vec{e}_\sigma, \vec{e}_\tau, \vec{e}_\phi\}$ similar to (2.8) as done for spherical coordinates.

2.1.7. Consider the coordinate system on $\mathbb{R}^2$ that employs the pair $(s, \alpha) \in [0, +\infty) \times [0, 2\pi)$ to represent the point on the ellipse

$$\frac{x^2}{4} + y^2 = s^2$$

that lies on the ray from the origin and through $(\cos \alpha, \sin \alpha)$.

(a) Determine change of coordinate system equations from and to Cartesian coordinates.

(b) Find the set $\mathcal{B}$ of basis vectors for the associated frame according to Equation (2.9).

(c) Prove that $\dfrac{\partial \vec{r}}{\partial s} \cdot \dfrac{\partial \vec{r}}{\partial \alpha} = \dfrac{3}{8} s \sin(2\alpha)$ and conclude that this coordinate system is not orthogonal.

(d) Calculate the rate of change matrix associated to this frame $\mathcal{B}$.

2.2 Frames Associated to Trajectories

In the study of trajectories, whether in physics or geometry, it is often convenient to use a frame that is different from the Cartesian frame. Changing types of frames sometimes makes difficult integrals tractable or makes certain difficult differential equations manageable. In the particular context of special relativity, one talks about a *momentarily comoving reference frame*, abbreviated to MCRF.[50]

In the study of plane curves, it is common to use the frame $\{\vec{T}, \vec{U}\}$ to study the local properties of a plane curve $\vec{x}(t)$. (See [5, Chapter 1].) The vector $\vec{T}(t)$ is the unit tangent vector $\vec{T}(t) = \vec{x}'(t)/\|\vec{x}'(t)\|$, and the unit normal vector $\vec{U}(t)$, is the result of rotating $\vec{T}(t)$ by $\pi/2$ in the counterclockwise direction. This is a moving frame that is defined in terms of a given regular curve $\vec{x}(t)$ and, at $t = t_0$, is viewed as based at the point $\vec{x}(t_0)$. To compare with applications in physics, it is important to note that the $\{\vec{T}, \vec{U}\}$ frame is not the same as the polar coordinate frame $\{\vec{e}_r, \vec{e}_\theta\}$. From Equation (2.4) (and ignoring the z-component), we know that

$$\vec{e}_r = (\cos\theta, \sin\theta) \qquad \text{and} \qquad \vec{e}_\theta = (-\sin\theta, \cos\theta).$$

Assuming that x, y, r, and θ are functions of t and since $x = r\cos\theta$ and $y = r\sin\theta$, we have

$$\vec{x}'(t) = (x'(t), y'(t)) = (r'\cos\theta - r\theta'\sin\theta, r'\sin\theta + r\theta'\cos\theta) = r'\vec{e}_r + r\theta'\vec{e}_\theta.$$

We then calculate the speed function to be

$$s'(t) = \|\vec{x}'(t)\| = \sqrt{(r')^2 + r^2(\theta')^2}$$

and find the unit tangent and unit normal vectors to be

$$\vec{T} = \frac{1}{\sqrt{(r')^2 + r^2(\theta')^2}} (r'\vec{e}_r + r\theta'\vec{e}_\theta),$$

$$\vec{U} = \frac{1}{\sqrt{(r')^2 + r^2(\theta')^2}} (-r\theta'\vec{e}_r + r'\vec{e}_\theta).$$

Therefore, the orthogonal matrix

$$\frac{1}{\sqrt{(r')^2 + r^2(\theta')^2}} \begin{pmatrix} r' & -r\theta' \\ r\theta' & r' \end{pmatrix}$$

is the transition matrix between the $\{\vec{T}, \vec{U}\}$ basis and the $\{\vec{e}_r, \vec{e}_\theta\}$ basis.

Parenthetically, it is now not difficult to obtain a formula for the plane curvature of $\vec{x}(t)$ in terms of the functions $r(t)$ and $\theta(t)$. We use either of the formulations

$$\kappa_g(t) = \frac{1}{s'(t)} \vec{T}' \cdot \vec{U} = \frac{1}{(s'(t))^3} (\vec{x}' \times \vec{x}'') \cdot \vec{k},$$

and we find that

$$\kappa_g(t) = \frac{-rr''\theta' + r^2(\theta')^3 + 2(r')^2\theta' + rr'\theta''}{\left((r')^2 + r^2(\theta')^2\right)^{3/2}}. \tag{2.12}$$

In general, a frame $\mathcal{F}$ in $\mathbb{R}^3$ that varies with respect to a parameter t consists of a quadruple of vector functions $(\vec{\alpha}(t), \vec{e}_1(t), \vec{e}_2(t), \vec{e}_3(t))$. The vector function $\vec{\alpha}(t)$ is a curve that traces out the motion of base (or origin) of the frame $\mathcal{F}$ and the set of vector functions $\{\vec{e}_1(t), \vec{e}_2(t), \vec{e}_3(t)\}$ are linearly independent for all t. We are not constrained to only consider frames in which $\{\vec{e}_1(t), \vec{e}_2(t), \vec{e}_3(t)\}$ form an orthonormal set for all t, but we will make that assumption for the remainder of this section and we will assume in addition that this basis is a positively oriented basis, i.e., it satisfies $\vec{e}_1 \times \vec{e}_2 = \vec{e}_3$. Now for all t,

$$\vec{e}_i \cdot \vec{e}_j = \begin{cases} 1, & \text{if } i = j, \\ 0, & \text{if } i \neq j, \end{cases}$$

so by a dot product rule,

$$\vec{e}_i' \cdot \vec{e}_j = \begin{cases} 0, & \text{if } i = j, \\ -\vec{e}_i \cdot \vec{e}_j', & \text{if } i \neq j. \end{cases} \tag{2.13}$$

Let $\mathcal{F} = (\vec{\alpha}, \vec{e}_1, \vec{e}_2, \vec{e}_3)$ be a moving positive orthonormal frame. Consider the vector function $\vec{\Omega}(t)$ defined by

$$\vec{\Omega} = (\vec{e}_2' \cdot \vec{e}_3)\vec{e}_1 + (\vec{e}_3' \cdot \vec{e}_1)\vec{e}_2 + (\vec{e}_1' \cdot \vec{e}_2)\vec{e}_3. \tag{2.14}$$

Using (2.13), it is easy to check that $\vec{e}_i' = \vec{\Omega} \times \vec{e}_i$ for all i and for all t.

Definition 2.2.1. The vector function $\vec{\Omega}(t)$ is called the *angular velocity vector* of the moving frame $\mathcal{F}$.

We now consider a particle following a trajectory $\vec{x} : I \to \mathbb{R}^3$ and we propose to determine the *perceived* position, velocity, and acceleration vectors in the moving frame $\mathcal{F}$ in terms of the true position, velocity, and acceleration. Label $(\vec{x})_{\mathcal{F}}$, $(\vec{x}')_{\mathcal{F}}$, and $(\vec{x}'')_{\mathcal{F}}$ as the perceived position, velocity, and acceleration vectors. First,

$$(\vec{x})_{\mathcal{F}} = \vec{x} - \vec{\alpha}. \tag{2.15}$$

However, the perceived velocity and acceleration of $\vec{x}$ are obtained by taking the derivatives of the components of $(\vec{x})_{\mathcal{F}}$ in $\{\vec{e}_1, \vec{e}_2, \vec{e}_3\}$. More explicitly,

$$(\vec{x})_{\mathcal{F}} = ((\vec{x} - \vec{\alpha}) \cdot \vec{e}_1)\vec{e}_1 + ((\vec{x} - \vec{\alpha}) \cdot \vec{e}_2)\vec{e}_2 + ((\vec{x} - \vec{\alpha}) \cdot \vec{e}_3)\vec{e}_3,$$

$$(\vec{x}')_{\mathcal{F}} = \frac{d}{dt}\Big((\vec{x} - \vec{\alpha}) \cdot \vec{e}_1\Big)\vec{e}_1 + \frac{d}{dt}\Big((\vec{x} - \vec{\alpha}) \cdot \vec{e}_2\Big)\vec{e}_2 + \frac{d}{dt}\Big((\vec{x} - \vec{\alpha}) \cdot \vec{e}_3\Big)\vec{e}_3, \tag{2.16}$$

$$(\vec{x}'')_{\mathcal{F}} = \frac{d^2}{dt^2}\Big((\vec{x} - \vec{\alpha}) \cdot \vec{e}_1\Big)\vec{e}_1 + \frac{d^2}{dt^2}\Big((\vec{x} - \vec{\alpha}) \cdot \vec{e}_2\Big)\vec{e}_2 + \frac{d^2}{dt^2}\Big((\vec{x} - \vec{\alpha}) \cdot \vec{e}_3\Big)\vec{e}_3.$$

We can now relate the perceived position, velocity, and acceleration in the moving frame $\mathcal{F}$ to the actual position, velocity, and acceleration. By (2.15),

$$\vec{x} = (\vec{x})_{\mathcal{F}} + \vec{\alpha}.$$

Then for the velocity,

$$\vec{x}' = \frac{d}{dt}(\vec{x})_{\mathcal{F}} + \vec{\alpha}'$$

$$= \Big(\sum_{i=1}^{3} \frac{d}{dt}((\vec{x} - \vec{\alpha}) \cdot \vec{e}_i)\vec{e}_i\Big) + \Big(\sum_{i=1}^{3}((\vec{x} - \vec{\alpha}) \cdot \vec{e}_i)\vec{e}_i'\Big) + \vec{\alpha}' \tag{2.17}$$

$$= (\vec{x}')_{\mathcal{F}} + \Big(\sum_{i=1}^{3}((\vec{x} - \vec{\alpha}) \cdot \vec{e}_i)\vec{\Omega} \times \vec{e}_i\Big) + \vec{\alpha}'$$

$$= (\vec{x}')_{\mathcal{F}} + \vec{\Omega} \times (\vec{x})_{\mathcal{F}} + \vec{\alpha}'.$$

For the acceleration,

$$\vec{x}'' = \frac{d}{dt}(\vec{x}')_{\mathcal{F}} + \frac{d}{dt}\Big(\vec{\Omega} \times (\vec{x})_{\mathcal{F}}\Big) + \vec{\alpha}''$$

$$= \frac{d}{dt}\Big(\sum_{i=1}^{3}((\vec{x} - \vec{\alpha}) \cdot \vec{e}_i)\vec{e}_i\Big) + \frac{d\vec{\Omega}}{dt} \times (\vec{x})_{\mathcal{F}} + \vec{\Omega} \times \frac{d}{dt}(\vec{x})_{\mathcal{F}} + \vec{\alpha}''$$

$$= \Big(\sum_{i=1}^{3}\frac{d^2}{dt^2}((\vec{x} - \vec{\alpha}) \cdot \vec{e}_i)\vec{e}_i\Big) + \Big(\sum_{i=1}^{3}\frac{d}{dt}((\vec{x} - \vec{\alpha}) \cdot \vec{e}_i)\vec{e}_i'\Big)$$

$$\qquad + \vec{\Omega}' \times (\vec{x})_{\mathcal{F}} + \vec{\Omega} \times \frac{d}{dt}(\vec{x})_{\mathcal{F}} + \vec{\alpha}''$$

$$= (\vec{x}'')_{\mathcal{F}} + \vec{\Omega} \times (\vec{x}')_{\mathcal{F}} + \vec{\Omega} \times (\vec{x})_{\mathcal{F}} + \vec{\Omega} \times \Big((\vec{x}')_{\mathcal{F}} + \vec{\Omega} \times (\vec{x})_{\mathcal{F}}\Big) + \vec{\alpha}'',$$

where the second-to-last term follows from Equation (2.17). Thus,

$$\vec{x}'' = (\vec{x}'')_{\mathcal{F}} + 2\vec{\Omega} \times (\vec{x}')_{\mathcal{F}} + \vec{\Omega}' \times (\vec{x})_{\mathcal{F}} + \vec{\Omega} \times (\vec{\Omega} \times (\vec{x})_{\mathcal{F}}) + \vec{\alpha}''. \qquad (2.18)$$

All of the above terms have names in physics (see [22, p. 118]):

- $(\vec{x}'')_{\mathcal{F}}$ is called the *perceived acceleration* or *acceleration with respect to* $\mathcal{F}$;

- $2\vec{\Omega} \times (\vec{x}')_{\mathcal{F}}$ is the *Coriolis acceleration*;

- $\vec{\Omega} \times (\vec{\Omega} \times (\vec{x})_{\mathcal{F}})$ is the *centripetal acceleration*;

- $\vec{\Omega}' \times (\vec{x})_{\mathcal{F}}$ is sometimes called the *transverse acceleration* because it is perpendicular to the perceived position vector $(\vec{x})_{\mathcal{F}}$;

- $\vec{\alpha}''$ is the *translational acceleration* of the frame.

The above discussion described the moving frame $\mathcal{F}$ in terms of some absolute (unmoving) frame. Though an absolute frame arises naturally in the mental framework of Cartesian coordinates, to assume the existence of an absolute frame in physical systems poses serious challenges. We may think of a point fixed to the Earth as the origin for an absolute frame, but taking into account that the Earth moves around the Sun, and the Sun moves around the galaxy and so on should disqualify this choice. Using Newton's second law of motion as a reference, classical mechanics defines an *inertial frame* as one in which the motion of a particle not subject to any forces travels in a straight line.

Now suppose that we have identified one inertial frame $\mathcal{F}_1$ and we consider another (moving) frame $\mathcal{F}_2$. From (2.18), $\mathcal{F}_2$ will also be an inertial frame if and only if $(\vec{x}'')_{\mathcal{F}_1} = (\vec{x}'')_{\mathcal{F}_2}$ for all trajectories $\vec{x}(t)$. This implies that $\vec{\Omega} = \vec{0}$ and that $\vec{\alpha}'' = \vec{0}$, expressed in reference to $\mathcal{F}_1$. Hence the unit vectors in $\mathcal{F}_2$ do not move with reference to the basis vectors of $\mathcal{F}_1$ and the origin of $\mathcal{F}_2$ moves with a constant velocity vector in reference to the frame $\mathcal{F}_1$.

Admittedly, the problem in practice of finding one inertial frame leads to a vicious circle. How do we know we have found a body free of external forces? We can only content ourselves with finding a frame in which Newton's laws of motion hold to a "satisfactory" degree.[21]

As an example of the application of differential geometry of curves to physics, we consider the notion of centripetal acceleration of a curve and its relation to the Frenet frame.

Example 2.2.2 (Centripetal Acceleration of Curves). One first encounters centripetal acceleration in the context of a particle moving around on a circle with constant speed v, and one defines it as the acceleration due to the change in the velocity vector. Phrasing the scenario mathematically, consider a particle moving along the trajectory with equations of motion

$$\vec{x}(t) = (R\cos(\omega t), R\sin(\omega t)),$$

where R is the radius of the circle and ω is the (constant) angular speed. The velocity, speed and acceleration are, respectively,

$$\vec{x}'(t) = (-R\omega\sin(\omega t), R\omega\cos(\omega t)),$$
$$s'(t) = v = R\omega,$$
$$\vec{x}''(t) = (-R\omega^2\cos(\omega t), -R\omega^2\sin(\omega t)).$$

Hence, the acceleration is

$$\vec{x}''(t) = -\omega^2\vec{x}(t) = -\omega^2 R\vec{e}_r = -\frac{v^2}{R}\vec{e}_r, \tag{2.19}$$

where $\vec{e}_r$ is the unit vector in the radial direction (see Equation (2.4)). This is the centripetal acceleration for circular motion, often written $\vec{a}_c$.

The angular velocity vector $\vec{\Omega}$ is the vector of magnitude ω that is perpendicular to the plane of rotation and with direction given by the right-hand rule. Thus, taking $\vec{k}$ as the direction perpendicular to the plane, we have in this simple setup $\vec{\Omega} = \omega\vec{k}$. Setting the radial vector $\vec{R} = r\vec{e}_r$, it is not hard to show that for this circular motion,

$$\vec{a}_c = \vec{\Omega} \times (\vec{\Omega} \times \vec{R}),$$

as expected from Equation (2.18).

Now consider a general curve in space $\vec{x} : I \to \mathbb{R}^3$, where I is an interval of $\mathbb{R}$. We recall a few differential geometric properties of space curves. The derivative $\vec{x}'(t)$ is called the velocity and $s'(t) = \|\vec{x}'(t)\|$ is called the speed. The curve $\vec{x}(t)$ is called regular at t if $\vec{x}'(t) \neq \vec{0}$. At all regular points of a curve, we define the unit tangent as $\vec{T}(t) = \vec{x}'(t)/\|\vec{x}'(t)\|$. Because $\vec{T}(t)$ is a unit vector for all t, $\vec{T}'(t)$ is perpendicular to $\vec{T}(t)$.

The curvature of the curve is the unique nonnegative function $\kappa(t)$ such that

$$\vec{T}'(t) = s'(t)\kappa(t)\vec{P}(t) \tag{2.20}$$

for some unit vector $\vec{P}(t)$. The vector function $\vec{P}(t)$ is called the *principal normal* vector. Finally, we define the *binormal* vector function $\vec{B}(t)$ by $\vec{B} = \vec{T} \times \vec{P}$. In so doing, we have defined an orthonormal set $\{\vec{T}, \vec{P}, \vec{B}\}$ associated to each point of the curve $\vec{x}(t)$. This set $\{\vec{T}, \vec{P}, \vec{B}\}$ is called the *Frenet frame*.

It is not hard to show that, by construction, the derivative $\vec{B}'(t)$ is perpendicular to $\vec{B}$ and to $\vec{T}$. We define the torsion function $\tau(t)$ of a space curve as the unique function such that

$$\vec{B}'(t) = -s'(t)\tau(t)\vec{P}(t). \tag{2.21}$$

Finally, from Equations (2.20) and (2.21) that

$$\vec{P}'(t) = -s'(t)\kappa(t)\vec{T}(t) + s'(t)\tau(t)\vec{B}(t). \tag{2.22}$$

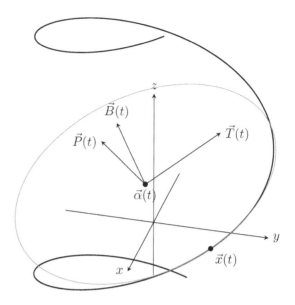

Figure 2.4: Center of curvature and osculating circle.

We summarize (2.20), (2.21) and (2.22) as

$$\frac{d}{dt}\begin{pmatrix} \vec{T} & \vec{P} & \vec{B} \end{pmatrix} = \begin{pmatrix} \vec{T} & \vec{P} & \vec{B} \end{pmatrix}\begin{pmatrix} 0 & -s'\kappa & 0 \\ s'\kappa & 0 & -s'\tau \\ 0 & s'\tau & 0 \end{pmatrix}. \tag{2.23}$$

(The above paragraphs only give the definitions of the concepts we will use below. A full treatment of these topics can be found in [5, Chapter 3].)

Since a space curve is not necessarily circular, one cannot use Equation (2.19) to determine the centripetal acceleration of $\vec{x}$. Instead, we view $\vec{x}$ in relation to an appropriate moving frame in which centripetal acceleration makes sense. The osculating circle is the unique circle of maximum contact with the curve $\vec{x}(t)$ at any point t, and hence, the appropriate frame $\mathcal{F}$ is based at the center of curvature

$$\vec{\alpha}(t) = \vec{x}(t) + \frac{1}{\kappa(t)}\vec{P}(t)$$

and has the vectors of the Frenet frame $\{\vec{T}, \vec{P}, \vec{B}\}$ as its basis. Figure 2.4 depicts a space curve along with the center of curvature $\vec{\alpha}(t)$ and the osculating circle associated to a point $\vec{x}(t)$ on the curve.

By Equation (2.14), the angular velocity vector of $\mathcal{F}$ is

$$\vec{\Omega} = (\vec{P}' \cdot \vec{B})\vec{T} + (\vec{B}' \cdot \vec{T})\vec{P} + (\vec{T}' \cdot \vec{P})\vec{B} = s'\tau\vec{T} + s'\kappa\vec{B}.$$

The relative position vector for the curve $\vec{x}$ with reference to its center of curvature is $\vec{R} = (\vec{x})_{\mathcal{F}} = -\frac{1}{\kappa}\vec{P}$. Therefore, the centripetal acceleration is

$$\vec{a}_c = \vec{\Omega} \times (\vec{\Omega} \times \vec{R}) = \vec{\Omega} \times \left((s'\tau\vec{T} + s'\kappa\vec{B}) \times (-\frac{1}{\kappa}\vec{P}) \right)$$

$$= (s'\tau\vec{T} + s'\kappa\vec{B}) \times (s'\vec{T} - s'\frac{\tau}{\kappa}\vec{B}) = (s')^2\kappa\vec{P} + (s')^2\frac{\tau^2}{\kappa}\vec{P}$$

$$= (s')^2\frac{\kappa^2 + \tau^2}{\kappa}\vec{P}. \tag{2.24}$$

It is interesting to note that if a curve happens to be planar, then $\tau = 0$, and the centripetal acceleration becomes $\vec{a}_c = (s')^2\kappa\vec{P}$, which matches Equation (2.19) exactly since $s' = v$ and κ is the reciprocal of the radius of curvature, $1/R$. However, Equation (2.24) shows that, for a curve in space, the "corkscrewing" effect, measured by τ, produces a greater centripetal acceleration than does simply rotating about the same axis. (Hence, on a rollercoaster a rider will experience more centrifugal force – the force that balances out centripetal acceleration – if the rollercoaster corkscrews than when it simply rotates around with the same radius of curvature.)

We finish this section with a classical example of how using a useful moving frame renders equations of motion tractable.

Example 2.2.3 (Radial Forces). As an application of cylindrical coordinate systems, we can study Newton's equation of motion applied to a particle under the influence of a radial force. By definition, a force is called radial if $\vec{F}(\vec{r}) = f(r)\vec{e}_r$, that is, if the force only depends on the distance from an origin and is parallel to the position vector $\vec{r}$. (The force of gravity between two point objects and the electric force between two charged point objects are radial forces, while the magnetic force on a charged particle is not.)

Newton's law of motion produces the following vector differential equation:

$$m\vec{r}'' = f(r)\vec{e}_r.$$

In order to solve this differential equation explicitly, one needs the initial position $\vec{r}_0$ and the initial velocity $\vec{v}_0$.

For convenience, choose a plane $\mathcal{P}$ that goes through the origin and is parallel to both $\vec{r}_0$ and $\vec{v}_0$. (If $\vec{r}_0$ and $\vec{v}_0$ are not parallel, then this information defines a unique plane in $\mathbb{R}^3$. If $\vec{r}_0$ and $\vec{v}_0$ are parallel, then any plane parallel to these vectors suffices.) Consider $\mathcal{P}$ to be the xy-plane, choose any direction for the ray $[Ox)$ and now use cylindrical coordinates in $\mathbb{R}^3$.

For radial forces, Newton's law of motion written in the cylindrical frame as three differential equations is

$$\begin{cases} \vec{e}_r : & m(r'' - r(\theta')^2) = f(r), \\ \vec{e}_\theta : & m(2r'\theta' + r\theta'') = 0, \\ \vec{e}_z : & 0 = 0. \end{cases} \tag{2.25}$$

Obviously, since $\vec{r}_0$ and $\vec{v}_0$ lie in the plane through the origin and parallel to $\vec{e}_r$ and $\vec{e}_\theta$, the equations show that $\vec{r}(t)$ never leaves the xy-plane. Thus $z(t) = 0$.

We can now solve the second differential equation in the above system to obtain a relationship between the functions r and θ. First write

$$\frac{2r'}{r} = -\frac{\theta''}{\theta'}.$$

Integrating both sides with respect to t, we then obtain $2 \ln |r| = -\ln |\theta'| + C$, where C is some constant of integration. Taking the exponential of both sides, one obtains the relationship $r^2\theta' = h$ where h is a constant. In terms of the initial conditions, we have

$$\vec{r} \times \vec{r}' = r^2\theta' \vec{e}_z$$

and therefore, for all time t, we have

$$h = (\vec{r}_0 \times \vec{v}_0) \cdot \vec{e}_z.$$

Thus, we conclude that the quantity $\vec{L} = \vec{r} \times (m\vec{v}) = m(\vec{r} \times \vec{v})$, which is called the *angular momentum* and in general depends on t, is a constant vector function for radial forces.

Finally, to solve the system in Equation (2.25) completely, it is convenient to substitute variables and write the first equation in terms of $u = 1/r$ and θ. Since $r = 1/u$, we have

$$\frac{dr}{dt} = -\frac{1}{u^2}\frac{du}{dt} = -\frac{1}{u^2}\frac{d\theta}{dt}\frac{du}{d\theta} = -h\frac{du}{d\theta}.$$

The second derivative of r gives

$$\frac{d^2r}{dt^2} = -h\frac{d}{dt}\left(\frac{du}{d\theta}\right) = -h\frac{d^2u}{d\theta^2}\frac{d\theta}{dt} = -h^2u^2\frac{d^2u}{d\theta^2},$$

where the last equality follows from fact that $r^2\theta'$ is the constant h. The first part of Equation (2.25) becomes

$$\frac{d^2u}{d\theta^2} + u = -\frac{1}{mh^2u^2}f(u^{-1}). \tag{2.26}$$

If the radial force in question is an inverse-square law (such as the force of gravity and the electrostatic force caused by a point charge), then the radial force is of the form

$$f(r) = -\frac{k}{r^2} = -ku^2.$$

In this case, Equation (2.26) becomes

$$\frac{d^2u}{d\theta^2} + u = \frac{k}{mh^2}.$$

Techniques from differential equations show that the general solution to this equation is

$$u(\theta) = \frac{k}{mh^2} + C\cos(\theta - \theta_0),$$

where C and θ_0 are constants of integration that depend on the original position and velocity of the point particle under the influence of this radial force. In polar coordinates, this gives the equation

$$r(\theta) = \frac{1}{\frac{k}{mh^2} + C\cos(\theta - \theta_0)}. \tag{2.27}$$

PROBLEMS

2.2.1. Provide the details for the proof of Equation (2.12).

2.2.2. Prove that Equation (2.14) is the correct vector to satisfy $\vec{e}_i' = \vec{\Omega} \times \vec{e}_i$ for all i.

2.2.3. Determine the transition matrix between the cylindrical coordinate frame and the Frenet frame.

2.2.4. Calculate the curvature and torsion of a space curve defined by the functions, in cylindrical coordinates, $(r, \theta, z) = (r(t), \theta(t), z(t))$.

2.2.5. Determine the transition matrix between the spherical coordinate frame and the Frenet frame.

2.2.6. Calculate the curvature and torsion of a space curve defined by the functions, in spherical coordinates, $(r, \theta, \phi) = (r(t), \theta(t), \phi(t))$.

2.2.7. Determine the transition matrix between the parabolic coordinate frame and the Frenet frame (see Problem 2.1.5).

2.2.8. Consider the solution $r(\theta)$ in Equation (2.27). Determine h, C, and θ_0 in terms of some initial conditions for position and velocity $\vec{r}(0)$ and $\vec{v}(0)$. Prove that for different initial conditions and different values of the constants, the locus of Equation (2.27) is a conic. State under what conditions the locus is a circle, ellipse, parabola, and hyperbola.

2.2.9. (ODE) Find the locus of the trajectory of a particle moving under the effect of a radial force that is an inverse cube, i.e., $f(r) = -k/r^3$. [Hint: There are three separate cases depending on whether $mh^2 > k$, $mh^2 = k$, or $mh^2 < k$.]

2.3 Variable Frames and Matrix Functions

In the preceding sections, we often described the rate of change of variable frames using a matrix function. Equations (2.1), (2.6), (2.8), and (2.23) established a matrix formula to describe the rate of change of the frame vectors for Cartesian, cylindrical, spherical, and Frenet frames respectively. This section generalizes this perspective for variable frames.

By a *matrix function*, we mean a function $F : U \to M_{m \times n}(\mathbb{R})$, where U is an open subset of $\mathbb{R}^p$. Identifying the set of $m \times n$ matrices $M_{m \times n}(\mathbb{R})$ with the Euclidean space $\mathbb{R}^{mn}$, the analysis on multivariable functions developed in Chapter 1 applies. A single variable matrix function can be viewed as a curve $\gamma : I \to M_{m \times n}(\mathbb{R})$, where I is an interval of $\mathbb{R}$. As with any parametrized curve, the derivative $\gamma'(t)$ is the $m \times n$ matrix of derivatives of component functions of $\gamma(t)$.

Proposition 2.3.1. *Let $\gamma_1(t)$ and $\gamma_2(t)$ be matrix functions defined over an interval I, and let A be any constant matrix. Assuming the operations are defined, the following identities hold:*

1. $\frac{d}{dt}(A)$ *is the 0-matrix of the same dimensions of A.*

2. $\frac{d}{dt}(A\gamma_1(t)) = A\gamma_1'(t)$ *and* $\frac{d}{dt}(\gamma_1(t)A) = \gamma_1'(t)A$.

3. $\frac{d}{dt}(\gamma_1(t) + \gamma_2(t)) = \gamma_1'(t) + \gamma_2'(t)$.

4. $\frac{d}{dt}\big(\gamma_1(t)^\top\big) = (\gamma_1'(t))^\top$.

5. $\frac{d}{dt}(\gamma_1(t)\gamma_2(t)) = \gamma_1'(t)\gamma_2(t) + \gamma_1(t)\gamma_2'(t)$.

6. *If $\gamma_1(t)$ is invertible for all t, then* $\frac{d}{dt}\big(\gamma_1(t)^{-1}\big) = -\gamma_1(t)^{-1}\gamma_1'(t)\gamma_1(t)^{-1}$.

Proof. (Left as exercises for the reader.) $\qquad\qquad\qquad\qquad\qquad\qquad\qquad\square$

A particularly useful matrix function involves the exponential of matrices. Let A be a $p \times p$ matrix and let $\vec{v} \in \mathbb{R}^p$. Consider the sequence $\{\vec{x}_n\}_{n=0}^\infty$ of vectors defined by

$$\vec{x}_n = \sum_{k=0}^n \frac{1}{k!} A^n \vec{v},$$

where by A^0, we mean the identity matrix I. We prove that this sequence is a Cauchy sequence. Denoting $|A|$ by the matrix norm of A, we have

$$\|\vec{x}_n - \vec{x}_m\| = \left\| \sum_{k=m+1}^n \frac{1}{k!} A^k \vec{v} \right\| \leq \sum_{k=m+1}^n \frac{1}{k!} \|A^k \vec{v}\|$$

$$\leq \sum_{k=m+1}^n \frac{1}{k!} |A|^k \|\vec{v}\| \leq \sum_{k=m+1}^\infty \frac{1}{k!} |A|^k \|\vec{v}\|,$$

where the last inequality follows since all the terms are nonnegative. Then since $(k - m - 1)!/k! \leq 1/(m+1)!$, we have

$$\|\vec{x}_n - \vec{x}_m\| \leq |A|^{m+1} \|\vec{v}\| \sum_{k=m+1}^\infty \frac{(k-m-1)!}{k!} \frac{1}{(k-m-1)!} |A|^{k-m-1}$$

$$\leq \frac{|A|^{m+1}}{(m+1)!} \|\vec{v}\| \sum_{j=0}^\infty \frac{1}{j!} |A|^j \leq \frac{|A|^{m+1}}{(m+1)!} \|\vec{v}\| e^{|A|},$$

where we substituted $j = k - m - 1$. For any positive real number $|A|$, the limit of $|A|^{m+1}/(m+1)!$ as $m \to \infty$ is 0. Hence, for any positive ε, there exists N large enough so that $m, n \geq N$ implies that

$$\|\vec{x}_n - \vec{x}_m\| \leq \frac{|A|^{m+1}}{(m+1)!}\|\vec{v}\|e^{|A|} \leq \varepsilon.$$

This establishes that the sequence $\{\vec{x}_n\}_{n=0}^{\infty}$ is a Cauchy sequence. Since $\mathbb{R}^p$ is a complete metric space, we conclude that this sequence converges. Since the sequence of vectors converges for all $\vec{v}$, we conclude that series of matrices in the following definition converges.

Definition 2.3.2. Let A be an $n \times n$ matrix. We define the *exponential* of A as

$$e^A = \sum_{k=0}^{\infty} \frac{1}{k!}A^k.$$

Proposition 2.3.3. *Let A and B be two matrices that commute, i.e., satisfying $AB = BA$. Then*

$$e^{A+B} = e^A e^B.$$

Proof. (Left as an exercise for the reader.) □

This proposition allows us to conclude the following interesting result.

Proposition 2.3.4. *For all $A \in M_{n \times n}(\mathbb{R})$, the exponential matrix e^A is invertible.*

Proof. The matrices A and $-A$ commute. Hence, by Proposition 2.3.3, $e^A e^{-A} = e^{A-A} = e^0 = I$. Thus e^A has an inverse. □

Now let $A \in M_{n \times n}(\mathbb{R})$ and consider the matrix function $\gamma(t) = e^{At}$. Note first that $\gamma(0) = I$. The derivative of $\gamma(t)$ is

$$\gamma'(t) = \frac{d}{dt}\left(\sum_{k=0}^{\infty}\frac{1}{k!}A^k t^k\right) = \sum_{k=0}^{\infty}\frac{1}{k!}\frac{d}{dt}(A^k t^k) = \sum_{k=1}^{\infty}\frac{1}{k!}A^k k t^{k-1}$$

$$= \sum_{k=1}^{\infty}\frac{1}{(k-1)!}A^k t^{k-1} = A\left(\sum_{k=1}^{\infty}\frac{1}{(k-1)!}A^k t^{k-1}\right) = Ae^{At}.$$

In particular $\gamma'(0) = A$.

We now connect these concepts to moving frames.

Any frame $\mathcal{F}$ of $\mathbb{R}^n$ consists of an origin and a basis $(\vec{u}_1, \vec{u}_2, \ldots, \vec{u}_n)$ of $\mathbb{R}^n$. Since the basis consists of n linearly independent vectors, the matrix

$$M = \begin{pmatrix} | & | & & | \\ \vec{u}_1 & \vec{u}_2 & \cdots & \vec{u}_n \\ | & | & & | \end{pmatrix},$$

where the ith column is the vector $\vec{u}_i$, is an invertible matrix. Similarly, for any invertible $n \times n$ matrix M, the columns form a basis of $\mathbb{R}^n$. If $\mathcal{F}$ is a moving frame, we can view the vectors of this moving frame as a matrix function $M(t)$ where $M(t)$ is invertible for all t. More precisely, M is a matrix function $M : I \to \mathrm{GL}_n(\mathbb{R})$, where I is an interval of $\mathbb{R}$ and $\mathrm{GL}_n(\mathbb{R}) = \{A \in M_{n \times n}(\mathbb{R}) \mid \det a \neq 0\}$ is the set of invertible matrices, also called the *general linear group*.

For any moving frame $\mathcal{F}$ with basis vectors $(\vec{u}_1(t), \ldots, \vec{u}_n(t))$, we can express each derivative $\vec{u}_i'(t)$ as a linear combination of these same basis vectors at a given t. As we did with Equations (2.1), (2.6), (2.8), and (2.23), if we consider the matrix function $M(t) = \begin{pmatrix} \vec{u}_1(t) & \cdots & \vec{u}_n(t) \end{pmatrix}$, then this decomposition can be expressed as

$$M'(t) = M(t)A(t)$$

for some matrix $A(t)$.

Proposition 2.3.5. *Let $B \in M_{n \times n}(\mathbb{R})$ be arbitrary. There exists a variable frame with matrix function $M : I \to \mathrm{GL}_n(\mathbb{R})$ with rate of change matrix $A(t)$ satisfying $M'(t) = M(t)A(t)$ such that $A(t_0) = B$ for some $t_0 \in I$.*

Proof. Consider $M(t) = e^{B(t-t_0)} = e^{Bt}e^{-Bt_0}$. We note that $M(t_0) = e^{\mathbf{0}} = I_n$. By the differentiation property of the matrix exponential, $M'(t) = Be^{Bt}e^{-Bt_0} = Be^{B(t-t_0)}$. Thus $A(t_0) = M(t_0)A(t_0) = M'(t_0) = Be^{\mathbf{0}} = B$. $\qquad\square$

Many of the examples that we discussed in the previous section involved orthonormal variable frames. An orthonormal basis in $\mathbb{R}^n$ is any n-tuple of vectors $(\vec{u}_1, \vec{u}_2, \ldots, \vec{u}_n)$ such that

$$\vec{u}_i \cdot \vec{u}_j = \begin{cases} 1, & \text{if } i = j, \\ 0, & \text{if } i \neq j. \end{cases} \tag{2.28}$$

Then using as usual $M = \begin{pmatrix} \vec{u}_1 & \vec{u}_2 & \cdots & \vec{u}_n \end{pmatrix}$, we note that the ijth entry of $M^\top M$ is precisely the dot product $\vec{u}_i \cdot \vec{u}_j$. Consequently, the vectors form an orthonormal frame if and only if $M^\top M = I_n$, i.e., M is an orthogonal matrix.

Since $\det(M^\top) = \det(M)$, an orthogonal matrix M satisfies $\det(M) = \pm 1$. An orthonormal basis $(\vec{u}_1, \vec{u}_2, \ldots, \vec{u}_n)$ is called *positively oriented* if $\det(M) = 1$ and *negatively oriented* if $\det(M) = -1$. The set of orthogonal $n \times n$ matrices is denoted by $\mathrm{O}(n)$, and the set of positive orthogonal matrices is denoted by

$$\mathrm{SO}(n) = \{M \in \mathrm{O}(n) \mid \det(M) = 1\}.$$

Both $\mathrm{O}(n)$ and $\mathrm{SO}(n)$ have a group structure, a property not discussed in this book, and are respectively called the *orthogonal group* and *special orthogonal group*.

(Note that the order of the basis vectors in the n-tuple $(\vec{u}_1, \vec{u}_2, \ldots, \vec{u}_n)$ matters since a permutation of these vectors may change the sign of the determinant of the corresponding matrix M. Consequently, we must talk about an n-tuple of vectors as opposed to just a set of vectors. One should also be aware that a permutation of vectors in the basis $\mathcal{B} = (\vec{u}_1, \vec{u}_2, \ldots, \vec{u}_n)$ would lead to another basis $\mathcal{B}'$ which consists of the same set of vectors but has a coordinate transition matrix which is a permutation matrix.)

We can therefore view an orthonormal moving frame in $\mathbb{R}^n$ as a map $M : I \to O(n)$, where $I \subset \mathbb{R}$ is an interval. In this case, if M is continuous, then $\det(M(t))$ is a continuous function from I to $\{-1, 1\}$. Consequently, $\det(M(t))$ is either 1 or -1 for all t. When $\det(M(t)) = 1$, we say that M corresponds to a positive orthonormal moving frame, and we view M as a function $M : I \to SO(n)$. The sets of matrices $O(n)$ and $SO(n)$ have the subset topology induced from the Euclidean topology on $\mathbb{R}^{n^2}$. Consequently, the notions of continuity and differentiability of a moving frame are familiar notions.

Proposition 2.3.6. *Let $I \subset \mathbb{R}$ be an interval, and let $M : I \to O(n)$ be a differentiable function. Then the matrix function $A(t)$ defined by*

$$M'(t) = M(t)A(t)$$

is antisymmetric for all $t \in I$. Furthermore, for any antisymmetric matrix B, there exists a matrix function $M : I \to O(n)$ such that $A(t_0) = B$ for some $t_0 \in I$.

Proof. Since $M(t) \in O(n)$, we have $M(t)^\top M(t) = I_n$ for all t, and similarly $M(t)M(t)^\top = I_n$. Hence, using the differentiation rules,

$$\mathbf{0} = M'(t)M(t)^\top + M(t)\frac{d}{dt}(M(t)^\top) = M'(t)M(t)^\top + M(t)(M'(t))^\top.$$

Thus $M(t)(M'(t))^\top = -M'(t)M(t)^\top$ so after multiplying on the right by $\left(M(t)^{-1}\right)^\top$ and on the left by $M(t)^{-1}$, we get

$$(M'(t))^\top \left(M(t)^{-1}\right)^\top = -M'(t)M(t)^{-1}$$
$$\implies \left(M(t)^{-1}M'(t)\right)^\top = -M(t)^{-1}M'(t).$$

However, from the definition of the matrix function $A(t)$, we have $A(t) = M(t)^{-1}M'(t)$. Hence, we deduce that $A(t)^\top = -A(t)$ and therefore that $A(t)$ is antisymmetric.

For the second part of the proof, let B be antisymmetric, i.e., $B^\top = -B$, and consider the matrix function $M(t) = e^{B(t-t_0)}$. Then

$$M(t)^\top = \left(\sum_{k=0}^{\infty} \frac{1}{k!}B^k(t - t_0)^k\right)^\top = \sum_{k=0}^{\infty} \frac{1}{k!}(B^\top)^k(t - t_0)^k$$

$$= \sum_{k=0}^{\infty} \frac{1}{k!}(-1)^k(B)^k(t - t_0)^k = \sum_{k=0}^{\infty} \frac{1}{k!}(B)^k(t_0 - t)^k$$

$$= e^{B(t_0 - t)} = M(t)^{-1}.$$

Hence, $M(t)$ is orthogonal for all $t \in I$. Then $M'(t) = Be^{B(t-t_0)} = BM(t)$ so

$$M'(t_0) = BM(t_0) = BI = B,$$

and the result follows. $\qquad\qquad\qquad\qquad\qquad\qquad\qquad\qquad\qquad\qquad\qquad\qquad\quad \square$

The four cases that motivated this section, namely Equations (2.1), (2.6), (2.8), and (2.23), illustrate the first part of Proposition 2.3.6. In all of those examples, the $A(t)$ matrix function was an antisymmetric matrix for all t.

PROBLEMS

2.3.1. Prove Proposition 2.3.3.

2.3.2. Let J be a constant $n \times n$ matrix and let $\gamma : (-\varepsilon, \varepsilon) \to \mathrm{GL}_n(\mathbb{R})$ be a differentiable matrix function, where $\varepsilon > 0$. Suppose that $\gamma(0) = I_n$ and suppose that $\gamma(t)^\top J \gamma(t) = J$ for all $t \in (-\varepsilon, \varepsilon)$. Prove that the matrix $\gamma'(0)$ satisfies

$$\gamma'(0)^\top J = -J\gamma'(0).$$

2.3.3. Find an example of an $n \times n$-matrix function $\gamma(t)$ such that $f'(t)$ is never 0, where $f(t) = \det(\gamma(t))$, but such that $\det(\gamma'(t)) = 0$ for all t.

2.3.4. Let $\gamma : I \to \mathbb{R}^{m \times n}$ be a differentiable matrix function, and let $f : J \to I$ be a differentiable function. Prove the chain rule for matrix functions, namely,

$$\frac{d}{dt}\big(\gamma(f(t))\big) = \gamma'(f(t)) \, f'(t).$$

2.3.5. Let $A(t)$ and $B(t)$ be two $n \times n$ matrix functions defined over an interval $I \subset \mathbb{R}$.

(a) Suppose that $A(t)$ and $B(t)$ are similar for all $t \in I$. Prove that $A'(t)$ and $B'(t)$ are not necessarily similar.

(b) Suppose that $A(t)$ and $B(t)$ are similar for all $t \in I$ and that $A(t_0) = \lambda I$. Prove that $A'(t_0)$ and $B'(t_0)$ are similar.

(c) Suppose that $A(t)$ and $B(t)$ are similar in that $B(t) = SA(t)S^{-1}$ for some fixed invertible matrix S. Prove that $A'(t)$ and $B'(t)$ are similar.

2.3.6. Prove Proposition 2.3.1.

2.3.7. Suppose that $\gamma : I \to \mathrm{GL}_n(\mathbb{R})$ be a matrix function and let $A(t) = e^{\gamma(t)}$. Is it true that $A'(t) = e^{\gamma(t)}\gamma(t)$.

2.3.8. Let A be a diagonalizable matrix with $A = PDP^{-1}$, where D is diagonal with

$$D = \begin{pmatrix} \lambda_1 & 0 & \cdots & 0 \\ 0 & \lambda_2 & \cdots & 0 \\ \vdots & \vdots & \ddots & \vdots \\ 0 & 0 & \cdots & \lambda_n \end{pmatrix},$$

and $\lambda_i \in \mathbb{R}$ for all i. Prove that

$$e^{At} = P \begin{pmatrix} e^{\lambda_1 t} & 0 & \cdots & 0 \\ 0 & e^{\lambda_2 t} & \cdots & 0 \\ \vdots & \vdots & \ddots & \vdots \\ 0 & 0 & \cdots & e^{\lambda_n t} \end{pmatrix} P^{-1}.$$

2.3.9. Show that if

$$A = \begin{pmatrix} 0 & -\omega \\ \omega & 0 \end{pmatrix}$$

then

$$e^{At} = \begin{pmatrix} \cos \omega t & -\sin \omega t \\ \sin \omega t & \cos \omega t \end{pmatrix}.$$

2.3.10. Let $\vec{\omega} \in \mathbb{R}^3$ be a nonzero vector.

(a) Show that for any $\vec{x} \in \mathbb{R}^3$, we can write the cross product as the matrix product

$$\vec{\omega} \times \vec{x} = \begin{pmatrix} 0 & -\omega_3 & \omega_2 \\ \omega_3 & 0 & -\omega_1 \\ -\omega_2 & \omega_1 & 0 \end{pmatrix} \vec{x}.$$

(b) Call W the 3×3 matrix in the above expression. Prove that e^{Wt} is the matrix of rotation about the axis with direction $\vec{\omega}$ and with angle $\|\vec{\omega}\| t$.

2.3.11. We define $\mathrm{SL}_n(\mathbb{R}) = \{A \in M_{n \times n}(\mathbb{R}) \mid \det(A) = 1\}$ and call this set the *special linear group*. Suppose that $M : (-\varepsilon, \varepsilon) \to \mathrm{SL}_n(\mathbb{R})$ with $M(0) = I_n$, the identity matrix, and suppose that $M'(t) = M(t)A(t)$ for all $t \in (-\varepsilon, \varepsilon)$. Prove that the trace of $A(0)$ is $\mathrm{Tr}\, A(0) = 0$. [Recall that the trace of a matrix is the sum of its diagonal elements. Hint: Use the definition of the determinant that if $M = (m_{ij})$, then

$$\det(M) = \sum_{\sigma \in S_n} (\mathrm{sign}\, \sigma) m_{1\sigma(1)} m_{2\sigma(2)} \cdots m_{n\sigma(n)},$$

where S_n is the set permutations on n elements.]

2.3.12. This exercise gives an interesting property about the derivative of determinants of square matrices of functions. Let $A = (a_{ij}(t))$ be an $n \times n$ matrix of functions.

(a) Use the formula for the determinant given in the previoius exercise to show that

$$\frac{d}{dt}(\det A) = \sum_{i=1}^{n} \sum_{j=1}^{n} (-1)^{i+j} \det(A_{ij}) \frac{da_{ij}}{dt},$$

where A_{ij} is the ijth minor of A.

(b) Conclude that if A is a symmetric matrix, then

$$\frac{d}{dt}(\det A) = (\det A) \sum_{i,j=1}^{n} a^{ij} \frac{da_{ij}}{dt},$$

where the a^{ij} are the entries of the inverse matrix A^{-1}.

CHAPTER 3

Differentiable Manifolds

In previous geometry or calculus courses, we studied curves and surfaces as subsets of some ambient Euclidean space $\mathbb{R}^n$. We defined parametrizations as vector functions of one (for a curve) or two (for a surface) variables into $\mathbb{R}^2$ or $\mathbb{R}^3$, without pointing out that many of our constructions relied on the fact that $\mathbb{R}^2$ and $\mathbb{R}^3$ are topological vector spaces. That we have only studied geometric objects that are subsets of $\mathbb{R}^3$ does not belie our intuition since the daily reality of human experience evolves (or at least appears to evolve) completely in three dimensions that we feel are flat. However, both in mathematics and in physics, one does not need to take such a large step in abstraction to realize the insufficiency of this intuition.

In geometry, one can easily define natural point sets that cannot be properly represented in only three dimensions. For example, the real projective plane $\mathbb{RP}^2$ can be defined as the set of equivalence classes of lines through the origin in $\mathbb{R}^3$ or also as the set of equivalence classes of points in $\mathbb{R}^3 - \{(0,0,0)\}$ under the equivalence relation

$$(x_0, x_1, x_2) \sim (y_0, y_1, y_2) \qquad \text{if and only if}$$
$$(y_0, y_1, y_2) = (\lambda x_0, \lambda x_1, \lambda x_2) \text{ for some } \lambda \in \mathbb{R} - \{0\}.$$

The projective plane plays a fundamental role in geometry, and also in topology and algebraic geometry. From the above construction, it appears that the projective plane (as its name suggests) should be a two-dimensional object since, from a topological viewpoint, it is the identification space (see Definition A.2.44) of a three-dimensional object by a one-dimensional object. Both in classical geometry and in algebraic geometry, there exist natural methods to study curves on the projective plane, thereby providing a language to "do analysis" on the projective plane. Nonetheless, it is not hard to show that no subset of $\mathbb{R}^3$ is homeomorphic to $\mathbb{RP}^2$. There does exist a subset of $\mathbb{R}^4$ that is in fact homeomorphic to $\mathbb{RP}^2$ but this fact is not obvious from the definition of the projective plane. Consequently, to provide definitions that include projective spaces and other more abstract geometric objects, we must avoid referring to some ambient Euclidean space.

In physics, the need for eliminating a Euclidean ambient space boasts a more colorful history. Inspired by evidence provided by scientists like Toricelli, explorers of the 15th and 16th centuries debunked the flat-earth theory by circumnavigating the globe. Though the normal Euclidean geometry remained valid on the small scale, namely, doing geometry on a flat surface (sheet of paper or plot of land), such methods no longer sufficed when considering the geometry of the earth as a whole. In particular, the science of cartography suddenly became far more mathematical in nature as navigators attempted to represent, with some degree of accuracy, coastlines of continents on a flat sheet of paper.

No less revolutionary was Einstein's theory of general relativity in which both space and time are connected as a single, four-dimensional space-time entity that could itself be curved. In fact, following from the postulate that nothing with mass travels faster than the speed of light, Einstein's theory purports that mass must distort space-time.

The practical need to do geometry or do physics in continuous point-set spaces that are not Euclidean leads us to generalize our concepts of curves and surfaces to higher-dimensional objects. We will call these objects of study differentiable manifolds. We will then define maps between manifolds and establish an analysis of maps between differentiable manifolds. Our definitions, which may seem a little weighty, attempt to retain sufficient restrictions to ensure that doing calculus on the sets is possible, while preserving enough freedom to incorporate the rich variety of geometric objects to which we wish to apply our techniques.

3.1   Definitions and Examples

As a motivating example for differentiable manifolds, we recall the definition of a regular surface in $\mathbb{R}^3$ (see [5, Chapter 5] for more background).

Definition 3.1.1. A subset $S \subset \mathbb{R}^3$ is a *regular surface* if for each $p \in S$, there exists an open set $U \subset \mathbb{R}^2$, an open neighborhood V of p in $\mathbb{R}^3$, and a surjective continuous function $\vec{X} : U \to V \cap S$ such that

1. $\vec{X}$ is differentiable: if we write $\vec{X}(u,v) = (x(u,v), y(u,v), z(u,v))$, then the functions $x(u,v)$, $y(u,v)$, and $z(u,v)$ have continuous partial derivatives of all orders;

2. $\vec{X}$ is a homeomorphism: $\vec{X}$ is continuous and has an inverse $\vec{X}^{-1} : V \cap S \to U$ such that $\vec{X}^{-1}$ is continuous;

3. $\vec{X}$ satisfies the regularity condition: for each $(u,v) \in U$, the differential $d\vec{X}_{(u,v)} : \mathbb{R}^2 \to \mathbb{R}^3$ is a one-to-one linear transformation.

This definition already introduces many of the subtleties that are inherent in the concept of a manifold. In the above definition, each function $\vec{X} : U \to V \cap S$ is called a *parametrization* of a *coordinate neighborhood*.

Now, as we set out to define differentiable manifolds and remove any reference to an ambient Euclidean space, we begin from the context of topological spaces.

(Appendix A gives a brief introduction to topological spaces.) Not every topological space can fit the bill of usefulness for differential geometry, so we require some additional properties of what types of topological spaces we will consider. We first impose the requirement of having a cover of open sets, each of which is homeomorphic to an open set in a Euclidean space.

Definition 3.1.2. A *topological manifold* of dimension n is a Hausdorff topological space M with a countable base such that for all $x \in M$, there exists an open neighborhood of x that is homeomorphic to an open set of $\mathbb{R}^n$.

The reader is encouraged to refer to Section A.2 in the appendices for definitions and discussions about the base of a topology and the Hausdorff property. A topological space that has a countable base is called *second countable*. The technical aspect of this definition attempts to define a category of objects as general as possible, while still remaining relevant for geometry that generalizes that on $\mathbb{R}^n$.

In the definition of a topological manifold, a given homeomorphism of a neighborhood of M with a subset of $\mathbb{R}^k$ provides a local coordinate system or coordinate patch. As one moves around on the manifold, one passes from one coordinate patch to another. In the overlap of coordinate patches, there exist change-of-coordinate functions that, by definition, are homeomorphisms between open sets in $\mathbb{R}^n$ (see Figure 3.1). However, in order to define a theory of calculus on the manifold, these functions must be differentiable. We make this clear in the following definition.

Definition 3.1.3. A *differentiable manifold* M of dimension n is a topological manifold along with a collection of functions $\mathcal{A} = \{\phi_\alpha : U_\alpha \to \mathbb{R}^n\}_{\alpha \in I}$ with U_α open in M called *charts*, satisfying

1. For each chart, $\phi_\alpha(U_\alpha) = V_\alpha$ is open in $\mathbb{R}^n$ and $\phi_\alpha : U_\alpha \to V_\alpha$ is a homeomorphism;

2. The collection of sets U_α, called *coordinate patches*, cover M, i.e.,

$$M = \bigcup_{\alpha \in I} U_\alpha;$$

3. For any pair of charts ϕ_α and ϕ_β, the change-of-coordinates

$$\phi_{\alpha\beta} \stackrel{\text{def}}{=} \phi_\alpha \circ \phi_\beta^{-1}|_{\phi_\beta(U_\alpha \cap U_\beta)} : \phi_\beta(U_\alpha \cap U_\beta) \longrightarrow \phi_\alpha(U_\alpha \cap U_\beta),$$

   called the *transition function*, is a function of class C^1 between open subsets of $\mathbb{R}^n$.

The collection of functions $\mathcal{A} = \{\phi_\alpha\}_{\alpha \in I}$ satisfying the above conditions is called an *atlas*.

A differentiable manifold is called a C^k *manifold*, a *smooth manifold*, or an *analytic manifold* if all the transition functions in the atlas are respectively C^k, C^∞, or analytic.

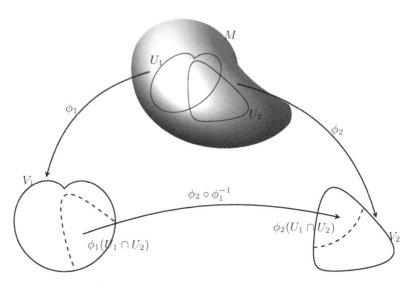

Figure 3.1: Change-of-coordinate maps.

A few comments about notation are in order here. Mimicking the notation habits for common sets (Euclidean space $\mathbb{R}^n$, or the n sphere as $\mathbb{S}^n$), if M is an n-dimensional differentiable manifold, we sometimes shorten the language by referring to the "differentiable manifold M^n." Also, though technically a chart is a function $\phi : U \to \mathbb{R}^n$, where U is an open subset of the manifold M, one sometimes refers to the chart (U, ϕ) to emphasize the letter to be used for the domain of ϕ. Though we use a single letter to designate a differentiable manifold, the atlas $\mathcal{A}$ is an essential part of the definition; consequently, we sometimes refer to the differentiable manifold as the pair $(M^n, \mathcal{A})$ to indicate the letter we are using to designate the atlas. Finally, since the domains U_α of the charts cover M, they satisfy the condition of a topological manifold that each $x \in M$ must have an open neighborhood that is homeomorphic to an open set in $\mathbb{R}^n$.

At first pass, the definition of a differentiable manifold may seem unnecessarily complicated. However, this definition removes any reference to an ambient space, a feature whose virtues we discussed in the introduction to this chapter. After all, from a geometric perspective, this is the safe thing to do: a priori we do not know whether a given manifold can be described as a subset of an ambient Euclidean space. The application to general relativity also gives a compelling reason: in general relativity, the universe is a spacetime whole that is not Euclidean, sometimes called "curved." However, it would be misleading to think of this curved spacetime as a subset of a larger Euclidean space. Removing any reference to an ambient space is the proper approach to presenting a mathematical structure that appropriately models a non-Euclidean space in which we wish to do calculus. The above definition

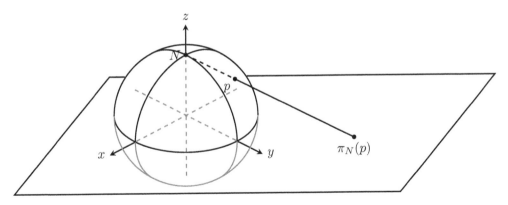

Figure 3.2: Stereographic projection.

and subsequent constructions have proven general enough and structured enough to be useful in geometry and in physics.

Many properties of manifolds that arise in analysis are local properties, in that we only need to know information about the manifold M in some neighborhood of a point $p \in M$. When this is the case, we can restrict our attention to a single coordinate chart $\phi_\alpha : U_\alpha \to \mathbb{R}^n$, where $p \in U_\alpha$. Saying that the coordinates of a point p (with respect to this chart) are $(x^1, x^2, \ldots, x^n)$ means that $\phi_\alpha(p) = (x^1, x^2, \ldots, x^n)$. For reasons that will only become clear later, it is convenient to follow the tensor notation convention of using superscripts for coordinates. This makes writing polynomial functions in the coordinates more tedious but this notation will provide a convenient way to distinguish between covariant and contravariant properties.

Example 3.1.4 (Sphere). Consider the unit sphere $\mathbb{S}^2 = \{(x, y, z) \in \mathbb{R}^3 \mid x^2 + y^2 + z^2 = 1\}$ and call $N = (0, 0, 1)$ the North pole and call $S = (0, 0, -1)$ the South pole. We define the stereographic projection from the North pole N as the function $\pi_N : \mathbb{S}^2 - \{N\} \to \mathbb{R}^2$, where $\pi_N(p)$ is the intersection of the line (Np) with the xy-plane. (See Figure 3.2). The definition for π_S, the stereographic projection from the South pole, is similar.

In Exercise 3.1.1, we prove the following results. The formula for stereographic projection:

1. from the north pole is $\pi_N(x, y, z) = \left(\dfrac{x}{1 - z}, \dfrac{y}{1 - z} \right)$, and

2. from the south pole is $\pi_S(x, y, z) = \left(\dfrac{x}{1 + z}, \dfrac{y}{1 + z} \right)$.

The inverses of stereographic projection are not hard to find either. In particular,

$$\pi_N^{-1}(u,v) = \left(\frac{2u}{u^2+v^2+1}, \frac{2v}{u^2+v^2+1}, \frac{u^2+v^2-1}{u^2+v^2+1} \right), \quad \text{and}$$

$$\pi_S^{-1}(u,v) = \left(\frac{2u}{u^2+v^2+1}, \frac{2v}{u^2+v^2+1}, -\frac{u^2+v^2-1}{u^2+v^2+1} \right)$$

The domain of π_N is $U_N = \mathbb{S}^2 - \{N\}$ and the domain of π_S is $U_S = \mathbb{S}^2 - \{S\}$, so these domains do cover the sphere $\mathbb{S}^2$. As fractions of polynomials, π_N and π_N^{-1} are both continuous, so π_N is a homeomorphism. The same holds for π_S.

For the transition function $\pi_s \circ \pi_N^{-1}$, we note first that the domain is $\pi_N(U_N \cap U_S)$. Since $\pi_N(S) = (0,0)$, we have $\pi_N(U_N \cap U_S) = \mathbb{R}^2 - \{(0,0)\}$. Furthermore, it is not hard to show that

$$\pi_S \circ \pi_N^{-1}(u,v) = \left(\frac{u}{u^2+v^2}, \frac{v}{u^2+v^2} \right). \tag{3.1}$$

By repeated application of the quotient rule, any repeated partial derivative of either component function of $\pi_S \circ \pi_N^{-1}$ is a polynomial in u and v divides by a power of $u^2 + v^2$. Since the domain of $\pi_S \circ \pi_N^{-1}$ is $\mathbb{R}^2 - \{(0,0)\}$, all of these partial derivatives exist and are continuous. Thus, $\pi_S \circ \pi_N^{-1}$ is C^∞. The same thing occurs for $\pi_N \circ \pi_S^{-1}$. Hence, the set $\{\pi_N, \pi_S\}$ provides an atlas that equips $\mathbb{S}^2$ with the structure of a smooth manifold.

Example 3.1.5 (Sphere with Another Atlas). We can prove that the unit sphere $\mathbb{S}^2$ is a smooth two-dimensional manifold using another atlas, this time using rectangular coordinates for the parametrizations.

Consider a point $p = (x,y,z) \in \mathbb{S}^2$, and let $V = \{(u,v) \mid u^2 + v^2 < 1\}$. If $z > 0$, then the mapping $\vec{X}_{(1)} : V \to \mathbb{R}^3$ defined by $(u, v, \sqrt{1-u^2-v^2})$ is clearly a bijection between V and $\mathbb{S}^2 \cap \{(x,y,z) \mid z > 0\}$. $\vec{X}_{(1)}$ is also a homeomorphism because it is continuous and its inverse $\vec{X}_{(1)}^{-1}$ is simply the vertical projection of the upper unit sphere onto $\mathbb{R}^2$, and since projection is a linear transformation, it is continuous.

We cover $\mathbb{S}^2$ with the following parametrizations $\vec{X}_{(i)} : V \to \mathbb{R}^3$:

$$\text{if } z > 0, \ \vec{X}_{(1)}(u,v) = (u, v, \sqrt{1-u^2-v^2}),$$

$$\text{if } z < 0, \ \vec{X}_{(2)}(u,v) = (u, v, -\sqrt{1-u^2-v^2}),$$

$$\text{if } y > 0, \ \vec{X}_{(3)}(u,v) = (u, \sqrt{1-u^2-v^2}, v),$$

$$\text{if } y < 0, \ \vec{X}_{(4)}(u,v) = (u, -\sqrt{1-u^2-v^2}, v),$$

$$\text{if } x > 0, \ \vec{X}_{(5)}(u,v) = (\sqrt{1-u^2-v^2}, u, v),$$

$$\text{if } x < 0, \ \vec{X}_{(6)}(u,v) = (-\sqrt{1-u^2-v^2}, u, v).$$

Figure 3.3 depicts an expanded view of these coordinate patches. The inverses for each of these parametrizations give coordinate charts $\phi_i = \vec{X}_{(i)}^{-1} : U_i \to V_i$, which together form an atlas on the sphere.

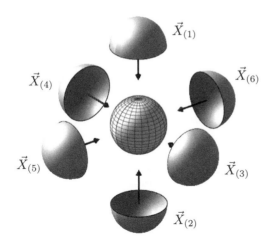

Figure 3.3: Six coordinate patches on the sphere.

We notice in this case that all $V_i = \{(u,v) \,|\, u^2 + v^2 < 1\}$. Also, not all U_i overlap; in particular, $U_1 \cap U_2 = \emptyset$, $U_3 \cap U_4 = \emptyset$, and $U_5 \cap U_6 = \emptyset$. To show that the sphere equipped with this atlas is a differentiable manifold, we must show that all transition functions are C^1. We illustrate this with $\phi_{31} = \phi_3 \circ \phi_1^{-1}$.

The identification $(\bar{u}, \bar{v}) = \phi_3 \circ \phi_1^{-1}(u,v)$ is equivalent to

$$(\bar{u}, \sqrt{1 - \bar{u}^2 - \bar{v}^2}, \bar{v}) = (u, v, \sqrt{1 - u^2 - v^2}).$$

This leads to

$$\phi_1(U_1 \cap U_3) = \{(u,v) \,|\, u^2 + v^2 < 1 \text{ and } v > 0\},$$
$$\phi_3(U_1 \cap U_3) = \{(\bar{u}, \bar{v}) \,|\, \bar{u}^2 + \bar{v}^2 < 1 \text{ and } \bar{v} > 0\},$$

and

$$\phi_{31}(u,v) = (u, \sqrt{1 - u^2 - v^2}).$$

It is now easy to verify that ϕ_{31} is of class C^1 over $\phi_1(U_1 \cap U_3)$. In fact, higher derivatives of ϕ_{31} involve polynomials in u and v possibly divided by powers of $\sqrt{1 - u^2 - v^2}$. Hence, over $\phi_1(U_1 \cap U_3)$, the function ϕ_{31} is of class C^∞. It is not hard to see that all other transition functions are similar. Thus, this atlas $\mathcal{A} = \{\phi_i\}_{i=1}^6$ equips $\mathbb{S}^2$ with the structure of a smooth manifold.

Example 3.1.6 (Projective Space). The n-dimensional real projective space $\mathbb{RP}^n$ is defined as the set of lines in $\mathbb{R}^{n+1}$ through the origin. No two lines through the origin intersect any place else and, for each point p in $\mathbb{R}^n$, there exists a unique line through the origin and p. Therefore, one can describe $\mathbb{RP}^n$ as the set of equivalence classes of points in $\mathbb{R}^{n+1} - \{(0, 0, \ldots, 0)\}$ under the equivalence relation

$$(x_0, x_1, \ldots, x_n) \sim (y_0, y_1, \ldots, y_n) \qquad \text{if and only if}$$
$$(y_0, y_1, \ldots, y_n) = (\lambda x_0, \lambda x_1, \ldots, \lambda x_n) \text{ for some } \lambda \in \mathbb{R} - \{0\}.$$

We designate the equivalence class of a point $(x_0, x_1, \ldots, x_n)$ by the notation $(x_0 : x_1 : \ldots : x_n)$. The set $\mathbb{RP}^n$ is a topological space with the quotient topology coming from the quotient map $\pi : \mathbb{R}^{n+1} - \{0\} \to \mathbb{RP}^n$ given by $\pi(x_0, x_1, \ldots, x_n) = (x_0 : x_1 : \ldots : x_n)$. Since $\mathbb{R}^{n+1} - \{0\}$ is second countable (has a countable base, namely the open balls in $\mathbb{R}^{n+1} - \{0\}$ of rational radius with centers of rational coordinates), the quotient space $\mathbb{RP}^n$ inherits a countable base.

General theorems in topology quickly establish that $\mathbb{RP}^n$ is Hausdorff but we give a direct proof here. Call $O = (0, 0, \ldots, 0)$ in $\mathbb{R}^{n+1}$. For $\alpha > 0$ and $A \in \mathbb{R}^{n+1} - \{0\}$, define $C_\alpha(A)$ as the double open cone

$$C_\alpha(A) = \{B \in \mathbb{R}^{n+1} \mid \angle AOB < \alpha \text{ or } \angle AOB > \pi - \alpha\},$$

with axis of revolution (OA) and opening angle of 2α. For all α, the cone $C_\alpha(A)$ is an open subset of $\mathbb{R}^{n+1}$.

Let $p, q \in \mathbb{RP}^n$ be distinct points. Let $p_1 \in \pi^{-1}(p)$ and let $q_1 \in \pi^{-1}(q)$ such that $\angle p_1 O q_1 \leq \pi/2$. Since $p \neq q$, the angle $\angle p_1 O q_1$ is positive. Define α to be an angle with $0 < \alpha < \frac{1}{2} \angle p_1 O q_1$. Then $C_\alpha(p_1) \cap C_\alpha(q_1) = \emptyset$.

Call $U = \pi(C_\alpha(p_1))$ and $V = \pi(C_\alpha(q_1))$. Since $C_\alpha(p_1)$ and $C_\alpha(q_1)$ are open, the topology on $\mathbb{RP}^n$ is defined so that U and V are open in $\mathbb{RP}^n$. Furthermore, $p \in U$ and $q \in V$. Also,

$$\pi^{-1}(U \cap V) = \pi^{-1}(U) \cap \pi^{-1}(V) = C_\alpha(p_1) \cap C_\alpha(q_1) = \emptyset,$$

where the middle equality holds because we used cones, namely unions of lines through O. However, the function π is surjective, so we deduce that $U \cap V = \emptyset$. Since p and q were arbitrarily chosen, we deduce that $\mathbb{RP}^n$ is Hausdorff.

We can define an atlas on $\mathbb{RP}^n$ as follows. Note that if $(x_0, x_1, \ldots, x_n) \sim (y_0, y_1, \ldots, y_n)$, then for any i, we have $x_i = 0$ if and only if $y_i = 0$. For $i \in \{0, 1, \ldots, n\}$, define $U_i = \{(x_0 : x_1 : \ldots : x_n) \in \mathbb{RP}^n \mid x_i \neq 0\}$ and define $\phi_i : U_i \to \mathbb{R}^n$ by

$$\phi_i(x_0 : x_1 : \ldots : x_n) = \left(\frac{x_0}{x_i}, \frac{x_1}{x_i}, \ldots, \widehat{\frac{x_i}{x_i}}, \ldots, \frac{x_n}{x_i} \right),$$

where the $\widehat{a}$ notation indicates deleting that entry from the $(n+1)$-tuple. It is easy to see that each ϕ_i is a homeomorphism between U_i and $\mathbb{R}^n$. Furthermore, $U_0 \cup \cdots \cup U_n$ includes all ratios $(x_0 : \ldots : x_n)$ for which not all $x_i = 0$. Thus, $U_0 \cup \cdots \cup U_n = \mathbb{RP}^n$.

So far, we have established that $\mathbb{RP}^n$ has the structure of a topological manifold and we have given it natural charts. We need to show that the transition functions between coordinate patches are differentiable.

Assume without loss of generality that $i < j$. Then $\phi_i(U_i \cap U_j) = \{(a_1, \ldots, a_n) \in \mathbb{R}^n \mid a_j \neq 0\}$ and $\phi_j(U_i \cap U_j) = \{(a_1, \ldots, a_n) \in \mathbb{R}^n \mid a_{i+1} \neq 0\}$. (The apparent difference comes from $i < j$.) Then the change-of-coordinate function $\phi_j \circ \phi_i^{-1}$ is

$$\phi_j \circ \phi_i^{-1}(a_1, a_2, \ldots, a_n) = \phi_j(a_1 : a_2 : \ldots : a_i : 1 : a_{i+1} : \ldots : a_n)$$
$$= \left(\frac{a_1}{a_j}, \frac{a_2}{a_j}, \ldots, \frac{a_i}{a_j}, \frac{1}{a_j}, \frac{a_{i+1}}{a_j}, \ldots, \underbrace{\widehat{\frac{a_j}{a_j}}}_{(j+1)\text{th}}, \ldots, \frac{a_n}{a_j} \right).$$

Note that we remove the $(j+1)$th entry from an $(n+1)$-tuple labeled, whose first index is 1.

It is not hard to see that $\phi_j \circ \phi_i^{-1}$ is indeed a bijection between $\phi_i(U_i \cap U_j)$ and $\phi_j(U_i \cap U_j)$. Furthermore, all higher partial derivatives of $\phi_j \circ \phi_i^{-1}$ exist over $\phi_i(U_i \cap U_j)$.

The same reasoning works if $i > j$. Therefore, this atlas satisfies the condition required to equip $\mathbb{RP}^n$ with the structure of a smooth manifold.

We point out that it is possible to define $\mathbb{RP}^n$ in a slightly different way. Consider the unit sphere $\mathbb{S}^n$ as a subset of $\mathbb{R}^{n+1}$ and consider the antipodal function $A : \mathbb{S}^n \to \mathbb{S}^n$ defined by $A(p) = -p$. We can define $\mathbb{RP}^n$ as $\mathbb{S}^n$ where antipodal points are identified. In other words, projective space is the set of equivalence classes of antipodal points

$$\mathbb{RP}^n = \{\{p, -p\} \mid p \in \mathbb{S}^n\}.$$

We define the projection $\pi : \mathbb{S}^2 \to \mathbb{RP}^2$ as the function $\pi(p) = [p]$, where $[p] = \{p, -p\}$ is the equivalence class. This function helps define the topology on $\mathbb{RP}^2$ (see Section A.2.3) but it is not as simple to define the manifold structure of $\mathbb{RP}^2$ from this quotient map.

Before providing more examples, we must emphasize a technical aspect of the definition of a differentiable manifold. If M is a topological manifold not inherently defined as the subset of a Euclidean space, we do not study whether M is or is not a differentiable manifold, but rather, we discuss whether it is possible to equip M with an atlas that equips it with the structure of a differentiable manifold. Also, as we saw in the above examples, since the domains of the charts cover M, these domains provide the open neighborhoods for each point that occur in the last condition of a topological manifold.

Definition 3.1.7. Two differentiable (respectively, C^k, smooth, analytic) atlases $\{\phi_\alpha\}$ and $\{\psi_i\}$ on a topological manifold M are said to be *compatible* if the union of the two atlases is again an atlas on M in which all the transition functions are differentiable (respectively, C^k, smooth, analytic).

Interestingly enough, not all atlases are compatible in a given category. It is also possible for the union of two atlases of class C^k to form an atlas of class C^l, with $l < k$. The notion of compatibility between atlases is an equivalence relation, and an equivalence class of differentiable (respectively, C^k, smooth, analytic) atlases is called a *differentiable (respectively, C^k, smooth, analytic) structure*. Proving that a given topological manifold has a unique differentiable structure or enumerating the differentiable structures on a given topological manifold involves techniques that are beyond the scope of this book. For example, in [29], published in 1963, Kervaire and Milnor prove that $\mathbb{S}^7$ has exactly 28 nondiffeomorphic smooth structures.

Example 3.1.8. We point out that for any integer $n \geq 1$, the Euclidean space $\mathbb{R}^n$ is an n-dimensional manifold. (The standard atlas consists of only one function, the identity function on $\mathbb{R}^n$.)

Example 3.1.9. A manifold M of dimension 0 is a set of points with the discrete topology, i.e., every subset of M is open. The notion of differentiability is vacuous over a 0-dimensional manifold.

Note that this example indicates that a manifold is not necessarily connected but may be a union of connected components, each of which is a manifold in its own right.

Example 3.1.10 (An Alternate Smooth Structure on $\mathbb{R}$). Let $M = \mathbb{R}$ and consider the function $\psi : M \to \mathbb{R}$ defined by $\psi(x) = x^3$. The function ψ is a homeomorphism so the singleton set $\{\psi\}$ forms an atlas on $\mathbb{R}$. The standard structure on $\mathbb{R}$, as described in Example 3.1.8, uses the atlas $\{\phi\}$, where $\phi : M \to \mathbb{R}$ is $\phi(x) = x$. However, though $\{\phi\}$ and $\{\psi\}$ define smooth structures on $\mathbb{R}$, these two atlases are incompatible. Consider the function $\phi \circ \psi^{-1}(x) = \sqrt[3]{x}$. It is a homeomorphism but it is not differentiable at 0. Hence, $\{\phi, \psi\}$ is not a differentiable atlas, let alone a smooth one.

Example 3.1.11 (Open Subsets of Manifolds). Let M^n be a differentiable manifold with atlas $\mathcal{A} = \{\phi_\alpha : U_\alpha \to \mathbb{R}^n\}_{\alpha \in I}$. Let V be an open subset of M. Consider the set of functions $\mathcal{A}' = \{\phi_\alpha|_V : U_\alpha \cap V \to \mathbb{R}^n\}_{\alpha \in I}$. We have

$$\bigcup_{\alpha \in I} U_\alpha \cap V = \left(\bigcup_{\alpha \in I} U_\alpha \right) \cap V = M \cap V = V.$$

Hence $U_\alpha \cap V$, for $\alpha \in I$, cover V. Because ϕ_α is a homeomorphism with its image $\phi_\alpha(U_\alpha)$, then $\phi_\alpha(U_\alpha \cap V)$ is open in $\mathbb{R}^n$ and $\phi_\alpha|_V$ is also a homeomorphism. Finally,

$$\phi_\alpha|_V \circ (\phi_\beta|_V)^{-1}\big|_{\phi_\beta(U_\alpha \cap U_\beta \cap V)} = \phi_\alpha \circ \phi_\beta^{-1}\big|_{\phi_\beta(U_\alpha \cap U_\beta \cap V)},$$

which is the same class of function as $\phi_\alpha \circ \phi_\beta^{-1}$. Hence, if V is any open subset of M, then it inherits the same class of structure of M.

When working with examples of manifolds that are subsets of $\mathbb{R}^k$, it is often easier to specify coordinate charts $x : U \subset M^n \to \mathbb{R}^n$ by providing a parametrization $x^{-1} : x(U) \to M^n$ that is homeomorphic with its image. Since the chart is a homeomorphism, this habit does not lead to any difficulties.

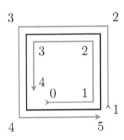

Figure 3.4: A square as a differentiable manifold.

Example 3.1.12. Consider a trefoil knot K in $\mathbb{R}^3$. One can realize K as the image of the parametric curve

$$\gamma(t) = \big((2 + \cos(3t))\cos(2t), (2 + \cos(3t))\sin(2t), \sin(3t)\big)$$

for $t \in \mathbb{R}$. We can choose an atlas of K as follows. Set one coordinate patch on K to be $U_1 = \gamma\big((0, 2\pi)\big)$ and another patch to be $U_2 = \gamma\big((\pi, 3\pi)\big)$. Use as charts the functions ϕ_1 and ϕ_2, which are the inverse functions of $\gamma : (0, 2\pi) \to K$ and $\gamma : (\pi, 3\pi) \to K$, respectively. Now

$$\phi_2(U_1 \cap U_2) = (\pi, 2\pi) \cup (2\pi, 3\pi) \qquad \text{and} \qquad \phi_1(U_1 \cap U_2) = (0, \pi) \cup (\pi, 2\pi),$$

and the coordinate transition functions are

$$\phi_1 \circ \phi_2^{-1}(t) = \begin{cases} t, & \text{if } t \in (\pi, 2\pi), \\ t - 2\pi, & \text{if } t \in (2\pi, 3\pi), \end{cases}$$

$$\phi_2 \circ \phi_1^{-1}(t) = \begin{cases} t + 2\pi, & \text{if } t \in (0, \pi), \\ t, & \text{if } t \in (\pi, 2\pi). \end{cases}$$

Both of these transition functions are differentiable on their domains. This shows that K, equipped with the given atlas, is a 1-manifold.

From the previous example, it is easy to see that any regular, simple, closed curve in $\mathbb{R}^k$ can be given an atlas that gives it the structure of a differentiable 1-manifold. Our intuition might tell us that, say, a square in the plane should not be a differentiable 1-manifold because of its corners. This idea, however, is erroneous, as we shall now explain.

Example 3.1.13. Consider the square with unit length side, and define two chart functions as follows. The function ϕ_1 measures the distance traveled as one travels around the square in a counterclockwise direction, starting with a value of 0 at $(0, 0)$. The function ϕ_2 measures the distance traveled as one travels around the square in the same direction, starting with a value of 1 at $(1, 0)$ (see Figure 3.4).

Figure 3.5: Not a bijection.

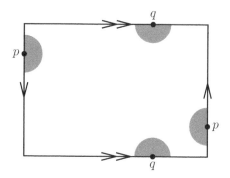

Figure 3.6: The Klein bottle.

The functions ϕ_1 and ϕ_2 are homeomorphisms, and the coordinate transition function is

$$\phi_1 \circ \phi_2^{-1} : (1,4) \cup (4,5) \to (0,1) \cup (1,4),$$

with

$$\phi_1 \circ \phi_2^{-1}(x) = \begin{cases} x, & \text{if } x \in (1,4), \\ x - 4, & \text{if } x \in (4,5). \end{cases}$$

This transition function (and its inverse, the other transition function) is differentiable over its domain. Therefore, the atlas $\{\phi_1, \phi_2\}$ equips the square with the structure of a differentiable manifold.

This example shows that, in and of itself, the square can be given the structure of a differentiable 1-manifold. However, this does not violate our intuition about differentiability and smoothness because one only perceives the "sharp" corners of the square in reference to the differential structure of $\mathbb{R}^2$. Once we have the appropriate definitions, we will say that the square is not a submanifold of $\mathbb{R}^2$ with the usual differential structure (see Definition 3.6.1). In fact, the atlases in Examples 3.1.12 and 3.1.13 bear considerable similarity, and, ignoring the structure of the ambient space, both the square and the knot resemble a circle. We develop these notions further when we consider functions between manifolds.

It is not hard to verify that a regular surface S in $\mathbb{R}^3$ (see Definition 3.1.1) is a differentiable 2-manifold. The only nonobvious part is showing that the properties of coordinate patches of a regular surface imply that the coordinate transition functions are differentiable. We leave this as an exercise for the reader (see Problem 3.1.6).

Parametrized surfaces that are not regular surfaces provide examples of geometric sets in $\mathbb{R}^3$ that are not differentiable manifolds. For example, with the surface in Figure 3.5, for any point along the line of self-intersection, there cannot exist an open set of $\mathbb{R}^2$ that is in bijective correspondence with any given neighborhood of p. However, the notion of a regular surface in $\mathbb{R}^3$ has more restrictions than that of a 2-manifold for two reasons. Applying the ideas behind Example 3.1.13, a circular (single) cone can be given the structure of a differentiable manifold even though it is not a regular surface. Furthermore, not every differentiable 2-manifold can be realized as a regular surface or even as a union of such surfaces in $\mathbb{R}^3$. A simple example is the Klein bottle, defined topologically as follows. Consider a rectangle, and identify opposite edges according to Figure 3.6. One pair of sides is identified directly opposite each other (in Figure 3.6, the horizontal edges), and the other pair of sides is identified in the reverse direction.

It is not hard to see that the Klein bottle can be given an atlas that makes it a differentiable 2-manifold. However, it turns out that the Klein bottle cannot be realized as a regular surface in $\mathbb{R}^3$.

We end the section by defining the product structure of two manifolds.

Definition 3.1.14. Let M^m and N^n be two differentiable (respectively, C^k, smooth, analytic) manifolds. Call their respective atlases $\{\phi_\alpha\}_{\alpha \in I}$ and $\{\psi_\beta\}_{\beta \in J}$. Consider the set $M \times N$ that is equipped with the product topology. If $\phi : U \to \mathbb{R}^m$ is a chart for M and $\psi : V \to \mathbb{R}^n$ is a chart for N, then define the function $\phi \times \psi : U \times V \to \mathbb{R}^{m+n}$ by $\phi \times \psi(p_1, p_2) = (\phi(p_1), \psi(p_2))$. The collection $\{\phi_\alpha \times \psi_\beta\}_{(\alpha,\beta) \in I \times J}$ defines a differentiable (respectively, C^k, smooth, analytic) structure on $M \times N$, called the *product structure*.

Consider, for example, the circle $\mathbb{S}^1$ with a smooth structure. The product $\mathbb{S}^1 \times \mathbb{S}^1$ is topologically equal to a (two-dimensional) torus, and the product structure defines a smooth structure on the torus. By extending this construction, we define the 3-torus as the manifold $T^3 = \mathbb{S}^1 \times \mathbb{S}^1 \times \mathbb{S}^1$ and inductively the n-torus as $T^n = T^{n-1} \times \mathbb{S}^1$.

PROBLEMS

3.1.1. *Stereographic Projection.* One way to define coordinates on the surface of the sphere $\mathbb{S}^2$ given by $x^2 + y^2 + z^2 = 1$ is to use the stereographic projection of $\pi : \mathbb{S}^2 - \{N\} \to \mathbb{R}^2$, where $N = (0, 0, 1)$, defined as follows. Given any point $p \in \mathbb{S}^2$, the line (pN) intersects the xy-plane at exactly one point, which is the image of the function $\pi(p)$. If (x, y, z) are the coordinates for p in $\mathbb{S}^2$, let us write $\pi(x, y, z) = (u, v)$ (see Figure 3.2).

(a) Prove that $\pi_N(x, y, z) = \left(\frac{x}{1-z}, \frac{y}{1-z}\right)$.

(b) Prove that

$$\pi_N^{-1}(u, v) = \left(\frac{2u}{u^2 + v^2 + 1}, \frac{2v}{u^2 + v^2 + 1}, \frac{u^2 + v^2 - 1}{u^2 + v^2 + 1}\right).$$

(c) Show that $\pi_S \circ \pi_N^{-1}(u, v) = \left(\frac{u}{u^2 + v^2}, \frac{v}{u^2 + v^2}\right)$.

3.1.2. Consider the n-dimensional sphere $\mathbb{S}^n = \{(x_1, \ldots, x_{n+1}) \in \mathbb{R}^{n+1} \,|\, x_1^2 + \cdots + x_{n+1}^2 = 1\}$. Exhibit an atlas that gives $\mathbb{S}^n$ the structure of a differentiable n-manifold. Explicitly show that the atlas you give satisfies the axioms of a manifold.

3.1.3. Let V be an open set in $\mathbb{R}^k$, and let $f : V \to \mathbb{R}^m$ be a continuous function. Find an atlas that equips the graph of f, defined as the subset

$$G = \{(x, f(x)) \in \mathbb{R}^{k+m} \,|\, x \in V\},$$

with the structure of a smooth k-manifold.

3.1.4. Describe an atlas for the 3-torus $T^3 = \mathbb{S}^1 \times \mathbb{S}^1 \times \mathbb{S}^1$. Find a parametric function $X : U \to \mathbb{R}^4$, where U is a subset of $\mathbb{R}^3$, such that the image of X is a 3-torus.

3.1.5. We revisit Example 3.1.10. Let $\phi_0(x) = x$ be the identity map on $\mathbb{R}$. The atlas $\{\phi_0\}$ equips $\mathbb{R}$ with its usual differentiable structure. Let $\phi_1 : \mathbb{R} \to \mathbb{R}$ defined by $\phi_1(x) = x^3 + x$. Prove that ϕ_1 is a homeomorphism and conclude that $\{\phi_1\}$ is a differentiable atlas on $\mathbb{R}$. Prove that $\{\phi_0\}$ and $\{\phi_1\}$ are compatible atlases.

3.1.6. Prove that a regular surface in $\mathbb{R}^3$ (see Definition 3.1.1) is a differentiable 2-manifold.

3.1.7. Consider the following two parametrizations of the circle $\mathbb{S}^1$ as a subset of $\mathbb{R}^2$:

$$X_1(t) = (\cos t, \sin t) \qquad \text{for } t \in (0, 2\pi),$$

$$Y_1(u) = \left(\frac{1 - u^2}{1 + u^2}, \frac{2u}{1 + u^2}\right) \qquad \text{for } u \in \mathbb{R}.$$

Find functions X_2 and Y_2 "similar" to X_1 and Y_1 respectively, to make $\{X_1, X_2\}$ and $\{Y_1, Y_2\}$ atlases that give $\mathbb{S}^1$ differentiable structures. Show that these two differentiable structures are compatible.

3.1.8. Consider the real projective plane $\mathbb{RP}^2$. The atlas described for $\mathbb{RP}^2$ has three coordinate charts. Calculate explicitly all six of the coordinate transition functions, and verify directly that $\phi_{ij} = \phi_{ji}^{-1}$.

3.1.9. Consider $\mathbb{S}^2$ to be the unit sphere in $\mathbb{R}^3$. Consider the parametrizations

$$f : (0, 2\pi) \times (0, \pi) \to \mathbb{S}^2, \quad \text{with} \quad f(u, v) = (\cos u \sin v, \sin u \sin v, \cos v),$$

$$g : (0, 2\pi) \times (0, \pi) \to \mathbb{S}^2, \quad \text{with} \quad g(\bar{u}, \bar{v}) = (-\cos \bar{u} \sin \bar{v}, \cos \bar{v}, \sin \bar{u} \sin \bar{v}).$$

We have seen that f is injective and so is a bijection onto its range.

    (a) Find the range U of f and the range V of g.

    (b) Determine $f^{-1}(x, y, z)$ and $g^{-1}(x, y, z)$, where $(x, y, z) \in \mathbb{S}^2$.

    (c) Show that the set of functions $\{(U, f^{-1}), (V, g^{-1})\}$ forms an atlas for $\mathbb{S}^2$ and equips $\mathbb{S}^2$ with a differentiable structure.

3.1.10. Let M^n be a topological manifold and let $D(M)$ be the collection of atlases on M that equip M with a differentiable structure. Prove that the relation of compatibility is an equivalence relation. [Recall that two atlases $\mathcal{A}$ and $\mathcal{B}$ are compatible when $\mathcal{A} \cup \mathcal{B} \in D(M)$.]

3.1.11. Let M^n be a topological manifold and let $D(M)$ be the collection of atlases on M that equip M with a differentiable structure. Consider the partial order of containment $\subset$ on $D(M)$. Show that every chain (totally ordered subsets) of $D(M)$ has an upper bound. Use Zorn's Lemma to conclude that $(D(M), \subset)$ has maximal elements. [Some authors reserve the expression *differentiable structure* for these maximal elements in $D(M)$.]

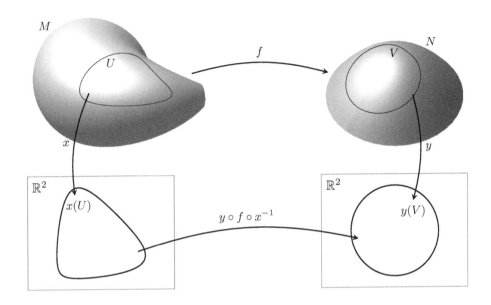

Figure 3.7: Differentiable map between manifolds.

3.2　Differentiable Maps between Manifolds

From a purely set-theoretic perspective, it is easy to define functions between manifolds. Since differentiable manifolds are topological manifolds to begin with, we can discuss continuous functions between manifolds just as we do in the context of topology. However, a differential structure on a manifold expressed by a specific atlas, allows us to make sense of the notion of differentiable maps between manifolds.

Definition 3.2.1. Let M^m and N^n be differentiable (respectively, C^k, smooth, analytic) manifolds. A continuous function $f : M^m \to N^n$ is said to be differentiable (respectively, C^k, smooth, analytic) if for any chart $y : V \to \mathbb{R}^n$ on N and for any chart $x : U \to \mathbb{R}^m$ on M, the map

$$y \circ f \circ x^{-1} : x(U \cap f^{-1}(V)) \subset \mathbb{R}^m \longrightarrow y(V) \subset \mathbb{R}^n \qquad (3.2)$$

is a differentiable (respectively, C^k, smooth, analytic) function. (See Figure 3.7.) We denote by $C^k(M^m, N^n)$ the set of C^k-differentiable maps from M to N.

In the above definition, the domain and codomain of $y \circ f \circ x^{-1}$ may seem complicated, but they are the natural ones for this composition of functions.

It follows from this definition that a function between two manifolds cannot have a stronger differentiability property than do the manifolds themselves. (See Exercise 3.2.9.) In particular, if M and N are C^k-differentiable manifolds, we cannot discuss functions of class C^{k+1} or higher between them. Restricting attention to smooth manifolds, removes this concern.

In linear algebra, we do not care about all functions between vector spaces but only linear transformations because, in an intuitive sense, linear transformations "preserve the structure" of vector spaces. Furthermore, two vector spaces V and W are considered the same (isomorphic) if there exists a bijective linear transformation between the two. In the same way, in the category of differentiable manifolds where we only consider differentiable (or perhaps C^k or smooth) maps between manifolds, we consider two manifolds the same if they are *diffeomorphic*.

Definition 3.2.2. Let M and N be two differentiable manifolds. A *diffeomorphism* (respectively, C^k diffeomorphism) between M and N is a bijective function $F : M \to N$ such that F is differentiable (respectively, C^k) and F^{-1} is differentiable (respectively, C^k). If a diffeomorphism exists between M and N, we say that M and N are *diffeomorphic*.

Example 3.2.3. Consider the projection map $\pi : \mathbb{S}^2 \to \mathbb{RP}^2$ that identifies antipodal points on the unit sphere

$$\pi(x, y, z) = (x : y : z)$$

for any $(x, y, z) \in \mathbb{S}^2$. For $\mathbb{S}^2$, we use the atlas $\{\pi_N, \pi_S\}$ as presented in Example 3.1.4, and for $\mathbb{RP}^2$, we use the atlas in Example 3.1.6, namely, $\phi_i : U_i \to \mathbb{R}^2$ for $0 \le i \le 2$, with

$$\phi_0(x_0 : x_1 : x_2) = \left(\frac{x_1}{x_0}, \frac{x_2}{x_0}\right), \quad \phi_1(x_0 : x_1 : x_2) = \left(\frac{x_0}{x_1}, \frac{x_2}{x_1}\right), \tag{3.3}$$

$$\text{and} \quad \phi_2(x_0 : x_1 : x_2) = \left(\frac{x_0}{x_2}, \frac{x_1}{x_2}\right). \tag{3.4}$$

For each pairing of coordinate charts we have

$$\phi_i \circ \pi \circ \pi_N^{-1}(u, v) = \phi_i\left(\frac{2u}{u^2 + v^2 + 1} : \frac{2v}{u^2 + v^2 + 1} : \frac{u^2 + v^2 - 1}{u^2 + v^2 + 1}\right)$$

and

$$\phi_i \circ \pi \circ \pi_S^{-1}(\bar{u}, \bar{v}) = \phi_i\left(\frac{2\bar{u}}{\bar{u}^2 + \bar{v}^2 + 1} : \frac{2\bar{v}}{\bar{u}^2 + \bar{v}^2 + 1} : \frac{1 - \bar{u}^2 - \bar{v}^2}{\bar{u}^2 + \bar{v}^2 + 1}\right).$$

For example,

$$\phi_0 \circ \pi \circ \pi_N^{-1}(u, v) = \left(\frac{v}{u}, \frac{u^2 + v^2 - 1}{2u}\right),$$

with domain $\{(u, v) \in \mathbb{R}^2 \mid u \ne 0\}$. In all six cases, the resulting functions are differentiable on their domains and in fact smooth. This shows that the projection map $\pi : \mathbb{S}^2 \to \mathbb{RP}^2$ is smooth.

Example 3.2.4. Similar to the real projective space $\mathbb{RP}^n$, we can also define the complex projective space $\mathbb{CP}^n$ as follows. Define the relation $\sim$ on nonzero $(n+1)$-tuples in $\mathbb{C}^{n+1}$ by $(z_0, z_1, \ldots, z_n) \sim (w_0, w_1, \ldots, w_n)$ if and only if there exists nonzero $\lambda \in \mathbb{C}$ such that $w_i = \lambda z_i$ for $0 \le i \le n$. This relation is an equivalence relation, and the complex projective space $\mathbb{CP}^n$ is the set of equivalence classes, written as $\mathbb{CP}^n = \left(\mathbb{C}^{n+1} - \{(0, \ldots, 0)\} \right) / \sim$ in the notation of quotient sets. We write $(z_0 : z_1 : \cdots : z_n)$ for the equivalence class of $(z_0, z_1, \ldots, z_n)$.

The stereographic projection π_N of the sphere onto the plane sets up a homeomorphism $h : \mathbb{CP}^1 \to \mathbb{S}^2$ defined by

$$h(z_0 : z_1) = \begin{cases} \pi_N^{-1}(z_1/z_0), & \text{if } z_0 \neq 0, \\ (0, 0, 1), & \text{if } z_0 = 0. \end{cases}$$

Note that if $z_0 \neq 0$, then there is a unique z' such that $(z_0 : z_1) = (1 : z')$, namely, $z' = z_1/z_0$, and that if $z_0 = 0$, then $(z_0 : z_1) = (0 : z)$ for all $z \neq 0$. Therefore, one sometimes says that $\mathbb{CP}^1$ is the complex plane $\mathbb{C}$ with a "point at infinity," where this point at infinity corresponds to the class of $(0 : z_1)$. The function h is a bijection that maps the point at infinity to the north pole of the sphere, but we leave it as an exercise for the reader to verify that this function is indeed a homeomorphism.

Complex analysis studies holomorphic (i.e., analytic) functions. This notion is tantamount to differentiable in the complex variable. Any holomorphic function $f : \mathbb{C} \to \mathbb{C}$ defines a map $p_f : \mathbb{S}^2 \to \mathbb{S}^2$ by identifying $\mathbb{R}^2$ with $\mathbb{C}$ and

$$p_f(q) = \begin{cases} \pi_N^{-1} \circ f \circ \pi_N(q) & \text{if } q \neq (0, 0, 1) \\ (0, 0, 1) & \text{if } q = (0, 0, 1) \end{cases}.$$

(That p_f must send $(0, 0, 1)$ to $(0, 0, 1)$ follows from a theorem in complex analysis, namely Liouville's Theorem.)

Consider $\mathbb{S}^2$ as a differentiable manifold with atlas $\{\pi_N, \pi_S\}$, with coordinates (u, v) and $(\bar{u}, \bar{v})$ respectively, as described in Example 3.1.4. It is interesting to notice that, according to Example 3.1.4, the change-of-coordinates map $\pi_S \circ \pi_N^{-1}$ corresponds to $z \mapsto 1/\bar{z}$ over $\mathbb{C} - \{0\}$, where $\bar{z}$ is the complex conjugate of z.

Take for example $f(z) = z^2$. The associated function p_f leaves $(0, 0, -1)$ and $(0, 0, 1)$ fixed and acts in a nonobvious manner on $\mathbb{S}^2$. According to Definition 3.2.1, in order to verify the differentiability of p_f as a function $\mathbb{S}^2 \to \mathbb{S}^2$, we need to determine explicitly the four combinations

$$(\pi_N \text{ or } \pi_S) \circ p_f \circ (\pi_N \text{ or } \pi_S)^{-1}$$

and show that they are differentiable on their appropriate domains.

Setting $z = u + iv$, we have $z^2 = (u^2 - v^2) + (2uv)i$. Since we are using the stereographic projection from the north pole to define p_f in the first place, we have $\pi_N \circ p_f \circ \pi_N^{-1}(u, v) = (u^2 - v^2, 2uv)$. Determining the other three combinations, we find that

$$\pi_N \circ p_f \circ \pi_N^{-1}(u, v) = (u^2 - v^2, 2uv),$$

$$\bar{\pi}_S \circ p_f \circ \pi_N^{-1}(u, v) = \left(\frac{u^2 - v^2}{(u^2 + v^2)^2}, \frac{2uv}{(u^2 + v^2)^2} \right),$$

$$\pi_N \circ p_f \circ \pi_S^{-1}(\bar{u}, \bar{v}) = \left(\frac{\bar{u}^2 - \bar{v}^2}{(\bar{u}^2 + \bar{v}^2)^2}, \frac{2\bar{u}\bar{v}}{(\bar{u}^2 + \bar{v}^2)^2} \right),$$

$$\pi_S \circ p_f \circ \pi_S^{-1}(\bar{u}, \bar{v}) = (\bar{u}^2 - \bar{v}^2, 2\bar{u}\bar{v}).$$

It is not hard to show that, with $f(z) = z^2$, the corresponding natural domains of these four functions are $\mathbb{R}^2$, $\mathbb{R}^2 - \{(0,0)\}$, $\mathbb{R}^2 - \{(0,0)\}$, and $\mathbb{R}^2$. Then it is an easy check that all these functions are differentiable on their domain and, hence, that p_f is a differentiable function from $\mathbb{S}^2$ to $\mathbb{S}^2$.

Since $\mathbb{R}$ is a one-dimensional manifold, if M is a differentiable manifold, we can discuss whether a real-valued function $f : M \to \mathbb{R}$ is differentiable by testing it against Definition 3.2.1. Suppose also that p is a point of M and that $x : U \to \mathbb{R}^m$ is a coordinate chart of a neighborhood of p. Then $f \circ x^{-1}$ is a differentiable function from the open set $x(U)$ in $\mathbb{R}^m$ to $\mathbb{R}$. Then we define the partial derivative of f at p in the x^i coordinate as

$$\left. \frac{\partial f}{\partial x^i} \right|_p \overset{\text{def}}{=} \left. \frac{\partial (f \circ x^{-1})}{\partial x^i} \right|_{x(p)}. \tag{3.5}$$

The notation on the left-hand side is defined by the partial derivative on the right-hand side, which is taken in the usual multivariable calculus sense.

The notion of a differentiable map between differentiable manifolds also allows us to easily define what we mean by a curve on a manifold.

Definition 3.2.5. Let M be a differentiable manifold. A *differentiable curve on* M is a differentiable function $\gamma : (a, b) \to M$, where the interval (a, b) is understood as a one-dimensional manifold with the differential structure inherited from $\mathbb{R}$. A *closed* differentiable curve on M is a differentiable function $\gamma : \mathbb{S}^1 \to M$, where $\mathbb{S}^1$ is the circle manifold.

PROBLEMS

3.2.1. Consider the antipodal identification map described in Example 3.2.3. Explicitly write out all six functions $\phi_i \circ f \circ \pi_N^{-1}$ and $\phi_i \circ f \circ \pi_S^{-1}$. Prove that each one is differentiable on its natural domain.

3.2.2. In Example 3.2.4, with $f(z) = z^2$, consider points on the unit sphere $\mathbb{S}^2$ with coordinates $(x, y, z) \in \mathbb{R}^3$. Express p_f on $\mathbb{S}^2 - \{(0, 0, 1)\}$ in terms of (x, y, z)-coordinates by calculating $\pi_N^{-1} \circ f \circ \pi_N(x, y, z)$.

3.2.3. Consider the torus T^2 in $\mathbb{R}^3$ parametrized by
$$X(u, v) = \big((2 + \cos v) \cos u, (2 + \cos v) \sin u, \sin v\big)$$
for $(u, v) \in [0, 2\pi]^2$. Consider the Gauss map of the torus $n : T^2 \to \mathbb{S}^2$ that sends each point of the torus to its outward unit normal vector as an element of $\mathbb{S}^2$. Using the stereographic projection of the sphere, explicitly show that this Gauss map is differentiable.

3.2.4. Consider the torus T^2 parametrized in the same way as in the previous exercise. The function X, restricted to $(0, 2\pi)^2$, gives a homeomorphism

(a) Prove that the function X, restricted to $(0, 2\pi)^2$, gives a homeomorphism between an open subset of this torus and an open square in $\mathbb{R}^2$. Define $\phi_1 = X^{-1}$.

(b) Show that if we defined ϕ_2 as the inverse of $(u, v) \to X\left(u + \frac{\pi}{2}, v + \frac{\pi}{2}\right)$ over $(0, 2\pi)^2$, and ϕ_3 as the inverse of $(u, v) \to X(u + \pi, v + \pi)$ over $(0, 2\pi)^2$, then $\{\phi_1, \phi_2, \phi_3\}$ is an atlas for the torus T^2. Show that no subset of this atlas is also an atlas of T^2.

(c) Define $f : T^2 \to \mathbb{S}^2$ in reference to the ϕ_1 chart as
$$f(u, v) = (\cos u \sin v, \sin u \sin v, \cos v).$$
Show that f is well-defined and can be continued continuously over all of T^2.

(d) Use the stereographic projection atlas $\{\pi_N, \pi_S\}$ of $\mathbb{S}^2$ to calculate df_p for $(y_1, y_2) = \pi_N \circ f \circ \phi_1^{-1}$.

3.2.5. Consider the (unit) sphere given with the atlas defined by stereographic projection $\mathcal{A} = \{\pi_N, \pi_S\}$ as in Example 3.1.4. Consider the function $f : \mathbb{S}^2 \to \mathbb{R}$ given by $f(x, y, z) = z$ in terms of Cartesian coordinates.

(a) Show that for points in the sphere in the coordinate chart of π_N, a formula for the partial derivatives of f is
$$\frac{\partial f}{\partial u} = \frac{4u}{(u^2 + v^2 + 1)^2} \quad \text{and} \quad \frac{\partial f}{\partial v} = \frac{4v}{(u^2 + v^2 + 1)^2}.$$

(b) Writing the coordinates on the π_S chart as $(\bar{u}, \bar{v})$, find a formula for the partial derivatives $\frac{\partial f}{\partial \bar{u}}$ and $\frac{\partial f}{\partial \bar{v}}$ over the coordinate chart π_S.

(c) Explain in what sense these partial derivatives are equal over $\mathbb{S}^2$, with the poles $\{(0, 0, 1), (0, 0, -1)\}$ removed. [Hint: Use the chain rule and the Jacobian matrix from Equation (3.22).]

3.2.6. Consider the function $f : \mathbb{RP}^3 \to \mathbb{RP}^2$ defined by
$$f(x_0 : x_1 : x_2 : x_3) = \left(x_0 x_3 - x_1 x_2 : x_0^2 - 10 x_1 x_2 : x_3^2 \cos\left(\frac{x_1^2}{x_0^2 + x_1^2 + x_2^2 + x_3^2}\right)\right).$$
Prove that f is a well-defined function. Prove also that f is a differentiable map.

3.2.7. Consider the 3-sphere described by $\mathbb{S}^3 = \{(z_1, z_2) \in \mathbb{C}^2 \,|\, |z_1|^2 + |z_2|^2 = 1\}$. Consider the function $h : \mathbb{S}^3 \to \mathbb{S}^2$ defined by $h(z_1, z_2) = (z_1 : z_2)$ where we identify $\mathbb{S}^2$ with $\mathbb{CP}^1$ as in Example 3.2.4. (This function is called the *Hopf map*.)

(a) Suppose that $z_1 = x_2 + iy_1$ and $z_2 = x_2 + iy_2$. Find an explicit formulation of $h(x_1, y_2, x_2, y_2)$.

(b) Prove that this function is a smooth map $\mathbb{S}^3 \to \mathbb{S}^2$. [Hint: Use atlases based on stereographic projection.]

3.2.8. Let $f : \mathbb{R}^{n+1} - \{0\} \to \mathbb{R}^{m+1} - \{0\}$ be a differentiable map. Let $d \in \mathbb{Z}$, and suppose that f is such that $f(\lambda x) = \lambda^d f(x)$ for all $\lambda \in \mathbb{R} - \{0\}$ and all $x \in \mathbb{R}^{n+1} - \{0\}$. Such a map is said to be *homogeneous* of degree d. For any $x \in \mathbb{R}^{k+1} - \{0\}$, denote by $\bar{x}$ the corresponding equivalence class in $\mathbb{RP}^k$. Show that the map $F : \mathbb{RP}^n \to \mathbb{RP}^m$ defined by $F(\bar{x}) = \overline{f(x)}$ is well defined and differentiable.

3.2.9. Let $f : M^m \to N^n$ be differentiable map between differentiable manifolds. Let (U_1, x) and $(U_2, \bar{x})$ be overlapping coordinate charts on M and let (V_1, y) and $(V_2, \bar{y})$ be overlapping coordinate charts on N. Since we can write

$$\bar{y} \circ f \circ \bar{x}^{-1} = (\bar{y} \circ y^{-1}) \circ (y \circ f \circ x^{-1}) \circ (x \circ \bar{x}^{-1}),$$

show why in order for a function $f : M \to N$ to be of class C^k, both manifolds must be C^k-differentiable manifolds.

3.3 Tangent Spaces

In the local theory of regular surfaces $S \subset \mathbb{R}^3$, the tangent plane plays a particularly important role. We define the first fundamental form on the tangent plane as the restriction of the dot product in $\mathbb{R}^3$ to the tangent. From the coefficients of the first fundamental form, one obtains all the concepts of intrinsic geometry, which include angles between curves, areas of regions, Gaussian curvature, geodesics, and even the Euler characteristic (see references to intrinsic geometry in [5]). The definition of a real differentiable manifold, however, makes no reference to an ambient Euclidean space, so we cannot imitate the theory of surfaces in $\mathbb{R}^3$ to define a tangent space to a manifold as a vector subspace of some $\mathbb{R}^n$.

From a physical perspective, we often think of a tangent vector to a surface $S \subset \mathbb{R}^3$ as the velocity vector at p of some curve on S through p. We understand this velocity vector to be an element in $\mathbb{R}^3$. Since we define manifolds without reference to an ambient Euclidean space, simply imagining the notion of a tangent vector poses serious conceptual challenges.

The reader can anticipate that to circumvent this difficulty, we must take a step in the direction of abstraction. We identify a tangent vector as a directional derivative at a point p of a real-valued function on a manifold M. Furthermore, since we cannot use vectors in an ambient Euclidean space to describe the notion of direction, we use curves on M through p to provide a notion of direction. The following construction makes this precise.

Definition 3.3.1. Let M^m be a differentiable manifold and let p be a point on M. Let $\varepsilon > 0$, and let $\gamma : (-\varepsilon, \varepsilon) \to M$ be a differentiable curve on M with $\gamma(0) = p$.

For any real-valued differentiable f defined on some neighborhood of p, we define the directional derivative of f along γ at p to be the number

$$D_\gamma(f) = \frac{d}{dt}(f(\gamma(t)))\bigg|_{t=0}. \qquad (3.6)$$

The operator D_γ is called the *tangent vector* to γ at p.

If γ_1 and γ_2 are two curves satisfying the conditions in the above definition, then $D_{\gamma_1} = D_{\gamma_2}$ if these operators have the same value at p for all differentiable functions defined in open neighborhoods of p.

Note that $f \circ \gamma$ is a function $(-\varepsilon, \varepsilon) \to \mathbb{R}$, so the derivative in Equation (3.6) is taken in the usual sense. It is also interesting to observe that the above definition does not explicitly refer to any particular chart on U. However, in order to calculate $D_\gamma(f)$ it may be necessary to refer to a chart around p.

The above definition of a tangent vector may initially come as a source of mental discomfort since it presents tangent vectors as operators instead of as the geometric objects with which we are used to working. However, any tangent vector (defined in the classical sense) to a regular surface S in $\mathbb{R}^3$ naturally defines a directional derivative of a function $S \to \mathbb{R}$ so Definition 3.3.1 generalizes the usual notion of a tangent vector (see [5, Section 5.2]).

As the name "tangent vector" suggests, the set of all tangent vectors forms a vector space, a fact that we show now.

Let U be an open neighborhood of p in M. Call $C^1(U, \mathbb{R})$ the set (vector space) of all differentiable functions from U to $\mathbb{R}$. A priori, the set of tangent vectors D_γ at p on M is a subset of all operators $\mathcal{W} = \{C^1(U, \mathbb{R}) \to \mathbb{R}\}$. By the differentiation properties

$$D_\gamma(f + g) = D_\gamma(f) + D_\gamma(g) \qquad \text{and} \qquad D_\gamma(cf) = cD_\gamma(f),$$

so D_γ is a linear transformation from $C^1(U, \mathbb{R})$ to $\mathbb{R}$. For readers who are familiar with the dual of a vector space, this latter result shows that D_γ is in the dual vector space $C^1(U, \mathbb{R})^*$. (We discuss the dual of a vector space in Section 4.1.) We would like to show that the set of tangent vectors is a subspace of $C^1(U, \mathbb{R})^*$, i.e., closed under addition and scalar multiplication.

Let $\gamma : (-\varepsilon, \varepsilon) \to M$ be a differentiable curve with $\gamma(0) = p$. If we define $\gamma_1(t) = \gamma(at)$, where a is some real number, then using the usual chain rule for any differentiable function $f \in C^1(U, \mathbb{R})$, we have

$$D_{\gamma_1}(f) = \frac{d}{dt}(f(\gamma(at)))\bigg|_{t=0} = a\frac{d}{dt}(f(\gamma(t)))\bigg|_{t=0} = aD_\gamma(f).$$

This shows that the set of tangent vectors is closed under scalar multiplication.

In order to prove that the set of tangent vectors is closed under addition, we make reference to a coordinate chart $x : U \to \mathbb{R}^m$, where U is an open neighborhood of p. Without loss of generality, we assume that $x(p) = (0, 0, \ldots, 0)$. We rewrite the composition $f \circ \gamma = f \circ x^{-1} \circ x \circ \gamma$ where $x \circ \gamma : (-\varepsilon, \varepsilon) \to \mathbb{R}^m$ and $f \circ x^{-1} :$ $x(U) \subset \mathbb{R}^m \to \mathbb{R}$. By the chain rule in multivariable analysis, Theorem (1.3.3), we have

$$
\begin{aligned}
D_\gamma(f) &= \frac{d}{dt}\left(f(\gamma(t))\right)\Big|_{t=0} \\
&= \frac{d}{dt}\left(f \circ x^{-1}(x \circ \gamma(t))\right)\Big|_{t=0} \\
&= d(f \circ x^{-1})_{\vec{0}}\, d(x \circ \gamma)\big|_{t=0},
\end{aligned}
$$

where we evaluate $d(f \circ x^{-1})$ at $\vec{0} = (0, 0, \dots, 0)$ because $x(p) = \vec{0}$.

Let α and β be two differentiable curves on M such that $\alpha(0) = \beta(0) = p$. Over the intersection of the domains of α and β, define the curve γ by

$$
\gamma(t) = x^{-1}\big(x \circ \alpha(t) + x \circ \beta(t)\big).
$$

Note that $\gamma(0) = x^{-1}(x(\alpha(0)) + x(\beta(0))) = x^{-1}(\vec{0} + \vec{0}) = x^{-1}(\vec{0}) = p$. Furthermore, for any function $f : U \to \mathbb{R}$, we have

$$
\begin{aligned}
D_\alpha(f) + D_\beta(f) &= d(f \circ x^{-1})_{\vec{0}}\, d(x \circ \alpha)|_{t=0} + d(f \circ x^{-1})_{\vec{0}}\, d(x \circ \beta)|_{t=0} \\
&= d(f \circ x^{-1})_{\vec{0}}\big(d(x \circ \alpha)|_{t=0} + d(x \circ \beta)|_{t=0}\big) \\
&= d(f \circ x^{-1})_{\vec{0}}\, d(x \circ \alpha + x \circ \beta)\big|_{t=0} \\
&= D_\gamma(f).
\end{aligned}
$$

Thus, the set of tangent vectors is closed under addition. This brings us in a position to prove the following foundational fact.

Proposition 3.3.2. *Let M be a differentiable manifold of dimension m, and let p be a point of M. The set of all tangent vectors to M at p is a vector space of dimension m with basis $\{\partial/\partial x^i \mid i = 1, \dots, m\}$, where $(x^1, x^2, \dots, x^m)$ are the coordinates on some chart around p.*

Definition 3.3.3. The vector space of tangent vectors is called the *tangent space* of M at p and is denoted by $T_p M$.

Proof of Proposition 3.3.2. The prior discussion has shown that the set $T_p M$ is a vector space. It remains to be shown that it has dimension m.

Let $x : U \to \mathbb{R}^m$ be a system of local coordinates at p. Write $x(q) = (x^1(q), \dots, x^m(q))$, and define the coordinate line curve $v_i : (-\varepsilon, \varepsilon) \to M$ by $v_i(t) = x^{-1}(0, \dots, 0, t, 0, \dots, 0)$ where the t occurs in the ith place. Then

$$
D_{v_i}(f) = \frac{d}{dt}\big(f \circ x^{-1}(0, \dots, 0, t, 0, \dots, 0)\big)\Big|_{t=0} = \frac{\partial f}{\partial x^i}\Big|_p
$$

according to the notation given in Equation (3.5). We can therefore write, as operators, $D_{v_i} = \dfrac{\partial}{\partial x^i}\Big|_p$.

For any differentiable curve γ on M with $\gamma(0) = p$, we can then write in coordinates $x \circ \gamma(t) = (\gamma^1(t), \ldots, \gamma^m(t))$, where $\gamma^i = x^i(\gamma(t))$. Then

$$D_\gamma(f) = \frac{d}{dt} f \circ x^{-1}(\gamma^1(t), \ldots, \gamma^m(t))\big|_{t=0}$$

$$= \sum_{i=1}^m \frac{\partial f}{\partial x^i}\bigg|_p \frac{d\gamma^i}{dt}\bigg|_{t=0}.$$

This presents the operator D_γ as a linear combination of the operators $\partial/\partial x^i\big|_p$.

It is also a trivial matter to show that for $1 \leq i \leq m$, the operators $\partial/\partial x^i\big|_p$ are linearly independent. Consequently, they form a basis of T_pM, which proves that $\dim T_pM = m$. □

Because the operators $\partial/\partial x^i$ occur so often in the theory of manifolds, one often uses an abbreviated notation. Whenever the coordinate system is understood by context, where one uses $x = (x^1, \ldots, x^n)$ or another letter, we write

$$\partial_i \stackrel{\text{def}}{=} \frac{\partial}{\partial x^i}, \tag{3.7}$$

whose explicit meaning is given by Equation (3.5). This notation shortens the standard partial derivative notation and makes it easier to write it in inline formulas.

From our definition of tangent vectors, if the manifold is of class C^2 we can give an alternate characterization of the tangent space T_pM.

Definition 3.3.4. A function from $X : C^1(M, \mathbb{R}) \to \mathbb{R}$ is called a *derivation* of $C^1(M, \mathbb{R})$ at p if it satisfies

1. Linearity: $X(af + bg) = aX(f) + bX(g)$ for all $f, g \in C^1(U, \mathbb{R})$ and $a, b \in \mathbb{R}$;
2. Leibniz's rule: $X(fg) = X(f)g(p) + f(p)X(g)$ for all $f, g \in C^1(U, \mathbb{R})$.

Note that $C^k(M, \mathbb{R})$ is an *algebra* that is, a vector space equipped a "multiplication" operation that is bilinear over the vector space. So, if $k \geq 1$, a derivation on $C^k(M, \mathbb{R})$ at p is a linear transformation from the algebra of $C^k(M, \mathbb{R})$ to $\mathbb{R}$, satisfying additionally what is tantamount to a product rule.

Proposition 3.3.5. *Let X be derivation of $C^1(M, \mathbb{R})$ at p and f a constant function on M. Then $X(f) = 0$.*

Proof. (Left as an exercise for the reader.) □

Theorem 3.3.6. *Let M^m be a C^2-differentiable manifold. The tangent space T_pM is the set of derivations of $C^2(M, \mathbb{R})$ at p.*

Proof. We have already seen that every tangent vector is derivation so T_pM is vector subspace of the set of derivations of $C^2(M, \mathbb{R})$ at p.

Conversely, let X be a derivation of $C^1(M, \mathbb{R})$ at p. Let U be a coordinate neighborhood of p with coordinates $x = (x^1, x^2, \ldots, x^m)$ and suppose that the coordinates of p are $x(p) = (a^1, a^2, \ldots, a^m)$. Without loss of generality, suppose that $x(U)$ is an open ball in $\mathbb{R}^m$ with radius $x(p)$. For $i = 1, \ldots, m$, let $X(x^i) = v^i$.

By Theorem 1.3.8, for any function $f \in C^2(M, \mathbb{R})$, setting

$$c_i = \partial f / \partial x^i(p) = \partial(f \circ x^{-1}) / \partial x^i \big|_{\vec{a}},$$

the first-order Taylor series of f at p is

$$f \circ x^{-1}(x_1, x_2, \ldots, x_m) = (f \circ x^{-1})(\vec{a}) + \sum_{i=1}^{m} c_i (x^i - a^i)$$

$$+ \sum_{i=1}^{m} (g_i \circ x^{-1})(x^1, x^2, \ldots, x^m)(x^i - a^i),$$

where $g_i \in C^1(U, \mathbb{R})$ with $(g_i \circ x^{-1})(\vec{a}) = 0$. Since g_i are of class C^1, we can take a derivation of it. Then by linearity and the Leibniz rule,

$$X(f) = X(f(p)) + \sum_{i=1}^{m} (X(c_i)(x^i - a^i) + c_i(X(x^i) - X(a^i)))$$

$$+ \sum_{i=1}^{m} X(g_i)(x^i - a^i)\big|_p + g_i(p)(X(x^i) - X(a^i)).$$

Then by Proposition 3.3.5 and the assumption that $X(x^i) = v^i$,

$$X(f) = \sum_{i=1}^{m} c_i v^i + \sum_{i=1}^{m} (X(g_i)0 + 0v^i) = \sum_{i=1}^{m} c_i v^i.$$

Thus $X = v^1 \partial_1 + v^2 \partial_2 + \cdots + v^m \partial_m$. Since $\partial_i \in T_p(M)$, we deduce that the set of derivations of $C^2(M, \mathbb{R})$ at p is also a subspace of $T_p M$. The result follows. $\qquad \square$

Example 3.3.7 (Tangent Space of $\mathbb{R}^n$). We consider the tangent space for the manifold $\mathbb{R}^n$ itself. We assume the standard differential structure.

Let p be a point in $\mathbb{R}^n$, and let $v = (v_1, \ldots, v_n)$ be a vector. Consider the line traced out by the curve $\gamma(t) = p + tv$. We wish to find the coordinates of the tangent vector D_γ with respect to the standard basis of $T_p M$, namely, $\{\partial/\partial x^i\}$ or, according to the notation of Equation (3.7), $\{\partial_i\}$. For any real function f defined over a neighborhood of p, we have

$$D_\gamma(f) = \frac{d}{dt} f(p_1 + tv_1, \ldots, p_n + tv_n) \bigg|_{t=0} = \sum_{i=1}^{n} \frac{\partial f}{\partial x^i} \bigg|_p v_i$$

So with respect to the basis $\{\partial/\partial x^i\}$, the coordinates of D_γ are $(v_1, \ldots, v_n)$. Therefore, at each $p \in \mathbb{R}^n$, the map $v \mapsto D_\gamma$ sets up an isomorphism between the vector spaces $\mathbb{R}^n$ and $T_p(\mathbb{R}^n)$ by identifying ∂_i with the ith standard basis vector. It is common to abuse the notation and view the tangent space $T_p(\mathbb{R}^n)$ as equal to $\mathbb{R}^n$.

Note that if $\vec{v}$ is a unit vector, the D_γ is equal to the directional derivative operator in the direction of $\vec{v}$.

More generally, for any differentiable curve $\gamma : (-\varepsilon, \varepsilon) \to \mathbb{R}^n$ with $\gamma(0) = p$, we have

$$D_\gamma(f) = \sum_{i=1}^n \frac{\partial f}{\partial x^i}\bigg|_p \gamma_i'(0).$$

Hence, as an operator, we can write

$$D_\gamma = \sum_{i=1}^n \gamma_i'(0) \frac{\partial}{\partial x^i}\bigg|_p,$$

which illustrates D_γ as a vector with the same components of the usual velocity vector $\gamma'(0)$ given with respect to the basis $\{\partial/\partial x^i\}$.

Example 3.3.8 (Regular Surfaces). Let S be a regular surface in $\mathbb{R}^3$. In Chapter 5 of [5], the authors define the tangent plane to S at p as the subspace of $\mathbb{R}^3$ consisting of all vectors $\gamma'(0)$, where $\gamma(t)$ is a curve on S with $\gamma(0) = p$. The correspondence $\gamma'(0) \leftrightarrow D_\gamma$ identifies the tangent space for regular surfaces with the tangent space of manifolds as defined above. This shows that the present definition directly generalizes the previous definition as a subspace of the ambient space $\mathbb{R}^n$.

In multivariable calculus, one shows that given a parametrization $\vec{X} : V \subset \mathbb{R}^2 \to \mathbb{R}^3$ of a coordinate patch of a regular surface, if $p = \vec{X}(u_0, v_0)$, then a basis for $T_p S$ is

$$\{\vec{X}_u(u_0, v_0), \vec{X}_v(u_0, v_0)\}.$$

The definition of the tangent plane given in calculus meshes with Definition 3.3.1 and Proposition 3.3.2 in the following way. A tangent vector in the classical sense, $\vec{w} \in T_p M$, is a vector such that $\vec{w} = \vec{\gamma}'(t_0)$, where $\vec{\gamma}(t)$ is a curve on S with $\vec{\gamma}(t_0) = p$. Write $\vec{\gamma}(t) = \vec{X}(\alpha(t))$, with $\alpha(t_0) = (u_0, v_0)$. Writing $\alpha(t) = (u(t), v(t))$, we have

$$\vec{w} = u'(t_0)\vec{X}_u(u_0, v_0) + v'(t_0)\vec{X}_v(u_0, v_0). \tag{3.8}$$

Now the corresponding coordinate chart x on S in the language of manifolds is the inverse of the parametrization $x = \vec{X}^{-1}$ defined over $U = \vec{X}(V) \subset S$. The tangent vector (in the phrasing of Definition 3.3.1) associated to γ at p is

$$D_\gamma(f) = \frac{d}{dt}\left(f(\vec{\gamma}(t))\right)\bigg|_{t_0} = \frac{d}{dt}\left(f(\vec{X}(\alpha(t)))\right)\bigg|_{t_0} = \frac{d}{dt}\left(f \circ x^{-1}(\alpha(t))\right)\bigg|_{t_0}$$

$$= u'(t_0)\frac{\partial f}{\partial u}\bigg|_p + v'(t_0)\frac{\partial f}{\partial v}\bigg|_p \tag{3.9}$$

where the partial derivatives $\left.\dfrac{\partial f}{\partial u}\right|_p$ and $\left.\dfrac{\partial f}{\partial v}\right|_p$ are in the sense of Equation (3.5). We can write as operators

$$D_\gamma = u'(t_0)\left.\frac{\partial}{\partial u}\right|_p + v'(t_0)\left.\frac{\partial}{\partial v}\right|_p. \tag{3.10}$$

Therefore, we see that the correspondence between the definition of the tangent space for manifolds and the definition for tangent spaces to regular surfaces in $\mathbb{R}^3$ identifies $\vec{X}_u(u_0, v_0)$ with $\left.\dfrac{\partial f}{\partial u}\right|_p$ and similarly for the v-coordinate.

Obviously, the bases for T_pM described in Proposition 3.3.2 are dependent on the coordinate charts. The following proposition shows how to change coordinates.

Proposition 3.3.9. *Let M^n be a differentiable manifold; let (U_1, ϕ_1) and (U_2, ϕ_2) be overlapping coordinate charts; and let $p \in U_1 \cap U_2$. Denote by (x^i) the coordinates of ϕ_1 and by $(\bar{x}^j)$ the coordinates of ϕ_2 Let $\mathcal{B} = \{\partial_1, \partial_2, \ldots, \partial_n\}$ and $\bar{\mathcal{B}} = \{\bar\partial_1, \bar\partial_2, \ldots, \bar\partial_n\}$ be the two bases for T_pM defined by Proposition 3.3.2 with respect to the coordinate systems (where by $\bar\partial_j$ we mean $\partial/\partial\bar{x}^j$). The coordinate change matrix from $\mathcal{B}$ to $\bar{\mathcal{B}}$ coordinates on T_pM is $d(\phi_2 \circ \phi_1^{-1})$, the differential of the transition function. In other words, for all $X \in T_pM$, if*

$$[X]_\mathcal{B} = \begin{pmatrix} v^1 \\ v^2 \\ \vdots \\ v^n \end{pmatrix} \quad and \quad [X]_{\bar{\mathcal{B}}} = \begin{pmatrix} \bar{v}^1 \\ \bar{v}^2 \\ \vdots \\ \bar{v}^n \end{pmatrix}, then \quad \bar{v}^j = \sum_{i=1}^n \frac{\partial \bar{x}^j}{\partial x^i} v^i.$$

Proof. (Left as an exercise for the reader.) □

PROBLEMS

3.3.1. Let M be a differentiable manifold. Let $p \in M$ and let $X \in T_pM$. Prove that if f is a constant function, then $X(f) = 0$.

3.3.2. Prove Proposition 3.3.9.

3.3.3. Consider $\mathbb{RP}^2$ with the usual atlas $\{\phi_0, \phi_1, \phi_2\}$. Let (u_1, u_2) be coordinates corresponding to ϕ_0 and (v_1, v_2) coordinates corresponding to ϕ_1. Let $p \in U_0 \cap U_1$. Calculate the change of coordinate matrix on $T_p(\mathbb{RP}^2)$ from u-coordinates to v-coordinates.

3.3.4. Let M be a differentiable manifold. A class C^k (resp. smooth) function element on M is a pair (f, U) where U is an open subset of M and $f : U \to \mathbb{R}$ that is of class C^k (resp. smooth). Recall that we cannot discuss functions of class C^k unless M is a C^k-differentiable manifold. Given a point $p \in M$, define the relation $\equiv$ on the set of function elements with $(f, U) \equiv (g, V)$ whenever $p \in U \cap V$ and there is a neighborhood W of p in $U \cap V$ such that $f|_W = g|_W$, i.e., the restrictions of f and g to W are equal as functions.

(a) Fix a k and a point $p \in M$. Prove that $\equiv$ is an equivalence relation. [The equivalence class $[(f, U)]$ of some element (f, U), where $p \in U$ is called a germ at p. The set of all germs at p of class C^k functions is denoted by $C_p^k(M, \mathbb{R})$.]

(b) Prove that the following addition and scalar multiplication

$$[(f, U))] + [(g, V)] \overset{\text{def}}{=} [(f + g, U \cap V)] \quad \text{and} \quad c[(f, U)] = [(cf, U)]$$

are well-defined and make $C_p^k(M, \mathbb{R})$ into a vector space.

(c) Prove that the multiplication on $C_p^k(M, \mathbb{R})$

$$[(f, U))][(g, V)] \overset{\text{def}}{=} [(fg, U \cap V)]$$

is associative, has an identity, and distributes over the addition. [This makes $C_p^k(M, \mathbb{R})$ into an associative algebra.]

(d) Let $\gamma : (-\varepsilon, \varepsilon) \to M$ be a curve on M with $\gamma(0) = p$. Prove that $D_\gamma([(f, U)]) \overset{\text{def}}{=} D_\gamma f$ is well-defined, i.e., that if $(f, U) \equiv (g, V)$, then $D_\gamma f = D_\gamma g$.

3.4 The Differential of a Differentiable Map

Having established the notion of a tangent space to a differentiable manifold at a point, we are in a position to define the differential of a differentiable map $f : M \to N$. Recall that in multivariable real analysis, we call a function $F : \mathbb{R}^m \to \mathbb{R}^n$ differentiable at a point $\vec{p} \in \mathbb{R}^m$ if there exists an $n \times m$ matrix A such that

$$F(\vec{p} + \vec{h}) = F(\vec{p}) + A\vec{h} + R(\vec{h}),$$

where $R(\vec{h})$ is a continuous function defined around $\vec{0}$ such that $\|R(\vec{h})\|/\|\vec{h}\| \to 0$ as $\|\vec{h}\| \to 0$. We refer to the matrix A as the differential $dF_{\vec{p}}$. Surprisingly, given our definition of the tangent space to a manifold, there exists a more natural way to define the differential.

Definition 3.4.1. Let $F : M^m \to N^n$ be a differentiable map between differentiable manifolds. We define the *differential* of F at $p \in M$ as the linear transformation between vector spaces

$$dF_p : T_pM \longrightarrow T_{F(p)}N,$$
$$D_\gamma \longmapsto D_{F \circ \gamma}.$$

The differential dF_p is also denoted by F_* with p is understood by context. If $X \in T_pM$, then $F_*(X)$ is also called the *push-forward* of X by F.

From this definition, it is not immediately obvious that dF_p is linear, but, as the following proposition shows, we can give an equivalent definition of the differential that makes it easy to show that the differential is linear. Figure 3.8 depicts the differential of a map between manifolds.

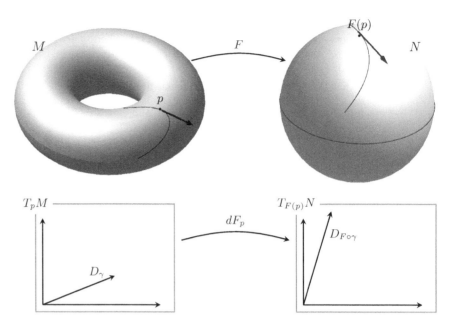

Figure 3.8: The differential of a map between manifolds.

Proposition 3.4.2. *Let $F : M \to N$ be a differentiable map between manifolds. Then at each $p \in M$, the function $F_* = dF_p$ satisfies*

$$F_*(X)(g) = X(g \circ F)$$

for every vector $X \in T_pM$ and every function g from N into $\mathbb{R}$ defined in a neighborhood of $F(p)$. Furthermore, F_* is linear.

Proof. Let $X \in T_pM$, with $X = D_\gamma$, for some curve γ on M with $\gamma(0) = p$. For all real-valued function g defined in a neighborhood of $F(p)$ on N,

$$F_*(X)(g) = dF_p(D_\gamma)(g) = D_{F \circ \gamma}(g)$$

$$= \frac{d}{dt}(g \circ F \circ \gamma)(t)\Big|_p = D_\gamma(g \circ F) = X(g \circ F).$$

To show linearity, let $X, Y \in T_pM$ and $a, b \in \mathbb{R}$. Then

$$F_*(aX + bY)(g) = aX(g \circ F) + bY(g \circ F) = aF_*(X)(g) + bF_*(Y)(g),$$

which shows that F_* is linear. $\qquad\square$

Note that this definition is independent of any coordinate system near p or $F(p)$. However, given specific coordinate charts $x : U \to \mathbb{R}^m$ and $y : V \to \mathbb{R}^n$ whose domains are, respectively, neighborhoods of p in M and $F(p)$ in N, with $F(U) \subset V$, we can define a matrix that represents dF_p. Set v_i as the coordinate line for the variable x^i in the chart x. In the usual basis of T_pM, we have

$$F_*(D_{v_i}) = F_*\left(\frac{\partial}{\partial x^i}\right) = D_{F \circ v_i}.$$

However, for any smooth function $g : N \to \mathbb{R}$,

$$D_{F \circ \gamma}(g) = \sum_{j=1}^{n} \frac{\partial g}{\partial y^j} \frac{d(y^j \circ F \circ \gamma)}{dt}\bigg|_{t=0} \tag{3.11}$$

$$= \sum_{j=1}^{n} \frac{\partial g}{\partial y^j} \left(\sum_{i=1}^{m} \frac{\partial(y^j \circ F)}{\partial x^i} \frac{d\gamma^i}{dt}\bigg|_{t=0} \right), \tag{3.12}$$

where $\gamma^i = x^i \circ \gamma$. Therefore, in terms of these coordinate patches, the matrix for F_* with respect to the standard bases $\{\partial/\partial x^i\}$ on T_pM and $\{\partial/\partial y^j\}$ on $T_{F(p)}N$ is

$$[dF_p] = \left(\frac{\partial F^j}{\partial x^i}\right)_p, \quad \text{with} \quad 1 \le i \le m, \text{ and } 1 \le j \le n, \tag{3.13}$$

by which we explicitly mean

$$\frac{\partial F^j}{\partial x^i}\bigg|_p \stackrel{\text{def}}{=} \frac{\partial(y^j \circ F \circ x^{-1})}{\partial x^i}\bigg|_{x(p)}. \tag{3.14}$$

Example 3.4.3 (Curves on a Manifold). We used the notion of a curve on a manifold to define tangent vectors in the first place. However, we can now restate the notion of a curve on a manifold as a differentiable map $\gamma : I \to M$, where I is an open interval of $\mathbb{R}$ and M is a differentiable manifold. The tangent vector $D_\gamma \in T_{\gamma(t_0)}M$ to the curve γ can be understood as

$$D_\gamma = \gamma_*\left(\frac{d}{dt}\bigg|_{t_0}\right). \tag{3.15}$$

Matching with notation from calculus courses, this tangent vector is sometimes denoted as $\gamma'(t_0)$. Then this tangent vector acts on differentiable functions $f : M \to \mathbb{R}$ by

$$\gamma'(t_0)(f) = \gamma_*\left(\frac{d}{dt}\bigg|_{t_0}\right)(f) = \frac{d(f \circ \gamma)}{dt}\bigg|_{t_0}. \tag{3.16}$$

Example 3.4.4 (Gauss Map). Consider a regular oriented surface S in $\mathbb{R}^3$ with orientation $n : S \to \mathbb{S}^2$. (Recall from calculus that the orientation is a choice of a unit normal vector to S at each point such that $n : S \to \mathbb{S}^2$ is a continuous function.) In the local theory of surfaces, the function n is often called the *Gauss map*. The differential of the Gauss map plays a central role in the differential geometry of surfaces. In that context, we define the differential of the Gauss map dn_p at a point $p \in S$ in the following way.

A parametrization $\vec{X}(u, v)$ of a coordinate patch U around p amounts to the inverse $\vec{X} = x^{-1}$ of a chart $x : U \to \mathbb{R}^2$. Similarly, on $\mathbb{S}^2$, the parametrization $\vec{N} = n \circ \vec{X}$ is the inverse of a chart y on $\mathbb{S}^2$ of a neighborhood of $n(p)$. Since $\vec{N} : U \to \mathbb{R}^3$ is a unit vector, by the comments in Section 2.2, we know that $\vec{N}_u$ and $\vec{N}_v$ are perpendicular to $\vec{N}$ and hence are in the tangent space $T_p S$. Hence, we often identify $T_p S = T_{n(p)}(\mathbb{S}^2)$. Let $\vec{X}(t) = \vec{X}(\vec{\alpha}(t))$ be any curve on the surface such that $\vec{X}(0) = p$. Then dn_p is the transformation on $T_p S$ that sends a tangent vector $\vec{X}'(0) \in T_p(S)$ to $\frac{d}{dt}(\vec{N}(\vec{\alpha}(t)))\big|_{t=0}$.

Via the association of $\gamma'(0) \to D_\gamma$ between the classical and the modern definition of the tangent space, we see that the classical definition of the differential of the Gauss map is precisely Definition 3.4.1. (Note that Figure 3.8 specifically illustrates the differential of the Gauss map.)

Over some neighborhood of $n(p)$, the function $\vec{N} : x(U) \to \mathbb{S}^2$ gives a parametrization of a coordinate neighborhood of $n(p)$ on $\mathbb{S}^2$. Write the coordinate functions as $x(q) = (x_1(q), x_2(q))$ and similarly for y. Then the associated bases on $T_p S$ and $T_{n(p)}(\mathbb{S}^2)$ are

$$\left\{ \frac{\partial}{\partial x^1}, \frac{\partial}{\partial x^2} \right\} \text{ identified as } \{\vec{X}_u, \vec{X}_v\} \text{ and}$$

$$\left\{ \frac{\partial}{\partial y^1}, \frac{\partial}{\partial y^2} \right\} \text{ identified as } \{\vec{N}_u, \vec{N}_v\}.$$

Thus, with respect to the coordinate charts x and y as described here, the matrix for dn_p is

$$[dn_p] = (a_j^i), \qquad \text{where} \qquad \vec{N}_j = a_j^1 \vec{X}_1 + a_j^2 \vec{X}_2,$$

where by $\vec{X}_i$, we mean $\partial \vec{X}/\partial x^i$. It is not hard to show that

$$\begin{pmatrix} a_1^1 & a_2^1 \\ a_1^2 & a_2^2 \end{pmatrix} = - \begin{pmatrix} g_{11} & g_{12} \\ g_{21} & g_{22} \end{pmatrix}^{-1} \begin{pmatrix} L_{11} & L_{12} \\ L_{21} & L_{22} \end{pmatrix}, \tag{3.17}$$

where $g_{ij} = \vec{X}_i \cdot \vec{X}_j$ and $L_{ij} = \vec{X}_{ij} \cdot \vec{N}$. In classical differential geometry, this matrix equation for the coefficients a_j^i is called the *Weingarten equations*. Equation (3.17) is written as

$$a_j^i = - \sum_{k=1}^n g^{ik} L_{kj},$$

where g^{ij} are the components of the inverse matrix $(g_{kl})^{-1}$.

Corollary 3.4.5 (The Chain Rule). *Let M, N, and S be differentiable manifolds, and consider $F : M \to N$ and $G : N \to S$ to be differentiable maps between them. Then*

$$(G \circ F)_* = G_* \circ F_*.$$

More specifically, at every point $p \in M$,

$$d(G \circ F)_p = dG_{F(p)} \circ dF_p \, .$$

Proof. By Proposition 3.4.2, for all functions h from a neighborhood of $G(F(p))$ on S to $\mathbb{R}$ and for all $X \in T_pM$, we have

$$(G \circ F)_*(X)(h) = X(h \circ G \circ F) = (F_*(X))(h \circ G)$$
$$= (G_*(F_*(X)))(h) = (G_* \circ F_*)(X)(h). \qquad \square$$

Definition 3.4.1 for the differential avoids referring to any coordinate neighborhood on M. In contrast to the matrix for the differential introduced in Chapter 1, the matrix for the differential df_p of maps between manifolds depends on the coordinate charts used around p and $f(p)$, according to Equations (3.13) and (3.14). We can, however, say the following about how the matrix of the differential changes under coordinate changes.

Proposition 3.4.6. *Let $f : M \to N$ be a differentiable map between differentiable manifolds. Let $x = (x^1, \ldots, x^m)$ and $\bar{x} = (\bar{x}^1, \ldots, \bar{x}^m)$ be two coordinate systems in a neighborhood of p, and let $y = (y^1, \ldots, y^n)$ and $\bar{y} = (\bar{y}^1, \ldots, \bar{y}^n)$ be two coordinate systems in a neighborhood of $f(p)$. Let $[df_p]$ be the matrix for df_p associated to the x- and y- coordinate systems and let $[d\bar{f}_p]$ be the matrix of the differential of f but expressed in $\bar{x}$-coordinates in the domain and $\bar{y}$-coordinates in the codomain. Then*

$$[d\bar{f}_p] = \left(\frac{\partial \bar{y}^i}{\partial y^j} \Big|_{f(p)} \right) [df_p] \left(\frac{\partial x^k}{\partial \bar{x}^l} \Big|_p \right).$$

Proof. (The proof is left as an exercise for the reader.) $\qquad \square$

PROBLEMS

3.4.1. Let $F : \mathbb{R}^m \to \mathbb{R}^n$ be a linear transformation. Show that under the identification of $T_p(\mathbb{R}^k)$ with $\mathbb{R}^k$ as described in Example 3.3.7, F_* is identified with F.

3.4.2. Consider a differentiable manifold M^m and a real-valued, differentiable function $h : M^m \to \mathbb{R}$. Apply Proposition 3.4.2 to show that $h_*(X)$ corresponds to the differential operator

$$h_*(X) = X(h) \frac{d}{dt}$$

on functions $g : \mathbb{R} \to \mathbb{R}$, where we assume we use the variable t on $\mathbb{R}$.

3.4.3. Let T^2 be the torus given as a subset of $\mathbb{R}^3$ with a parametrization

$$\vec{X}(u, v) = \big((2 + \cos v) \cos u, (2 + \cos v) \sin u, \sin v \big).$$

Consider the sphere $\mathbb{S}^2$ given as a subset of $\mathbb{R}^3$, and use the stereographic atlas $\{\pi_N, \pi_S\}$ as the coordinate patches of $\mathbb{S}^2$. Consider the map $f : T^2 \to \mathbb{S}^2$ defined by

$$x \mapsto \frac{x}{\|x\|}.$$

Explicitly calculate the matrix of the differential df_p, with p given in terms of (u, v)-coordinates for $(u, v) \in (0, 2\pi)^2$ and using the stereographic atlas on the sphere.

3.4.4. Let $\mathbb{S}^3$ be the 3-sphere given in $\mathbb{R}^4$ by $\mathbb{S}^3 = \{u \in \mathbb{R}^4 : \|u\| = 1\}$, and let $\mathbb{S}^2$ be the unit sphere in $\mathbb{R}^3$, where we use coordinates (x_1, x_2, x_3). Consider the Hopf map $h : \mathbb{S}^3 \to \mathbb{S}^2$ given by

$$h(u_1, u_2, u_3, u_4) = \left(2(u_1 u_2 + u_3 u_4), 2(u_1 u_4 - u_2 u_3), (u_1^2 + u_3^2) - (u_2^2 + u_4^2)\right).$$

(Note: the description of h is equivalent to the one given in Problem 3.2.7.)

(a) Show that this map indeed surjects $\mathbb{S}^3$ onto $\mathbb{S}^2$.

(b) Show that the preimage $h^{-1}(q)$ of any point $q \in \mathbb{S}^2$ is a circle on $\mathbb{S}^3$. [Using the notation of 3.2.7, show that $h^{-1}(1 : z_2)$ is the circle in $\mathbb{C}^2$ parametrized by $\gamma(t) = (Re^{it}, Rz_2 e^{it})$ with $R = 1/\sqrt{1 + |z_2|^2}$ and $h^{-1}(0 : 1)$ is the circle in $\mathbb{C}^2$ parametrized by $(0, e^{it})$.]

(c) For a coordinate patch of your choice on $\mathbb{S}^3$ and also on $\mathbb{S}^2$, calculate the differential dh_p for points p on $\mathbb{S}^3$.

3.4.5. Consider the map $F : \mathbb{RP}^3 \to \mathbb{RP}^2$ defined by

$$F(x : y : z : w) = (x^3 - y^3 : xyz - 2xw^2 + z^3 : z^3 + 2yz^2 - 6y^2 z - w^3).$$

This function is homogeneous, and the result of Problem 3.2.8 ensures that this map is differentiable. Let $p = (1 : 2 : -1 : 3) \in \mathbb{RP}^3$.

(a) After choosing standard coordinate neighborhoods of $\mathbb{RP}^3$ and $\mathbb{RP}^2$ that contain, respectively, p and $F(p)$, calculate the matrix of dF_p with respect to these coordinate neighborhoods.

(b) Choose a different pair of coordinate neighborhoods for p and $F(p)$ and repeat the above calculation.

(c) Explain how these two matrices are related.

3.4.6. Let $f : U \to \mathbb{R}$ be a function of class C^2 over an open set $U \subset \mathbb{R}^2$. Use the coordinates (u, v) that arise from the parametrization of the graph by $(u, v, f(u, v))$. Define $n : M \to \mathbb{S}^2$ to be the function that returns that upward pointing unit normal vector of M, as a subset of $\mathbb{R}^3$.

(a) Find the matrix of dn_p for an arbitrary point $p \in M$ where we use the atlas $\{\pi_N, \pi_S\}$ for charts on $\mathbb{S}^2$.

(b) Find the matrix of dn_p for an arbitrary point $p \in M$ where we use the inverse of the parametrization $\vec{N}(u, v)$ as a coordinate chart for the image set $n(U)$ in $\mathbb{S}^2$.

3.4.7. Example 3.1.6 shows that if we give $\mathbb{R}^3$ the coordinates (x_0, x_1, x_2), there is a natural surjection $f : \mathbb{R}^3 - \{(0, 0, 0)\} \to \mathbb{RP}^2$ via $\pi(x_0, x_1, x_2) = (x_0 : x_1 : x_2)$. Consider the unit sphere $\mathbb{S}^2$ (centered at the origin), and consider the map $g : \mathbb{S}^2 \to \mathbb{RP}^2$ given as the restriction of f to $\mathbb{S}^2$. Using the oriented atlas on the sphere given in Example 3.7.3 and the coordinate patches for $\mathbb{RP}^2$ as described in Example 3.1.6, give the matrix for dg_p between the north pole patch π_N and U_0. Do the same between the north pole patch and U_1 and explicitly verify Proposition 3.4.6.

3.4.8. Prove Proposition 3.4.6.

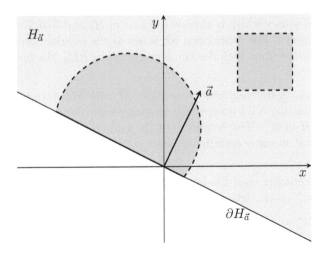

Figure 3.9: Open subsets in a half-space of $\mathbb{R}^2$.

3.4.9. Let M_1 and M_2 be two differentiable manifolds, and consider their product manifold $M_1 \times M_2$. Call $\pi_i : M_1 \times M_2 \to M_i$ for $i = 1, 2$ the projection maps. Show that for all points $p_1 \in M_1$ and $p_2 \in M_2$, the linear transformation

$$S : T_{(p_1, p_2)}(M_1 \times M_2) \longrightarrow T_{p_1} M_1 \oplus T_{p_2} M_2,$$
$$X \longmapsto (\pi_{1*}(X), \pi_{2*}(X))$$

is an isomorphism.

3.5 Manifolds with Boundaries

Despite the flexibility of the definition of a differentiable manifold, it does not allow for a boundary. In many applications, it is useful to have the notion of a manifold with a boundary. This notion relies on the concept of a Euclidean half-space.

Definition 3.5.1. Let $\vec{a}$ be a unit vector in $\mathbb{R}^n$, i.e., $\vec{a} \in \mathbb{S}^n$. The *half-space* $H_{\vec{a}}$ is

$$H_{\vec{a}} = \{\vec{x} \in \mathbb{R}^n \mid \vec{x} \cdot \vec{a} \geq 0\}.$$

The boundary of the half-space is $\partial H_{\vec{a}} = \{\vec{x} \in \mathbb{R}^n \mid \vec{x} \cdot \vec{a} = 0\}$.

Note that for distinct unit vectors $\vec{a}$ and $\vec{b}$, the half-spaces $H_{\vec{a}}$ and $H_{\vec{b}}$ are not equal.

Since the topology on $H_{\vec{a}}$ is the subset topology inherited from $\mathbb{R}^n$, a set is open in $H_{\vec{a}}$ if and only if it is equal to $U \cap H_{\vec{a}}$ for some open set $U \subset \mathbb{R}^n$. Figure 3.9 depicts a Euclidean half-space of of $\mathbb{R}^2$ along with two open subsets. One open set arises as an open square which is already a subset of $H_{\vec{a}}$ and so does not include any point of its boundary. The other open set arises as the intersection of an open disk with $H_{\vec{a}}$. This intersection includes the segment along $\partial H_{\vec{a}}$, the line perpendicular to $\vec{a}$.

Definition 3.5.2. A differentiable n-manifold M *with boundary* has the same definition as in Definition 3.1.3 except that the ranges for the charts are open subsets of a half-space H of $\mathbb{R}^n$. The *boundary* of the manifold, written ∂M, is the set of points p such that in some coordinate chart $\phi : U \to H$, where H is a half-space, $\phi(p) \in \partial H$.

The most commonly used Euclidean half-spaces in $\mathbb{R}^n$ are the *upper half-space* and the *lower half-space*, defined respectively as

$$\mathbb{R}^n_+ = \{(x_1, x_2, \ldots, x_n) \in \mathbb{R}^n \mid x_n \geq 0\} = H_{(0,\ldots,0,1)}$$
$$\mathbb{R}^n_- = \{(x_1, x_2, \ldots, x_n) \in \mathbb{R}^n \mid x_n \leq 0\} = H_{(0,\ldots,0,-1)}.$$

For any unit vector $\vec{a}$, define the projection function $\pi_{\vec{a}} : H_{\vec{a}} \to \partial H_{\vec{a}}$ as

$$\pi_{\vec{a}}(\vec{x}) = \vec{x} - \mathrm{proj}_{\vec{a}} \vec{x} = \vec{x} - (\vec{a} \cdot \vec{x})\vec{a}. \tag{3.18}$$

Since $\partial H_{\vec{a}}$ is a $(n-1)$-dimensional subspace of $\mathbb{R}^n$, assigning coordinates to elements of $\partial H_{\vec{a}}$ gives an isomorphism (and homeomorphism) between $\mathbb{R}^{n-1}$ and $\partial H_{\vec{a}}$.

Proposition 3.5.3. *Let M be a differentiable (respectively, C^k, smooth, analytic) n-manifold with boundary. Its boundary ∂M is a differentiable (respectively, C^k, smooth, analytic) $(n-1)$-manifold without boundary.*

Proof. Let $\mathcal{A} = \{\phi_\alpha\}_{\alpha \in I}$ be an atlas for M. Let I' be the subset of the indexing set I such that the domain of ϕ_α contains points of ∂M. For $\alpha \in I'$ with $\phi_\alpha : U_\alpha \to H$ for some half-space H, we consider the projection $\pi : H \to \partial H$ as a mapping into $\mathbb{R}^{n-1}$.

By definition, the restricted function $\psi_\alpha = (\pi \circ \phi_\alpha)|_{\partial M}$ is a bijection. It is continuous, as the restriction of the composition of two continuous maps. The projection function π is an open function, so maps open sets to open sets. Hence, since ϕ_α^{-1} is continuous and maps open sets to open sets, then so does ψ_α. This shows that ψ_α is a homeomorphism from $U_\alpha \cap \partial M$ to its image.

Furthermore, the domains of ψ_α for $\alpha \in I'$ cover ∂M.

Finally, consider two overlapping charts $\phi_\alpha : U_\alpha \to H_{\vec{a}}$ and $\phi_\beta : U_\beta \to H_{\vec{b}}$ with $\alpha, \beta \in I'$. Consider the transition function

$$\psi_\alpha \circ \psi_\beta^{-1} = \pi_{\vec{a}} \circ (\phi_\alpha \circ \phi_\beta^{-1}) \circ \pi_{\vec{b}}^{-1} : \pi_{\vec{b}} \circ \phi_\alpha(U_\alpha \cap U_\beta) \to \pi_{\vec{a}} \circ \phi_\beta(U_\alpha \cap U_\beta).$$

Then $\pi_{\vec{b}}^{-1} : \pi_{\vec{b}} \circ \phi_\alpha(U_\alpha \cap U_\beta) \to \phi_\alpha(U_\alpha \cap U_\beta)$ is a smooth injection and $\pi_{\vec{a}} : \phi_\alpha(U_\alpha \cap U_\beta) \to \mathbb{R}^{n-1}$ is a smooth projection. Hence, the differentiability class of $\psi_\alpha \circ \psi_\beta^{-1}$ is the same as the differentiability class of $\phi_\alpha \circ \phi_\beta^{-1}$.

This shows that the collection $\mathcal{A}' = \{\psi_\alpha\}_{\alpha \in I'}$ equips ∂M with the same differential (respectively, C^k or smooth) structure as M. $\square$

Example 3.5.4 (Closed Interval). Let $M = [a, b]$ be a closed and bounded real interval. Define $\phi_1 : [a, b) \to \mathbb{R}_+$ by $\phi_1(x) = x - a$ and $\phi_2 : (a, b] \to \mathbb{R}_-$ as $\phi_2(\bar{x}) = \bar{x} - b$. The set $\{\phi_1, \phi_2\}$ equips $[a, b]$ with the structure of a manifold with boundary. Note that the boundary $\partial M = \{a, b\}$ is a discrete manifold, consisting of exactly two points.

Example 3.5.5 (Closed Ball). Example 3.1.5 inspires a relatively easy way to equip the closed unit ball $\mathbb{B}^3 = \{(x, y, z) \in \mathbb{R}^3 \mid x^2 + y^2 + z^2 \leq 1\}$ with the structure of a manifold with boundary. Consider first the function $\phi_1 : \mathbb{B}^3 \cap \{(x, y, z) \in \mathbb{R}^3 \mid z > 0\} \to \mathbb{R}^3$ defined by

$$\phi_1(x, y, z) = (x, y, \sqrt{1 - x^2 - y^2} - z).$$

We can visualize in Figure 3.3 the domain of this function as the portion of the closed ball inside the dome corresponding to $\vec{X}_{(1)}$. Since $z \leq \sqrt{1 - x^2 - y^2}$ in the domain of ϕ_1, then the codomain of ϕ_1 is $\mathbb{R}^3_+$. Furthermore, $\phi_1(x, y, z) \in \partial \mathbb{R}^3_+$ if and only if $\sqrt{1 - x^2 - y^2} - z = 0$, which is precisely the portion of the unit sphere in $\{(x, y, z) \in \mathbb{R}^3 \mid z > 0\}$. It is easy to see that ϕ_1 is continuous and also a homeomorphism with its image.

We can create in a similar manner charts ϕ_2, ϕ_3, ϕ_4, ϕ_5, and ϕ_6 corresponding to $\vec{X}_{(2)}$, $\vec{X}_{(3)}$, $\vec{X}_{(4)}$, $\vec{X}_{(5)}$, and $\vec{X}_{(6)}$.

As constructed, the union of the domains of ϕ_i for $i = 1, 2, \ldots, 6$ is not all of $\mathbb{B}^3$ but only $\mathbb{B}^3 - \{(0, 0, 0)\}$. To remedy this situation, it suffices to enlarge the domains of at least one of the ϕ_i functions to include $(0, 0, 0)$. For example, using as the domain of ϕ_1 as the set

$$U_1 = \mathbb{B}^3 \cap \{(x, y, z) \in \mathbb{R}^3 \mid x^2 + y^2 + (z - 1)^2 < 2\}$$

suffices. The open set U_1 in $\mathbb{B}^3$ includes the same portion of the manifold's boundary as the open set $\mathbb{B}^3 \cap \{(x, y, z) \in \mathbb{R}^3 \mid z > 0\}$, but also includes $(0, 0, 0)$.

We leave it as an exercise for the reader to show that the transition functions between coordinate charts are differentiable. Therefore, the atlas $\{\phi_1, \phi_2, \ldots, \phi_6\}$, equips the closed ball $\mathbb{B}^3$ with the structure of a manifold with boundary. The boundary $\partial \mathbb{B}^3$ is the unit sphere $\mathbb{S}^2$.

Example 3.5.6. As an example of a manifold with boundary, consider the half-torus in $\mathbb{R}^3$ given as the image of $\vec{X} : [0, \pi] \times [0, 2\pi] \to \mathbb{R}^3$, with

$$\vec{X}(u, v) = \big((2 + \cos v) \cos u, (2 + \cos v) \sin u, \sin v\big).$$

The image of $\vec{X}$ is a half-torus M with $y \geq 0$, which, to conform to Definition 3.5.2, is easily covered by four coordinate patches. The boundary ∂M is the manifold consisting of two connected components:

$$R_+ = \{(x, y, z) \mid (x + 2)^2 + z^2 = 1, y = 0\},$$
$$R_- = \{(x, y, z) \mid (x - 2)^2 + z^2 = 1, y = 0\}.$$

We leave it as an exercise for the reader to decide on precise patches that make this half-torus into a manifold with boundary. (See Exercise 3.5.2.)

Since manifolds with boundaries are topological spaces, the concept of a continuous map between them is still the same as Definition A.2.26. Furthermore, the concept of a continuous or differentiable map between manifolds with or without boundaries remains essentially the same the original Definition 3.2.1 but with one clarification.

Let $f : M^m \to N^n$ be a continuous function from a manifold M with boundary to any other manifold. Deciding the limit or the differentiability of f a point $p \in M$ that is not on the boundary ∂M is the same as always. However, suppose that $p \in \partial M$, with (U, x) where x has for codomain a half-space of $\mathbb{R}^n$, a coordinate neighborhood of p, where H is a half-space of $\mathbb{R}^m$, and (V, y) a coordinate neighborhood of $f(p)$. Then the function $y \circ f \circ x^{-1}$ described in Definition 3.2.1 has the domain $x(U \cap f^{-1}(V))$, which is an open subset of a half-space H. Since $p \in \partial M$, then $x(p) \in \partial H$. In order to decide on the differentiability of $y \circ f \circ x^{-1}$ at $x(p)$, we only consider the condition and the limit in Definition 1.2.14 for $\vec{h}$ such that $x(p) + \vec{h} \in H$. This is a restricted limit, which generalizes a one-sided limit from calculus of a single variable.

The concept of a manifold with boundary, allows us to generalize the notion of a curve on a manifold.

Definition 3.5.7. A *differentiable curve* on a manifold M, possibly with boundary, is a differentiable function $\gamma : J \to M$, where J is any interval of the real line and is understood as a one-dimensional manifold, possibly with boundary.

This definition allows for curves with endpoints on a manifold.

Since curves on manifolds served an essential role in defining the tangent space to a manifold at a point, curves with endpoints help us define the tangent space to a manifold with boundary M at any point p, even if $p \in \partial M$. We now restate Definition 3.3.1 to accommodate manifolds with boundary.

Definition 3.5.8. Let M^m be a differentiable manifold (possibly with boundary) and let p be a point on M. Let J be some interval of $\mathbb{R}$ containing 0 and let $\gamma : J \to M$ be a differentiable curve on M with $\gamma(0) = p$. For any real-valued differentiable f defined on some neighborhood of p, we define the directional derivative of f along γ at p to be the number

$$D_\gamma(f) = \frac{d}{dt}\left(f(\gamma(t))\right)\Big|_{t=0}, \tag{3.19}$$

where this derivative is understood as a one-sided derivative, if 0 is an endpoint of the interval J. The operator D_γ is called the *tangent vector* to γ at p.

Though this definition adds nothing new for points $p \in M$ that are not on the boundary ∂M, including curves with endpoints allows us to consider curves whose endpoints are on the boundary.

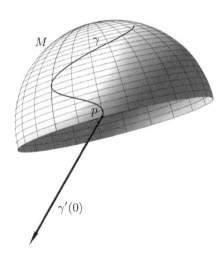

Figure 3.10: Tangent vectors to point on the boundary.

Figure 3.10 depicts a manifold with boundary M, specifically a half sphere, along with a curve $\gamma : [-1, 0] \to M$ such that $\gamma(0) = p$. The figure illustrates what occurs in regular calculus, visualizing a tangent vector as a vector in $\mathbb{R}^3$. However, even in this context, the tangent vector $\gamma'(0)$ must be understood as a one-side derivative, namely,

$$\gamma'(0) = \lim_{h \to 0^-} \frac{1}{h}(\gamma(h) - \gamma(0)).$$

It is not surprising then that in the more abstract setting of manifolds with boundary, (3.19) should involve a one-sided derivative.

Section 3.3 showed that the set of tangent vectors to a manifold M at a point p is a vector space.

It takes a little more work to show that set of tangent vectors to M at p, even when $p \in \partial M$, is a vector space. For example, showing that the set of tangent vectors is closed under scalar multiplication breaks into two cases. Suppose that $\varepsilon > 0$ and that $\gamma : [-\varepsilon, 0] \to M$ is a curve on M with $\gamma(0) = p$. Then if $a \in \mathbb{R}_+$, defining $\gamma_1 : [-\varepsilon/a, 0] \to M$ by $\gamma_1(t) = \gamma(at)$ will give $D_{\gamma_1} = D_\gamma$. However, if $a \in \mathbb{R}_-$, defining $\gamma_2 : [0, \varepsilon/a] \to M$ by $\gamma_2(t) = \gamma(t)$ leads to $D_{\gamma_2} = D_\gamma$. The issue here is that we needed to change the domain of γ_2 when $a < 0$. However, combining cases, we see that the set of tangent vectors is closed under scalar multiplication.

We leave as an exercise for the reader the technical details for the following proposition.

Proposition 3.5.9. *Let M be a manifold with boundary ∂M and suppose that $p \in \partial M$. The set of tangent vectors to M at p forms a vector space.*

As with manifolds without boundary, we call the set of tangent vectors to M at p the *tangent space* to M at p and denote it by $T_p M$.

With the notions developed in this section, Definition 3.4.1 for the differential of a map between manifolds does not need to change for the generalized context of manifolds with boundary.

PROBLEMS

3.5.1. Explicitly show that the solid ball $B^n = \{(x_1, \ldots, x_n) \in \mathbb{R}^n \mid x_1^2 + \cdots + x_n^2 \leq 1\}$ is a smooth n-manifold with boundary and show that its boundary is the sphere $\mathbb{S}^{n-1}$.

3.5.2. Referring to the parametrization $\vec{X}(u, v)$ in Example 3.5.6, give four coordinate patches that equip the half-torus with the structure of a manifold with boundary.

3.5.3. Prove that all the transition functions in Example 3.5.5 are differentiable.

3.5.4. This exercise shows how to equip the closed ball $B^3 = \{(x, y, z) \in \mathbb{R}^3 \mid z^2 + y^2 + z^2 \leq 1\}$ with the structure of a manifold with boundary in a manner inspired by stereographic projection. Let $N = (0, 0, 1)$ and define the function $\Pi_N : B^3 - \{N\} \to \mathbb{R}^3_+$ as follows. For all $A \in B^3 - \{N\}$, let B be the intersection of the line $\overleftrightarrow{AN}$ with the unit sphere $\mathbb{S}^3$. The $\Pi_N(A) = (\pi_N(B), \lambda)$ where π_N is the stereographic projection as in Example 3.1.4 and where as vectors $\overrightarrow{BA} = \lambda \overrightarrow{BN}$.

(a) Calculate $\Pi_N(x, y, z)$ explicitly.

(b) Show that the third component of $\Pi_N(x, y, z)$ is equal to 0 if and only if $(x, y, z) \in \mathbb{S}^2 - \{N\}$.

(c) Show that Π_N is a homeomorphism between $B^3 - \{N\}$ and $\{(x, y, z) \in \mathbb{R}^3 \mid 0 \leq z < 1\}$.

(d) Let $S = (0, 0, -1)$ and define Π_S in a similar fashion as Π_N. If we call $(\bar{u}, \bar{v}, \bar{w}) = \Pi_S \circ \Pi_N^{-1}(u, v, w)$, show that

$$(\bar{u}, \bar{v}, \bar{w}) = \left(\frac{u(1-w)}{u^2 + v^2 + w}, \frac{v(1-w)}{u^2 + v^2 + w}, \frac{w(1-w)}{u^2 + v^2 + w} \right),$$

and deduce that $\Pi_S \circ \Pi_N^{-1}$ is differentiable on its domain.

[These steps show that the atlas $\{\Pi_N, \Pi_S\}$ equips B^3 with the structure of a differentiable manifold with boundary.]

3.5.5. Show that if M is a compact manifold, then so is ∂M. [Hint: See Definition A.2.51.]

3.5.6. Modify the approach in Section 3.3 to prove Proposition 3.5.9.

3.6   Immersions, Submersions, and Submanifolds

The linear transformation F_*, which is implicitly local to p, and the associated matrix $[dF_p]$ allow us to discuss the relation of one manifold to another. A number of different situations occur frequently enough to warrant their own terminologies.

Definition 3.6.1. Let $F : M \to N$ be a differentiable map between differentiable manifolds.

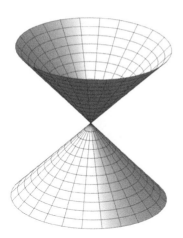 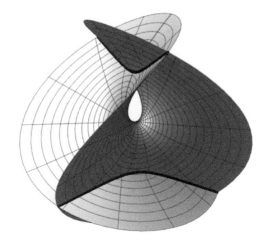

Figure 3.11: Double cone.　　　　Figure 3.12: Enneper's surface.

1. If F_* is an injection at all points $p \in M$, then F is called an *immersion*.

2. If F_* is a surjection at all points $p \in M$, then F is called a *submersion*.

3. If F is an immersion and one-to-one, then the pair (M, F) is called a *submanifold* of N.

4. If (M, F) is a submanifold and $F : M \to F(M)$ is a homeomorphism for the topology on $F(M)$ induced from N, then F is called an *embedding* and $F(M)$ is called an *embedded submanifold*.

It is important to give examples of the above four situations. Clearly, every embedded submanifold is a submanifold and every submanifold is an immersion. In fact, in the theory of differentiable manifolds, it is only in the context of Definition 3.6.1 that we can discuss how a manifold "sits" in an ambient Euclidean space by considering a differentiable function $f : M \to \mathbb{R}^n$, where $\mathbb{R}^n$ is viewed as a manifold with its usual differential structure.

These three categories represent different situations that we addressed when studying regular surfaces in $\mathbb{R}^3$. The cylinder $\mathbb{S}^1 \times \mathbb{R}$ is a differentiable manifold. We can consider the double cone in Figure 3.11 as the image of a map $f : \mathbb{S}^1 \times \mathbb{R} \to \mathbb{R}^3$ given by

$$f(u, v) = (v \cos u, v \sin u, v),$$

where we use u as an angle. We note that

$$[df] = \begin{pmatrix} -v \sin u & \cos u \\ v \cos u & \sin u \\ 0 & 1 \end{pmatrix}.$$

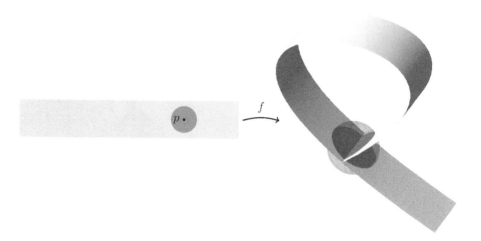

Figure 3.13: Not a homeomorphism.

Clearly, at all points $(u, 0) \in \mathbb{S}^1 \times \mathbb{R}$, the differential df_p is not injective. Thus, the cone is not an immersion in $\mathbb{R}^3$.

Enneper's surface (see Figure 3.12) is the locus of the parametrization

$$\vec{X}(u, v) = \left(u - \frac{u^3}{3} + uv^2, v - \frac{v^3}{3} + vu^2, u^2 - v^2 \right) \quad \text{for } (u, v) \in \mathbb{R}^2. \qquad (3.20)$$

Enneper's surface can be considered a differentiable map of $\vec{X} : \mathbb{R}^2 \to \mathbb{R}^3$. It is not hard to check (Exercise 3.6.1) that according to Definition 3.6.1, Enneper's surface is an immersion, but because $\vec{X}$ is not one-to-one, the surface is not a submanifold. (In Figure 3.12, the locus of self-intersection of the parametrized surface is indicated in thick black or thick white.)

To illustrate the idea of a submanifold that is not an embedded submanifold, consider the ribbon surface in Figure 3.13. We can consider this surface to be a function between manifolds in the following sense. Consider the two-dimensional manifold without boundary $M = (0, 5) \times (0, 1)$ with the natural product topology and differential structure inherited from $\mathbb{R}^2$. Then the ribbon surface can be viewed as the image of a differentiable map $f : (0, 5) \times (0, 1) \to \mathbb{R}^3$ between manifolds. One of the (open) ends of the ribbon comes arbitrarily close to the surface of the ribbon ("touching" but not intersecting). The pair (M, f) is a submanifold but not an embedded submanifold because, as Figure 3.13 shows, open sets on M might not be open sets on $f(M)$ with the topology induced from $\mathbb{R}^3$. Note that no open set V of $\mathbb{R}^3$ around $f(p)$ can intersect $f(M)$ in to obtain the set $f(U)$, where U is the open neighborhood around p that is depicted by a darker gray circle.

Some authors (usually out of sympathy for their readers) introduce the theory of "manifolds in $\mathbb{R}^n$." By this we mean manifolds that are embedded submanifolds of

$\mathbb{R}^n$. Though not as general as Definition 3.6.1, that approach has some merit as it more closely mirrors Definition 3.1.1 for regular surfaces in $\mathbb{R}^3$. However, our current approach is more general. Admittedly, it might seem strange to call a differentiable map a submanifold, but, as the above examples show, this tactic generalizes the various situations of interest for subsets of $\mathbb{R}^3$. Furthermore, this approach again removes the dependence on an ambient Euclidean space. Consequently, it is not at all strange to discuss submanifolds of $\mathbb{RP}^n$ or any other space of interest.

We now wish to discuss specifically embedded submanifolds of a differentiable manifold since they occupy an important role in subsequent sections and allow us to quickly determine certain classes of manifolds.

Proposition 3.6.2. *Let M^m be a differentiable manifold. An open subset S of M is an embedded submanifold of dimension m.*

Proof. Let $\{\phi_i : U_i \to \mathbb{R}^m\}_{i \in I}$ be the atlas of M. Equip S with the atlas $\{\phi_i|_S\}_{i \in I}$. The inclusion map $\iota : S \to M$ is a one-to-one immersion. The topology of S is induced from M, so S, with the given atlas, is an embedded submanifold of M. $\square$

Example 3.6.3. Consider the set $M_{n \times n}$ of $n \times n$ matrices with real coefficients. We can equip $M_{n \times n}$ with a Euclidean topology by identifying $M_{n \times n}$ with $\mathbb{R}^{n^2}$. In particular, $M_{n \times n}$ is a differentiable manifold. Consider the subset $\mathrm{GL}_n(\mathbb{R})$ of invertible matrices in $M_{n \times n}$. We claim that, with the topology induced from $M_{n \times n}$, $\mathrm{GL}_n(\mathbb{R})$ is an embedded submanifold. We can see this by the fact that an $n \times n$ matrix A is invertible if and only if $\det A \neq 0$. However, the function $\det : M_{n \times n} \to \mathbb{R}$ is continuous, and therefore,

$$\mathrm{GL}_n(\mathbb{R}) = \det{}^{-1}(\mathbb{R} - \{0\})$$

is an open subset of M_n. Proposition 3.6.2 proves the claim.

The proof of Proposition 3.6.2 is deceptively simple. If $S \subset M$, though the inclusion map $\iota : S \to M$ is obviously one-to-one, one cannot use it to show that any subset is an embedded submanifold. Consider the subset S of $\mathbb{R}^2$ defined by the equation $y^2 - x^3 = 0$ (see Figure 3.14). The issue is that in order to view S as a manifold, we must equip it with an atlas. In this case, the atlas of $\mathbb{R}^2$ consists of one coordinate chart, the identity map. The restriction of the identity map $\mathrm{id}|_S$ is not a homeomorphism into an open subset of $\mathbb{R}$ or $\mathbb{R}^2$ so cannot serve as a chart. In fact, if we put any atlas $\{\phi_i\}$ of coordinate charts on S, the inclusion $\iota : S \to \mathbb{R}^2$ is such that $\iota \circ \phi_i$ will be some regular reparametrization of $t \mapsto (t^2, t^3)$, i.e.,

$$\iota \circ \phi_i(t) = (g(t)^2, g(t)^3),$$

where $g'(t) \neq 0$. Hence,

$$[d\iota_t] = \begin{pmatrix} 2g(t)g'(t) \\ 3g(t)^2 g'(t) \end{pmatrix},$$

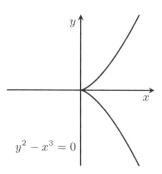

Figure 3.14: Not an embedded submanifold of $\mathbb{R}^2$.

and thus $d\iota_t$ fails to be an immersion at the point where $g(t) = 0$, which corresponds to $(0,0) \in S$, the cusp of the curve.

Having a clear definition of the differential of a function between manifolds, we can now imitate Definition 1.4.1 to give a definition for regular points and for critical points of functions between manifolds.

Definition 3.6.4. Let $f : M^m \to N^n$ be a differentiable map between differentiable manifolds. Then any point $p \in M$ is called a *critical point* if $\mathrm{rank}(f_*) < \min(m,n)$, i.e., df_p is not of maximal rank. If p is a critical point, then the image $f(p)$ is called a *critical value*. Furthermore, any element $q \in N$ that is not a critical value is called a *regular value* (even if $q \notin f(M)$).

We remind the reader that this definition for critical point directly generalizes all the previous definitions for critical points of functions (see the discussion following Definition 1.4.1). The only novelty here from the discussion in Chapter 1 was to adapt the definition for functions from $\mathbb{R}^m$ to $\mathbb{R}^n$ to functions between manifolds.

Our main point in introducing the above definition is to introduce the Regular Value Theorem. A direct generalization to a similar theorem for regular surfaces (see Proposition 5.2.13 in [5]), the Regular Value Theorem provides a class of examples of manifolds for which it would otherwise take a considerable amount of work to verify that these sets are indeed manifolds. However, we need a few supporting theorems first.

Theorem 3.6.5. *Let $f : M^m \to N^n$ be a differentiable function between differentiable manifolds, and assume that df_p is injective. Then there exist charts ϕ around p and ψ around $f(p)$ that are compatible with the respective differential structures on M and N and such that $\bar{f} = \psi \circ f \circ \phi^{-1}$ corresponds to the standard inclusion*

$$(x_1, \ldots, x_m) \longmapsto (x_1, \ldots, x_m, 0, \ldots, 0)$$

of $\mathbb{R}^m$ into $\mathbb{R}^n$.

Proof. Let ϕ and ψ be charts on M and N, respectively, for neighborhoods of p and $f(p)$. If necessary, translate ϕ and ψ so that $\phi(p)$ and $\psi(f(p))$ correspond to the origin $\vec{0}$. Then, with respect to these charts, as matrices $[df]_p = [d\bar{f}]_{\vec{0}}$, so $d\bar{f}_{\vec{0}}$ is injective by assumption. The image of the linear transformation $d\bar{f}_{\vec{0}} : \mathbb{R}^m \to \mathbb{R}^n$ is a subspace of $\mathbb{R}^n$. Thus, by a rotation in $\mathbb{R}^n$ (applied to the chart ψ), we can assume that $\operatorname{Im}(d\bar{f}_{\vec{0}})$ is $\{(x_1, \ldots, x_m, 0, \ldots, 0) \in \mathbb{R}^n\}$.

We wish to change coordinates on $\mathbb{R}^n$ via some diffeomorphism $h : \mathbb{R}^n \to \mathbb{R}^n$ that would make $\bar{f}$ the standard inclusion. We view $\mathbb{R}^n$ as $\mathbb{R}^m \times \mathbb{R}^{n-m}$ and define the function $h : \mathbb{R}^n \to \mathbb{R}^n$ by

$$h(x, y) = (\bar{f}(x)) + (0, \ldots, 0, y) = (\pi \circ \bar{f}(x), y), \tag{3.21}$$

where π is the orthogonal projection of $\mathbb{R}^n$ onto the subspace of its first m-dimensional components. Note that the differential of h at $\vec{0}$ is $dh_{\vec{0}} = d\bar{f}_{\vec{0}} \oplus \operatorname{id}$ or, as matrices,

$$[dh_{\vec{0}}] = \begin{pmatrix} [d\bar{f}_{\vec{0}}] & 0 \\ 0 & I_{n-m} \end{pmatrix}.$$

Since $d\bar{f}_{\vec{0}}$ is injective, we see that $dh_{\vec{0}}$ is invertible.

Now, by the Inverse Function Theorem (Theorem 1.4.5), there exists some open neighborhood V of $\vec{0}$, such that h is injective on V, $h(V)$ is open, and the inverse function h^{-1} exists and is differentiable. Thus, h is a diffeomorphism between open neighborhoods of the origin in $\mathbb{R}^n$. We reparametrize the neighborhood of $f(p)$ with the chart $h^{-1} \circ \psi$. By Equation (3.21), replacing ψ with $\psi' = h^{-1} \circ \psi$ leads to an atlas that is compatible with the atlas on N. Furthermore, by construction, the new $\bar{f}$ satisfies

$$\bar{f}'(x) = \psi' \circ f \circ \phi^{-1}(x) = h^{-1} \circ \bar{f}(x) = (x, 0, \ldots, 0)$$

as desired. $\qquad\square$

The functional relationship discussed in the above proof is often depicted using the following diagram:

$$
\begin{array}{ccc}
M^m & \xrightarrow{\ f\ } & N^n \\
\phi \downarrow & & \downarrow \psi \\
\mathbb{R}^m & \xrightarrow{\ \bar{f}\ } & \mathbb{R}^n
\end{array}
$$

We say that the diagram is *commutative* if, when one takes different directed paths from one node to another, the different compositions of the corresponding functions are equal. In this simple case, to say that the above diagram is commutative means that

$$\psi \circ f = \bar{f} \circ \phi.$$

These kinds of diagrams are often used in algebra and in geometry as a schematic to represent the kind of relationship illustrated by Figure 3.7.

Theorem 3.6.5 offers a strategy to prove a number of theorems about embedded submanifolds. We mention a few of these here and leave some others for the reader to prove.

Corollary 3.6.6. *Let M^m be a differentiable manifold, and let $S \subset M$. The subset S is an embedded submanifold of M of dimension k if and only if, for all $p \in S$, there is a coordinate neighborhood (U, ϕ) of p compatible with the atlas on M such that*

$$\phi(U \cap S) = \{(x^1, \ldots, x^k, x^{k+1}, \ldots, x^m) \in \phi(U) \,|\, x^{k+1} = \cdots = x^m = 0\}.$$

Proof. The implication ($\Rightarrow$) follows immediately from Theorem 3.6.5.

For the converse ($\Leftarrow$), assume that for all $p \in S$, there is a coordinate neighborhood (U, ϕ) in M compatible with the atlas of M satisfying the condition for $U \cap S$. We cover S with a collection of such open sets $\{U_\alpha \cap S\}_{\alpha \in I}$. Let $\pi : \mathbb{R}^m \to \mathbb{R}^k$ be the projection that ignores the last $m - k$ variables. Then on each coordinate neighborhood, $\psi_\alpha = \pi \circ \phi_\alpha : U_\alpha \cap S \to \mathbb{R}^k$ is a coordinate chart for S. Since $\phi_\alpha : U_\alpha \to \phi_\alpha(U_\alpha)$ is a homeomorphism and since

$$\phi_\alpha(U_\alpha \cap S) \subset \{(x^1, \ldots, x^k, x^{k+1}, \ldots, x^m) \in \mathbb{R}^m \,|\, x^{k+1} = \cdots = x^m = 0\},$$

then $\pi \circ \phi_\alpha$ is a homeomorphism onto its image. By definition, $\phi_\beta \circ \phi_\alpha^{-1}$ is differentiable for any pair of indices α and β. However, $\psi_\beta \circ \psi_\alpha^{-1} = \pi \circ (\phi_\beta \circ \phi_\alpha^{-1})|_{\mathbb{R}^k}$ and so is differentiable as well. Consequently, $\{(U_\alpha \cap S, \psi_\alpha)\}_{\alpha \in I}$ forms an atlas on S that gives S the structure of a differentiable manifold. Furthermore, the inclusion map satisfies all the requirements of an embedded submanifold. $\square$

The following theorem is similar to Theorem 3.6.5 but applies to local submersions.

Theorem 3.6.7. *Let $f : M^m \to N^n$ be a differentiable map such that $f_* : T_p M \to T_{f(p)} N$ is onto. Then there are charts ϕ at p and ψ at $f(p)$ compatible with the differentiable structures on M and N such that*

$$\psi \circ f = \pi \circ \phi,$$

where π is the standard projection of $\mathbb{R}^m$ onto $\mathbb{R}^n$ by ignoring the last $m - n$ variables.

Proof. (The proof mimics the proof of Theorem 3.6.5 and is left to the reader.) $\square$

Theorem 3.6.8 (Regular Value Theorem). *Suppose that $m \geq n$, let $f : M^m \to N^n$ be a differentiable map, and let q be a regular value of f. Then $f^{-1}(q)$ is an embedded submanifold of M of dimension $m - n$.*

Proof. Since $m \geq n$ and q is a regular value, then for all $p \in f^{-1}(q)$, df_p has rank n.

We first prove that the set of points $p \in M$, where $\operatorname{rank} df_p = n$ is an open subset of M. Let $\{(U_\alpha, \phi_\alpha)\}$ be an atlas for M. For all α, define $g_\alpha : U_\alpha \to \mathbb{R}$, with $g_\alpha(p)$ as the sum of squares of all $n \times n$ minors of $[df_p]$. Note that there are $\binom{m}{n}$ minors in $[df_p]$ and that, for all α, each function g_α is well defined on the coordinate patch U_α. The functions g_α need not induce a well defined function $g : M \to \mathbb{R}$. (We would need $g_\alpha|_{U_\alpha \cap U_\beta} = g_\beta|_{U_\alpha \cap U_\beta}$ for all pairs (α, β).) The equation $g_\alpha(p) = 0$ holds if and only if all the maximal minors of $[df_p]$ are 0, which is equivalent to $\operatorname{rank} df_p$. However, $\operatorname{rank} df_p$ is independent of any coordinate system, so regardless of choices made in the construction of of g_α, we have $g_\alpha^{-1}(0) \cap U_\beta = g_\beta^{-1}(0) \cap U_\alpha$. Define $V_\alpha = g_\alpha^{-1}(\mathbb{R} - \{0\})$. Since each g_α is continuous, V_α is open and the set

$$V = \bigcup_{\alpha \in I} V_\alpha$$

is an open subset of M. By construction, V is precisely the set of points in which $\operatorname{rank} df_p = n$. Since V is open in M, by Proposition 3.6.2, V is an embedded submanifold of dimension m.

We consider now the differentiable map $f|_V : V \to N$. Let $p \in f^{-1}(q)$. By construction of V, the differential df_p is surjective for all $p \in V$. Applying Theorem 3.6.7, we can assume that there is a coordinate chart (U, ϕ) of p with coordinates $(x^1, \ldots, x^m)$ and a chart ψ of q such that $\psi \circ f = \pi \circ \phi$, where π is the projection of $\mathbb{R}^m$ onto $\mathbb{R}^n$ by ignoring the last $m - n$ coordinates. Furthermore, by taking a translation if necessary, we can assume that $\psi(q) = (0, 0, \ldots, 0)$. Consequently, $\psi(q) = \pi \circ \phi(p)$ and, so in coordinates,

$$f^{-1}(q) \cap U = \{(x^1, \ldots, x^n, x^{n+1}, \ldots, x^m) \mid x^1 = \cdots = x^n = 0\}.$$

By Corollary 3.6.6, $f^{-1}(q)$ is an embedded submanifold of M. $\qquad\square$

The Regular Value Theorem is also called the Regular Level Set Theorem because any subset of the form $f^{-1}(q)$, where $q \in N$, is called a *level set* of f.

Example 3.6.9 (Spheres). With the Regular Value Theorem at our disposal, it is now easy to show that certain objects are differentiable manifolds. We consider the sphere $\mathbb{S}^n$ as the subset of $\mathbb{R}^{n+1}$, with

$$(x^1)^2 + (x^2)^2 + \cdots + (x^{n+1})^2 = 1.$$

The Euclidean spaces $\mathbb{R}^{n+1}$ and $\mathbb{R}$ are differentiable manifolds with trivial coordinate charts. Consider the differentiable map $f : \mathbb{R}^{n+1} \to \mathbb{R}$ defined by

$$f(x) = \|x\|^2.$$

In the standard coordinates, the differential of f is

$$[df_x] = \begin{pmatrix} 2x^1 \\ 2x^2 \\ \vdots \\ 2x^{n+1} \end{pmatrix}.$$

We note that the only critical point of f is $(0, \ldots, 0)$ and that the only critical value is 0. Thus, $\mathbb{S}^n = f^{-1}(1)$ is an embedded submanifold of $\mathbb{R}^{n+1}$, and hence, $\mathbb{S}^n$ is a differentiable manifold in its own right when equipped with the subspace topology of $\mathbb{R}^{n+1}$. Notice that this establishes $\mathbb{S}^2$ as an embedded submanifold of $\mathbb{R}^3$ without reference to any charts.

PROBLEMS

3.6.1. Prove that $[d\vec{X}]_{(u,v)}$ for the parametrization of Enneper's surface in (3.20) is injective for all $(u, v) \in \mathbb{R}^2$. [This confirms that the parametrization of Enneper's surface is an immersion of $\mathbb{R}^2$ in $\mathbb{R}^3$.]

3.6.2. Let M be a differentiable manifold, and suppose that $f : M \to \mathbb{R}$ is a differentiable map. Prove that if $f_* = 0$ at all points of M, then f is constant on each connected component of M.

3.6.3. Let $M_{m \times n}$ be the set of $m \times n$ matrices. Show that the subset of $M + m \times n$ of matrices of rank less than or equal to r is an embedded submanifold.

3.6.4. Show that the square as described in Example 3.1.13 is not an embedded submanifold of $\mathbb{R}^2$.

3.6.5. Show that the function $f : \mathbb{R}^2 \to \mathbb{R}^3$ defined by $f(u, v) = (u^3 + u + v, u^2 + uv, v^3)$ is smooth and injective but is not an immersion.

3.6.6. Let N be an embedded submanifold of a differentiable manifold M. Prove that at all points $p \in N$, the space $T_p N$ is a subspace of $T_p M$.

3.6.7. Let M^m be a differentiable manifold that is embedded in $\mathbb{R}^n$. By Exercise 3.6.6, $T_p M$ is a subspace of $T_p(\mathbb{R}^n) \cong \mathbb{R}^n$.) Let $f : \mathbb{R}^n \to \mathbb{R}$ be a differentiable function defined in a neighborhood of $p \in M$. Show that if f is constant on M, then $f_*(v) = 0$ for all $v \in T_p M$. Conclude that, viewed as a vector in $T_p(\mathbb{R}^n)$, the differential df_p is perpendicular to $T_p M$.

3.6.8. Let M be a differentiable manifold, and let U be an open set in M. Define $\iota : U \to M$ as the inclusion map. Prove that for any $p \in U$, the differential $\iota_* : T_p U \to T_p M$ is an isomorphism.

3.6.9. Let M and N be k-manifolds in $\mathbb{R}^n$, in the sense that they are both embedded submanifolds. Show that the set $M \cup N$ is not necessarily an embedded submanifold of $\mathbb{R}^n$. Give sufficient conditions for $M \cup N$ to be a manifold.

3.6.10. Suppose that the defining rectangle of the Klein bottle, as illustrated in Figure 3.6 or 3.16, is $[0, 2\pi] \times [0, 2\pi]$. It is a well-known fact that it is impossible to embed the Klein bottle in $\mathbb{R}^3$. Show that the parametrization

$$X(u, v) = \left((2 + \cos v) \cos u, (2 + \cos v) \sin u, \sin v \cos(u/2), \sin v \sin(v/2) \right)$$

gives an embedding of the Klein bottle in $\mathbb{R}^4$. (Remark: This parametrization is similar to the standard parametrization of the torus in $\mathbb{R}^3$ as the union of circles traced in the normal planes of a planar circle of larger radius. A planar circle in $\mathbb{R}^4$ admits a normal three-space. The parametrization X is the locus of a circle in the normal three-space that rotates in the fourth coordinate dimension by half a twist as one travels around the circle of larger radius.)

3.6.11. Define $O(n)$ as the set of all orthogonal $n \times n$ matrices.

   (a) Prove that $O(n)$ is a smooth manifold of dimension $\frac{1}{2}n(n-1)$.

   (b) Consider the tangent space to $O(n)$ at the identity matrix, $T_I(O(n))$, as a subspace of the tangent space to $M_{n \times n}$ (which is $M_{n \times n}$ itself). Prove that $A \in M_{n \times n}$ is a tangent vector in $T_I(O(n))$ if and only if A is skew-symmetric, i.e., $A^T = -A$.

3.6.12. Prove Theorem 3.6.7.

3.6.13. Let M^m and N^n be embedded submanifolds of a differentiable manifold S^s, and suppose that $m + n > s$. Let $p \in M \cap N$. We say that M and N *intersect transversally* at p in S if $T_p M + T_p N = T_p S$, viewed as subspaces of $T_p S$ by virtue of Problem 3.6.6. In this exercise, you will show that if M and N intersect transversally at each point of $M \cap N$, then $M \cap N$ is a differentiable manifold.

   (a) Let $p \in M \cap N$. Prove that there is a coordinate chart (U, ϕ) of p and a function $f_1 : U \to \mathbb{R}^{s-m}$ such that $U \cap M = f_1^{-1}(\vec{0})$. [This also shows that there is a coordinate chart (V, ψ) of p and a function $f_2 : V \to \mathbb{R}^{s-n}$ such that $V \cap N = f_2^{-1}(\vec{0})$.]

   (b) Consider the function $F : U \cap V \to \mathbb{R}^{s-m} \times \mathbb{R}^{s-n}$ defined by $F(x) = (f_1(x), f_2(x))$. Prove that $(\vec{0}, \vec{0})$ is a regular value.

   (c) Deduce that $M \cap N$ is an embedded submanifold of S.

3.7   Orientability

In the final section in this chapter we introduce the notion of orientability. We usually first encounter this notion with the example of the Möbius strip M. See Figure 3.15. Consider the Möbius strip as a 2-manifold with boundary embedded in $\mathbb{R}^3$, and consider trying to assign a unit normal vector to every point on the Möbius strip in a continuous fashion. Such an assignment corresponds to a continuous function $n : M \to \mathbb{S}^2$. It is not hard to see that this is impossible: Once a unit normal vector goes around the strip once, it will be pointing in the other direction.

Since manifolds do not necessarily exist in some ambient Euclidean space, we cannot talk about unit normal vectors to generalize the notion of orientability. The Klein bottle gives us another way of visualizing non-orientability. Consider a curve on the Klein bottle as shown in Figure 3.16. We point out that the depicted curve is differentiable: In this diagram for the Klein bottle, at the point p, the "vertical" orientation changes as the curve passes through the vertical boundary of

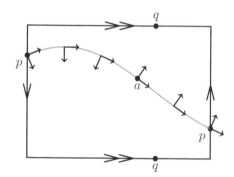

Figure 3.15: Möbius strip.

Figure 3.16: The Klein bottle.

the diagram. Attached to the curve, we show a unit tangent vector to the curve and a unit normal as well, which form a variable frame. As we move along the curve from left to right starting from the point a, the unit tangent and the unit normal change continuously. As we cycle around the diagram of the Klein bottle, passing through the vertical boundary on the right, vectors (and directions) are reflected vertically. returning to the point a, we do not recover the frame we started with, but one with opposite orientation. We point out the we did not have to use a this particular frame involving a unit tangent and a unit normal to the curve, or even an orthonormal one; this observation would remain the same with any frame.

Suppose that $\gamma(t)$ continuously parametrizes the curve on the Klein bottle. If we call $\vec{T}(t)$ and $\vec{U}(t)$ the tangent and normal vectors in this diagram, we see that $\det(\vec{T}(t) \ \ \vec{U}(t))$, which measures the sign of the angle swept from $\vec{T}(t)$ to $\vec{U}(t)$ is not a continuous function; it must change sign in a discontinuous fashion. This observation gives us another way to think about orientability without reference to an ambient space. However, this still is not quite enough to motivate a definition. Notice that if we had done the same construction with a vertical line connecting the two instances of q in the diagram, the same function $\det(\vec{T}(t) \ \ \vec{U}(t))$ would be continuous along the curve.

The key to the notion of orientability that we do consider from the example of the Klein bottle is that of variable frames on the manifold whose determinant does not change sign.

Definition 3.7.1. Let M^n be a differentiable n-manifold equipped with an atlas $\mathcal{A} = \{\phi_\alpha\}_{\alpha \in I}$. Suppose that for any two charts ϕ_α and ϕ_β of the atlas $\mathcal{A}$, the Jacobian of the transition function $\phi_{\beta\alpha} = \phi_\beta \circ \phi_\alpha^{-1}$ is positive at all points in its domain. Then $(M, \mathcal{A})$ is called an *oriented manifold*.

As we saw in Proposition 3.3.9, at any point $p \in M$ the matrix of the differential $[d\phi_{\beta\alpha}]$ coordinate change matrix on T_pM between the basis derived from the ϕ_α coordinate charts and the basis derived from the ϕ_β chart. As a coordinate change matrix, $[d\phi_{\beta\alpha}]$ is invertible and hence, its determinant is never 0.

At present, in our development of the theory of manifolds, we do not have a way to connect a tangent space at one point to the tangent space at some other nearby point of the manifold. Consequently, we cannot (currently) think of the the the bases on T_pM derived from the ϕ_α coordinates as a variable frame, since each basis is in a different vector space. Nonetheless, the function $\det(d\phi_{\beta\alpha})$ is defined over $\phi_\alpha(U_\alpha \cap U_\beta)$.

Definition 3.7.2. Let M^n be a differentiable n-manifold equipped with an atlas $\mathcal{A}$. Then M with $\mathcal{A}$ is called *orientable* if there is an atlas $\mathcal{B}$ on M that is compatible with $\mathcal{A}$ such that M equipped with $\mathcal{B}$ is an oriented manifold.

Example 3.7.3. Consider Example 3.1.4 of the sphere, with the atlas $\mathcal{A} = \{\pi_N, \pi_S\}$ of stereographic projection from the North and South poles. From the change of coordinates $(\bar{u}, \bar{v}) = \pi_S \circ \pi_N^{-1}(u, v)$ in (3.1), we find that

$$\det(d(\pi_S \circ \pi_N^{-1})) = \frac{\partial(\bar{u}, \bar{v})}{\partial(u, v)} = \begin{vmatrix} \dfrac{-u^2 + v^2}{(u^2 + v^2)^2} & -\dfrac{2uv}{(u^2 + v^2)^2} \\[2mm] -\dfrac{2uv}{(u^2 + v^2)^2} & \dfrac{u^2 - v^2}{(u^2 + v^2)^2} \end{vmatrix}$$

$$= \frac{(-u^2 + v^2)(u^2 - v^2) - 4u^2v^2}{(u^2 + v^2)^4} = -\frac{1}{(u^2 + v^2)^2}.$$

Consequently, the atlas $\{\pi_N, \pi_S\}$ on $\mathbb{S}^2$ does not equip $\mathbb{S}^2$ with the structure of an oriented manifold.

Consider instead the atlas $\mathcal{B} = \{\pi_N, \bar{\pi}_S\}$, where $\bar{\pi}_S$ is the composition of π_S with the reflection $(u, v) \mapsto (u, -v)$ so that $\bar{\pi}_S(x, y, z) = \left(\frac{x}{1+z}, -\frac{y}{1+z}\right)$. It is easy to tell that $\{\pi_N, \bar{\pi}_S\}$ is an atlas for $\mathbb{S}^2$ that it is compatible with $\{\pi_N, \pi_S\}$. Thus this atlas gives $\mathbb{S}^2$ the same differentiable structure as the original atlas. Furthermore, writing $(\tilde{u}, \tilde{v}) = \bar{\pi}_S \circ \pi_N^{-1}(u, v)$, we get

$$\tilde{u} = \frac{u}{u^2 + v^2} \qquad \text{and} \qquad \tilde{v} = -\frac{v}{u^2 + v^2}.$$

We easily find that the Jacobian is

$$\frac{\partial(\tilde{u}, \tilde{v})}{\partial(u, v)} = \begin{vmatrix} \dfrac{u^2 - v^2}{(u^2 + v^2)^2} & -\dfrac{2uv}{(u^2 + v^2)^2} \\[2mm] \dfrac{2uv}{(u^2 + v^2)^2} & \dfrac{u^2 - v^2}{(u^2 + v^2)^2} \end{vmatrix} = \frac{1}{(u^2 + v^2)^2}. \tag{3.22}$$

This shows that the atlas $\{\pi_N, \bar{\pi}_S\}$ gives the sphere the structure of an oriented smooth manifold. So though $(\mathbb{S}^2, \mathcal{A})$ is not an oriented manifold, it is orientable, and $(\mathbb{S}^2, \mathcal{B})$ is an oriented manifold.

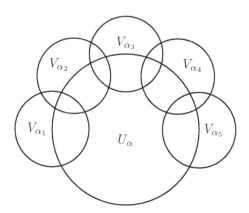

Figure 3.17: Sequence in the proof of Proposition 3.7.5.

In the study of surfaces in $\mathbb{R}^3$, an orientation of an orientable surface meant choosing one of the two possible directions for a unit normal vector function that is continuous over the whole surfaces. With manifolds, this notion of choice is more subtle but still exists.

Lemma 3.7.4. *Let M be a connected differentiable manifold with atlas $\mathcal{A} = \{(U_\alpha, \phi_\alpha)\}_{\alpha \in I}$. For any pair $\alpha, \alpha' \in I$, there exists a finite sequence $\alpha = \alpha_1, \alpha_2, \ldots, \alpha_n = \alpha'$ of indices in I such that $U_{\alpha_i} \cap U_{\alpha_{i+1}} \neq \emptyset$, for $i = 1, 2, \ldots, n-1$.*

Proof. If $U_\alpha \cap U_{\alpha'} \neq \emptyset$, then we are done. However, this need not be true. Let C_α be the set of indices $\alpha'' \in I$ such that there exists a finite sequence $\alpha = \alpha_1, \alpha_2, \ldots, \alpha_n = \alpha''$ such that $U_{\alpha_i} \cap U_{\alpha_{i+1}} \neq \emptyset$, for $i = 1, 2, \ldots, n-1$. Then

$$U = \bigcup_{a \in C} U_a \qquad \text{and} \qquad V = \bigcup_{b \in I - C} U_b$$

are both open as union of open sets and $U \cup V = M$ by definition of an atlas. We claim that $U \cap V = \emptyset$. If not, if $x \in U \cap V$, then there exists $x \in U_a \cap U_b$ for some $a \in C$ and $b \in I - C$. But this is a contradiction because in this case whatever finite sequence of indices that gave a chain of nonempty intersections from U_α to U_a can be extended by one more so that $b \in C$ and not in $I - C$. Since M is connected (see Definition A.2.62), $V = \emptyset$ and $C = I$ and the result follows. $\qquad\square$

Proposition 3.7.5. *Let M^n be an orientable connected differentiable manifold and let $\mathcal{A} = \{(U_\alpha, \phi_\alpha)\}_{\alpha \in I}$ and $\mathcal{B} = \{(V_\beta, \psi_\beta)\}_{\beta \in J}$ be two compatible atlases on M both of which separately make M into an oriented manifold. Then either $\det(d(\psi_\beta \circ \phi_\alpha^{-1})) > 0$ for all $(\alpha, \beta) \in I \times J$ or $\det(d(\psi_\beta \circ \phi_\alpha^{-1})) < 0$ for all $(\alpha, \beta) \in I \times J$ such that $\psi_\beta \circ \phi_\alpha^{-1}$ have nonempty domains.*

Proof. Let $\alpha \in I$ be arbitrary and let $\beta \in J$. Then, as a continuous function, $\det(d(\psi_\beta \circ \phi_\alpha^{-1}))$ is either always positive or always negative on its domain. Suppose that $\det(d(\psi_\beta \circ \phi_\alpha^{-1})) > 0$ and let $\beta' \in J$. By Lemma 3.7.4, there is a finite sequence $\beta = \beta_1, \beta_2, \ldots, \beta_n = \beta'$ such that $V_{\beta_i} \cap V_{\beta_{i+1}} \neq \emptyset$. Furthermore, because we can think of U_α as a submanifold of M, we can assume that $V_{\beta_i} \cap U_\alpha \neq \emptyset$ for all V_{β_i} in the sequence. (See Figure 3.17.)

Since $\mathcal{B}$ equips M with an oriented differentiable manifold structure, then $\det(d(\psi_{\beta_{i+1}} \circ \psi_{\beta_i}^{-1})) > 0$ for all $i = 1, 2, \ldots, n-1$. By the chain rule,

$$\det(d(\psi_{\beta'} \circ \psi_\alpha^{-1}))$$
$$= \det(d(\psi_{\beta_n} \circ \psi_{\beta_{n-1}}^{-1}) \circ \cdots \circ d(\psi_{\beta_2} \circ \psi_{\beta_1}^{-1}) \circ d(\psi_{\beta_1} \circ \phi_\alpha^{-1}))$$
$$= \det(d(\psi_{\beta_n} \circ \psi_{\beta_{n-1}}^{-1})) \cdots \det(d(\psi_{\beta_2} \circ \psi_{\beta_1}^{-1})) \det(d(\psi_\beta \circ \phi_\alpha^{-1}))$$
$$> 0.$$

Suppose alternatively that $\det(d(\psi_\beta \circ \phi_\alpha^{-1})) < 0$. Then the same composition holds but the product of determinants leads to $\det(d(\psi_{\beta'} \circ \phi_\alpha^{-1})) < 0$. The result follows. $\qquad\square$

Definition 3.7.6. Let M^n be an orientable connected differentiable n-manifold. Two compatible atlases $\mathcal{A} = \{\phi_\alpha\}_{\alpha \in I}$ and $\mathcal{B} = \{\psi_\beta\}_{\beta \in J}$ are said to have equivalent orientations if the atlas $\mathcal{A} \cup \mathcal{B}$ also makes M an oriented manifold. An equivalence class of oriented atlases is called an *orientation*.

Proposition 3.7.5 shows that on a connected differentiable maniofld there can only be two orientations. More generally, if M is a manifold with c connected components, there are 2^c possible orientations on M. This includes the degenerate case of 0-manifolds that correspond to a set of points equipped with the discrete topology. In this case, each point can have an orientation of $+1$ or -1.

Definition 3.7.7. Let M^n be an oriented manifold and let $p \in M$. Any ordered basis on $T_p M$ is called positively oriented if its change of coordinate matrix with $(\partial_1, \partial_2, \ldots, \partial_n)$ has positive determinant.

We now discuss how orientations on manifolds with boundary M induce orientations on the boundary manifold ∂M. We must make a choice in how to induce an orientation on the boundary but do so to conform with Green's Theorem, Stokes' Theorem, the divergence theorem, and even the fundamental theorem of calculus. Recall that Green's Theorem states that for a compact region $\mathcal{R} \subset \mathbb{R}^2$ with a boundary $\partial \mathcal{R}$ that consists of a finite number of regular curves

$$\iint_\mathcal{R} \left(\frac{\partial F_2}{\partial x} - \frac{\partial F_1}{\partial y} \right) dA = \int_{\partial \mathcal{R}} \vec{F} \cdot d\vec{r},$$

for any differentiable vector field $\vec{F} = (F_1, F_2)$ on $\mathcal{R}$. The integral on the right breaks into the sum of k integrals if $\partial \mathcal{R}$ has k boundary components. Furthermore,

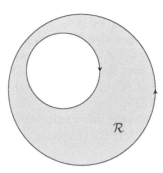

Figure 3.18: Orientation of the boundary of a plane region.

we require that each boundary component be oriented so that as "someone travels along the curve" the interior of the region is to the left. Another way to state this choice of orientation of the boundary components is that the ordered pair of vectors consisting of an outward pointing normal and the direction of travel along the curve is a positive frame for $\mathbb{R}^2$. (See Figure 3.18.)

Let M^n be an oriented differentiable manifold with boundary. Let $p \in \partial M$, and let (U, ϕ) be a coordinate neighborhood of p with coordinates $(x^1, x^2, \ldots, x^n)$. Recall that since $p \in \partial M$, the coordinate chart ϕ is a homeomorphism onto an open subset of half-space $H_{\vec{a}}$ of $\mathbb{R}^n$. If $\vec{a} = (a^1, a^2, \ldots, a^n) \in \mathbb{R}^n$, the tangent vector

$$X_{\vec{a}} = a^1 \partial_1 + a^2 \partial_2 + \cdots + a^n \partial_n,$$

where $\partial_i = \partial/\partial x^i$, is in $T_p M$ but not in $T_p(\partial M)$. We say that $X_{\vec{a}}$ is a tangent vector that *points inward* from ∂M, while $-X_{\vec{a}}$ is outward pointing. The induced coordinate chart on ∂M is $\pi_{\vec{a}} \circ \phi : U \cap \partial M \to \mathbb{R}^{n-1}$ with coordinates $(u^1, u^2, \ldots, u^{n-1})$.

Definition 3.7.8. Then the ordered basis $(\partial/\partial u^1, \ldots, \partial/\partial u^{n-1})$ of $T_p(\partial M)$ gives the *induced orientation* on ∂M if

$$\left(X_{\vec{a}}, \frac{\partial}{\partial u^1}, \ldots, \frac{\partial}{\partial u^{n-1}} \right)$$

is positively oriented on $T_p M$.

If the coordinate chart of a point on the boundary of an oriented manifold is $\phi : U \to \mathbb{R}^n_+$, i.e., with $\vec{a} = (0, \ldots, 0, 1)$, then $-\partial_n = -\partial/\partial x^n$ is a tangent vector that points outward from ∂M. The ordered basis $(\partial_1, \ldots, \partial_{n-1})$ of $T_p(\partial M)$ gives the *induced orientation* on ∂M if $(-\partial_n, \partial_1, \ldots, \partial_{n-1})$ is positively oriented on $T_p M$.

Example 3.7.9. Consider the half-torus M shown in Figure 3.19. The boundary ∂M has two components. The point q is a generic point in a neighborhood that contains the boundary component where p is. If (x^1, x^2) is a coordinate system in a neighborhood of p, the boundary component that contains p is given by $x^2 = 0$. The

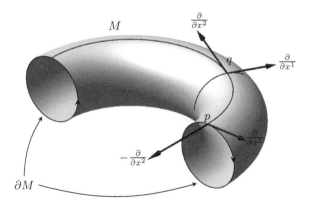

Figure 3.19: Half-torus with boundary.

figure depicts the ordered basis (∂_1, ∂_2) at the generic point q and also the ordered basis $(-\partial_2, \partial_1)$ at p. Since these two bases have the same orientation (imagine moving the standard basis at q over to p), then ∂_1 determines the induced orientation on ∂M (as opposed to $-\partial_1$).

For the other boundary component, the reasoning is the same except that we must use at least one other coordinate chart $(\bar{x}^1, \bar{x}^2)$ where the boundary is given by $\bar{x}^2 = 0$ and the portion of M that is not on ∂M has $\bar{x}^2 > 0$. Intuitively speaking, in order for $(\bar{x}^1, \bar{x}^2)$ to have an orientation compatible with (x^1, x^2), one must switch the direction of the basis vector $\partial/\partial\bar{x}^2$ (from what one would obtain from moving $\partial/\partial x^2$ over along a line of $x^1 =$const.). We must then also switch the sign of $\partial/\partial x^1$ to get the equivalent $\partial/\partial\bar{x}^1$ in order to keep a positively oriented atlas. The induced orientation on the second boundary component is shown with an arrow.

Example 3.7.10 (Closed Interval). We set a convention for use later concerning 1-manifolds. Let $\gamma : [a, b] \to M$ be a 1-manifold with two boundary points $p_1 = \gamma(a)$ and $p_2 = \gamma(b)$. Example 3.5.4 gives two coordinate charts that explicitly define $[a, b]$ as a manifold with boundary. If $p \in [a, b)$, then the basis on T_pM with respect to the chart ϕ_1 is $\{d/dx\}$, while if $p \in (a, b]$, then the basis on T_pM with respect to the chart ϕ_2 is $\{d/d\bar{x}\}$. It is easy to tell that for all $p \in (a, b)$, on T_pM, we have $d/dx = d/d\bar{x}$. Clearly, $[a, b]$ with the atlas $\{\phi_1, \phi_2\}$ is an oriented manifold.

The outward pointing vector at p_2 is in the same orientation as d/dx, so we say that p_2 is equipped with a positive orientation. In contrast, the outward pointing vector at p_1 is $-d/d\bar{x}$, which is negative with respect to the induced orientation.

This association of -1 and $+1$ to the endpoints as shown above is, by convention, the induced orientation of γ onto $\partial\gamma$.

PROBLEMS

3.7.1. Prove that a manifold that has a single chart is orientable.

3.7.2. Prove that every one-dimensional manifold is orientable.

3.7.3. Let M be a differentiable manifold of dimension 2 or greater that has an atlas of exactly 2 charts. Prove that M is orientable.

3.7.4. Show that the closed ball $\mathbb{B}^3$ equipped with the atlas $\{\Pi_N, \Pi_S\}$ described in Exercise 3.5.4 does not make $\mathbb{B}^3$ into an oriented manifold. Modify the atlas to explicitly show that $\mathbb{B}^3$ is orientable. Sketch the ball and indicate with a frame in the tangent plane to a point on the surface, the orientation that is induced on the surface $\mathbb{S}^2 = \partial\mathbb{B}^3$.

3.7.5. Prove that if M is an orientable manifold with boundary, then ∂M is also orientable.

3.7.6. Show that $\mathbb{RP}^2$ is not orientable.

3.7.7. Let M and N be two orientable differentiable manifolds. Show that $M \times N$ with the product structure is an orientable manifold.

CHAPTER 4

Multilinear Algebra

Many of the objects of interest in differential geometry on manifolds are expressed properly in the context of multilinear algebra. Consequently, this chapter introduces linear algebraic concepts that are not commonly included in a first linear algebra course. The underlying field for all objects outside this chapter is the set of reals $\mathbb{R}$, but this chapter introduces the concepts for an arbitrary field K of characteristic 0 (e.g., $\mathbb{Q}$, $\mathbb{R}$, or $\mathbb{C}$).

Before jumping in, we mention our habit of notation for components associated to certain linear algebraic objects. Let V be a vector space over K with $\dim V = n$. If $\mathcal{B} = (e_1, e_2, \ldots, e_n)$ is an ordered basis of V, the coordinates of $v \in V$ with respect to $\mathcal{B}$ are

$$[v]_\mathcal{B} = \begin{pmatrix} v^1 \\ v^2 \\ \vdots \\ v^n \end{pmatrix}, \quad \text{where} \quad v = v^1 e_1 + v^2 e_2 + \cdots + v^n e_n.$$

If the basis of V is understood from the problem or if we use a standard basis of V, we write $[v]$. It is common to abuse the notation and say that a vector is equal to the $n \times 1$ matrix of its coordinates but we must always be careful to understand that components are given with respect to some basis.

If V is a vector space of dimension n and if $\mathcal{B} = \{e_1, e_2, \ldots, e_n\}$ and $\mathcal{B}' = \{f_1, f_2, \ldots, f_n\}$ are two bases, there is an $n \times n$ matrix $P_{\mathcal{B}'}^\mathcal{B}$ that converts the $\mathcal{B}$-coordinates of a vector to $\mathcal{B}'$-coordinates. In particular, for all $v \in V$,

$$[v]_{\mathcal{B}'} = P_{\mathcal{B}'}^\mathcal{B} [v]_\mathcal{B}. \tag{4.1}$$

This matrix is found by $P_{\mathcal{B}'}^\mathcal{B} = ([e_1]_{\mathcal{B}'} \quad [e_2]_{\mathcal{B}'} \quad \cdots \quad [e_n]_{\mathcal{B}'})$. Writing the components of $P_{\mathcal{B}'}^\mathcal{B}$ as (p_j^i), where i is the row index and j is the column index, and the $\mathcal{B}'$ coordinate of v as $(\bar{v}^i)$, we can write (4.1) as

$$\bar{v}^i = \sum_{j=1}^n p_j^i v^j. \tag{4.2}$$

As we introduce constructions in multilinear algebra and refer to how the components of objects change under a change of basis, we will refer to (4.2) repeatedly to see appropriate generalizations.

4.1   Hom Space and Dual

Definition 4.1.1. Let V and W be two vector spaces over K. Denote the set of linear transformations from V to W by $\mathrm{Hom}_K(V,W)$, or simply $\mathrm{Hom}(V,W)$ if the field K is understood by context.

We can define addition and multiplication by a K-scalar on $\mathrm{Hom}(V,W)$ in the following way. If $T_1, T_2 \in \mathrm{Hom}(V,W)$, then $T_1 + T_2$ is the linear transformation given by

$$(T_1 + T_2)(v) \stackrel{\mathrm{def}}{=} T_1(v) + T_1(v) \qquad \text{for all } v \in V.$$

Also, if $\lambda \in K$ and $T \in \mathrm{Hom}(V,W)$, define the linear transformation λT by

$$(\lambda T)(v) \stackrel{\mathrm{def}}{=} \lambda(T(v)) \qquad \text{for all } v \in V.$$

These definitions lead us to the following foundational proposition.

Proposition 4.1.2. *Let V and W be vector spaces over K of dimension m and n, respectively. Then $\mathrm{Hom}(V,W)$ is a vector space over K, with $\dim \mathrm{Hom}(V,W) = mn$.*

Proof. We leave it to the reader to check that $\mathrm{Hom}(V,W)$ satisfies all the axioms of a vector space over K.

To prove that $\dim \mathrm{Hom}(V,W) = mn$, first choose an ordered basis $\mathcal{B} = (e_1, e_2, \ldots, e_m)$ of V and an ordered basis $\mathcal{B}' = (f_1, f_2, \ldots, f_n)$ of W. Define $T_{ij} \in \mathrm{Hom}(V,W)$ as the linear transformations defined by

$$T_{ij}(e_k) = \begin{cases} f_i, & \text{if } j = k, \\ 0, & \text{if } j \neq k, \end{cases}$$

and extended by linearity over all V. We show that the set $\{T_{ij}\}$ for $1 \leq i \leq m$ and $1 \leq j \leq n$ forms a basis of $\mathrm{Hom}(V,W)$.

Because of linearity, any linear transformation $L \in \mathrm{Hom}(V,W)$ is completely defined given the knowledge of $L(e_j)$ for all $1 \leq j \leq m$. Suppose that for each j, there exist mn constants a^i_j in K, indexed by $i = 1, 2, \ldots, m$ and $j = 1, 2, \ldots, n$, such that

$$L(e_j) = \sum_{i=1}^{n} a^i_j f_i. \tag{4.3}$$

Then

$$L = \sum_{i=1}^{n} \sum_{j=1}^{m} a^i_j T_{ij},$$

and hence, $\{T_{ij}\}$ spans $\text{Hom}(V, W)$. Furthermore, suppose that for some constants c_{ij}

$$\sum_{i=1}^{n} \sum_{j=1}^{m} c_{ij} T_{ij} = \mathbf{0},$$

the trivial linear transformation. Then for all $1 \le k \le m$,

$$\sum_{i=1}^{n} \sum_{j=1}^{m} c_{ij} T_{ij}(e_k) = \mathbf{0} \iff \sum_{i=1}^{n} c_{ik} f_i = \mathbf{0}.$$

However, since $\{f_1, f_2, \ldots, f_n\}$ is a linearly independent set, given any k, we have $c_{ik} = 0$ for all $1 \le 1 \le m$. Hence, for all i and j, the constants $c_{ij} = 0$, which shows that the linear transformations T_{ij} are linearly independent.

We conclude that the set $\{T_{ij} \mid$ for $1 \le i \le n$ and $1 \le j \le m\}$, forms a basis of $\text{Hom}(V, W)$. Consequently, $\dim \text{Hom}(V, W) = mn$. $\qquad\square$

The proof of Proposition 4.1.2 provides the set of linear transformations $\{T_{ij}\}$ as a standard basis of $\text{Hom}(V, W)$. Furthermore, with respect to these bases, $\left[T_{ij}\right]_{\mathcal{B}'}^{\mathcal{B}} = E_{ij}$, the $n \times m$ matrices where the entries are all 0 except for a 1 in the (i, j)th entry.

Recall that the matrix (a_j^i) described in (4.3), where i is the row index and j is the column index, is called matrix representing L with respect to $\mathcal{B}$ and $\mathcal{B}'$. Using the notation from this chapter's introduction, we denote this by $[L]_{\mathcal{B}'}^{\mathcal{B}} = (a_j^i)$. A standard result from linear algebra is that

$$[L]_{\mathcal{B}'}^{\mathcal{B}} = \left([L(e_1)]_{\mathcal{B}'} \quad [L(e_2)]_{\mathcal{B}'} \quad \cdots \quad [L(e_m)]_{\mathcal{B}'}\right).$$

Clearly, the matrix that represents a linear transformation with respect to certain bases will change if the ordered bases change. Let $T : V \to W$ be a linear transformation. Suppose that $\dim V = m$ and $\dim W = n$. Let $\mathcal{A}$ and $\mathcal{A}'$ be two bases of V, and let $P = P_{\mathcal{A}'}^{\mathcal{A}}$ be the change of coordinate matrix from $\mathcal{A}$ to $\mathcal{A}'$. Let $\mathcal{B}$ and $\mathcal{B}'$ be two bases of W and let $Q = Q_{\mathcal{B}'}^{\mathcal{B}}$ be the corresponding change of coordinate matrix from $\mathcal{B}$ to $\mathcal{B}'$. We point out that the process of taking, say, the $\mathcal{A}$ coordinates of V is a linear transformation $[\,]_{\mathcal{A}} : V \to \mathbb{R}^m$. With this in mind, the following diagram depicts how the linear transformations, coordinates, and matrix multiplications are all related.

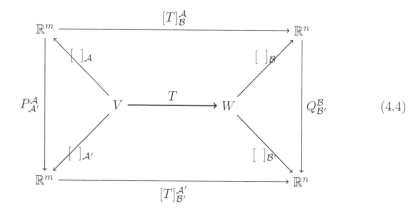

$$\text{(4.4)}$$

In particular, we read off the matrix relationship

$$[T]_{\mathcal{B}'}^{\mathcal{A}'} = Q_{\mathcal{B}'}^{\mathcal{B}} [T]_{\mathcal{B}}^{\mathcal{A}} P_{\mathcal{A}}^{\mathcal{A}'} = Q_{\mathcal{B}'}^{\mathcal{B}} [T]_{\mathcal{B}}^{\mathcal{A}} \left(P_{\mathcal{A}'}^{\mathcal{A}} \right)^{-1} . \tag{4.5}$$

To write this relationship as a sum similar to (4.2), we suppose that $P^{-1} = (\breve{p}_j^i)$, $Q = (q_j^i)$, $[T]_{\mathcal{B}}^{\mathcal{A}} = (a_j^i)$ and $[T]_{\mathcal{B}'}^{\mathcal{A}'} = (\bar{a}_j^i)$, then

$$\bar{a}_\ell^k = \sum_{i=1}^{m} \sum_{j=1}^{n} q_j^k a_i^j \breve{p}_\ell^i . \tag{4.6}$$

In the particular case of a linear transformation $T : V \to V$, we always assume that the same basis change occurs simultaneously on both the domain and codomain. Hence, $P = Q$. The relationship in (4.5) becomes

$$[T]_{\mathcal{A}'}^{\mathcal{A}'} = P_{\mathcal{A}'}^{\mathcal{A}} [T]_{\mathcal{A}}^{\mathcal{A}} \left(P_{\mathcal{A}'}^{\mathcal{A}} \right)^{-1} ,$$

making $[T]_{\mathcal{A}}^{\mathcal{A}}$ and $[T]_{\mathcal{A}'}^{\mathcal{A}'}$ similar matrices, and (4.6) changes to

$$\bar{a}_\ell^k = \sum_{i=1}^{m} \sum_{j=1}^{m} p_j^k \breve{p}_\ell^i a_i^j . \tag{4.7}$$

Though a particular case of a Hom-space, the dual to a vector space plays a critical role in multilinear algebra. We devote the remainder of this section to the dual space.

Definition 4.1.3. Let V be a vector space over K. The vector space $\mathrm{Hom}_K(V, K)$ is called the *dual space* to V and is denoted by V^*. Elements of V^* are called *covectors* to V.

By Proposition 4.1.2, if V is finite-dimensional, then $\dim V^* = \dim V$ and, by well-known facts from linear algebra, V and V^* are isomorphic. Knowing that two vector spaces are isomorphic may seem like there is not much difference between them. However, the difference in how coordinates of covectors change versus how coordinates of vectors change under a basis change is of foundational importance with implications reaching into many areas of mathematics.

Proposition 4.1.4. *If $\mathcal{B} = (e_1, e_2, \ldots, e_n)$ is an ordered basis for V, then the linear functions $e^{*i} : V \to K$, with $1 \leq i \leq n$ such that*

$$e^{*i}(v) = v^i, \quad whenever \quad [v]_\mathcal{B} = \begin{pmatrix} v^1 \\ \vdots \\ v^n \end{pmatrix}, \tag{4.8}$$

form a basis of V^*. In particular, $\dim V = \dim V^*$.

Proof. This follows from the basis of $\mathrm{Hom}_K(V, W)$ exhibited in Proposition 4.1.2. $\qquad\square$

We can give an alternate characterization of the functions e^{*i}. They are linear and satisfy the property that

$$e^{*i}(e_j) = \delta^i_j = \begin{cases} 1 & \text{if } i = j \\ 0 & \text{otherwise.} \end{cases}$$

This δ^i_j symbol is called the *Kronecker delta*. The Kronecker delta appears repeatedly in multilinear algebra and represents the components of the identity matrix.

We point out that the map $\varphi : V \to V^*$ that sends e_i to e^{*i} for all $1 \leq i \leq n$ (and that is completed by linearity) provides an explicit isomorphism between V and V^*. This isomorphism depends on the choice of ordered basis.

Definition 4.1.5. The ordered basis $\mathcal{B}^* = (e^{*1}, \ldots, e^{*n})$ is called the *dual basis* or *cobasis* to $\mathcal{B} = (e_1, e_2, \ldots, e_n)$.

It is important to remember that each e^{*i} depends on the whole basis $\mathcal{B}$. There is no canonical (without reference to a basis of V) way to define a function $v^* \in \mathrm{Hom}(V, K)$ in reference to a single vector $v \in V$.

Proposition 4.1.6. *Let V be a vector space with two bases $\mathcal{A}$ and $\mathcal{B}$ and suppose that $\lambda \in V^*$ has coordinates (λ_i) with respect to the dual basis $\mathcal{A}^*$ and has coordinates $(\bar\lambda_i)$ with respect to $\mathcal{B}^*$. If $Q = Q^\mathcal{A}_\mathcal{B}$ then*

$$\bar\lambda_i = \sum_{j=1}^n \breve{q}^j_i \lambda_j,$$

where $\breve{q}^j_i$ are the components of the inverse matrix Q^{-1}.

Proof. This is a particular case of (4.5) and (4.6).   □

In the context of a dual space, it is possible to give a natural interpretation of the transpose of a matrix. Suppose that V and W are two vector spaces over K and that $T \in \mathrm{Hom}_K(V,W)$. There is a natural way to define an associated linear transformation $W^* \to V^*$ as follows. Given a linear function $g \in W^*$, the composition $v \mapsto g(T(v))$ is an element of V^*. Therefore, we call $T^* : W^* \to V^*$ the transformation such that $T^*(g)$ is the unique element of V^* that satisfies

$$T^*(g)(v) = g(T(v)). \tag{4.9}$$

As the composition of linear transformations, T^* is again linear, and hence, $T^* \in \mathrm{Hom}(W^*,V^*)$. This transformation T^* is called the *dual* of T.

Proposition 4.1.7. *Let V and W be finite-dimensional K-vector spaces with ordered bases $\mathcal{A}$ and $\mathcal{B}$, respectively. Let $T : V \to W$ be a linear transformation. The matrix representing the dual $T^* : W^* \to V^*$ with respect to the cobases $\mathcal{B}^*$ and $\mathcal{A}^*$ is the transpose of the matrix representing T with respect to $\mathcal{A}$ and $\mathcal{B}$. In other words,*

$$[T^*]_{\mathcal{A}^*}^{\mathcal{B}^*} = \left([T]_{\mathcal{B}}^{\mathcal{A}}\right)^{\top}.$$

Proof. Suppose that $\mathcal{A} = (e_1, e_2, \ldots, e_m)$ and $\mathcal{B} = (f_1, f_2, \ldots, f_n)$, respectively, and let $A = (a_j^i)$ be the matrix representing T with respect to these bases, so that

$$T(e_i) = \sum_{k=1}^{n} a_i^k f_k.$$

For all $v \in V$, we can write v as $v = v^1 e_1 + v^2 e_2 + \cdots + v^m e_m$. Then

$$T^*(f^{*j})(v) = f^{*j}(T(v)) = f^{*j}\left(T\left(\sum_{i=1}^{m} v^i e_i\right)\right)$$

$$= f^{*j}\left(\sum_{i=1}^{m} v^i T(e_i)\right) = f^{*j}\left(\sum_{i=1}^{m} v^i \sum_{k=1}^{n} a_i^k f_k\right)$$

$$= \sum_{i=1}^{m} v^i\left(\sum_{k=1}^{n} a_i^k f^{*j}(f_k)\right) = \sum_{i=1}^{m} v^i\left(\sum_{k=1}^{n} a_i^k \delta_k^j\right)$$

$$= \sum_{i=1}^{m} v^i a_i^j = \sum_{i=1}^{m} a_i^j e^{*i}(v).$$

Thus, as covectors,

$$T^*(f^{*j}) = \sum_{i=1}^{m} a_i^j e^{*i}.$$

Hence, the (i, j)-entry of the matrix representing T^* with respect to $\mathcal{B}^*$ and $\mathcal{A}^*$ is (a_i^j), which is the same of the matrix representing T, but with the role of rows and columns reversed. The result follows. $\qquad\square$

In the previous paragraphs, we emphasized the role of a basis in establishing an isomorphism between V and V^*. Operations on a vector space that can be described without reference to any particular basis are called *canonical*. For example, the definition of the dual of a vector space and the definition of a dual of a linear transformation in (4.9) are canonical definitions. On the other hand, when a vector space V has an ordered basis $\mathcal{B}$, the isomorphism $\varphi : V \to V^*$ defined in (4.8) is not canonical.

We now consider the double-dual of V, namely the dual of V^*. Given any vector $v \in V$, we define the co-covector $\lambda_v \in V^{**} = \operatorname{Hom}(V^*, K)$ by

$$\lambda_v(f) = f(\vec{v}). \tag{4.10}$$

This defines a function $\Lambda : V \to V^{**}$ by $\Lambda(v) = \lambda_v$.

Proposition 4.1.8. *The function Λ defined by (4.10) is an injective linear transformation. Furthermore, if V is finite-dimensional, then Λ is a canonical isomorphism between a vector space V and its double dual V^{**}.*

Proof. We defined Λ without reference to any basis so it is canonical.

We first prove that Λ is a linear transformation. Let $v, w \in V$, and let $c \in K$. For all $f \in V^*$,

$$\Lambda(v + w)(f) = \lambda_{v+w}(f) = f(v + w) = f(v) + f(w) = \lambda_v(f) + \lambda_w(f),$$
$$= \Lambda(v)(f) + \Lambda(w)(f), \tag{4.11}$$

so as co-covectors, $\Lambda(v + w) = \Lambda(v) + \Lambda(w)$. Similarly,

$$\Lambda(cv)(f) = \lambda_{cv}(f) = f(cv) = cf(v) = c\lambda_v(f) = c\Lambda(v)(f), \tag{4.12}$$

so again, as co-covectors, $\Lambda(cv) = c\Lambda(v)$.

Next, we show that Λ is injective. Let $u_1, u_2 \in V$ be vectors, and suppose that $\Lambda(u_1) = \Lambda(u_2)$. Thus, $f(u_1) = f(u_2)$ for all $f \in \operatorname{Hom}(V, K)$. Therefore, $f(u_1 - u_2) = 0$ for all $f \in V^*$, hence $u_1 - u_2 = 0$, and thus, $u_1 = u_2$, proving that Λ is injective.

Finally, we prove that if V is finite-dimensional, then Λ is an isomorphism. Since $\dim W = \dim W^*$ for all finite dimensional vector spaces, we also have $\dim V = \dim V^{**}$. Since Λ is injective, by the Rank-Nullity Theorem, $\dim(\operatorname{Im} \Lambda) = \operatorname{rank} \Lambda = \dim V = \dim V^{**}$. Thus Λ is surjective. Hence, Λ is an isomorphism. $\qquad\square$

We point out that requiring V to be finite-dimensional for V and V^{**} to be canonically isomorphic is not a limitation of the above proof. There exist infinite-dimensional vector spaces V with a basis $\mathcal{B}$ such that a basis of V^* has a strictly greater cardinality than $|\mathcal{B}|$. This suffices to show that V and V^* and by extension V^{**} cannot be isomorphic.

However, the benefit of the existence of a canonical isomorphism between V and V^{**} when V is finite dimensional arises especially in regards to how coordinates change under a change of basis. Proposition 4.1.6 shows that coordinates of covectors change by the inverse of the coordinate change matrix under a basis change on V. However, since there is a canonical isomorphism between V and V^{**} coordinates of co-covectors change by the regular coordinate change matrix and hence behave as regular vectors under a basis change on V.

As the reader has hopefully noticed, it is standard to use superscripts for the index of coordinates of a vector with respect to a basis $\mathcal{B}$ and to use subscripts for the index of coordinates of a covector with respect to $\mathcal{B}^*$. Superscript indices are called *contravariant* indices and subscript indices are called *covariant* indices. Proposition 4.1.8 shows that we do not need three or worse, a countable number of, types of index. This distinction between types of indices dovetails with our habits of matrix notation: we write coordinates of a vector as a column matrix and coordinates of a covector as a row matrix.

We can now introduce the Einstein summation convention, which shortens calculations involving components of objects in multilinear algebra. In any expression involving the product of components of vectors or matrices (or eventually tensors), we will assume that we sum over any index that is repeated in the superscript and in the subscript. For example, if $1 \le j \le n$, then

$$a_j^i v^j \quad \text{means} \quad \sum_{j=1}^{n} a_j^i v^j.$$

This equation shows the components of a matrix-vector multiplication, Av. Also if (c^i) are real numbers and (e_i) is a list of vectors, then under the Einstein summation convention

$$c^i e_i \quad \text{means the linear combination} \quad c^1 e_1 + c^2 e_2 + \cdots + c^n e_n.$$

As a third example, we could rewrite (4.7) using Einstein summation convention as

$$\bar{a}_\ell^k = q_j^k \breve{q}_\ell^i a_i^j.$$

Except for summations involving T_{ij} in the proof of Proposition 4.1.2, we properly used superscript and subscript indices in every equation in this section so that the summation symbol could be removed and the expression be correct using the Einstein summation convention. Through the remainder of the book, we will use the Einstein summation convention, and occasionally write (ESC) to remind the reader of this convention.

We conclude this section with a brief comment about the direct sum of two vector spaces. Though not immediately connected to the Hom-space or dual space, we mention this construction here since some exercises involve the direct sum.

Definition 4.1.9. Let V and W be two vector spaces over a field K. The direct sum $V \oplus W$ consists of the set $V \times W$ equipped with operations of addition and scalar multiplication defined as:

1. $(v_1, w_1) + (v_2, w_2) \overset{\text{def}}{=} (v_1 + v_2, w_1 + w_2)$.

2. $c \cdot (v, w) \overset{\text{def}}{=} (cv, cw)$ for all $c \in K$.

It is a simple exercise to show that $V \oplus W$ is a vector space over K. In the direct sum $V \oplus W$, the subset $\{(v, 0) \mid v \in V\}$ is a subspace isomorphic to V and the subset $\{(0, w) \mid w \in W\}$ is a subspace isomorphic to W. By an abuse of terminology, we will often say that V and W are subspaces of $V \oplus W$, with V and W identified according to these natural isomorphisms.

PROBLEMS

4.1.1. Let V be a vector space with basis $\{e_1, e_2, \ldots, e_n\}$. Clearly prove that the set of functions $\{e^{*1}, e^{*2}, \ldots, e^{*n}\}$ defined in (4.8) form a basis of V^*.

4.1.2. Prove that $\dim(V \oplus W) = \dim V + \dim W$.

4.1.3. Let U, V, and W be vector spaces over a field K. Prove that there exist canonical (vector space) isomorphisms

$$\text{Hom}(U \oplus V, W) \approx \text{Hom}(U, W) \oplus \text{Hom}(V, W), \qquad (4.13)$$
$$\text{Hom}(U, V \oplus W) \approx \text{Hom}(U, V) \oplus \text{Hom}(U, W). \qquad (4.14)$$

4.1.4. Let V_1, V_2, W_1, and W_2 be vector spaces over a field K. Suppose that $L : V_1 \to V_2$ and $T : W_1 \to W_2$ are linear transformations with respect to given bases. Define the function

$$f : V_1 \oplus W_1 \to V_2 \oplus W_2$$
$$(v, w) \longmapsto (L(v), T(w)).$$

Suppose that $\mathcal{A}_1$, $\mathcal{A}_2$, $\mathcal{B}_1$, and $\mathcal{B}_2$ are ordered bases on V_1, V_2, W_1, and W_2 respectively. Prove that the matrix of f with respect to the basis $\mathcal{A}_1 \cup \mathcal{B}_1$ on $V_1 \oplus W_1$ and to the basis $\mathcal{A}_2 \cup \mathcal{B}_2$ on $V_2 \oplus W_2$ is block diagonal

4.1.5. Let V and W be finite-dimensional vector spaces over a field K with dimensions m and n, respectively, and let $f : V \to W$ be a linear transformation. Define the linear transformation

$$T : V \oplus W \to V \oplus W$$
$$(v, w) \longmapsto (v, f(v) + w).$$

(a) Prove that the only eigenvalue of T is 1 (with multiplicity $m + n$).

(b) Prove that the eigenspace of 1 is $E_1 = \text{Ker} f \oplus W$, and conclude that the geometric multiplicity of 1 is $m + n - \text{rank} f$.

4.1.6. Let $T : V \to V$ be a linear transformation and suppose that T with respect to some basis $\mathcal{B}$ of V, it has the matrix $A = (a_j^i)$. Using Einstein summation convention, prove that a_i^i, which is the trace $\text{Tr}(A)$ of A, is independent of the basis.

4.1.7. Let V be a vector space with an ordered basis $\mathcal{B}$. Let $v \in V$ and let $f \in V^*$. Suppose that the coordinates of v with respect to $\mathcal{B}$ are (v^i) and that the coordinates of f with respect to $\mathcal{B}^*$ are (f_i). Prove that $f(v)$ is equal to $f_i v^i$ (ESC) and show that this quantity is independent of the basis.

4.1.8. Let U, V, and W be vector spaces. Prove that $\psi : \mathrm{Hom}(U, \mathrm{Hom}(v, W)) \to \mathrm{Hom}(V, \mathrm{Hom}(U, W))$ defined by $\psi(T) = L$, where for all $u \in U$ and for all $v \in V$ the linear transformation L satisfies $L(v)(u) = T(u)(v)$, is a canonical isomorphism.

4.1.9. Let $V = C^0([a, b], \mathbb{R})$ be the vector space of continuous real valued functions defined on the interval $[a, b]$. For all $f \in C^0([a, b], \mathbb{R})$, define λ_f as the covector satisfying.

$$\lambda_f(g) = \int_a^b f(x)g(x)\, dx \qquad \text{for all } g \in C^0([a, b], \mathbb{R}).$$

(a) Prove that $\Lambda : V \to V^*$ defined by $\Lambda(f) = \lambda_f$ is an injective linear transformation.

(b) Let $c \in [a, b]$ and define the evaluation at c as $\mathrm{ev}_c(g) = g(c)$. Prove that $\mathrm{ev}_c \in V^*$.

(c) Prove that there does not exist $f \in C^0([a, b], \mathbb{R})$ such that $\lambda_f = \mathrm{ev}_c$. [This shows that Λ is strictly injective.]

4.1.10. Let V be a vector space, and let W be a subspace. Define the relation $\sim$ on vectors of V by

$$v_1 \sim v_2 \iff v_1 - v_2 \in W.$$

(a) Prove that $\sim$ is an equivalence relation.

(b) Denote by V/W the set of equivalence classes. Prove that V/W has the structure of a vector space under the operations: $[v_1] + [v_2] \overset{\text{def}}{=} [v_1 + v_2]$ and $c \cdot [v] \overset{\text{def}}{=} [cv]$.

(c) Suppose that V is finite-dimensional. Prove that $\dim V/W = \dim V - \dim W$.

(The vector space V/W is called the *quotient vector space* of V with respect to W.)

4.2 Bilinear Forms and Inner Products

4.2.1 Bilinear Forms on $V \times W$

Definition 4.2.1. Let V and W be vector spaces over a field K. A *bilinear form* $\langle \cdot, \cdot \rangle$ on $V \times W$ is a function $V \times W \to K$ such that for all $v \in V$, $w \in W$, and $\lambda \in K$,

$$\langle v_1 + v_2, w \rangle = \langle v_1, w \rangle + \langle v_2, w \rangle, \quad \langle \lambda v, w \rangle = \lambda \langle v, w \rangle, \tag{4.15}$$

$$\langle v, w_1 + w_2 \rangle = \langle v, w_1 \rangle + \langle v, w_2 \rangle \quad \langle v, \lambda w \rangle = \lambda \langle v, w \rangle. \tag{4.16}$$

If $V = W$, then we say $\langle \cdot, \cdot \rangle$ is a bilinear form on V.

We can restate this definition to say that for any fixed $w_0 \in W$, the function $x \mapsto \langle x, w_0 \rangle$ is in V^* (corresponding to (4.15)) and that for any fixed $v_0 \in V$, the function $x \mapsto \langle v_0, x \rangle$ is in W^* (corresponding to (4.16)).

The notation used for a bilinear form varies widely in the literature because of the many areas in which it is used. In terms of function notation, we might encounter the functional notation $f : V \times W \to K$ or perhaps $w : V \times W \to K$ for a bilinear form and $\langle \cdot, \cdot \rangle$ or $(\cdot, \cdot)$ for the "product" notation. If $V = W$, we sometimes write the pair (V, f) to denote the vector space V equipped with the bilinear form f.

Example 4.2.2. In elementary linear algebra, the most commonly known example of a bilinear form on $\mathbb{R}^n$ is the dot product between two vectors defined in terms of standard coordinates by

$$\vec{v} \cdot \vec{w} = v^1 w^1 + v^2 w^2 + \cdots + v^n w^n.$$

The following functions $\mathbb{R}^n \times \mathbb{R}^n \to \mathbb{R}$ are also bilinear forms:

$$
\begin{aligned}
\langle \vec{v}, \vec{w} \rangle_1 &= v_1 w_2 + v_2 w_1 + v_3 w_3 \cdots + v_n w_n, \\
\langle \vec{v}, \vec{w} \rangle_2 &= 2 v_1 w_1 + v_2 w_2 + \cdots + v_n w_n + v_1 w_n, \\
\langle \vec{v}, \vec{w} \rangle_3 &= v_1 w_2 - v_2 w_1.
\end{aligned}
\tag{4.17}
$$

Despite the variety depicted in the above example, bilinear forms on finite dimensional vector spaces can be completely characterized by a single matrix.

Proposition 4.2.3. *Let V and W be finite-dimensional vector spaces, with $\dim V = m$ and $\dim W = n$. Let $\langle \cdot, \cdot \rangle$ be a bilinear form on $V \times W$. Given ordered bases $\mathcal{A}$ of V and $\mathcal{B}$ of W, there exists a unique $m \times n$ matrix M such that*

$$\langle v, w \rangle = [v]_{\mathcal{A}}^{\top} C \, [w]_{\mathcal{B}}.$$

Furthermore, if $\mathcal{A} = (e_1, e_2, \ldots, e_m)$ and $\mathcal{B} = (u_1, u_2, \ldots, u_n)$, then the entries of C are $c_{ij} = \langle e_i, u_j \rangle$ for $1 \le i \le m$ and $1 \le j \le n$.

Proof. Let $v \in V$ and $w \in W$ be vectors with coordinates

$$
[v] = \begin{pmatrix} v^1 \\ \vdots \\ v^m \end{pmatrix}
\quad \text{and} \quad
[w] = \begin{pmatrix} w^1 \\ \vdots \\ w^n \end{pmatrix}.
$$

Then since $\langle \cdot, \cdot \rangle$ is bilinear,

$$\langle v, w \rangle = \langle v^i e_i, w^j u_j \rangle = v^i w^j \langle e_i, u_j \rangle. \quad \text{(ESC)} \tag{4.18}$$

Setting $c_{ij} = \langle e_i, u_j \rangle$ and the matrix $C = (c_{ij})$, for $1 \le i \le m$, we have $\langle e_i, u_j \rangle w^j$ (ESC), which are the coordinates of $C[w]$. Then (4.18) shows that $\langle v, w \rangle = [v]^{\top} C[w]$, where $[v]^{\top}$ means the coordinates of v are written in a row vector as opposed to a column vector. $\qquad \square$

Definition 4.2.4. The mn constants (c_{ij}) described in Proposition 4.2.3 are called the components of $\langle \cdot, \cdot \rangle$ with respect to the ordered bases $\mathcal{A}$ and $\mathcal{B}$.

We emphasize that it is appropriate to use two subscript indices for (c_{ij}) in light of the comments at the end of the previous section about contravariant and covariant indices. Using the Einstein summation convention, we would write the evaluation of $\langle v, w \rangle$ as

$$\langle v, w \rangle = c_{ij} v^i w^j.$$

This hints that both indices are covariant. This is the content of the following proposition.

Example 4.2.5. Let $V = \mathbb{R}^n$, use the standard basis, and consider the bilinear forms in Example 4.2.2. First, note that the matrix for the dot product is just the identity matrix

$$\vec{v} \cdot \vec{w} = \vec{v}^\top \vec{w} = \vec{v}^\top I_n \vec{w}.$$

For the other forms, it is easy to see that

$$\langle \vec{v}, \vec{w} \rangle_1 = \vec{v}^\top \begin{pmatrix} 0 & 1 & 0 & \cdots & 0 \\ 1 & 0 & 0 & \cdots & 0 \\ 0 & 0 & 1 & \cdots & 0 \\ \vdots & \vdots & \vdots & \ddots & \vdots \\ 0 & 0 & 0 & \cdots & 1 \end{pmatrix} \vec{w}$$

$$\langle \vec{v}, \vec{w} \rangle_2 = \vec{v}^\top \begin{pmatrix} 2 & 0 & \cdots & 1 \\ 0 & 1 & \cdots & 0 \\ \vdots & \vdots & \ddots & \vdots \\ 0 & 0 & \cdots & 1 \end{pmatrix} \vec{w}$$

$$\langle \vec{v}, \vec{w} \rangle_3 = \vec{v}^\top \begin{pmatrix} 0 & 1 & 0 & \cdots & 0 \\ -1 & 0 & 0 & \cdots & 0 \\ \vdots & \vdots & & \ddots & \vdots \\ 0 & 0 & 0 & \cdots & 0 \end{pmatrix} \vec{w}$$

Proposition 4.2.6. *Let V and W be finite-dimensional vector spaces. Suppose that $\mathcal{A}$ and $\mathcal{A}'$ are ordered bases on V and that $\mathcal{B}$ and $\mathcal{B}'$ are ordered bases on W. Let $P = P_{\mathcal{A}'}^{\mathcal{A}}$ be the coordinate change matrix on V from $\mathcal{A}$ and $\mathcal{A}'$ and let $Q = Q_{\mathcal{B}'}^{\mathcal{B}}$ be the coordinate change matrix on W from $\mathcal{B}$ and $\mathcal{B}'$. Let $\langle \cdot, \cdot \rangle$ be a bilinear form on $V \times W$ with components (c_{ij}) with respect to $\mathcal{A}$ and $\mathcal{B}$ and with components $(\bar{c}_{kl})$ with respect to $\mathcal{A}'$ and $\mathcal{B}'$. Then*

$$\bar{c}_{k\ell} = \breve{p}_k^i \breve{q}_\ell^j c_{ij},$$

where $(\breve{p}_k^i)$ are the components of P^{-1} and $(\breve{q}_\ell^j)$ are the components of Q^{-1}.

Proof. (The proof is left as an exercise for the reader.)   □

Definition 4.2.7. Let V and W be vector spaces over K, and let $\langle \cdot, \cdot \rangle$ be a bilinear form on $V \times W$. Then $\langle \cdot, \cdot \rangle$ is called

1. *nondegenerate on the left* if for all nonzero $v \in V$, there exists $w \in W$ such that $\langle v, w \rangle \neq 0$;

2. *nondegenerate on the right* if for all nonzero $w \in W$, there exists $v \in V$ such that $\langle v, w \rangle \neq 0$;

3. *nondegenerate* if it is nondegenerate on the right and on the left.

Furthermore, the *rank* of $\langle \cdot, \cdot \rangle$ is the rank of its associated matrix with respect to any basis on V and W.

Basic facts about the rank of a matrix imply that if a form is nondegenerate on the left, then the number of rows of its associated matrix C is equal to the rank of the form. If a form is nondegenerate on the right, then the number of columns of C is equal to the rank of the form. Hence, a form can only be nondegenerate if $\dim V = \dim W$.

4.2.2   Bilinear Forms on V

Many applications of bilinear forms involve a bilinear form $\langle \cdot, \cdot \rangle$ on V.

When we consider the components of a bilinear form on V with respect to bases, we always assume that $\mathcal{A} = \mathcal{B}$. The components (c_{ij}) described in Proposition 4.2.3 can be written as an $n \times n$ matrix. In Proposition 4.2.6, we also suppose that $\mathcal{A}' = \mathcal{B}'$ so that change of coordinate matrices are equal, $P = Q$. Then

$$\bar{c}_{k\ell} = \breve{p}_k^i \breve{p}_\ell^j c_{ij}.$$

The matrices of components (c_{ij}) and $(\bar{c}_{k\ell})$ are not necessarily similar. If they were, they would satisfy (4.7). Consequently, though we do depict the components of a bilinear form according to Proposition 4.2.3, the matrix does not behave under coordinate changes like a matrix that represents a linear transformation. We leave it as an exercise for the reader to prove that

$$\overline{C} = (P^{-1})^\top C P^{-1}, \tag{4.19}$$

where C is the matrix (c_{ij}) and $\overline{C}$ is the matrix with components $(\bar{c}_{k\ell})$. Hence, C and $\overline{C}$ are similar only if P is orthogonal.

Definition 4.2.8. A bilinear form $\langle \cdot, \cdot \rangle$ on a vector space V is called

1. *symmetric* if $\langle y, x \rangle = \langle x, y \rangle$ for all $x, y \in V$; and

2. *antisymmetric* (or *skew-symmetric*) if $\langle y, x \rangle = -\langle x, y \rangle$ for all $x, y \in V$.

From Proposition 4.2.3, we see that $\langle \cdot, \cdot \rangle$ is symmetric if and only if its component matrix C is symmetric, and is antisymmetric if and only if C is antisymmetric.

By way of example, referring to the three bilinear forms on $\mathbb{R}^n$ in Example 4.2.2, $\langle \cdot, \cdot \rangle_1$ is symmetric and nondegenerate; $\langle \cdot, \cdot \rangle_2$ is nondegenerate but neither symmetric nor antisymmetric; $\langle \cdot, \cdot \rangle_3$ is antisymmetric and degenerate.

Proposition 4.2.10 below gives a key characterization of both symmetric and antisymmetric bilinear forms. Its proof repeatedly uses the notion of a perpendicular subspace.

Definition 4.2.9. Let V be a vector space with a bilinear form $\langle \cdot, \cdot \rangle$. If W is a subspace of V, the set

$$W^\perp = \{v \in V \mid \langle v, w \rangle \text{ for all } w \in W\}$$

is called the $\langle \cdot, \cdot \rangle$-*orthogonal* subspace to W.

Proposition 4.2.10. *Let* $\langle \cdot, \cdot \rangle$ *be a bilinear form on a vector space* V *with* $\dim V = n$. *Let* I_k *denote the* $k \times k$ *identity matrix.*

1. *If* $\langle \cdot, \cdot \rangle$ *is symmetric, there exists a basis* $\mathcal{B}$ *relative to which the component matrix is*

$$\begin{pmatrix} I_p & 0 & 0 \\ 0 & -I_q & 0 \\ 0 & 0 & 0 \end{pmatrix}, \tag{4.20}$$

   for some nonnegative integers p *and* q.

2. *If* $\langle \cdot, \cdot \rangle$ *is antisymmetric, there exists a basis* $\mathcal{B}$ *relative to which the component matrix is*

$$\begin{pmatrix} 0 & I_k & 0 \\ -I_k & 0 & 0 \\ 0 & 0 & 0 \end{pmatrix}. \tag{4.21}$$

Proof. (1) It is easy to check that a symmetric form $\langle \cdots, \cdot \rangle$ satisfies

$$\langle v, w \rangle = \frac{1}{4} \left(\langle v + w, v + w \rangle - \langle v - w, v - w \rangle \right). \tag{4.22}$$

Consequently, whenever the restriction $\langle \cdot, \cdot \rangle|_W$ to a subspace W is not trivial, there is a $w \in W$ with $\langle w, w \rangle \neq 0$.

Suppose that $\langle \cdot, \cdot \rangle|_V \neq 0$. Then there exists $e'_1 \in V$ with $\langle e'_1, e'_1 \rangle \neq 0$. Defining $e_1 = e'_1 / \sqrt{|\langle e'_1, e'_1 \rangle|}$, we have $\varepsilon_1 = \langle e_1, e_1 \rangle = \pm 1$. Let $V_1 = \text{Span}(e_1)$ and $W_1 = V_1^\perp$. Then W_1 is a subspace with $V_1 \cap W_1 = \{0\}$. Furthermore, for all $v \in V$, we have $v - \varepsilon_1 \langle v, e_1 \rangle e_1 \in W_1$, so $V_1 + W_1 = V$. Hence, V_1 and W_1 are complementary subspaces.

If $\langle \cdot, \cdot \rangle|_{W_1}$ is not trivial, then there exists some $e_2 \in W_1$ with $\varepsilon_2 = \langle e_2, e_2 \rangle = \pm 1$. We then define $V_2 = \text{Span}(e_1, e_2)$ and $W_2 = V_2^\perp$. By the same reasoning as above, V_2

and W_2 are complementary subspaces. We repeat this process until $\langle \cdot, \cdot \rangle$ restricted to some W_k is trivial.

Let $\mathcal{B}$ be an ordered basis consisting of $(e_1, e_2, \ldots, e_k)$ permuted so that all the vectors with $\varepsilon_i > 0$ come first, followed by any basis of W_k. With respect to $\mathcal{B}$, the form has the matrix described in (4.20).

(2) Since $\langle \cdot, \cdot \rangle$ is antisymmetric $\langle v, v \rangle = 0$ for all $v \in V$. If $\langle \cdot, \cdot \rangle$ is not trivial on V, there exist two linear independent vectors e_1, u_1 such that $\langle e_1, u_1 \rangle \neq 0$. By rescaling one of them, we can assume that $\langle e_1, u_1 \rangle = 1$. Define $V_1 = \mathrm{Span}(e_1, u_1)$. The matrix of $\langle \cdot, \cdot \rangle|_{V_1}$ with respect to the ordered basis (e_1, u_1) is

$$\begin{pmatrix} 0 & 1 \\ -1 & 0 \end{pmatrix}.$$

Define $W_1 = V_1^\perp$. We note that $V_1 \cap W_1 = \{0\}$. Clearly, for all $v \in V$,

$$v - \langle v, u_1 \rangle e_1 + \langle v, e_1 \rangle u_1 \in W_1.$$

Thus $V_1 + W_1 = V$, so since $V_1 \cap W_1 = \{0\}$, V_1 and W_1 are complementary subspaces in V.

As in part (1), if $\langle \cdot, \cdot \rangle$ restricted to W_1 is not trivial, then we can repeat the procedure on W_1 and construct e_2 and u_2 such that $\langle e_2, u_2 \rangle = 1$, and so forth. We repeat this until $\langle \cdot, \cdot \rangle$ restricted to W_k is trivial. Then define $\mathcal{B}$ as the ordered basis consisting first of $(e_1, \ldots, e_k, u_1, \ldots, u_k)$, followed by an basis of W_k. Then the matrix of $\langle \cdot, \cdot \rangle$ with respect to $\mathcal{B}$ is (4.21). $\qquad\square$

Three specific types of bilinear forms play important roles in this text: inner products, symplectic forms, and Minkowski metric. Each leads to a different kind of geometry. We mention them here together to show their similar origins.

Definition 4.2.11. Let V be a vector space over $\mathbb{R}$. An *inner product* on V is a bilinear form $\langle \cdot, \cdot \rangle$ on V that is symmetric and positive-definite, i.e., $\langle v, v \rangle > 0$ for all $v \in V - \{0\}$. We call the pair $(V, \langle \cdot, \cdot \rangle)$ an *inner product space*.

Inner products are often introduced in elementary linear algebra courses. We remind the reader that from any inner product space $(V, \langle \cdot, \cdot \rangle)$, we can generalize geometric concepts that originally arise in connection to the dot product on $\mathbb{R}^n$.

- We defined the *magnitude* of an elements $v \in V$ as $\|v\| = \sqrt{\langle v, v \rangle}$.

- The Cauchy-Schwartz inequality holds: $|\langle v, w \rangle| \leq \|v\| \, \|w\|$ for all $v, w \in V$.

- The *angle* between two vectors is θ, satisfying $0 \leq \theta \leq \pi$ and

$$\cos \theta = \frac{\langle v, w \rangle}{\|v\| \, \|w\|}.$$

- Two elements v and w are *orthogonal* to each other if $\langle v, w \rangle = 0$.

- We can perform Gram-Schmidt orthonormalization on V.

- If we define the function $d : V \times V \to \mathbb{R}^{\geq 0}$ by $d(x, y) = \|x - y\|$, then d satisfies the triangle inequality and is a metric on V, making (V, d) into a metric space. (See Section A.1.)

Definition 4.2.12. Let V be a vector space over $\mathbb{R}$. A *symplectic* form on V is a nondegenerate, antisymmetric bilinear form.

Definition 4.2.13. Let V be a finite-dimensional vector space over $\mathbb{R}$. A *Minkowski metric*, sometimes called a *Lorentz metric*, on V is a symmetric bilinear form for which there exists a basis that has a component matrix of either

$$
\begin{pmatrix}
-1 & 0 & \cdots & 0 \\
0 & 1 & \cdots & 0 \\
\vdots & \vdots & \ddots & \vdots \\
0 & 0 & \cdots & 1
\end{pmatrix}
\quad \text{or} \quad
\begin{pmatrix}
1 & 0 & \cdots & 0 \\
0 & -1 & \cdots & 0 \\
\vdots & \vdots & \ddots & \vdots \\
0 & 0 & \cdots & -1
\end{pmatrix}.
$$

4.2.3   Signature of a Symmetric Bilinear Form

Theorem 4.2.14 (Sylvester's Law of Inertia). *Let $\langle \cdot, \cdot \rangle$ be a symmetric bilinear form on V with $\dim V = n$. Setting $r = n - (p + q)$, the triple of nonnegative integers (p, q, r) arising in (4.20) is independent of the basis.*

Proof. Let $\mathcal{B} = (e_1, e_2, \ldots, e_n)$ be an ordered basis of V with respect to which the component matrix of $\langle \cdot, \cdot \rangle$ is given in (4.20). The rank of $\langle \cdot, \cdot \rangle$, which is independent of any basis, is $p + q$.

Let V_1 be a subspace of V of maximal dimension such that $\langle \cdot, \cdot \rangle$ restricted to V_1 is positive-definite. Then $\dim V_1 = p' \geq p$ because $\langle \cdot, \cdot \rangle$ is positive-definite over $\mathrm{Span}(e_1, \ldots, e_p)$. By Proposition 4.2.10, there is a basis $\mathcal{B}_1$ of V_1 with respect to which that matrix of the form is $I_{p'}$.

Let V_2 be a subspace of V of maximal dimension such that $\langle \cdot, \cdot \rangle$ restricted to V_1 is is negative-definite, i.e., for all $v \in V_2 - \{0\}$, $\langle v, v \rangle < 0$. Over $\mathrm{Span}(e_{p+1}, \ldots, e_{p+q})$, the form is positive-definite, so $\dim V_2 = p' \geq p$. By Proposition 4.2.10, there is a basis $\mathcal{B}_2$ of V_2 with respect to which that matrix of the form is $-I_{q'}$.

Clearly, $V_1 \cap V_2 = \{0\}$ so the subspace $V_1 + V_2$ has dimension $p' + q'$. Assume that $p' > p$ or $q' > q$. Then with respect to $\mathcal{B}_1 \cup \mathcal{B}_2$, the restriction of $\langle \cdot, \cdot \rangle$ to $V_1 + V_2$ is

$$
\begin{pmatrix}
I_{p'} & 0 \\
0 & -I_{q'}
\end{pmatrix},
$$

which implies that the rank of $\langle \cdot, \cdot \rangle$ on V is greater than $p + q$. This is a contradiction. We deduce that $\dim V_1 = p$ and $\dim V_2 = q$. Since V_1 and V_2 were defined without reference to a basis, the theorem follows. $\square$

The traditional statement of Sylvester's Law of Inertia is slightly different: If A is a symmetric matrix and S is any invertible matrix such that $D = SAS^\top$ is diagonal, then the number of negative elements in D is the same regardless of S.

Definition 4.2.15. Let $\langle \cdot, \cdot \rangle$ be a symmetric bilinear form on a finite-dimensional real vector space. The triple of nonnegative integers (p, q, r) is called the *signature* of $\langle \cdot, \cdot \rangle$.

We point out the following properties and their relation to the signature (p, q, r). A symmetric bilinear form is:

- nondegenerate if and only if $r = 0$;

- an inner product if and only if $(p, q, r) = (n, 0, 0)$.

- a Minkowski metric if (p, q, r) is either $(n - 1, 1, 0)$ or $(1, n - 1, 0)$.

PROBLEMS

4.2.1. Prove Proposition 4.2.6.

4.2.2. Prove Equation (4.19).

4.2.3. Prove that every inner product on a real vector space is nondegenerate.

4.2.4. Let $\langle \cdot, \cdot \rangle$ be a bilinear form on V. Fix $v \in V$ and define $i_v \in V^*$ as the element such that $i_v(w) = \langle v, w \rangle$. Let $\mathcal{B} = \{e_1, e_2 \ldots, e_n\}$ be a basis of V, and let $\mathcal{B}^* = \{e^{*1}, e^{*2} \ldots, e^{*n}\}$ be the cobasis of V^*.

(a) Prove that $\psi : V \to V^*$ defined by $\psi(v) = i_v$ is a linear transformation.

(b) Prove that in coordinates $\left[i_v\right]_{\mathcal{B}^*} = C^T \left[v\right]_{\mathcal{B}}$, where $c_{jk} = \langle e_j, e_k \rangle$. [Comment on notation: we think of $\left[i_v\right]_{\mathcal{B}^*}$ as a column vector, whereas we think of $\left[i_v\right]_{\mathcal{B}}$ as a row vector with $\left[i_v\right]_{\mathcal{B}} = \left(\left[i_v\right]_{\mathcal{B}^*}\right)^\top$.]

(c) Prove that ψ is invertible if and only if $\langle \cdot, \cdot \rangle$ is nondegenerate.

[If $\langle \cdot, \cdot \rangle$ is an inner product, we denote i_v by $v^\flat$ since it lowers the indices of the components of v, i.e., turns a vector into a covector. The components of $v^\flat$ with respect to $\mathcal{B}^*$ are $c_{jk}v^k$.]

4.2.5. Let $\langle \cdot, \cdot \rangle$ be a nondegenerate bilinear form on V and refer to the previous exercise for notations. Let C be the component matrix of $\langle \cdot, \cdot \rangle$ with respect to some basis $\mathcal{B}$. Define the function $\langle \cdot, \cdot \rangle^*$ on V^* by $\langle i_u, i_v \rangle^* = \langle v, u \rangle$, or in other words $\langle \eta, \tau \rangle^* = \langle i^{-1}(\tau), i^{-1}(\eta) \rangle$.

(a) Prove that $\langle \cdot, \cdot \rangle^*$ with respect to $\mathcal{B}^*$ is a bilinear form on V^*. Prove also that if $\langle \cdot, \cdot \rangle$ is an inner product, then so is $\langle \cdot, \cdot \rangle^*$.

(b) Prove that the component matrix of $\langle \cdot, \cdot \rangle^*$ with respect to $\mathcal{B}^*$ is C^{-1}.

(c) Define $\Psi_\lambda : V^* \to V^{**}$ by $\Psi_\lambda(\mu) = \langle \lambda, \mu \rangle^*$. Under the canonical isomorphism between V and V^{**}, show that $\Psi_{i_v} = v$ for all $v \in V$.

[In parallel with the previous exercise, if $\langle \cdot, \cdot \rangle$ is an inner product, we denote Ψ_λ by $\lambda^\sharp$ since it raises the indices of the components of λ, i.e., turns a covector into a vector. Writing (c^{jk}) as the components of C^{-1}, the components of $\lambda^\sharp$ with respect to $\mathcal{B}$ are $c^{jk}\lambda_k$.]

4.2.6. Let $\langle \cdot, \cdot \rangle$ be a bilinear form on V. Let W be a subspace of V. Consider the orthogonal subspace from Definition 4.2.9.

    (a) Prove that $W^\perp$ is indeed a subspace of V.

    (b) Prove that $W \subset W^\perp$ if and only if the form $\langle \cdot, \cdot \rangle$ restricted to W is identically 0. [When this is holds, W is called an *isotropic* subspace of V.]

    (c) Prove that if $\langle \cdot, \cdot \rangle$ is symmetric, then $W^\perp \cap W = \{0\}$.

    (d) Prove that if $\langle \cdot, \cdot \rangle$ is antisymmetric, it is not necessarily true that $W^\perp \cap W = \{0\}$.

4.2.7. Let V with $\dim V = 2k$ be equipped with a symplectic form. A *Lagrangian* subspace of V is one in which $L^\perp = L$. Prove that $\dim L = k$.

4.2.8. Let $\langle \cdot, \cdot \rangle$ be a bilinear form on V and let W, W_1 and W_2 be subspaces of V. Prove the following.

    (a) $W_1 \subset W_2$ implies $W_2^\perp \subset W_1^\perp$.

    (b) $(W_1 + W_2)^\perp = W_1^\perp \cap W_2^\perp$.

    (c) $(W_1 \cap W_2)^\perp = W_1^\perp + W_2^\perp$.

    (d) If $\langle \cdot, \cdot \rangle$ is nondegerate, then $(W^\perp)^\perp = W$.

4.2.9. Let V be a vector space over $\mathbb{C}$. An inner product $\langle \cdot, \cdot \rangle$ over V is a function $V \times V \to \mathbb{C}$ that is (1) conjugate symmetric: $\langle x, y \rangle = \overline{\langle y, x \rangle}$ for all $x, y \in V$; (2) linear in the first entry; (3) positive-definite. Prove that there is a basis $\mathcal{B}$ of V with respect to which, for all $x, y \in V$,

$$\langle x, y \rangle = x^1 \overline{y^1} + x^2 \overline{y^2} + \cdots + x^n \overline{y^n},$$

where $[x]_\mathcal{B} = (x^i)$ and $[y]_\mathcal{B} = (y^i)$.

4.3 Adjoint, Self-Adjoint, and Automorphisms

In applications of bilinear forms to geometry, linear transformations that preserve the form play a key role.

Suppose that V is a finite dimensional vector space and $\mathcal{B}$ is a basis. Let $\langle \cdot, \cdot \rangle$ be a nondegenerate bilinear form with component matrix C with respect to $\mathcal{B}$. If $L : V \to V$ is a linear transformation with $[L]_\mathcal{B}^\mathcal{B} = A$, then

$$\langle L(v), w \rangle = (A[v])^\top C[w] = [v]^\top A^\top C[w].$$

There exists a unique linear transformation $L^\dagger : V \to V$ such that

$$\langle L(v), w \rangle = \langle v, L^\dagger(w) \rangle \quad \text{for all } v, w \in V.$$

We find the associated matrix $A^\dagger$ of $L^\dagger$ by remarking that if

$$[v]^\top A^\top C[w] = [v]^\top C(A^\dagger[w])$$

for all $v, w \in V$, then $A^\top C = C A^\dagger$ as matrices. Hence,

$$A^\dagger = C^{-1} A^\top C. \tag{4.23}$$

Definition 4.3.1. The linear transformation $L^\dagger$ such that $\langle L(v), w \rangle = \langle v, L^\dagger(w) \rangle$ for all $v, w \in V$ is called the *adjoint operator* to L with respect to $\langle \cdot, \cdot \rangle$.

More generally, let V and W be vector spaces equipped with nondegenerate bilinear forms $\langle \cdot, \cdot \rangle_V$ and $\langle \cdot, \cdot \rangle_W$. Let $L : V \to W$ be a linear transformation. Then there exists a unique linear map $L^\dagger : W \to V$ such that

$$\langle L(v), w \rangle_W = \langle v, L^\dagger(w) \rangle_V.$$

We also call $L^\dagger$ the adjoint of L with respect to these forms. If C_1 is the matrix corresponding to $\langle \cdot, \cdot \rangle_V$ and C_2 is the matrix corresponding to $\langle \cdot, \cdot \rangle_W$ with respect to specific bases on V and W, and if A is the matrix representing L, then the adjoint matrix $A^\dagger$ of $L^\dagger$ is

$$A^\dagger = C_1^{-1} A^\top C_2.$$

Example 4.3.2. Let $L : \mathbb{R}^n \to \mathbb{R}^m$ be a linear transformation between Euclidean spaces, with matrix A with respect to the standard bases. For all $\vec{v}, \vec{w} \in \mathbb{R}^n$,

$$L(\vec{v}) \cdot \vec{w} = (A\vec{v}) \cdot \vec{w} = (A\vec{v})^\top \vec{w} = \vec{v}^\top A^\top \vec{w} = \vec{v} \cdot (A^\top \vec{w}).$$

Therefore, the transpose $A^\top$ is the matrix corresponding to the adjoint of L when we assume $\mathbb{R}^n$ and $\mathbb{R}^m$ are equipped with the usual dot product.

Proposition 4.3.3. *Let V, W, and U be vector spaces equipped with nondegenerate bilinear forms. Then the following formulas hold for the adjoint:*

1. *$(L_1 + L_2)^\dagger = L_1^\dagger + L_2^\dagger$ for all $L_1, L_2 \in \mathrm{Hom}(V, W)$.*
2. *$(cL)^\dagger = cL^\dagger$ for all $L \in \mathrm{Hom}(V, W)$ and all $c \in K$.*
3. *$(L_2 \circ L_1)^\dagger = L_1^\dagger \circ L_2^\dagger$ for all $L_1 \in \mathrm{Hom}(V, W)$ and all $L_2 \in \mathrm{Hom}(W, U)$.*

Proof. (Left as an exercise for the reader.) $\qquad\square$

We are often lead to consider two particular types of linear transformations associated to the adjoint: automorphisms with respect to the form and self-adjoint transformations. We describe these in the following paragraphs.

Definition 4.3.4. Let V be a vector space with a nondegenerate bilinear form $f = \langle \cdot, \cdot \rangle$. A linear transformation $L : V \to V$ is called *self-adjoint* with respect to this form if $L = L^\dagger$. We use the same term for the matrix representing L.

Example 4.3.5. Consider $V = \mathbb{R}^n$ equipped with the dot product. A matrix A is self-adjoint with respect to the dot product if $A = A^\top$, hence it is symmetric.

Because of this example, some authors refer to $L^\dagger$ as defined above as the transpose of L with respect to a form (or forms) and use the word adjoint of a linear transformation only in the cases when V and W are vector spaces over $\mathbb{C}$ and when the form $\langle \cdot, \cdot \rangle$ is sesquilinear. (A sesquilinear form on a complex vector space is one that satisfies conditions (1) and (2) in Exercise 4.2.9. See [31] for a discussion on sesquilinear forms.)

Definition 4.3.6. Let V be a vector space with a nondegenerate bilinear form $f = \langle \cdot, \cdot \rangle$. An *automorphism* of (V, f) is an invertible linear transformation $L : V \to V$ such that

$$\langle L(v_1), L(v_2) \rangle = \langle v_1, v_2 \rangle \qquad \text{for all } v_1, v_2 \in V. \tag{4.24}$$

The property (4.24) shows that automorphisms preserve the bilinear form. This condition is equivalent to $\langle v_1, L^\dagger(L(v_2)) \rangle = \langle v_1, v_2 \rangle$ for all $v_1, v_2 \in V$. Since f is nondegenerate, then $L^\dagger \circ L = \mathrm{id}_V$, where id_V is the identity on V. This gives the following proposition.

Proposition 4.3.7. *A linear transformation $L : V \to V$ is an automorphism of (V, f) if and only if L is invertible with $L^\dagger \circ L = \mathrm{id}_V$.*

If V is a finite dimensional vector space, we could simplify Definition 4.3.6. Suppose the V is finite dimensional and that L is a linear transformation satisfying (4.24). We can still conclude that $L^\dagger \circ L = \mathrm{id}_V$. By properties of functions, we deduce that L is injective. By the Rank-Nullity Theorem, an injective linear transformation between vector spaces of the same finite dimension is invertible. Hence, when V is finite dimensional condition (4.24) implies that L is invertible.

Proposition 4.3.8. *Let (V, f) be a vector space equipped with a nondegenerate bilinear form. Then the set S of automorphisms of (V, f) satisfies the following:*

1. S is closed under composition: $L_1 \circ L_2 \in S$ for all $L_1, L_2 \in S$.

2. The identity id_V is in S.

3. If $L \in S$, then L is invertible and $L^{-1} \in S$.

Proof. We have already discussed the first property, and the second is obvious. For the third property, note that for all $L \in S$, we have $L^\dagger \circ L = \mathrm{id}$ and L is invertible. Thus $L^{-1} = L^\dagger$ so $L \circ L^\dagger = \mathrm{id}_V$ as well. Furthermore, for all $v, w \in V$,

$$\langle L^\dagger(v), L^\dagger(w) \rangle = \langle L \circ L^\dagger(v), L \circ L^\dagger(w) \rangle \qquad \text{because } L \text{ is an automorphism}$$
$$= \langle v, w \rangle \qquad .$$

Thus, $L^\dagger$ is an automorphism. $\square$

(Using the language of modern algebra, Proposition 4.3.8, along with the associativity of linear transformations, shows that the set of automorphisms of (V, f) is a group. This group is denoted by $\mathrm{Aut}(V, f)$.)

If for a vector space V has an ordered basis $\mathcal{B} = \{e_1, e_2, \ldots, e_n\}$, then (4.23) leads to a characterization of matrices of automorphisms. Let C be the matrix associated to the bilinear form f, and let A be the matrix of a linear transformation $L : V \to V$ in reference to $\mathcal{B}$. Then by Proposition 4.3.7, L is an automorphism if and only if

$$A^{-1} = C^{-1}A^\top C. \tag{4.25}$$

Example 4.3.9. Example 4.2.5 indicates that the dot product is a symmetric, nondegenerate bilinear transformation with associated matrix I_n, and Example 4.3.2 shows that the transpose of a matrix is the adjoint of a matrix with respect to the dot product. However, consider the symmetric bilinear forms f_1 and f_2 on $\mathbb{R}^4$ given by the matrices

$$M_1 = \begin{pmatrix} 0 & 1 & 0 & 0 \\ 1 & 0 & 0 & 0 \\ 0 & 0 & 1 & 0 \\ 0 & 0 & 0 & 1 \end{pmatrix} \quad \text{and} \quad M_2 = \begin{pmatrix} 0 & 0 & 0 & 1 \\ 0 & 0 & 1 & 0 \\ 0 & 1 & 0 & 0 \\ 1 & 0 & 0 & 0 \end{pmatrix}.$$

Then a simple calculation using (4.23) shows that the adjoint of $A = (a_{ij})$ with respect to f_1 is

$$A^\dagger = \begin{pmatrix} a_{22} & a_{12} & a_{32} & a_{42} \\ a_{21} & a_{11} & a_{31} & a_{41} \\ a_{23} & a_{13} & a_{33} & a_{43} \\ a_{24} & a_{14} & a_{34} & a_{44} \end{pmatrix},$$

and the adjoint of A with respect to f_2 is

$$A^\dagger = \begin{pmatrix} a_{44} & a_{34} & a_{24} & a_{14} \\ a_{43} & a_{33} & a_{23} & a_{13} \\ a_{42} & a_{32} & a_{22} & a_{12} \\ a_{41} & a_{31} & a_{21} & a_{11} \end{pmatrix}.$$

Now just consider $\langle \cdot, \cdot \rangle_1$, the matrix A correspond to a self-adjoint linear transformation if

$$A = \begin{pmatrix} a & b & c & d \\ e & a & f & g \\ f & c & h & i \\ g & d & i & j \end{pmatrix}.$$

A matrix A corresponds to an automorphism if and only if $A^\dagger A = I_4$, which is a system of 16 quadratic equations in the 16 variables of the entries of A. Many of these equations will be redundant.

Proposition 4.3.10. *If $(V, \langle \cdot, \cdot \rangle)$ is a vector space with a nondegenerate symmetric form, then $L^{\dagger\dagger} = L$ for all $L \in \mathrm{Hom}(V, V)$.*

Proof. By definition, $\langle L(v), w \rangle = \langle v, L^{\dagger}(w) \rangle$ for all $v, w \in V$. Since the form is symmetric, $\langle L(v), w \rangle = \langle w, L(v) \rangle$. Thus,

$$\langle w, L(v) \rangle = \langle v, L^{\dagger}(w) \rangle = \langle L^{\dagger}(w), v \rangle = \langle w, L^{\dagger\dagger}(v) \rangle.$$

Since these equalities hold for all $v, w \in V$ and since $\langle \cdot, \cdot \rangle$ is nondegenerate, we conclude that $L = L^{\dagger\dagger}$. $\qquad\square$

Example 4.3.11. Consider $V = \mathbb{R}^n$ and consider the dot product as a symmetric bilinear form. We know from elementary linear algebra that an $n \times n$ matrix is an automorphism (of the dot product) if $A^{\top} = A^{-1}$, i.e., if A is orthogonal. The set of $n \times n$ orthogonal matrices is denoted by $\mathrm{O}(n)$. This is the set of isometries of $\mathbb{R}^n$ that fix the origin. The special orthogonal group $\mathrm{SO}(n)$ defined in Section 2.3, which are all orthogonal matrices with determinant 1, is the set of rotations in $\mathbb{R}^n$. In particular, if $n = 2$, all orthogonal matrices have the form

$$\begin{pmatrix} \cos\theta & -\varepsilon\sin\theta \\ \sin\theta & \varepsilon\cos\theta \end{pmatrix},$$

for some angle θ and some $\varepsilon = \pm 1$. Such a matrix is in $\mathrm{SO}(2)$ when $\varepsilon = 1$.

Example 4.3.12 (Lorenztian Transformations). Minkowski spacetime is $\mathbb{R}^4$ equipped with the Minkowski metric. It is common to denote Minkowski space by $\mathbb{R}^{3,1}$. Points are located by the quadruple (x^0, x^1, x^2, x^3), with (x^1, x^2, x^3) serving as space variables and x^0 represents time.

With respect to the standard basis of $\mathbb{R}^4$, the Minkowski metric is

$$g((x^0, x^1, x^2, x^3), (y^0, y^1, y^2, y^3)) = -x^0 y^0 + x^1 y^1 + x^2 y^2 + x^3 y^3,$$

so the representing matrix is

$$C = \begin{pmatrix} -1 & 0 \\ 0 & I_3 \end{pmatrix}.$$

Note that $C^{-1} = C$.

We propose to study the automorphisms of the Minkowski metric. Using block diagonal properties, it is not hard to see that any linear transformation with matrix

$$\begin{pmatrix} 1 & 0 \\ 0 & R \end{pmatrix},$$

where $R \in \mathrm{O}(3)$, is an automorphism of this form. This corresponds to an orthogonal transformation in the space variables. Also using block diagonal properties, we can see that

$$\begin{pmatrix} \pm 1 & 0 \\ 0 & I_3 \end{pmatrix}$$

are the only two matrices corresponding to automorphism that fix all the space variables.

To understand automorphisms that intermingle the space and time variables, we consider the situation on $\mathbb{R}^2$ where the Minkowski metric has the matrix

$$\begin{pmatrix} -1 & 0 \\ 0 & 1 \end{pmatrix}.$$

For a generic matrix A, the adjoint with respect to this bilinear form is

$$A^\dagger = C^{-1}A^\top C = \begin{pmatrix} -1 & 0 \\ 0 & 1 \end{pmatrix} \begin{pmatrix} a & b \\ c & d \end{pmatrix} \begin{pmatrix} -1 & 0 \\ 0 & 1 \end{pmatrix} = \begin{pmatrix} a & -c \\ -b & d \end{pmatrix}.$$

The matrix A represents an automorphism when $A^\dagger A = I_2$, which is equivalent to

$$\begin{cases} a^2 - c^2 = 1 \\ ab - cd = 0 \\ -b^2 + d^2 = 1. \end{cases}$$

As the first equation parametrizes a hyperbola, there exist $\varepsilon_1 = \pm 1$ and $u \in \mathbb{R}$ such $a = \varepsilon_1 \cosh u$ and $c = \sinh u$. By the second equation, we deduce that $b = cd/a = d\varepsilon_1 \tanh u$. Then from $-b^2 + d^2 = 1$, we deduce that

$$-d^2 \tanh^2 u + d^2 = d^2 \operatorname{sech}^2 u = 1,$$

so $d = \varepsilon_2 \cosh u$ for some $\varepsilon_3 = \pm 1$, from which we also deduce $b = \varepsilon_1 \varepsilon_2 \sinh u$. Hence, the matrix A has the form

$$A = \begin{pmatrix} \varepsilon_1 \cosh u & \varepsilon_1 \varepsilon_2 \sinh u \\ \sinh u & \varepsilon_2 \cosh u \end{pmatrix}.$$

This gives uniquely the all the matrices representing automorphisms of the Minkowski metric on $\mathbb{R}^{1,1}$. The set of automorphisms on $\mathbb{R}^{3,1}$ is the smallest subset of $\mathrm{GL}_4(\mathbb{R})$ closed under multiplication and taking inverses that includes every matrix of the form

$$\begin{pmatrix} \varepsilon_1 \cosh u & \varepsilon_1 \varepsilon_2 \sinh u & 0 \\ \sinh u & \varepsilon_2 \cosh u & \\ 0 & & I_2 \end{pmatrix} \quad \text{and} \quad \begin{pmatrix} 1 & 0 \\ 0 & R \end{pmatrix},$$

where $R \in O(3)$. This describes all automorphisms on $\mathbb{R}^{3,1}$.

To apply this to special relativity, set $t = x^0$ and $x = x^1$. We imagine that one observer $\mathcal{O}$ has frame axes t and x, and a second observer $\overline{\mathcal{O}}$ with axes $\bar{t}$ and $\bar{x}$ travels with respect to $\mathcal{O}$ along the x-axis with velocity v. In the $\bar{t}\bar{x}$-frame, the $\bar{t}$-axis, namely $\bar{x} = 0$, is the trajectory of $\overline{\mathcal{O}}$, namely the line with equation $vt = x$ in the $\mathcal{O}$ frame. Thus, there are nonzero constants λ and μ such that

$$\mu \begin{pmatrix} 1 \\ 0 \end{pmatrix} = \begin{pmatrix} \varepsilon_1 \cosh u & \varepsilon_1 \varepsilon_2 \sinh u \\ \sinh u & \varepsilon_2 \cosh u \end{pmatrix} \begin{pmatrix} \lambda \\ \lambda v \end{pmatrix} = \lambda \begin{pmatrix} \varepsilon_1 \cosh u + \varepsilon_1 \varepsilon_2 v \sinh u \\ \sinh u + \varepsilon_2 v \cosh u \end{pmatrix}.$$

From the $\bar{x}$ component, we deduce that $v = -\varepsilon_2 \tanh u$. Since $\operatorname{sech}^2 u = 1 - \tanh^2 u$, we get $\cosh u = 1/\sqrt{1 - v^2}$ and $\sinh u = -\varepsilon_2 v/\sqrt{1 - v^2}$. Using the variable v, we have

$$A = \begin{pmatrix} \dfrac{\varepsilon_1}{\sqrt{1 - v^2}} & -\dfrac{\varepsilon_1 v}{\sqrt{1 - v^2}} \\ -\dfrac{\varepsilon_2 v}{\sqrt{1 - v^2}} & \dfrac{\varepsilon_2}{\sqrt{1 - v^2}} \end{pmatrix}. \tag{4.26}$$

Since the range of $\tanh u$ is $(-1, 1)$, we still have all the automorphisms of the Minkowski metric, assuming that $\varepsilon_1 = \pm 1$, $\varepsilon_2 = \pm 1$, and $v \in (-1, 1)$. In $\operatorname{GL}_2(\mathbb{R}^2)$, the subset of matrices of this form consists of 4 connected components, each being a curve parametrized by $v \in (-1, 1)$ and designated by the four possible values of the pair $(\varepsilon_1, \varepsilon_2)$.

Returning to the full context of Minkowski space $\mathbb{R}^{3+1}$, the automorphisms include

$$\Lambda(v) = \begin{pmatrix} \dfrac{\varepsilon_1}{\sqrt{1 - v^2}} & -\dfrac{\varepsilon_1 v}{\sqrt{1 - v^2}} & 0 & 0 \\ -\dfrac{\varepsilon_2 v}{\sqrt{1 - v^2}} & \dfrac{\varepsilon_2}{\sqrt{1 - v^2}} & 0 & 0 \\ 0 & 0 & 1 & 0 \\ 0 & 0 & 0 & 1 \end{pmatrix}. \tag{4.27}$$

This is an example of a Lorentz transformation associated to the vector $(0, v, 0, 0)^\top$. It is also clear that the Minkowski metric is invariant under any orthogonal transformation in the (x^1, x^2, x^3) variables. The group of automorphisms of the Minkowski metric is called the *Lorentz group* and, as a subset of $\operatorname{GL}_4(\mathbb{R})$, consists of any finite product of matrices of the form (4.27) and orthogonal matrices in the space variables. We denote the Lorentz group by $\operatorname{SO}(3, 1)$.

Generalizing (4.27), if $\vec{v}$ is some vector in the space variables with $\|\vec{v}\| < 1$, then the Lorentz transformation associated to $\vec{v}$ is the linear map $\mathbb{R}^4 \to \mathbb{R}^4$ that has the matrix $\Lambda(\vec{v})$ obtained by the composition $M\Lambda(\|v\|)M^{-1}$, where M is some rotation matrix that sends the unit x-direction vector to $\vec{v}/\|\vec{v}\|$. Exercise 4.3.9 gives the exact value of this matrix.

As in the case of Minkowski space $\mathbb{R}^{1,1}$, the freedom of choosing values of ε_1 and ε_2 implies that $\operatorname{O}(3, 1)$ has 4 connected components. Only the component corresponding to $\varepsilon_1 = \varepsilon_2 = 1$ contains the identity matrix I_4. This subset is called the *restricted Lorentz group* and is denoted $\operatorname{SO}^+(3, 1)$. Matrices in the restricted Lorentz group are called *Lorentz transformations*. As we will see in Section 7.2, Lorentz transformations play a central role in special relativity.

PROBLEMS

4.3.1. Let $\langle \cdot, \cdot \rangle$ be a nondegenerate bilinear form on a finite vector space V. Prove that the set of self-adjoint linear transformations is a vector subspace of $\operatorname{Hom}(V, V)$.

4.3.2. Let $\langle \cdot, \cdot \rangle$ be a nondegenerate form on V and let $L \in \operatorname{Hom}(V, V)$. Let C be the component matrix of $\langle \cdot, \cdot \rangle$ with respect to a basis $\mathcal{B}$ and let A be the matrix representing L. Determine the matrix representing $L^{\dagger\dagger}$ and conclude that if $\langle \cdot, \cdot \rangle$ is not symmetric, then it is not necessarily true that $L^{\dagger\dagger} = L$.

4.3.3. Let (V, f) be a vector space equipped with a nondegenerate bilinear form. Prove that the set of automorphisms is not closed under addition.

4.3.4. Let (V, f) be a vector space equipped with a nondegenerate bilinear form. Prove that the set of self-adjoint transformations is closed under composition.

4.3.5. Let L be an automorphism of an inner product space. Prove that the eigenvalues of L are 1 or -1.

4.3.6. Let V and W be finite vector spaces over a field K. Suppose that V and W are equipped with nondegenerate bilinear forms denoted by $\langle\,,\,\rangle_V$ and $\langle\,,\,\rangle_W$, respectively. Let $L : V \to W$ be a surjective linear transformation, and let $L^\dagger$ be its adjoint, namely, $L^\dagger : W \to V$ satisfies

$$\langle L(v), w\rangle_W = \langle v, L^\dagger(w)\rangle_V$$

for all $v \in V$ and $w \in W$.

(a) Show that $L^\dagger$ is injective.

(b) Assume in addition that for all $v \in V$ with $v \neq 0$, $\langle v, v\rangle_V \neq 0$. Then show that $\operatorname{Ker} L$ and $\operatorname{Im} L^\dagger$ are orthogonal complements in V, that is:

    (i) $\operatorname{Ker} L \cap \operatorname{Im} L^\dagger = \{0\}$;

    (ii) for all $v_1 \in \operatorname{Ker} L$ and $v_2 \in \operatorname{Im} L^\dagger$, we have $\langle v_1, v_2\rangle_V = 0$; and

    (iii) all $v \in V$ can be written as $v = v_1 + v_2$, where $v_1 \in \operatorname{Ker} L$ and $v_2 \in \operatorname{Im} L^\dagger$. [Hint: Let $\phi = L \circ L^\dagger : W \to W$. Show that ϕ is invertible. For all $v \in V$, let $v_2 = (L^\dagger \circ \phi^{-1} \circ L)(v)$ and set $v_1 = v - v_2$; show that $v_1 \in \operatorname{Ker} L$.]

4.3.7. The definition for an *isometry* on $\mathbb{R}^n$ is any function $f : \mathbb{R}^n \to \mathbb{R}^n$ satisfying $d(f(\vec{x}), f(\vec{y})) = d(\vec{x}, \vec{y})$ for all $\vec{x}, \vec{y} \in \mathbb{R}^n$. Prove that any isometry on $\mathbb{R}^n$ that fixes the origin is an orthogonal transformation. [Hint: If f is an isometry, first prove that f preserves the dot product between any two vectors; prove that f maps parallelograms to parallelograms; deduce that f is linear.]

4.3.8. Let $\vec{u} = (0, u, 0, 0)^\top$ and $\vec{v} = (0, v, 0, 0)^\top$ be two vectors in Minkowski spacetime $\mathbb{R}^{1+3}$. The matrix $\Lambda(v)\Lambda(u)$ represents the Lorentz transformation from an observer $\mathcal{O}$ to $\mathcal{O}''$ in which $\mathcal{O}''$ moves relative to an observer $\mathcal{O}'$ along the x-axis with velocity v and $\mathcal{O}'$ moves relative to an observer $\mathcal{O}$ along the x-axis with velocity u. Prove that

$$\Lambda(v)\Lambda(u) = \Lambda\left(\frac{u + v}{1 + uv}\right).$$

[This is the velocity addition law in special relativity.]

4.3.9. Consider the velocity vector $\vec{v}$ in $\mathbb{R}^{1+3}$ with $\vec{v} = (0, v_1, v_2, v_3)^\top$ and $v = \|\vec{v}\| < 1$. Call $\gamma = 1/\sqrt{1 - v^2}$ and let $\vec{u} = \vec{v}/\|\vec{v}\| = (0, u_1, u_2, u_3)$. Show by direct verification that

$$\Lambda(\vec{v}) = \begin{pmatrix} \gamma & -v_1\gamma & -v_2\gamma & -v_3\gamma \\ -v_1\gamma & 1 + u_1^2(\gamma - 1) & u_1 u_2(\gamma - 1) & u_1 u_3(\gamma - 1) \\ -v_2\gamma & u_1 u_2(\gamma - 1) & 1 + u_2^2(\gamma - 1) & u_2 u_3(\gamma - 1) \\ -v_3\gamma & u_1 u_3(\gamma - 1) & u_2 u_3(\gamma - 1) & 1 + u_3^2(\gamma - 1) \end{pmatrix}$$

is an automorphism of the Minkowski metric, has $\Lambda(\vec{0}) = I_4$, and satisfies

$$\Lambda(\vec{v}) \begin{pmatrix} 1 \\ v_1 \\ v_2 \\ v_3 \end{pmatrix} = \mu \begin{pmatrix} 1 \\ 0 \\ 0 \\ 0 \end{pmatrix} \qquad \text{for some } \mu \in \mathbb{R}.$$

This is the matrix described at the end of Example 4.3.12.

4.4 Tensor Product

So far in this chapter, we have studied the dual of a vector space, the Hom-space, and bilinear forms on vector spaces. This section generalizes those constructions through what is called the tensor product. The order of presentation clearly betrays this book's mathematical bias in contrast to a physicist's approach: We first present the structure of a tensor product abstractly and only in the subsequent section discuss the components of a tensor and how they change under a change of basis.

(The following construction is a little abstract. The casual reader may feel free to focus attention on the explanations and propositions following Definition 4.4.2.)

Let U, V and W be vector spaces over a field K. Recall that a function $f : V \times W \to U$ is called a bilinear transformation if f is linear in both of its input variables. More precisely, f satisfies

$$f(v_1 + v_2, w) = f(v_1, w) + f(v_2, w), \qquad f(\lambda v, w) = \lambda f(v, w),$$
$$f(v, w_1 + w_2) = f(v, w_1) + f(v, w_2), \qquad f(v, \lambda w) = \lambda f(v, w),$$

for all $v_1, v_2, v \in V$, for all $w_1, w_2, w \in W$, and all $\lambda \in K$.

It is crucial to point out that a bilinear transformation $V \times W \to U$ is not equivalent to a linear transformation $V \oplus W \to U$. A linear transformation $T : V \oplus W \to U$ satisfies

$$T(v_1, w) + T(v_2, w) = T((v_1, w) + (v_2, w)) = T(v_1 + v_2, 2w).$$

for all $v_1, v_2 \in V$ and $w \in W$. The differs from the first axiom of a bilinear transformation. This observation motivates the following important proposition, which leads to the concept of tensor product.

Proposition 4.4.1. *Let U, V, and W be vector spaces over a field K. There exists a unique vector space Z over K and a bilinear transformation $\psi : V \times W \to Z$ such that for any bilinear transformation $f : V \times W \to U$, there exists a unique linear transformation $\bar{f} : Z \to U$ such that $f = \bar{f} \circ \psi$.*

Proof. We first prove the existence of the vector space Z. Consider the set $\bar{Z}$ of formal finite linear combinations

$$c_1(v_1, w_1) + c_2(v_2, w_2) + \cdots + c_l(v_l, w_l),$$

where $v_i \in V$, $w_i \in W$, and $c_i \in K$ for $1 \leq i \leq l$. It is not hard to see that $\bar{Z}$ is a vector space over K. Consider now the subspace $\bar{Z}_{\text{lin}}$ spanned by vectors of the form

$$(v_1 + v_2, w) - (v_1, w) - (v_2, w), \quad (\lambda v, w) - \lambda(v, w),$$
$$(v, w_1 + w_2) - (v, w_1) - (v, w_2), \quad (v, \lambda w) - \lambda(v, w). \tag{4.28}$$

Define Z as the quotient vector space $Z = \bar{Z}/\bar{Z}_{\text{lin}}$. The elements of Z are equivalence classes of elements of $\bar{Z}$ under the equivalence relation $u \sim v$ if and only if $v - u \in \bar{Z}_{\text{lin}}$. (See Exercise 4.1.10 for a description of the vector space quotient.)

Define $\psi : V \times W \to Z$ as the composition $\psi = \pi \circ i$, where $\pi : \bar{Z} \to Z$ is the canonical projection and $i : V \times W \to \bar{Z}$ is the inclusion. The space $\bar{Z}_{\text{lin}}$ is defined in such a way that the canonical projection π turns ψ into a bilinear transformation.

Now given any bilinear transformation $f : V \times W \to U$, we can complete f by linearity to define a linear transformation $\tilde{f}$ from $\bar{Z}$ to U via

$$\tilde{f}\big(c_1(v_1, w_1) + \cdots + c_l(v_l, w_l)\big) = c_1 f(v_1, w_1) + \cdots + c_l f(v_l, w_l).$$

If $z_0 \in \bar{Z}_{\text{lin}}$, then z_0 is a linear combination of elements of the form in Equation (4.28). However, every element of the form given in Equation (4.28) maps to 0 under $\tilde{f}$, so $\tilde{f}(z_0)$. Therefore, if $z_1, z_2 \in \bar{Z}$ are such that $z_1 \sim z_2$, then $z_1 - z_2 = z_0 \in \bar{Z}_{\text{lin}}$, so $\tilde{f}(z_1 - z_2) = 0$ and $\tilde{f}(z_1) = \tilde{f}(z_2)$. Hence, $\tilde{f}$ induces a function $\bar{f} : \bar{Z}/\bar{Z}_{\text{lin}} \to U$. It is easy to check that $\bar{f}$ is a linear transformation and that $f = \bar{f} \circ \psi$. Since the image of ψ spans $\bar{Z}/\bar{Z}_{\text{lin}}$, it follows that the induced map $\bar{f}$ is uniquely determined. This proves the existence of Z.

To prove uniqueness of Z, suppose there is another vector space Z' and a bilinear transformation $\psi' : V \times W \to Z'$ with the desired property. Then there exist $\bar{\psi}$ and $\bar{\psi}'$ such that $\psi' = \bar{\psi}' \circ \psi$ and $\psi = \bar{\psi} \circ \psi'$. Then we have $\psi = \bar{\psi} \circ \bar{\psi}' \circ \psi$. However, $\psi = \text{id}_Z \circ \psi$, and since we know that ψ factors through Z with a unique map, then $\bar{\psi} \circ \bar{\psi}' = \text{id}_Z$. Similarly, we can show that $\bar{\psi}' \circ \bar{\psi} = \text{id}_{Z'}$. Thus, $Z \cong Z'$, and so Z is unique up to a canonical isomorphism. $\qquad\square$

As a first comment about the proof, we observe that $\bar{Z}$ is a "large" vector space; its basis consists of every pair (v, w) with $v \in V$ and $w \in W$, so has cardinality $|V \times W|$.

We can depict the functional relationships described in this proposition by the following commutative diagram.

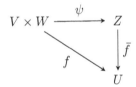

Observe that f and ψ are bilinear transformations while $\bar{f}$ is linear. Consequently, though a bilinear transformation $V \times W \to U$ is not equivalent to a linear transformation $V \oplus W \to U$, it is equivalent to a linear transformation $Z \to U$.

Definition 4.4.2. The vector space Z in the above proposition is called the *tensor product* of V and W and is denoted by $V \otimes W$. The element $\psi(v, w)$ in $V \otimes W$ is denoted by $v \otimes w$.

Elements of $V \otimes W$ are linear combinations of vectors of the form $v \otimes w$, with $v \in V$ and $w \in W$. With this notation, the following identities hold:

$$(v_1 + v_2) \otimes w = v_1 \otimes w + v_2 \otimes w, \qquad (\lambda v) \otimes w = \lambda(v \otimes w),$$
$$v \otimes (w_1 + w_2) = v \otimes w_1 + v \otimes w_2, \qquad v \otimes (\lambda w) = \lambda(v \otimes w). \tag{4.29}$$

Definition 4.4.3. Any element of $V \otimes W$ is often simply called a tensor. A tensor in $V \otimes W$ that can be written as $v \otimes w$ for $v \in V$ and $w \in W$ is called a *pure tensor*.

From the identity $c(v \otimes w) = (cv) \otimes w$, a linear combination of pure tensors is just a sum of pure tensors. We remind the reader that even if V and W are not finite dimensional, linear combinations always consist of a finite number of terms. So every tensor in $V \otimes W$ is a finite sum of pure tensors.

Definition 4.4.4. The *rank* of a tensor $t \in V \otimes W$ is the least integer r such that

$$t = v_1 \otimes w_1 + v_2 \otimes w_2 + \cdots + v_r \otimes w_r$$

for some $v_i \in V$ and $w_i \in W$.

Example 4.4.5. Let $V = \mathbb{R}^3$. Let $t \in V \otimes V$ be $t = \vec{\imath} \otimes \vec{\jmath} + \vec{\imath} \otimes \vec{k}$. Though currently written as a sum of two pure tensors, it is in fact not a tensor of rank 2 but is a pure tensor because $t = \vec{\imath} \otimes (\vec{\jmath} + \vec{k})$.

Proposition 4.4.6. *A pure tensor $v \otimes w$ in $V \otimes W$ is the 0 tensor if and only if $v = 0$ or $w = 0$.*

Proof. By linearity $0 = 0(v' \otimes w) = (0v') \otimes w$ so $0 \otimes w = 0$ and the same is true for if $w = 0$.

Conversely, assume that $v \neq 0$ and $w \neq 0$. Let $\mathcal{B}_1$ be a basis of V containing v and let $\mathcal{B}_2$ be a basis of W containing w. Consider the function $f : V \times W \to K$ defined by $f(x, y) = v^*(x)w^*(y)$, where $v^* \in V^*$ is the dual basis vector to v in $\mathcal{B}_1^*$ and similarly for w^*. The function f is bilinear and nontrivial since $f(v, w) = 1$. Using the construction from the proof of Proposition 4.4.1, $\bar{f} : V \otimes W \to K$ satisfies $\bar{f}(v \otimes w) = f(v, w) = v^*(v)w^*(w) = 1$. Hence, $v \otimes w$ is not the 0 element in $V \otimes W$. $\qquad\square$

Proposition 4.4.7. *Let U, V, and W be three vector spaces over a field K. There exists a unique isomorphism*

$$(U \otimes V) \otimes W \cong U \otimes (V \otimes W)$$

such that

$$(u \otimes v) \otimes w \mapsto u \otimes (v \otimes w)$$

for all $u \in U$, $v \in V$, and $w \in W$.

Proof. By the identities (4.29) in

$$(u_1 \otimes v_1 + u_2 \otimes v_2) \otimes w = (u_1 \otimes v_1) \otimes w + (u_2 \otimes v_2) \otimes w,$$

so every tensor in $(U \otimes V) \otimes W$ is the sum of tensors of the form $(u \otimes v) \otimes w$ with $u \in U$, $v \in V$ and $w \in W$. Hence any linear transformation from $(U \otimes V) \otimes W$ is uniquely determined by how it maps tensors of the form $(u \otimes v) \otimes w$.

Define the function $f : (U \otimes V) \times W \to U \otimes (V \otimes W)$ by

$$f\left(\sum_{i=1}^{s}(u_i \otimes v_i), w_i\right) = \sum_{i=1}^{s} u_i \otimes (v_i \otimes w_i).$$

By distributivity properties of finite sums, it is easy to see that this is bilinear. By Proposition 4.4.1, there exists a unique linear transformation $\bar{f} : (U \otimes V) \otimes W \to U \otimes (V \otimes W)$ satisfying

$$\bar{f}\left(\left(\sum_{i=1}^{s}(u_i \otimes v_i)\right) \otimes w_i\right) = \sum_{i=1}^{s} u_i \otimes (v_i \otimes w_i).$$

Clearly, $\bar{f}((u \otimes v) \otimes w) = u \otimes (v \otimes w)$.

We can construct the inverse linear transformation $\bar{f}^{-1}$ in the same way. We already showed uniqueness, but this establishes the existence of an isomorphism satisfying the desired property. □

In light of Proposition 4.4.7, the notation $U \otimes V \otimes W$ without parentheses is uniquely defined. More generally, this allows us to consider the tensor product $V_1 \otimes V_2 \otimes \cdots \otimes V_k$ of k vector spaces $V_1, V_2, \ldots, V_k$ over the same field. In this general context, we call a pure tensor in $V_1 \otimes V_2 \otimes \cdots \otimes V_k$ any element of the form $v_1 \otimes v_2 \otimes \cdots \otimes v_k$. Again all elements of $V_1 \otimes V_2 \otimes \cdots \otimes V_k$ are finite sums of pure tensors. We also denote by $V^{\otimes k}$ the k-fold tensor product of a vector space V with itself.

Proposition 4.4.8. *Let V and W be two vector spaces over a field K. There exists a unique isomorphism*

$$V \otimes W \cong W \otimes V$$

such that $v \otimes w \mapsto w \otimes v$ for all $v \in V$ and $w \in W$.

Proof. (Left as an exercise for the reader.) □

Proposition 4.4.9. *If V and W are finite-dimensional vector spaces over a field K. If $\mathcal{B}_1 = \{e_1, \ldots, e_m\}$ is a basis of V and $\mathcal{B}_2 = \{f_1, \ldots, f_n\}$ is a basis of W, then*

$$\mathcal{B} = \{e_i \otimes f_j \mid 1 \leq i \leq m \text{ and } 1 \leq j \leq n\}$$

is a basis of $V \otimes W$ and therefore $\dim(V \otimes W) = (\dim V)(\dim W)$.

Proof. For every pure tensor $v \otimes w \in V \otimes W$, using the coordinates of v and of w, we have

$$\begin{aligned}
v \otimes w &= (v^1 e_1 + \cdots + v^m e_m) \otimes w = v^1(e_1 \otimes w) + \cdots + v^m(e_m \otimes w) \\
&= v^1(e_1 \otimes (w^1 f_1 + \cdots + w^n f_n)) + \cdots + v^m(e_m \otimes (w^1 f_1 + \cdots + w^n f_n)) \\
&= v^i w^j e_i \otimes f_j \quad \text{(ESC)}.
\end{aligned}$$

Thus, $\mathcal{B}$ spans $V \otimes W$.

Now suppose that (using ESC), $c^{ij} e_i \otimes f_j = 0$. Let $\mathcal{B}_2^* = \{f^{*1}, \ldots, f^{*n}\}$ be the cobasis of V^* associated to $\mathcal{B}_2$. For each j with $1 \leq j \leq n$, the function $\varphi_j : V \times W \to V$ defined by $\varphi_j(v, w) = f^{*j}(w)v$ is bilinear so uniquely defines a linear transformation $\bar{\varphi}_j : V \otimes W \to V$ satisfying $\bar{\varphi}_j(v \otimes w) = f^{*j}(w)v$ on all pure tensors. Since $c^{ij} e_i \otimes f_j = 0$, then for all j_0,

$$0 = \bar{\varphi}_{j_0}(c^{ij} e_i \otimes f_j) = c^{ij} f^{*j_0}(f_j) e_i c^{ij} \delta_j^{j_0} e_i = c^{ij_0} e_i$$

as an element of V. Since $\mathcal{B}$ is a basis of V, then $c^{ij_0} = 0$ for all i, and this is for all j_0. We conclude that $\mathcal{B}$ is linearly independent in $V \otimes W$. Hence, $\mathcal{B}$ is a basis of $V \otimes W$ and so $\dim(V \otimes W) = mn$. $\qquad\square$

Because of Proposition 4.4.9, if $t \in V \otimes W$, it is common to use two indices to index the components of t with respect to the basis $\mathcal{B} = \{e_i \otimes f_j\}$. Saying that the tensor t has components (c^{ij}) with respect to $\mathcal{B}$ means that

$$t = \sum_{i=1}^{m} \sum_{j=1}^{n} c^{ij} e_i \otimes f_j.$$

We have used the superscript notation for the components of t to be consistent with the Einstein summation convention.

The next propositions illustrate how the tensor product construction directly generalizes the Hom space and the space of bilinear forms.

Proposition 4.4.10. *Let V and W be finite-dimensional vector spaces over a field K. The space $V^* \otimes W$ is canonically isomorphic to $\mathrm{Hom}(V, W)$.*

Proof. Consider the function $\varphi : V^* \times W \longrightarrow \mathrm{Hom}(V, W)$ defined by $\varphi(\lambda, w) = (v \mapsto \lambda(v)w)$. Since it is bilinear, by Proposition 4.4.1 there is a unique linear transformation $\bar{\varphi} : V^* \otimes W \to \mathrm{Hom}(V, W)$ that maps the pure tensor $\lambda \otimes w$ to the linear transformation $v \mapsto \lambda(v)w$.

The kernel of φ consists of all linear combinations $\lambda_1 \otimes w_1 + \cdots + \lambda_m \otimes w_m$ such that the function in $\mathrm{Hom}(V, W)$ defined by

$$\lambda_1(\cdot)w_1 + \cdots + \lambda_m(\cdot)w_m$$

is identically 0. By the identities (4.29), we can assume that $\{w_1, \ldots, w_m\}$ is a linear independent set of vectors in W. Thus for each $v \in V$, we have $\lambda_i(v) = 0$ for all $1 \le i \le m$. Hence each λ_i is the 0-map in V^*. From this we conclude that $\mathrm{Ker}\,\varphi = \{0\}$.

To show surjectivity, let $T \in \mathrm{Hom}(V, W)$. Let $\{v_1, \ldots, v_n\}$ be a basis of V, and consider the linear functions $\{v_1^*, \ldots, v_n^*\}$ (see Equation (4.8) and the subsequent explanation). Then the element

$$\sum_{i=1}^{n} v_i^* \otimes T(v_i)$$

maps to T under φ. Therefore, φ is also surjective. $\qquad \square$

Proposition 4.4.11. *Let V and W be vector spaces over a field K. The set of bilinear forms on $V \times W$ is a vector space with*

$$(\omega_1 + \omega_2)(v, w) \overset{\text{def}}{=} \omega_1(v, w) + \omega_2(v, w)$$

$$(c\omega)(v, w) \overset{\text{def}}{=} c\omega(v, w),$$

for all bilinear forms ω_1, ω_2, and ω and for all $c \in K$. Furthermore, the vector space of bilinear forms on $V \times W$ is canonically isomorphic to $V^* \otimes W^*$.

Proof. (The proof is left as an exercise for the reader.) $\qquad \square$

PROBLEMS

4.4.1. Let V be a vector space over a field K. Let $v_1, v_2 \in V$. Show that in $V \otimes V$, we have $v_1 \otimes v_2 = v_2 \otimes v_1$ if and only if v_1 and v_2 are collinear.

4.4.2. Let V be a vector space over the field K. Prove that $V \otimes K$ is canonically isomorphic to V.

4.4.3. Let V and W be finite dimensional vector spaces over a field K with respective bases $\mathcal{B} = \{e_1, \ldots, e_n\}$ and $\mathcal{B}' = \{f_1, \ldots, f_m\}$. Let $T : V \to W$ be a linear transformation with matrix A with respect to the bases $\mathcal{B}$ and $\mathcal{B}'$. T determines a linear transformation $T^{\otimes 2} : V \otimes V \to W \otimes W$ defined on pure tensors by

$$T^{\otimes 2}(v_1 \otimes v_2) = T(v_1) \otimes T(v_2)$$

and completed for other elements of $V \otimes V$ by linearity.

(a) If $V = W = \mathbb{R}^2$ and the matrix of a linear transformation T with respect to the standard basis is

$$A = \begin{pmatrix} 2 & 3 \\ 5 & 7 \end{pmatrix},$$

find the matrix of $T^{\otimes 2}$.

(b) In general, for any finite dimensional vector spaces V and W and linear transformation T, if the coefficients of A are (a_{ij}), find the coefficients of the matrix for $T^{\otimes 2}$.

4.4.4. Let V and W be vector spaces over $\mathbb{C}$, and let $S : V \to V$ and $T : W \to W$ be linear transformations. Consider the linear transformation $S \otimes T : V \otimes W \longrightarrow V \otimes W$ defined on pure tensors by

$$(S \otimes T)(v \otimes w) = S(v) \otimes T(w).$$

(a) Suppose that $\dim V = 2$ and that $\dim W = 3$, with bases $\{e_1, e_2\}$ and $\{f_1, f_2, f_3\}$ respectively. Suppose also that with respect to these bases, the matrices for S and T are

$$\begin{pmatrix} 1 & 3 \\ 5 & 2 \end{pmatrix} \quad \text{and} \quad \begin{pmatrix} -1 & 0 & 2 \\ 1 & 3 & -2 \\ 0 & 1 & 4 \end{pmatrix}.$$

Find the matrix for $S \otimes T$ with respect to the basis for $V \otimes W$ defined in Proposition 4.4.9.

(b) Suppose that S and T are diagonalizable with eigenvalues $\lambda_1, \ldots, \lambda_m$ and $\mu_1, \ldots, \mu_n$, respectively. Prove that $S \otimes T$ is diagonalizable and that the eigenvalues of $S \otimes T$ are $\lambda_i \mu_j$ for $1 \leq i \leq m$ and $1 \leq j \leq n$.

4.4.5. Let V be a vector space over $\mathbb{C}$, and let $T : V \to V$ be a linear transformation.

(a) Suppose that the Jordan canonical form of T is $J = \lambda I$. Find the Jordan canonical form of $T^{\otimes 2}$.

(b) Suppose that the Jordan canonical form of T is

$$J = \begin{pmatrix} \lambda & 1 & 0 & \cdots & 0 & 0 \\ 0 & \lambda & 1 & \cdots & 0 & 0 \\ 0 & 0 & \lambda & \cdots & 0 & 0 \\ \vdots & \vdots & \vdots & \ddots & \vdots & \vdots \\ 0 & 0 & 0 & \cdots & \lambda & 1 \\ 0 & 0 & 0 & \cdots & 0 & \lambda \end{pmatrix}$$

Find the Jordan canonical form of $T^{\otimes 2}$.

4.4.6. Prove Proposition 4.4.8.

4.4.7. Let V, W_1, and W_2 be finite dimensional vector spaces over a field K. Show that there exists a canonical (independent of a given basis) isomorphism

$$V \otimes (W_1 \oplus W_2) \cong (V \otimes W_1) \oplus (V \otimes W_2).$$

4.4.8. Prove Proposition 4.4.11. [Hint: Call $\mathrm{Bil}(V, W)$ the set of bilinear forms on $V \times W$. Show that $\Psi : V^* \otimes W^* \to \mathrm{Bil}(V, W)$ defined on pure tensors by $\Psi(\lambda \otimes \mu)(v, w) = \lambda(v)\mu(w)$ gives the canonical isomorphism.]

4.4.9. Let U, V, and W be vector spaces over a field K. Prove that $V^* \otimes W^* \otimes U$ is canonically isomorphic to the vector space of bilinear transformations $V \times W \to U$.

4.4.10. In the identification $V^* \otimes W \cong \mathrm{Hom}(V, W)$ described in Proposition 4.4.10, show that tensors of rank r in $V^* \otimes W^*$ precisely correspond to linear transformations in $\mathrm{Hom}(V, W)$ of rank r.

4.4.11. Consider the linear transformation $\mathrm{Tr} : V^* \otimes V \to K$ defined on pure tensors by $\mathrm{Tr}(\lambda \otimes v) = \lambda(v)$. Under the isomorphism $\mathrm{Hom}(V, V) \cong V^* \otimes V$, show that Tr corresponds to the trace of a linear transformation.

4.5 Components of Tensors over V

Let V be a vector space over a field K. Many applications of multilinear algebra, in particular to differential geometry, involve tensor products in

$$V^{\otimes r} \otimes V^{*\otimes s} \overset{\text{def}}{=} \overbrace{V \otimes V \otimes \cdots \otimes V}^{r \text{ times}} \otimes \overbrace{V^* \otimes V^* \otimes \cdots \otimes V^*}^{s \text{ times}}.$$

For example, $\mathrm{Hom}(V, V)$ is $V \otimes V^*$ and the vector space of bilinear forms on V is $V^{*\otimes 2}$. We will see in Section 4.7 that the set (vector space) of all bilinear products on V is $V \otimes V^{*\otimes 2}$.

Definition 4.5.1. A tensor *over* V of type (r, s) is an element of $V^{\otimes r} \otimes V^{*\otimes s}$. A scalar in K is called a tensor of type $(0, 0)$.

Suppose that V has an ordered basis $\mathcal{B} = (e_1, e_2, \ldots, e_n)$ and that the associated ordered cobasis is $\mathcal{B}^* = (e^{*1}, e^{*2}, \ldots, e^{*n})$. By Proposition 4.4.9 the basis of $V^{\otimes r} \otimes V^{*\otimes s}$ associated to $\mathcal{B}$ consists of all the pure tensors

$$e_{i_1} \otimes \cdots \otimes e_{i_r} \otimes e^{*j_1} \otimes \cdots \otimes e^{*j_s}$$

for $i_k = 1, \ldots n$ and $j_\ell = 1, \ldots, n$, for all $1 \leq k \leq r$ and $1 \leq \ell \leq s$. This basis confirms that $\dim(V^{\otimes r} \otimes V^{*\otimes s}) = (\dim V)^{r+s}$. The components of a tensor $A \in V^{\otimes r} \otimes V^{*\otimes s}$ with respect to $\mathcal{B}$ are the n^{r+s} values

$$A^{i_1 i_2 \cdots i_r}_{j_1 j_2 \cdots j_s} \in K \tag{4.30}$$

such that

$$A = A^{i_1 i_2 \cdots i_r}_{j_1 j_2 \cdots j_s} e_{i_1} \otimes e_{i_2} \otimes \cdots \otimes e_{i_r} \otimes e^{*j_1} \otimes e^{*j_2} \otimes \cdots \otimes e^{*j_s}. \quad \text{(ESC)} \tag{4.31}$$

Note that this formula involves summations over $r + s$ indices, all from 1 to n. Following the explanation at the end of Section 4.1, the superscript indices are called *contravariant* indices, while the subscript indices are called *covariant* indices. As in the contrast between a vector space and its dual, the difference between contravariant and covariant indices lies in how they affect the transformation of components of a tensor under a change of basis on V.

4.5.1   Coordinate Changes

Proposition 4.5.2. *Let $\mathcal{B}$ and $\mathcal{B}'$ be two bases on a finite dimensional vector space V. Let (a_q^p) be the components of the coordinate change matrix A from $\mathcal{B}$ to $\mathcal{B}'$ and let $(\breve{a}_\beta^\alpha)$ be the components of A^{-1}. Let $T_{j_1 j_2 \cdots j_s}^{i_1 i_2 \cdots i_r}$ be the components of a tensor of type (r,s) with respect to $\mathcal{B}$, and let $\bar{T}_{j_1 j_2 \cdots j_s}^{i_1 i_2 \cdots i_r}$ be the components of the same tensor T with respect to $\mathcal{B}'$. Then*

$$\bar{T}_{\ell_1 \ell_2 \cdots \ell_s}^{k_1 k_2 \cdots k_r} = a_{i_1}^{k_1} a_{i_2}^{k_2} \cdots a_{i_r}^{k_r} \breve{a}_{\ell_1}^{j_1} \breve{a}_{\ell_2}^{j_2} \cdots \breve{a}_{\ell_r}^{j_r} T_{j_1 j_2 \cdots j_r}^{i_1 i_2 \cdots i_s}. \tag{4.32}$$

Proof. Suppose that $\mathcal{B} = (e_1, e_2, \ldots, e_n)$ and $\mathcal{B}' = (f_1, f_2, \ldots, f_n)$. By definition of the coordinate change matrix, $e_i = a_i^k f_k$ for all i and by Proposition 4.1.6 $e^{*j} = \breve{a}_\ell^j f^{*\ell}$. Thus

$$\bar{T}_{\ell_1 \ell_2 \cdots \ell_s}^{k_1 k_2 \cdots k_r} f_{k_1} \otimes f_{k_2} \otimes \cdots \otimes f_{k_r} \otimes f^{*\ell_1} \otimes f^{*\ell_2} \otimes \cdots \otimes f^{*\ell_s}$$

$$= T_{j_1 j_2 \cdots j_s}^{i_1 i_2 \cdots i_r} e_{i_1} \otimes e_{i_2} \otimes \cdots \otimes e_{i_r} \otimes e^{*j_1} \otimes e^{*j_2} \otimes \cdots \otimes e^{*j_s}$$

$$= T_{j_1 j_2 \cdots j_s}^{i_1 i_2 \cdots i_r} (a_{i_1}^{k_1} f_{k_1}) \otimes (a_{i_2}^{k_2} f_{k_2}) \otimes \cdots (a_{i_r}^{k_r} f_{k_r}) \otimes (\breve{a}_{\ell_1}^{j_1} f^{*\ell_1}) \otimes (\breve{a}_{\ell_2}^{j_2} f^{*\ell_2}) \otimes \cdots \otimes (\breve{a}_{\ell_s}^{j_s} f^{*\ell_s})$$

$$= \left(a_{i_1}^{k_1} a_{i_2}^{k_2} \cdots a_{i_r}^{k_r} \breve{a}_{\ell_1}^{j_1} \breve{a}_{\ell_2}^{j_2} \cdots \breve{a}_{\ell_s}^{j_s} T_{j_1 j_2 \cdots j_s}^{i_1 i_2 \cdots i_r} \right) f_{k_1} \otimes f_{k_2} \otimes \cdots \otimes f_{k_r} \otimes f^{*\ell_1} \otimes f^{*\ell_2} \otimes \cdots \otimes f^{*\ell_s}.$$

By identifying coordinates, the proposition follows. $\square$

Physicists often introduce tensors by saying that an n^{r+s} set of quantities indexed as in (4.30) that change according to (4.32) under a basis change on V "form the components of a tensor." This perspective may be sufficient for various calculations but it does not elucidate what a tensor over V is.

We comment now on the linear algebraic meaning of a few common operations on tensors, when viewed from their components' perspective.

If $A_{j_1 j_2 \cdots j_s}^{i_1 i_2 \cdots i_r}$ form the components of a (r,s)-tensor A and $B_{j_1 j_2 \cdots j_s}^{i_1 i_2 \cdots i_r}$ form the components of a (r,s)-tensor B, then the term-by-term addition

$$C_{j_1 j_2 \cdots j_s}^{i_1 i_2 \cdots i_r} = A_{j_1 j_2 \cdots j_s}^{i_1 i_2 \cdots i_r} + B_{j_1 j_2 \cdots j_s}^{i_1 i_2 \cdots i_r}.$$

also satisfies (4.32), so form the components of a tensor. This operation corresponds to the usual addition of A and B as elements in the vector space $V^{\otimes r} \otimes V^{*\otimes s}$. Similarly, given the components $A_{j_1 j_2 \cdots j_s}^{i_1 i_2 \cdots i_r}$ of a tensor of type (r,s), the operation of multiplying all the components by a given scalar c in the base field K corresponds to multiplying the tensor A by the scalar c again as an operation in the vector space $V^{\otimes r} \otimes V^{*\otimes s}$.

It is not hard to check that if $S_{j_1 j_2 \cdots j_s}^{i_1 i_2 \cdots i_r}$ and $T_{l_1 l_2 \cdots l_u}^{k_1 k_2 \cdots k_t}$ are components of tensors of type (r,s) and (t,u), respectively, then the quantities obtained by multiplying these components

$$W_{j_1 j_2 \cdots j_s l_1 l_2 \cdots l_u}^{i_1 i_2 \cdots i_r k_1 k_2 \cdots k_t} = S_{j_1 j_2 \cdots j_s}^{i_1 i_2 \cdots i_r} T_{l_1 l_2 \cdots l_u}^{k_1 k_2 \cdots k_t}$$

form the components of another tensor but of type $(r + t, s + u)$. This operation called *tensor multiplication* or the product of two tensors, corresponds to the bilinear transformation

$$V^{\otimes r} \otimes V^{*\otimes s} \times V^{\otimes t} \otimes V^{*\otimes u} \longrightarrow V^{\otimes(r+t)} \otimes V^{*\otimes(s+u)},$$

defined by $(\alpha, \beta) \mapsto \alpha \otimes \beta$. Therefore, this tensor multiplication utilizes the isomorphism

$$(V^{\otimes r} \otimes V^{*\otimes s}) \otimes (V^{\otimes t} \otimes V^{*\otimes u}) \cong V^{\otimes r+t} \otimes V^{*\otimes s+u}.$$

Finally, the contraction operation on the components of a tensor

$$B^{i_1 i_2 \cdots i_{r-1}}_{j_1 j_2 \cdots j_{s-1}} = A^{i_1 i_2 \cdots i_{r-1} k}_{j_1 j_2 \cdots j_{s-1} k}$$

corresponds to setting one contravariant and one covariant index to be the same and then summing over that index. (The contraction operation does not have to occur on the last indices as in the above equation.) On the indices involved, this corresponds to the linear transformation $C : V \otimes V^* \longrightarrow K$ defined on a pure tensor by $v \otimes \lambda \longmapsto \lambda(v)$. Exercise 4.4.11 showed that the contraction operation is similar to the operation of taking the trace of a matrix along certain specified indices.

If $v \in V$ is a vector and $A \in V^{\otimes r} \otimes V^{*s\otimes}$ be a tensor of type (r, s) with $s \geq 1$, then some writers use the symbol

$$v \lrcorner A$$

to indicate the $(r, s - 1)$ tensor that corresponds to the contraction along the index of v and the first covariant index of A.

4.5.2 Examples

Example 4.5.3 (Cross Product). Consider $V = \mathbb{R}^3$. The cross product between two vectors is a bilinear transformation $\times : V \times V \to V$, so is a linear transformation $V \otimes V \to V$. In this way of considering it, the cross product is a particular element of $V^* \otimes V^* \otimes V$. We can describe it through its components expressed in reference to the standard ordered basis $\mathcal{E} = (\vec{\imath}, \vec{\jmath}, \vec{k})$. We write its components as C^i_{jk} satisfying

$$C^3_{12} = 1, \quad C^1_{23} = 1, \quad C^2_{31} = 1,$$
$$C^3_{21} = -1, \quad C^1_{32} = -1, \quad C^2_{13} = -1,$$

and all other components are 0. Suppose that $\mathcal{B} = (u_1, u_2, u_3)$ is some other ordered basis with the change of coordinate matrix $P = P^{\mathcal{E}}_{\mathcal{B}}$ with components (p^i_j), then the cross product expressed with respect to $\mathcal{B}$ has the components

$$\bar{C}^r_{st} = p^r_i \breve{p}^j_s \breve{p}^k_t C^i_{jk}.$$

Example 4.5.4 (Inverse of a $(0, 2)$-Tensor). As a more involved example, consider the components C_{ij} of a $(0, 2)$-tensor over V with respect to some basis $\mathcal{B}$. Recall that a $(0, 2)$-tensor represents a bilinear form $\langle \cdot, \cdot \rangle$ on V. Suppose in addition that $\langle \cdot, \cdot \rangle$ is nondegenerate. This is equivalent to the fact that if the C_{ij} are organized into an $n \times n$ matrix, then this matrix is invertible. Denote by C^{ij} the coefficients of the inverse matrix of (C_{ij}). We prove that C^{ij} form the components of a $(2, 0)$-tensor.

Let $P = (p^i_j)$ be a coordinate change matrix from $\mathcal{B}$-coordinates to some other system of coordinates. If $(\bar{C}_{rs})$ are the components of the same object with respect to the other basis, then

$$C^{ij} C_{jk} = \delta^i_k, \qquad \text{and} \qquad \bar{C}^{rs} \bar{C}_{st} = \delta^r_t. \tag{4.33}$$

Equation (4.33) gives $\bar{C}^{rs} \breve{p}^i_s \breve{p}^j_t C_{ij} = \delta^r_t$. Multiplying both sides by p^t_α and summing over t, we obtain

$$\bar{C}^{rs} \breve{p}^i_s \breve{p}^j_t C_{ij} p^t_\alpha = \delta^r_t p^t_\alpha \implies \bar{C}^{rs} \breve{p}^i_s \delta^j_\alpha C_{ij} = \bar{C}^{rs} \breve{p}^i_s C_{i\alpha} = p^r_\alpha.$$

Multiplying both sides by $C^{\alpha\beta}$ and then summing over α, we get

$$\bar{C}^{rs} p^i_s \delta^\beta_i = \bar{C}^{rs} p^\beta_s = p^r_\alpha C^{\alpha\beta}.$$

Finally, multiplying the rightmost equality by p^s_β and summing over β, we conclude that

$$\bar{C}^{rs} = p^r_\alpha p^s_\beta C^{\alpha\beta}.$$

This shows that the quantities C^{ij} satisfy Proposition 4.5.2 and hence form the components of a $(2, 0)$-tensor.

We should ask ourselves whether we can understand this tensor in a coordinate-independent way. In fact, we already presented this object in Exercise 4.2.5. The components C^{ij} represent the bilinear form $\langle \cdot, \cdot \rangle^*$ on V^* defined in that exercise. Using notations from there, we see that the components of λ_u are $C_{ij} u^i$ and the components of λ_v are $C_{k\ell} v^k$. Then

$$C_{ij} u^i C^{j\ell} C_{k\ell} v^k = \delta^\ell_i u^i C_{k\ell} v^k = u^\ell C_{kl} v^k.$$

This last expression are the components of $\langle v, u \rangle$, which confirms that Exercise 4.2.5 $C^{j\ell}$ are the components of $\langle \cdot, \cdot \rangle^*$.

4.5.3   Numerical Tensors

Definition 4.5.5. A *numerical tensor* is a tensor that is not a scalar whose components are the same given with respect to any basis.

As a first example, consider the Kronecker delta

$$\delta^i_j = \begin{cases} 1, & \text{if } i = j, \\ 0, & \text{if } i \neq j. \end{cases}$$

Under a coordinate change with matrix (p^i_j) it transforms according to

$$\bar{\delta}^r_s = p^r_i \breve{p}^j_s \delta^i_j = p^r_i \breve{p}^i_s.$$

Since this last expression represents the product of a matrix with its inverse, the we see that again $\bar{\delta}^r_s$ is 1 if $r = s$ and 0 otherwise. This should make sense because δ^i_j represents the identity function on a vector space, and with respect to any basis the components of the identity transformation is the identity matrix. Therefore, δ^i_j is a $(1,1)$-tensor in a tautological way.

The *generalized Kronecker delta* of order r is a tensor of type (r,r), with components denoted by $\delta^{i_1 \cdots i_r}_{j_1 \cdots j_r}$ defined as the following determinant:

$$\delta^{i_1 \cdots i_r}_{j_1 \cdots j_r} = \begin{vmatrix} \delta^{i_1}_{j_1} & \delta^{i_1}_{j_2} & \cdots & \delta^{i_1}_{j_r} \\ \delta^{i_2}_{j_1} & \delta^{i_2}_{j_2} & \cdots & \delta^{i_2}_{j_r} \\ \vdots & \vdots & \ddots & \vdots \\ \delta^{i_r}_{j_1} & \delta^{i_r}_{j_2} & \cdots & \delta^{i_r}_{j_r} \end{vmatrix}. \tag{4.34}$$

For example, the components of the generalized Kronecker delta of order 2 as

$$\delta^{ij}_{kl} = \delta^i_k \delta^j_l - \delta^i_l \delta^j_k,$$

which presents δ^{ij}_{kl} as the difference between two $(2,2)$-tensors, which shows that δ^{ij}_{kl} is indeed a tensor. More generally, expanding out Equation (4.34) by the Laplace expansion of a determinant gives the generalized Kronecker delta of order r as a sum of $r!$ components of tensors of type (r,r), proving that $\delta^{i_1 \cdots i_r}_{j_1 \cdots j_r}$ are the components of an (r,r)-tensor.

Properties of the determinant imply that $\delta^{i_1 \cdots i_r}_{j_1 \cdots j_r}$ is antisymmetric in the superscript indices and also antisymmetric in the subscript indices. Equivalently, $\delta^{i_1 \cdots i_r}_{j_1 \cdots j_r} = 0$ if any of the superscript indices are equal or if any of the subscript indices are equal, and the value of a component is negated if any two superscript indices are interchanged and similarly for subscript indices. We also note that if $r > n$, where we assume $\delta^{i_1 \cdots i_r}_{j_1 \cdots j_r}$ are the components of a tensor over an n-dimensional vector space, then $\delta^{i_1 \cdots i_r}_{j_1 \cdots j_r} = 0$ for all choices of indices since at least two superscript (and at least two subscript) indices would be equal.

We introduce one more symbol that is commonly used in calculations with tensor components, the permutation symbol. Define

$$\varepsilon^{i_1 \cdots i_n} = \delta^{i_1 \cdots i_n}_{1 \cdots n},$$
$$\varepsilon_{j_1 \cdots j_n} = \delta^{1 \cdots n}_{j_1 \cdots j_n}. \tag{4.35}$$

Note that the maximal index n in Equation (4.35) as opposed to r is intentional.

Recall that a permutation of $\{1, 2, \ldots, n\}$ is a bijection on that set and a transposition is a permutation that interchanges to elements and leaves the rest fixed.

A fact in modern algebra (see [25, Theorem 5.5]) states that given a permutation σ on $\{1, 2, \ldots, n\}$, if we have two ways to write σ as a composition of transpositions, e.g.,

$$\sigma = \tau_1 \circ \tau_2 \circ \cdots \circ \tau_a = \tau'_1 \circ \tau'_2 \circ \cdots \circ \tau'_b,$$

then a and b have the same parity.

Definition 4.5.6. We call a permutation *even* (respectively *odd*) if this common parity is even (respectively odd) and the *sign* of σ is

$$\text{sign}(\sigma) = \begin{cases} 1, & \text{if } \sigma \text{ is even}, \\ -1, & \text{if } \sigma \text{ is odd}. \end{cases}$$

Because of the properties of the determinant, it is not hard to see that

$$\varepsilon^{i_1 \cdots i_n} = \varepsilon_{i_1 \cdots i_n} = \begin{cases} 1, & \text{if } (i_1, \ldots, i_n) \text{ is an even permutation of } (1, 2, \ldots, n), \\ -1, & \text{if } (i_1, \ldots, i_n) \text{ is an odd permutation of } (1, 2, \ldots, n), \\ 0, & \text{if } (i_1, \ldots, i_n) \text{ is not a permutation of } (1, 2, \ldots, n). \end{cases}$$

The permutation symbol is an example for which, despite the apparently proper notation, the collection of quantities is not a numerical tensor. Instead, we have the following proposition.

Proposition 4.5.7. *Let $\mathcal{B}$ and $\mathcal{B}'$ be two bases on a finite dimensional vector space V. Let $A = (a^i_j)$ be the components of the coordinate change matrix from $\mathcal{B}$ to $\mathcal{B}'$ coordinates. The permutation symbols transform according to*

$$\det(A)\bar{\varepsilon}^{j_1 \cdots j_n} = a^{j_1}_{i_1} a^{j_2}_{i_2} \cdots a^{j_n}_{i_n} \varepsilon^{i_1 \cdots i_n},$$

$$(\det(A))^{-1}\bar{\varepsilon}_{k_1 \cdots k_n} = \check{a}^{h_1}_{k_1} \check{a}^{h_2}_{k_2} \cdots \check{a}^{h_n}_{k_n} \varepsilon_{h_1 \cdots h_n}.$$

Proof. (Left as an exercise for the reader.) $\qquad\qquad\qquad\qquad\qquad\qquad\square$

The generalized Kronecker delta has a close connection to determinants which, we will elucidate here. Note that if the superscript indices are exactly equal to the subscript indices, then $\delta^{i_1 \cdots i_r}_{j_1 \cdots j_r}$ is the determinant of the identity matrix. Thus, the contraction over all indices, $\delta^{j_1 \cdots j_r}_{j_1 \cdots j_r}$ counts the number of permutations of r indices taken from the set $\{1, 2, \ldots, n\}$. Thus,

$$\delta^{j_1 \cdots j_r}_{j_1 \cdots j_r} = \frac{n!}{(n-r)!}. \tag{4.36}$$

Another property of the generalized Kronecker delta is that

$$\varepsilon^{j_1 \cdots j_n} \varepsilon_{i_1 \cdots i_n} = \delta^{j_1 \cdots j_n}_{i_1 \cdots i_n},$$

the proof of which is left as an exercise for the reader (Problem 4.5.5). Now let a^i_j be the components of a $(1,1)$-tensor, which we can view as the matrix of a linear transformation from $\mathbb{R}^n$ to $\mathbb{R}^n$. By definition of the determinant,

$$\det(a^i_j) = \varepsilon^{j_1 \cdots j_n} a^1_{j_1} \cdots a^n_{j_n}.$$

Then, by properties of the determinant related to rearranging rows or columns, we have

$$\varepsilon^{i_1 \cdots i_n} \det(a^i_j) = \varepsilon^{j_1 \cdots j_n} a^{i_1}_{j_1} \cdots a^{i_n}_{j_n}.$$

Multiplying by $\varepsilon_{i_1 \cdots i_n}$ and summing over all the indices $i_1, \ldots, i_n$, we have

$$\varepsilon_{i_1 \cdots i_n} \varepsilon^{i_1 \cdots i_n} \det(a^i_j) = \delta^{j_1 \cdots j_n}_{i_1 \cdots i_n} a^{i_1}_{j_1} \cdots a^{i_n}_{j_n},$$

and since $\varepsilon_{i_1 \cdots i_n} \varepsilon^{i_1 \cdots i_n}$ counts the number of permutations of $\{1, \ldots, n\}$, we have

$$n! \det(a^i_j) = \delta^{j_1 \cdots j_n}_{i_1 \cdots i_n} a^{i_1}_{j_1} \cdots a^{i_n}_{j_n}. \tag{4.37}$$

4.5.4   Tensor Fields

Later in this book, tensor fields on manifolds will play a key role in describing structures of interest on manifolds. Before facing that full generality, we briefly consider tensor fields on $\mathbb{R}^n$ from the component perspective.

Let U be an open region of $\mathbb{R}^n$ equipped with two coordinate systems $(x^1, x^2, \ldots, x^n)$ and $(\bar{x}_1, \bar{x}_2, \ldots, \bar{x}_n)$ and let $p \in U$. For this section, we think of a tensor field over U as expressed by a collection of components $T^{i_1 i_2 \cdots i_r}_{j_1 j_2 \cdots j_s}$, where each of these is a function $U \to \mathbb{R}$. At a given point $p \in U$, these are tensors over the vector space $T_p \mathbb{R}^n$. The ordered basis associated to the $(x^1, x^2, \ldots, x^n)$ coordinates is $(\partial/\partial x^1, \ldots, \partial/\partial x^n)$, while the ordered basis associated to the $(\bar{x}^1, \bar{x}^2, \ldots, \bar{x}^n)$ coordinates is $(\partial/\partial \bar{x}^1, \ldots, \partial/\partial \bar{x}^n)$. The change of coordinate matrix between these is

$$\left(\left. \frac{\partial \bar{x}^i}{\partial x^j} \right|_p \right)$$

Example 4.5.8 (Gradient). Let $f : U \to \mathbb{R}$ be a differentiable function and consider the gradient ∇f_p. It has components $\partial f / \partial x^i$, evaluated at p. This is a tensor of type $(0,1)$ because in the $(\bar{x}^1, \bar{x}^2, \ldots, \bar{x}^n)$ coordinates, its components are

$$\frac{\partial \bar{f}}{\partial \bar{x}^j} = \frac{\partial f}{\partial \bar{x}^j} = \frac{\partial f}{\partial x^i} \frac{\partial x^i}{\partial \bar{x}^j},$$

by the chain rule. This satisfies (4.32) for a $(0,1)$. Consequently, in the expression $\partial f / \partial x^i$, though the i appears as a superscript index of the variable, we understand it as a covariant index instead of a contravariant index because it appears on the "denominator" of a partial derivative.

The following example illustrates some of the subtlety required when working with tensor fields in component form.

Example 4.5.9. Let B_i be the components of a covariant vector field. We prove that the collection of functions

$$A_{ij} = \frac{\partial B_i}{\partial x^j} - \frac{\partial B_j}{\partial x^i}$$

form the components of a $(0,2)$-tensor. In the $(\bar{x}^1, \bar{x}^2, \ldots, \bar{x}^n)$ coordinates, we have

$$\bar{A}_{k\ell} = \frac{\partial \bar{B}_k}{\partial \bar{x}^\ell} - \frac{\partial \bar{B}_\ell}{\partial \bar{x}^r k}$$

$$= \frac{\partial}{\partial \bar{x}^\ell}\left(\frac{\partial x^i}{\partial \bar{x}^k} B_i\right) - \frac{\partial}{\partial \bar{x}^k}\left(\frac{\partial x^j}{\partial \bar{x}^\ell} B_j\right)$$

$$= \frac{\partial^2 x^i}{\partial \bar{x}^\ell \partial \bar{x}^k} B_i + \frac{\partial x^i}{\partial \bar{x}^k}\frac{\partial B_i}{\partial \bar{x}^\ell} - \frac{\partial^2 x^j}{\partial \bar{x}^k \partial \bar{x}^\ell} B_j - \frac{\partial x^j}{\partial \bar{x}^\ell}\frac{\partial B_j}{\partial \bar{x}^k}. \tag{4.38}$$

Because we sum over variables repeated in superscript and the subscript, the first and third terms cancel out. So applying the chain rule on $\partial B_i/\partial \bar{x}^\ell$ and similarly in the fourth term,

$$\bar{A}_{k\ell} = \frac{\partial x^i}{\partial \bar{x}^k}\frac{\partial B_i}{\partial x^u}\frac{\partial x^u}{\partial \bar{x}^\ell} - \frac{\partial x^j}{\partial \bar{x}^\ell}\frac{\partial B_j}{\partial x^v}\frac{\partial x^v}{\partial \bar{x}^k}$$

$$= \frac{\partial x^i}{\partial \bar{x}^k}\frac{\partial x^j}{\partial \bar{x}^\ell}\frac{\partial B_i}{\partial x^j} - \frac{\partial x^j}{\partial \bar{x}^\ell}\frac{\partial x^i}{\partial \bar{x}^k}\frac{\partial B_j}{\partial x^i} \quad \text{by setting } u = j \text{ and } v = i$$

$$= \frac{\partial x^i}{\partial \bar{x}^k}\frac{\partial x^k}{\partial \bar{x}^\ell} A_{ij}.$$

We should also observe that the component functions $\partial B_i/\partial x^j$ do not describe a tensor field of type $(0,2)$ because of the mixed second partial derivative that appears (4.38).

PROBLEMS

4.5.1. Prove that (a) $\delta^i_j \delta^j_k \delta^k_l = \delta^i_l$; (b) $\delta^i_j \delta^j_k \delta^k_i = n$.

4.5.2. Let $T^{i_1 i_2 \cdots i_r}_{j_1 j_2 \cdots j_s}$ be a tensor of type (r,s). Prove that the quantities $T^{i i_2 \cdots i_r}_{i j_2 \cdots j_s}$, obtained by contracting over the first two indices, form the components of a tensor of type $(r-1, s-1)$. Explain in a coordinate-free way why we still obtain a tensor when we contract over any superscript and subscript index.

4.5.3. Let S_{ijk} be the components of a tensor, and suppose they are antisymmetric in $\{i,j\}$. Find a tensor with components T_{ijk} that is antisymmetric in j,k satisfying

$$-T_{ijk} + T_{jik} = S_{ijk}.$$

4.5.4. Prove Proposition 4.5.7.

4.5.5. Prove that $\varepsilon^{i_1 \cdots i_n} \varepsilon_{j_1 \cdots j_n} = \delta^{i_1 \cdots i_n}_{j_1 \cdots j_n}$.

4.5.6. Let A_{ij} be the components of an antisymmetric (i.e., $A_{ji} = -A_{ij}$) tensor field of type $(0,2)$, and define the quantities

$$B_{rst} = \frac{\partial A_{st}}{\partial x^r} + \frac{\partial A_{tr}}{\partial x^s} + \frac{\partial A_{rs}}{\partial x^t}.$$

(a) Prove that B_{rst} are the components of a tensor of type $(0,3)$.

(b) Prove that the components B_{rst} are antisymmetric in all their indices.

(c) Determine the number of independent components of antisymmetric tensors of type $(0,3)$ over $\mathbb{R}^n$.

(d) Would the quantities B_{rst} still be the components of a tensor if A_{ij} were symmetric?

4.5.7. Let a^i_j be the components of a $(1,1)$-tensor, or in other words the matrix of a linear transformation from $\mathbb{R}^n$ to $\mathbb{R}^n$ given with respect to some basis. Recall that the characteristic equation for the matrix is

$$\det(a^i_j - \lambda \delta^i_j) = 0. \tag{4.39}$$

Prove that Equation (4.39) is equivalent to

$$\lambda^n + \sum_{r=1}^{n} (-1)^r a_{(r)} \lambda^{n-r} = 0,$$

where

$$a_{(r)} = \frac{1}{r!} \delta^{i_1 \cdots i_r}_{j_1 \cdots j_r} a^{i_1}_{j_1} \cdots a^{i_r}_{j_r}.$$

[Hint: The solutions to Equation (4.39) are the eigenvalues of the matrix (a^i_j).]

4.5.8. *Moment of Inertia Tensor.* Suppose that $\mathbb{R}^3$ is given a basis that is not necessarily the standard one. Let g_{ij} be the components of the standard inner product corresponding to this basis, which means that the scalar product between two (contravariant) vectors A^i and B^j is given by

$$\vec{A} \cdot \vec{B} = g_{ij} A^i B^j.$$

In the rest of the problem, call (x^1, x^2, x^3) the coordinates of the position vector $\vec{r}$. Let S be a solid in space with a density function $\rho(\vec{r})$, and suppose that it rotates about an axis ℓ through the origin. The angular velocity vector $\vec{\omega}$ is defined as the vector along the axis ℓ, pointing in the direction that makes the rotation a right-hand corkscrew motion with magnitude $\omega = \|\vec{\omega}\|$ that is equal to the radians per second swept out by the motion of rotation. Let $(\omega^1, \omega^2, \omega^3)$ be the components of $\vec{\omega}$ in the given basis. The moment of inertia of the solid S about the direction $\vec{\omega}$ is defined as the quantity

$$I_\ell = \iiint_S \rho(\vec{r}) r_\perp^2 \, dV,$$

where $r_\perp$ is the distance from a point $\vec{r}$ with coordinates (x^1, x^2, x^3) to the axis ℓ. The moment of inertia tensor of a solid is often presented using cross products, but we define it here using a characterization that is equivalent to the usual definition but avoids cross products. We define the moment of inertia tensor as the unique $(0,2)$-tensor with components I_{ij} such that

$$\frac{1}{2}I_{ij}\omega^i\omega^j = \frac{1}{2}I_\ell\omega^2. \tag{4.40}$$

Note that this the kinetic energy of the rotating object.

(a) Prove that

$$r_\perp^2 = g_{ij}x^ix^j - \frac{(g_{kl}\omega^kx^l)^2}{g_{rs}\omega^r\omega^s}.$$

(b) Prove that, using the metric g_{ij}, the moment of inertia tensor is given by

$$I_{ij} = \iiint_S \rho(x^1,x^2,x^3)(g_{ij}g_{kl} - g_{ik}g_{jl})x^kx^l\, dV. \tag{4.41}$$

(c) Show that

$$(g_{ij}g_{kl} - g_{ik}g_{jl})x^kx^l = g_{ip}g_{ql}\delta^{pq}_{jk}x^kx^l,$$

where δ^{pq}_{jk} is the generalized Kronecker delta of order 2.

(d) Prove that $I_{ij} = I_{ji}$ for all $1 \le i,j \le n$.

(e) Prove that if the basis of $\mathbb{R}^3$ is orthonormal (which means that (g_{ij}) is the identity matrix), we recover the following familiar formulas:

$$I_{11} = \iiint_S \rho((x^2)^2 + (x^3)^2)\, dV, \qquad I_{12} = -\iiint_S \rho x^1x^2\, dV, \tag{4.42}$$

$$I_{22} = \iiint_S \rho((x^1)^2 + (x^3)^2)\, dV, \qquad I_{13} = -\iiint_S \rho x^1x^3\, dV, \tag{4.43}$$

$$I_{33} = \iiint_S \rho((x^1)^2 + (x^2)^2)\, dV, \qquad I_{23} = -\iiint_S \rho x^2x^3\, dV. \tag{4.44}$$

(We took the relation in (4.40) as the defining property of the moment of inertia tensor because of the theorem that $I_\ell\omega$ is the component of the angular moment vector along the axis of rotation that is given by $(I_{ij}\omega^i)\frac{\omega^j}{\omega}$. See [22] p. 221–222 and, in particular, Equation (9.7) for an explanation.

The interesting point about this approach is that it avoids the use of an orthonormal basis and provides a formula for the moment of inertia tensor when one has an affine metric tensor that is not the identity.)

4.6   Symmetric and Alternating Products

In the tensor product $V \otimes V$, in general $v_1 \otimes v_2 \ne v_2 \otimes v_1$. It is sometimes useful to have a tensor-like product that is either commutative or anticommutative.

For example, we have seen that every bilinear form on V is an element of $V^* \otimes V^*$. However, in geometry, many useful applications involve symmetric bilinear forms.

If A_{ij} are the components of an element in $A \in V^* \otimes V^*$ with respect to a given basis on V, then the condition that A be a symmetric bilinear form means that $A_{ij} = A_{ji}$ for all $1 \leq i, j \leq \dim V$. The set of symmetric bilinear forms is a linear subspace of $V^{*\otimes 2}$. Other applications may involve a higher type of tensor and have symmetry across more than two indices.

Let V be a vector space of dimension n. Let S_k be the set of permutations on k elements (i.e., bijections on $\{1, 2, \ldots, k\}$). This set of permutations acts on $V^{\otimes k}$ by doing the following on pure tensors:

$$\sigma \cdot (v_1 \otimes v_2 \otimes \cdots \otimes v_k) = v_{\sigma^{-1}(1)} \otimes v_{\sigma^{-1}(2)} \otimes \cdots \otimes v_{\sigma^{-1}(k)} \qquad (4.45)$$

and extending by linearity on nonpure tensors. (Taking σ^{-1} on the indices means that σ sends the vector in the ith position in the tensor product $v_1 \otimes v_2 \otimes \cdots \otimes v_k$ to the $\sigma(i)$th position.)

Definition 4.6.1. We say that tensor $\alpha \in V^{\otimes k}$ is *symmetric* (resp. *antisymmetric*) if $\sigma \cdot \alpha = \alpha$, (resp. $\sigma \cdot \alpha = \text{sign}(\sigma)\alpha$) for all $\sigma \in S_k$.

4.6.1 Symmetric Product

Definition 4.6.2. Let $\alpha \in V^{\otimes k}$. We define the *symmetrization* of α to be

$$\mathsf{S}(\alpha) = \sum_{\sigma \in S_k} \sigma \cdot \alpha.$$

Example 4.6.3. Let V be a vector space. We consider tensors in $V \otimes V \otimes V$. We will consider permutations in S_3, which has $3! = 6$ elements.

$$\mathsf{S}(e_1 \otimes e_2 \otimes e_3) = e_1 \otimes e_2 \otimes e_3 + e_2 \otimes e_1 \otimes e_3 + e_3 \otimes e_2 \otimes e_1$$
$$+ e_1 \otimes e_3 \otimes e_2 + e_2 \otimes e_3 \otimes e_1 + e_3 \otimes e_1 \otimes e_2.$$

In contrast,

$$\mathsf{S}(e_1 \otimes e_1 \otimes e_2) = e_1 \otimes e_1 \otimes e_2 + e_1 \otimes e_1 \otimes e_2 + e_2 \otimes e_1 \otimes e_1$$
$$+ e_1 \otimes e_2 \otimes e_1 + e_1 \otimes e_1 \otimes e_1 + e_2 \otimes e_1 \otimes e_1$$
$$= 2(e_1 \otimes e_1 \otimes e_2 + e_1 \otimes e_2 \otimes e_1 + e_2 \otimes e_1 \otimes e_1).$$

By construction, the symmetrization S defines a linear transformation $\mathsf{S} : V^{\otimes k} \to V^{\otimes k}$.

Definition 4.6.4. The subspace of $V^{\otimes k}$ given as the image of $\mathsf{S} : V^{\otimes k} \to V^{\otimes k}$ is called the *kth symmetric product* of V and is denoted by $\text{Sym}^k V$.

Proposition 4.6.5. *The subspace $\text{Sym}^k V$ is invariant under the action of S_k on $V^{\otimes k}$.*

Proof. Let $\tau \in S_k$ be a permutation. Then on any pure tensor $v_{i_1} \otimes v_{i_2} \otimes \cdots \otimes v_{i_k}$, the action of τ on $\mathsf{S}(v_{i_1} \otimes v_{i_2} \otimes \cdots \otimes v_{i_k})$ gives

$$\tau \cdot \mathsf{S}(v_{i_1} \otimes v_{i_2} \otimes \cdots \otimes v_{i_k}) = \tau \cdot \left(\sum_{\sigma \in S_k} \sigma \cdot v_{i_1} \otimes v_{i_2} \otimes \cdots \otimes v_{i_k} \right)$$

$$= \sum_{\sigma \in S_k} \tau \cdot (\sigma \cdot v_{i_1} \otimes v_{i_2} \otimes \cdots \otimes v_{i_k}) = \sum_{\sigma \in S_k} (\tau\sigma) \cdot v_{i_1} \otimes v_{i_2} \otimes \cdots \otimes v_{i_k}$$

$$= \sum_{\sigma' \in S_k} \sigma' \cdot v_{i_1} \otimes v_{i_2} \otimes \cdots \otimes v_{i_k} = \mathsf{S}(v_{i_1} \otimes v_{i_2} \otimes \cdots \otimes v_{i_k}),$$

where we obtain the second-to-last line because as σ runs through all the permutations in S_k, for any fixed $\tau \in S_k$, the compositions $\tau\sigma$ also run through all the permutations of S_k. $\square$

Corollary 4.6.6. *For all symmetric tensors $\alpha \in \mathrm{Sym}^k V$, we have $\mathsf{S}(\alpha) = k!\,\alpha$.*

Proof. This follows immediately from Proposition 4.6.5 and Definition 4.6.2. $\square$

Proposition 4.6.7. *Let $\{e_1, e_2, \ldots, e_n\}$ be a basis of V. Then the set*

$$\{\mathsf{S}(e_{i_1} \otimes e_{i_2} \otimes \cdots \otimes e_{i_k}) \mid 1 \le i_1 \le i_2 \le \cdots \le i_k \le n\}$$

is a basis of $\mathrm{Sym}^k V$.

Proof. Define $T(k,n) = \{(i_1, i_2, \ldots, i_k) \in \mathbb{N}^k \mid 1 \le i_1 \le i_2 \le \cdots \le i_k\}$. For this proof, if $\beth = (i_1, i_2, \ldots, i_k) \in \{1, 2, \ldots, n\}^k$, denote $e_{\mathbf{i}} \overset{\text{def}}{=} e_{i_1} \otimes e_{i_2} \otimes \cdots \otimes e_{i_k}$. In the action of S_k on $\{1, 2, \ldots, n\}^k$ defined by

$$\sigma \cdot (i_1, i_2, \ldots, i_k) = (i_{\sigma^{-1}(1)}, i_{\sigma^{-1}(2)}, \ldots, i_{\sigma^{-1}(k)})$$

the set $T(k,n)$ contains exactly one representative from each orbit of this action. This implies that $\{\mathsf{S}(e_{\mathbf{i}}) \mid \mathbf{i} \in T(k,n)\}$ spans $\mathrm{Im}\,\mathsf{S} = \mathrm{Sym}^k V$.

Now we show that $\{\mathsf{S}(e_{\mathbf{i}}) \mid \mathbf{i} \in T(k,n)\}$ is linearly independent. For any $\sigma \in S_k$ and for any $\mathbf{i} \in T(k,n)$, the permuted pure tensor $\sigma \cdot e_{\mathbf{i}}$ is another pure tensor with the same number of e_i basis vectors of a given index i. For any $\mathbf{j} \in \{1, 2, \ldots, n\}^k$ let $g(\mathbf{j})$ be the same k-tuple $\mathbf{j}$, but reorganized into nondecreasing order. If the k-tuple $\mathbf{j}$ consists of m_1 1s, m_2 2s, and so on, define $f(\mathbf{j}) = m_1! m_2! \cdots m_n!$. As the above examples illustrate, for all $\mathbf{i} \in T(k,n)$,

$$(S)(e_{\mathbf{i}}) = f(\mathbf{i}) \sum_{\mathbf{j} \in \{1, \ldots, n\}^k \,:\, g(\mathbf{j}) = \mathbf{i}} e_{\mathbf{j}}.$$

Thus, in a linear combination

$$0 = \sum_{\mathbf{i} \in T(k,n)} c_{\mathbf{i}} \mathsf{S}(e_{\mathbf{i}}) = \sum_{\mathbf{j} \in \{1, \ldots, n\}^k} c_{g(\mathbf{j})} f(\mathbf{j}) e_{\mathbf{j}}$$

we have $c_{g(\mathbf{j})} f(\mathbf{j}) = 0$ for all $\mathbf{j} \in \{1, \ldots, n\}^k$ because $\{e_{\mathbf{j}} \mid \mathbf{j} \in \{1, \ldots, n\}^k\}$ is a basis of $V^{\otimes k}$. However, $f(\mathbf{j}) \geq 1$, so we deduce that $c_{g(\mathbf{j})} = 0$ for all $\mathbf{j} \in \{1, \ldots, n\}^k$. In particular, $c_{\mathbf{i}} = 0$ for all $\mathbf{i} \in T(k, n)$. This establishes the linear independence.

We conclude that $\{S(e_{\mathbf{i}}) \mid \mathbf{i} \in T(k, n)\}$ is a basis of $\operatorname{Sym}^k V$. $\qquad \square$

Corollary 4.6.8. *Let V be a vector space of dimension n. Then*

$$\dim \operatorname{Sym}^k V = \binom{n + k - 1}{k}.$$

Proof. From Proposition 4.6.7, $\dim \operatorname{Sym}^k V$ is the cardinality of $T(k, n)$. This particular enumeration problem, of counting the number of nondecreasing sequences of length k with values in $\{1, 2, \ldots, n\}$, has a standard solution. Consider $n + k$ slots. We have a bag of n Xs and k Ys. Put an X in the first slot. Fill the remaining slots with Xs and Ys. Because we insist on an X in the first slot, there are $\binom{n+k-1}{k}$ ways to fill the slots. However, the set of fillings as described is in bijection with our desired set of sequences in the following way. For any filling, let i_t be the number of Xs that occur before the t'th Y. With this definition, the resulting k-tuple $(i_1, i_2, \ldots, i_k)$ is nondecreasing with $1 \leq i_t \leq n$. (Placing an X in the first slot ensured that $i_t \geq 1$.) Conversely, any nondecreasing sequence $(i_1, i_2, \ldots, i_k)$ leads to a unique filling of slots that satisfies our parameters. The result follows. $\qquad \square$

Given $\alpha \in \operatorname{Sym}^k V$ and $\beta \in \operatorname{Sym}^l V$, the tensor product $\alpha \otimes \beta$ is of course an element of $V^{\otimes(k+l)}$ but is not necessarily an element of $\operatorname{Sym}^{k+l} V$. However, it is possible to construct a new product that satisfies this deficiency.

Definition 4.6.9. Let $\alpha \in \operatorname{Sym}^k V$ and $\beta \in \operatorname{Sym}^l V$. Define the symmetric product between α and β as

$$\alpha\beta = \frac{1}{k!\, l!} S(\alpha \otimes \beta).$$

Note that if α and β are tensors of rank 1, then the product $\alpha\beta$ is precisely the symmetrization of $\alpha \otimes \beta$. However, a few other properties, which we summarize in Proposition 4.6.11, of this symmetric product also hold. We need a lemma first.

Lemma 4.6.10. *Let $\alpha \in V^{\otimes k}$. If $S(\alpha) = 0$, then $S(\alpha \otimes \beta) = S(\beta \otimes \alpha) = 0$ for all tensors β. Furthermore, if $S(\alpha) = S(\alpha')$, then $S(\alpha \otimes \beta) = S(\alpha' \otimes \beta)$ for all tensors β.*

Proof. We first prove that if $S(\alpha) = 0$, then $S(\alpha \otimes \beta) = 0$ for all tensors β of rank l, and the result for $S(\beta \otimes \alpha)$ follows similarly.

Let S_k be the subset of permutations in S_{k+l} that only permute the first k elements of $\{1, 2, \ldots, k + l\}$ and leave the remaining l elements unchanged. Define the relation $\sim$ on S_{k+l} as $\tau_1 \sim \tau_2$ if and only if $\tau_2^{-1}\tau_1 \in S_k$. Since S_k is closed under taking inverse functions and composition of functions, it is easy to see that $\sim$ is an equivalence relation on S_{k+l}.

Let C be a set of representatives of distinct equivalence classes of $\sim$. Then we have

$$\mathsf{S}(\alpha \otimes \beta) = \sum_{\sigma \in S_{k+l}} \sigma \cdot (\alpha \otimes \beta) = \sum_{\tau \in C} \sum_{\sigma' \in S_k} \tau \sigma' \cdot (\alpha \otimes \beta)$$

$$= \sum_{\tau \in C} \tau \cdot \left(\left(\sum_{\sigma' \in S_k} \sigma' \cdot \alpha \right) \otimes \beta \right) = \sum_{\tau \in C} \tau \cdot ((\mathsf{S}\alpha) \otimes \beta) = 0.$$

For the second part of the lemma, suppose that $\mathsf{S}(\alpha) = \mathsf{S}(\alpha')$. Then $\mathsf{S}(\alpha - \alpha') = 0$. Thus, for all tensors β we have $\mathsf{S}((\alpha - \alpha') \otimes \beta) = 0$. Hence, $\mathsf{S}(\alpha \otimes \beta) - \mathsf{S}(\alpha' \otimes \beta) = 0$ and the result follows. $\qquad\square$

Proposition 4.6.11. *Let V be a vector space of dimension n. The following hold:*

1. *The symmetric product is bilinear: for all $\alpha, \alpha_1, \alpha_2 \in \operatorname{Sym}^k V$, for all $\beta, \beta_1, \beta_2 \in \operatorname{Sym}^l V$, and λ in the base field,*

$$(\alpha_1 + \alpha_2)\beta = \alpha_1\beta + \alpha_2\beta, \qquad (\lambda\alpha)\beta = \lambda(\alpha\beta),$$
$$\alpha(\beta_1 + \beta_2) = \alpha\beta_1 + \alpha\beta_2, \qquad \alpha(\lambda\beta) = \lambda(\alpha\beta).$$

2. *The symmetric product is commutative: for all $\alpha \in \operatorname{Sym}^k V$ and $\beta \in \operatorname{Sym}^l V$,*

$$\alpha\beta = \beta\alpha.$$

3. *The symmetric product is associative: for all $\alpha \in \operatorname{Sym}^r V$, $\beta \in \operatorname{Sym}^s V$, and $\gamma \in \operatorname{Sym}^t V$, as an element of $\operatorname{Sym}^{r+s+t} V$, we have*

$$(\alpha\beta)\gamma = \alpha(\beta\gamma) = \frac{1}{r!\,s!\,t!}\mathsf{S}(\alpha \otimes \beta \otimes \gamma).$$

Proof. We leave part 1 of the proposition as an exercise for the reader.

For part 2, by Proposition 4.6.5, $\mathsf{S}(\alpha \otimes \beta)$ is invariant under the action of S_{k+l}. Consider the permutation $\sigma_0 \in S_{k+l}$ that maps the n-tuple $(1, 2, \ldots, k + l)$ to $(k + 1, \ldots, k + l, 1, \ldots, k)$. In each pure tensor in an expression of $\alpha \otimes \beta$, the action $\sigma_0(\alpha \otimes \beta)$ moves (and keeps in the proper order) the vector terms coming from β in front of the terms coming from α. Hence, we see that $\sigma_0(\alpha \otimes \beta) = \beta \otimes \alpha$. Thus, we conclude that

$$\beta\alpha = \sigma_0(\alpha\beta) = \sigma_0\left(\frac{1}{k!\,l!}\mathsf{S}(\alpha \otimes \beta)\right) = \frac{1}{k!\,l!}\mathsf{S}(\alpha \otimes \beta) = \alpha\beta.$$

Thus, the symmetric product is commutative.

For part 3, by Corollary 4.6.6, since $\alpha\beta$ is symmetric,

$$\mathsf{S}(\alpha\beta) = (r + s)!\,\alpha\beta = \frac{(r + s)!}{r!\,s!}\mathsf{S}(\alpha \otimes \beta).$$

Therefore, by Lemma 4.6.10, for all tensors γ of rank t,

$$\mathsf{S}(\alpha\beta \otimes \gamma) = \mathsf{S}\left(\frac{(r+s)!}{r!\,s!}(\alpha \otimes \beta) \otimes \gamma\right).$$

Consequently,

$$(\alpha\beta)\gamma = \frac{1}{(r+s)!\,t!}\mathsf{S}(\alpha\beta \otimes \gamma) = \frac{1}{(r+s)!\,t!}\frac{(r+s)!}{r!\,s!}\mathsf{S}((\alpha \otimes \beta) \otimes \gamma)$$

$$= \frac{1}{r!\,s!\,t!}\mathsf{S}(\alpha \otimes \beta \otimes \gamma).$$

It is easy to follow the same calculation and find that

$$\alpha(\beta\gamma) = \frac{1}{r!\,s!\,t!}\mathsf{S}(\alpha \otimes \beta \otimes \gamma),$$

which shows that $(\alpha\beta)\gamma = \alpha(\beta\gamma)$ for all tensors α, β, and γ. □

By virtue of associativity, the symmetrization of a pure tensor $\mathsf{S}(v_1 \otimes v_2 \otimes \cdots \otimes v_k)$ is in fact

$$v_1 v_2 \cdots v_k.$$

We think of this element as a commutative "product" between vectors, which is linear in each term. With this notation in mind, one usually thinks of $\mathrm{Sym}^k V$ as a vector space in its own right, independent of $V^{\otimes k}$, with basis

$$\{e_{i_1} e_{i_2} \cdots e_{i_k} \mid 1 \le i_1 \le i_2 \le \cdots \le i_k \le n\}.$$

Furthermore, analogous to polynomials in multiple variables where the monomial $xyx^2 z^3 y = x^3 y^2 z^3$, any symmetric product vector $e_{i_1} e_{i_2} \cdots e_{i_k}$ is equal to another expression on which the particular vectors in the product are permuted.

4.6.2 Alternating Product

We turn now to the alternating product, also called the wedge product. Many of the results for the alternating product parallel the symmetric product.

Let V be a vector space of dimension n and let us continue to consider the action of S_k on $V^{\otimes k}$ as described in Equation (4.45). Recall the sign of a permutation described in Definition 4.5.6.

Definition 4.6.12. Let $\alpha \in V^{\otimes k}$ be a tensor. We define the *alternation* of α to be

$$\mathsf{A}(\alpha) = \sum_{\sigma \in S_k} \mathrm{sign}(\sigma)(\sigma \cdot \alpha).$$

Example 4.6.13. Let V be a vector space. We consider tensors in $V \otimes V \otimes V$. We will consider permutations in S_3, which has $3! = 6$ elements. The identity permutation has a sign of 1, permutations that interchange only two elements have a sign of -1, and the permutations that cycle through the three indices have a sign of 1.

$$\mathsf{A}(e_1 \otimes e_2 \otimes e_3) = e_1 \otimes e_2 \otimes e_3 - e_2 \otimes e_1 \otimes e_3 - e_3 \otimes e_2 \otimes e_1$$
$$- e_1 \otimes e_3 \otimes e_2 + e_2 \otimes e_3 \otimes e_1 + e_3 \otimes e_1 \otimes e_2$$

In contrast,

$$\mathsf{A}(e_1 \otimes e_1 \otimes e_2) = e_1 \otimes e_1 \otimes e_2 - e_1 \otimes e_1 \otimes e_2 - e_2 \otimes e_1 \otimes e_1$$
$$- e_1 \otimes e_2 \otimes e_1 + e_1 \otimes e_1 \otimes e_1 + e_2 \otimes e_1 \otimes e_1 = 0$$

Proposition 4.6.14. *Let $v_1 \otimes v_2 \otimes \cdots \otimes v_k$ be a pure tensor in $V^{\otimes k}$. If $v_i = v_j$ for some pair (i, j), where $i \neq j$, then*

$$\mathsf{A}(v_1 \otimes v_2 \otimes \cdots \otimes v_k) = 0.$$

Proof. Suppose that in the pure tensor $v_1 \otimes v_2 \otimes \cdots \otimes v_k$, we have $v_i = v_j$ for some pair $i \neq j$. Let $f \in S_k$ be the permutation that interchanges the ith and jth entry and leaves all others fixed. This permutation f is a transposition so $\mathrm{sign}(f) = -1$. Define the relation $\sim$ on S_k by $\sigma \sim \tau$ if and only if $\tau^{-1}\sigma \in \{1, f\}$. Note that $f^2 = f \circ f = 1$ is the identity permutation, and hence, $f = f^{-1}$. Because of these properties of f, we can easily check that the relation $\sim$ is an equivalence relation on S_k.

Let C be a set of representatives for all of the equivalence classes of $\sim$. If $\alpha = v_1 \otimes v_2 \otimes \cdots \otimes v_k$, then $f \cdot \alpha = \alpha$ because $v_i = v_j$. Thus,

$$\mathsf{A}(\alpha) = \sum_{\sigma \in C} \Big(\mathrm{sign}(\sigma)(\sigma \cdot \alpha) + \mathrm{sign}(\sigma f)((\sigma f) \cdot \alpha) \Big)$$
$$= \sum_{\sigma \in C} \Big(\mathrm{sign}(\sigma)(\sigma \cdot \alpha) - \mathrm{sign}(\sigma)(\sigma \cdot (f \cdot \alpha)) \Big)$$
$$= \sum_{\sigma \in C} \Big(\mathrm{sign}(\sigma)(\sigma \cdot \alpha) - \mathrm{sign}(\sigma)(\sigma \cdot \alpha) \Big) = 0 \qquad \square$$

By construction, the alternation A defines a linear transformation $\mathsf{A} : V^{\otimes k} \to V^{\otimes k}$.

Definition 4.6.15. The subspace of $V^{\otimes k}$ given as the image of $\mathsf{A} : V^{\otimes k} \to V^{\otimes k}$ is called the *kth alternating product* or the *kth wedge product* of V and is denoted by $\bigwedge^k V$.

Proposition 4.6.16. *The subspace $\bigwedge^k V$ is skew-invariant under the action of S_k on $V^{\otimes k}$, i.e., for all $\sigma \in S_k$ and for all tensors $\alpha \in \bigwedge^k V$, we have $\sigma \cdot \alpha = \mathrm{sign}(\sigma)\alpha$.*

Proof. (Left as an exercise for the reader.) □

Corollary 4.6.17. *For all alternating tensors* $\alpha \in \bigwedge^k V$, *we have* $\mathsf{A}(\alpha) = k!\,\alpha$.

Proof. By Proposition 4.6.16,

$$\mathsf{A}(\alpha) = \sum_{\sigma \in S_k} \mathrm{sign}(\sigma)\sigma \cdot \alpha = \sum_{\sigma \in S_k} \mathrm{sign}(\sigma)^2 \alpha = k!\,\alpha. \qquad \square$$

Proposition 4.6.18. *Let* $\{e_1, e_2, \ldots, e_n\}$ *be a basis of* V. *Then the set*

$$\{\mathsf{A}(e_{i_1} \otimes e_{i_2} \otimes \cdots \otimes e_{i_k}) \mid 1 \leq i_1 < i_2 < \cdots < i_k \leq n\}$$

is a basis of $\bigwedge^k V$.

Proof. (The proof of this proposition is similar to the proof of Proposition 4.6.7 with the exception that $\mathsf{A}(e_{i_1} \otimes e_{i_2} \otimes \cdots \otimes e_{i_k}) = 0$ if $i_s = i_t$ for some $s \neq t$. We leave the proof as an exercise.) □

Corollary 4.6.19. *Let* V *be a vector space of dimension* n. *Then*

$$\dim \bigwedge^k V = \binom{n}{k}.$$

Proof. The proof of this corollary is similar to that of Corollary 4.6.8. However, we need to devise a counting argument that enumerates all strictly increasing sequences of length k with entries in $\{1, 2, \ldots, n\}$. Consider the scenario where we have n slots and a bag of $n - k$ Xs and k Ys. There are $\binom{n}{k}$ ways to fill n slots with the Xs and Ys, by choosing the slots in which we put Ys. However, there is a bijection between such fillings and our desired set of increasing sequences. For a given filling, define i_t as the number of Xs or Ys before or including the tth Y. This is clearly an increasing sequence with entries in $\{1, 2, \ldots, n\}$. Furthermore, any such increasing sequence gives us a unique filling of the slots with Xs and Ys. The result follows. □

As in the case of the symmetric product, it is not hard to see that the tensor product of alternating tensors is, in general, not another alternating tensor. However, it is possible to define a product between alternating tensors that produces another alternating tensor.

Definition 4.6.20. Let V be a vector space, and let $\alpha \in \bigwedge^k V$ and $\beta \in \bigwedge^l V$. We define

$$\alpha \wedge \beta = \frac{1}{k!\,l!}\mathsf{A}(\alpha \otimes \beta)$$

so that $\alpha \wedge \beta \in \bigwedge^{k+l} V$. We call this operation $\alpha \wedge \beta$ the *exterior product* or the *wedge product* of α and β.

Similar properties hold for the exterior product as for the symmetric product.

Proposition 4.6.21. *Let V be a vector space of dimension n. The following hold:*

1. *The exterior product is bilinear: for all $\alpha, \alpha_1, \alpha_2 \in \bigwedge^k V$, $\beta, \beta_1, \beta_2 \in \bigwedge^l V$, and λ in the base field,*

$$(\alpha_1 + \alpha_2) \wedge \beta = \alpha_1 \wedge \beta + \alpha_2 \wedge \beta, \qquad (\lambda\alpha) \wedge \beta = \lambda(\alpha \wedge \beta),$$
$$\alpha \wedge (\beta_1 + \beta_2) = \alpha \wedge \beta_1 + \alpha \wedge \beta_2, \qquad \alpha \wedge (\lambda\beta) = \lambda(\alpha \wedge \beta).$$

2. *The exterior product is anticommutative in the sense that for all $\alpha \in \bigwedge^k V$ and $\beta \in \bigwedge^l V$,*

$$\beta \wedge \alpha = (-1)^{kl} \alpha \wedge \beta.$$

3. *The exterior product is associative: for all $\alpha \in \bigwedge^r V$, $\beta \in \bigwedge^s V$, and $\gamma \in \bigwedge^t V$, as an element of $\bigwedge^{r+s+t} V$, we have*

$$(\alpha \wedge \beta) \wedge \gamma = \alpha \wedge (\beta \wedge \gamma) = \frac{1}{r!\, s!\, t!} \mathsf{A}(\alpha \otimes \beta \otimes \gamma).$$

Proof. Again we leave part 1 as an exercise for the reader.

For part 2, by Proposition 4.6.16, $\mathsf{A}(\alpha \otimes \beta)$ is skew-invariant under the action of S_{k+l}. As in the proof of Proposition 4.6.11, consider the permutation $\sigma_0 \in S_{k+l}$ that maps the n-tuple $(1, 2, \ldots, k+l)$ to $(k+1, \ldots, k+l, 1, \ldots, k)$. In each pure tensor in an expression of $\alpha \otimes \beta$, the action $\sigma_0 \cdot (\alpha \otimes \beta)$ moves (and keeps in the proper order) the vector terms coming from β in front of the terms coming from α. Hence we see that $\sigma_0 \cdot (\alpha \otimes \beta) = \beta \otimes \alpha$. Also, it is not difficult to see how σ_0 can be expressed using kl transpositions (permutations that interchange only two elements), and therefore, $\mathrm{sign}(\sigma_0) = (-1)^{kl}$. Thus, we conclude that

$$\beta \wedge \alpha = \frac{1}{k!\, l!} \mathsf{A}(\beta \otimes \alpha) = \frac{1}{k!\, l!} \mathsf{A}(\sigma_0 \cdot (\alpha \otimes \beta))$$
$$= \mathrm{sign}(\sigma_0) \frac{1}{k!\, l!} \mathsf{A}(\alpha \otimes \beta) = (-1)^{kl} \alpha \wedge \beta.$$

Part 3 follows in a similar manner to the proof of Proposition 4.6.11 with appropriate modifications, including an adaptation of Lemma 4.6.10 and using Corollary 4.6.17. □

By virtue of the associativity of the exterior product, the alternation of a pure tensor $\mathsf{A}(v_1 \otimes v_2 \otimes \cdots \otimes v_k)$ is denoted by

$$v_1 \wedge v_2 \wedge \cdots \wedge v_k,$$

where we often think of this element as an *anticommutative* "product" between vectors. This means that

$$\underset{\text{interchange } i,j}{v_1 \wedge v_2 \wedge \cdots \wedge v_k} = -v_1 \wedge v_2 \wedge \cdots \wedge v_k \qquad (4.46)$$

and also that for all $\sigma \in S_k$,

$$\sigma \cdot (v_1 \wedge v_2 \wedge \cdots \wedge v_k) = \text{sign}(\sigma) v_1 \wedge v_2 \wedge \cdots \wedge v_k. \qquad (4.47)$$

With this notation in mind, we often think of $\bigwedge^k V$ as a vector space in its own right, independent of $V^{\otimes k}$, with basis

$$\{e_{i_1} \wedge e_{i_2} \wedge \cdots \wedge e_{i_k} \mid 1 \le i_1 < i_2 < \cdots < i_k \le n\}.$$

Example 4.6.22. Let $V = \mathbb{R}^3$, and let

$$\vec{v} = \begin{pmatrix} 1 \\ -1 \\ 2 \end{pmatrix} \quad \text{and} \quad \vec{w} = \begin{pmatrix} 3 \\ 0 \\ 2 \end{pmatrix}$$

be vectors in $\mathbb{R}^3$. Recall that $\dim \bigwedge^2 V = 3$. With respect to the standard basis in $\bigwedge^2 V$, we have

$$\begin{aligned}
\vec{v} \wedge \vec{w} &= (\vec{e}_1 - \vec{e}_2 + 2\vec{e}_3) \wedge (3\vec{e}_1 + 2\vec{e}_3) \\
&= 3\vec{e}_1 \wedge \vec{e}_1 + 2\vec{e}_1 \wedge \vec{e}_3 - 3\vec{e}_2 \wedge \vec{e}_1 - 2\vec{e}_2 \wedge \vec{e}_3 + 6\vec{e}_3 \wedge \vec{e}_1 + 4\vec{e}_3 \wedge \vec{e}_3 \\
&= 2\vec{e}_1 \wedge \vec{e}_3 + 3\vec{e}_1 \wedge \vec{e}_2 - 2\vec{e}_2 \wedge \vec{e}_3 - 6\vec{e}_1 \wedge \vec{e}_3 \\
&= -4\vec{e}_1 \wedge \vec{e}_3 + 3\vec{e}_1 \wedge \vec{e}_2 - 2\vec{e}_2 \wedge \vec{e}_3.
\end{aligned}$$

For the symmetric product, recall that $\dim \text{Sym}^2 V = 6$. With respect to the standard basis in $\text{Sym}^2 V$, we have

$$\begin{aligned}
\vec{v}\vec{w} &= (\vec{e}_1 - \vec{e}_2 + 2\vec{e}_3)(3\vec{e}_1 + 2\vec{e}_3) \\
&= 3\vec{e}_1\vec{e}_1 + 2\vec{e}_1\vec{e}_3 - 3\vec{e}_2\vec{e}_1 - 2\vec{e}_2\vec{e}_3 + 6\vec{e}_3\vec{e}_1 + 4\vec{e}_3\vec{e}_3 \\
&= 3\vec{e}_1^2 + 4\vec{e}_3^2 - 3\vec{e}_1\vec{e}_2 + 8\vec{e}_1\vec{e}_3 - 2\vec{e}_2\vec{e}_3.
\end{aligned}$$

Proposition 4.6.23. *Let V be an n-dimensional vector space over a field. Let v_i for $i = 1, \ldots, m$ be m vectors in V where $m \le n$. Let w_j for $j = 1, \ldots, m$ be another set of vectors, with $w_j \in \text{Span}(v_i)$ given by $w_j = \sum_i c_{ji} v_i$. Then*

$$w_1 \wedge w_2 \wedge \cdots \wedge w_m = (\det c_{ji}) v_1 \wedge v_2 \wedge \cdots \wedge v_m.$$

Proof. This is a simple matter of calculation, as follows:

$$w_1 \wedge w_2 \wedge \cdots \wedge w_m = \left(\sum_{i_1=1}^{m} c_{1i_1} v_{i_1} \right) \wedge \left(\sum_{i_2=1}^{m} c_{2i_2} v_{i_2} \right) \wedge \cdots \wedge \left(\sum_{i_m=1}^{m} c_{mi_m} v_{i_m} \right). \quad (4.48)$$

In any wedge product, if there is a repeated vector, the wedge product is 0. Therefore, when distributing out the m summations, the only nonzero terms are those in which all the $i_1, i_2, \ldots, i_m$ are distinct. Furthermore, by Equation (4.47), any nonzero term can be rewritten as

$$v_{i_1} \wedge v_{i_2} \wedge \cdots \wedge v_{i_m} = \text{sign}(\sigma)\, v_1 \wedge v_2 \wedge \cdots \wedge v_m,$$

where σ is the permutation given as a table by

$$\sigma = \begin{pmatrix} 1 & 2 & \cdots & m \\ i_1 & i_2 & \cdots & i_m \end{pmatrix}.$$

Furthermore, by selecting which integer is chosen for each i_k in each term on the right side of Equation (4.48), we see that every possible permutation is used exactly once. Thus, we have

$$w_1 \wedge w_2 \wedge \cdots \wedge w_m = \left(\sum_{\sigma \in S_m} \text{sign}(\sigma) c_{1\sigma^{-1}(1)} c_{2\sigma^{-1}(2)} \cdots c_{m\sigma^{-1}(m)} \right) v_1 \wedge v_2 \wedge \cdots \wedge v_m.$$

The content of the parantheses in the above equation is precisely the determinant of the matrix (c_{ij}) and the proposition follows. □

Example 4.6.24. Let $V = \mathbb{R}^n$ with standard basis e_i, where $i = 1, 2, \ldots, n$. By Proposition 4.6.23, we have

$$\vec{v}_1 \wedge \vec{v}_2 \wedge \cdots \wedge \vec{v}_n = \det \begin{pmatrix} | & | & & | \\ \vec{v}_1 & \vec{v}_2 & \cdots & \vec{v}_n \\ | & | & & | \end{pmatrix} \vec{e}_1 \wedge \vec{e}_2 \wedge \cdots \wedge \vec{e}_n.$$

By a standard result, the determinant $\det \begin{pmatrix} \vec{v}_1 & \vec{v}_2 & \cdots & \vec{v}_n \end{pmatrix}$ is the volume of the parallelepiped spanned by $\{\vec{v}_1, \vec{v}_2, \cdots, \vec{v}_n\}$.

Furthermore, if we consider the element $\vec{e}^{*1} \wedge \cdots \wedge \vec{e}^{*n} \in \bigwedge^n V^*$ as an alternating multilinear function on V, we have

$$\vec{e}^{*1} \wedge \cdots \wedge \vec{e}^{*n}(\vec{v}_1, \ldots, \vec{v}_n) = \left(\sum_{\sigma \in S_k} \text{sign}(\sigma)\sigma \cdot (\vec{e}^{*1} \otimes \cdots \otimes \vec{e}^{*n}) \right)(\vec{v}_1, \ldots, \vec{v}_n)$$

$$= \det \begin{pmatrix} | & | & & | \\ \vec{v}_1 & \vec{v}_2 & \cdots & \vec{v}_n \\ | & | & & | \end{pmatrix}.$$

Therefore, the element $\vec{e}^{*1} \wedge \cdots \wedge \vec{e}^{*n}$ is often called the *signed volume form* on V.

PROBLEMS

4.6.1. Let $V = \mathbb{R}^3$, and consider the linear transformation $T : V \to V$ given by

$$T(\vec{v}) = \begin{pmatrix} 1 & 2 & 3 \\ 4 & 5 & 6 \\ 7 & 8 & 9 \end{pmatrix} \vec{v}$$

with respect to the standard basis of $\mathbb{R}^3$.

(a) Prove that the function $S : \bigwedge^2 V \to \bigwedge^2 V$ that satisfies

$$S(\vec{v}_1 \wedge \vec{v}_2) = T(\vec{v}_1) \wedge T(\vec{v}_2)$$

extends to a linear transformation.

(b) Determine the matrix of S with respect to the associated basis $\{\vec{i} \wedge \vec{j}, \vec{j} \wedge \vec{k}, \vec{k} \wedge \vec{i}\}$.

4.6.2. Repeat the above exercise but with $\mathrm{Sym}^2 V$ and changing the question accordingly.

4.6.3. Prove Proposition 4.6.16.

4.6.4. Prove Proposition 4.6.18.

4.6.5. Prove part 1 of Proposition 4.6.11.

4.6.6. Let V be a vector space over $\mathbb{C}$ of dimension n, and let T be a linear transformation $T : V \to V$ with eigenvalues λ_i, where $1 \le i \le n$. Let $S : \bigwedge^2 V \to \bigwedge^2 V$ be defined by $S(v_1 \wedge v_2) = T(v_1) \wedge T(v_2)$.

(a) Prove that the eigenvalues of S are $\lambda_i \lambda_j$ for $1 \le i < j \le n$.

(b) Prove that $\det S = (\det T)^{n-1}$.

(c) Prove that the trace of S is

$$\frac{1}{2} \left(\left(\sum_{i=1}^{n} \lambda_i \right)^2 - \sum_{i=1}^{n} \lambda_i^2 \right).$$

4.6.7. Let V be a vector space over $\mathbb{C}$ of dimension n, and let T be a linear transformation $T : V \to V$ with eigenvalues λ_i, where $1 \le i \le n$. Let $S : \mathrm{Sym}^2 V \to \mathrm{Sym}^2 V$ be defined by $S(v_1 v_2) = T(v_1)T(v_2)$.

(a) Prove that the eigenvalues of S are $\lambda_i \lambda_j$ for $1 \le i \le j \le n$.

(b) Prove that $\det S = (\det T)^{n+1}$.

(c) Prove that the trace of S is

$$\frac{1}{2} \left(\left(\sum_{i=1}^{n} \lambda_i \right)^2 + \sum_{i=1}^{n} \lambda_i^2 \right).$$

4.7 Algebra over a Field

We conclude this chapter by introducing the concept of an algebra over a field. If the reader is not familiar with the technical term of "algebra," she has already encountered this algebraic structure both in this book and in previous study. The reader surely is familiar with the word "algebra" used in a variety of contexts; the precise definition for an algebra aligns with the casual use of the word. Furthermore, introducing this notion here allows us to give a broader perspective on multilinear algebra. In addition, we present the concept of a derivation, which plays a central role in analysis on manifolds.

4.7.1 Algebras

Definition 4.7.1. Let K be a field. An *algebra* over K is a vector space A over K equipped with a bilinear transformation $A \times A \to A$. The bilinear transformation is usually called a product.

It is not uncommon to change the terminology slightly and refer to an algebra on A. We say that an algebra is commutative (resp. associative) depending on whether the product is commutative (resp. associative). Note that the bilinear property implies that the product distributes over the addition.

A few common vectors spaces are in fact algebras. We consider a few examples.

Example 4.7.2. One of the first nontrivial examples of an algebra that mathematics students encounter is the vector space of $\mathbb{R}^3$ equipped with the cross product. The properties that

$$(\vec{u} + \vec{v}) \times \vec{w} = \vec{u} \times \vec{w} + \vec{v} \times \vec{w} \qquad (c\vec{u}) \times \vec{w} = c(\vec{u} \times \vec{w})$$
$$\vec{u} \times (\vec{v} + \vec{w}) = \vec{u} \times \vec{v} + \vec{u} \times \vec{w} \qquad \vec{u} \times (c\vec{w}) = c(\vec{u} \times \vec{w})$$

for all $\vec{u}, \vec{v}, \vec{w} \in \mathbb{R}^3$ and for all $c \in \mathbb{R}$ establish that the cross product $\times$ is a bilinear transformation. This algebra is neither commutative nor associative.

Example 4.7.3. In linear algebra, the operations of addition and scalar multiplication on the set $M_n(\mathbb{R})$ of $n \times n$ matrices with coefficients in $\mathbb{R}$ makes it a vector space. However, the operation of matrix multiplication on $M_n(\mathbb{R})$ is bilinear in each entry, giving $M_n(\mathbb{R})$ the stucture of an algebra. This algebra is not commutative but it is associative.

Example 4.7.4. The set of polynomials of degree n or less, or more generally the set $\mathbb{R}[x]$ of all polynomials with coefficients in $\mathbb{R}$, equipped with scalar multiplication, addition, polynomial multiplication, is an algebra. That is why the expression "polynomial algebra" makes sense. Polynomial algebra over a field is both commutative and associative.

Example 4.7.5. Let I be an interval of $\mathbb{R}$. The set of continuous function $C^0(I, \mathbb{R})$, and more generally the set of functions of any differentiability class, forms a vector space when we consider scalar multiplication and addition of functions. However, by virtue of distributivity, multiplication of functions equips these sets with the structures of an algebra over $\mathbb{R}$.

The concept of an algebra allows us to recast some of our constructions concerning tensor products into a broader perspective that will be useful later on. We introduce the tensor, symmetric and alternating algebra on a vector space in tandem. Let V be a vector space over a field K. In all the following cases, the product of an element in K with anything else corresponds to scalar multiplicaiton.

1. The *tensor algebra* on V, denoted $T^\bullet V$ is the infinite direct sum

$$T^\bullet V = \bigoplus_{j=0}^{\infty} V^{\otimes j} = K \oplus V \oplus (V \otimes V) \oplus (V \otimes V \otimes V) \oplus \cdots$$

with the bilinear product on $T^\bullet V$ induced from $\otimes : V^{\otimes s} \times V^{\otimes t} \to V^{\otimes (s+t)}$ and extended by linearity.

2. The *symmetric algebra* on V, denoted $\mathrm{Sym}\, V$, is the infinite direct sum

$$\mathrm{Sym}\, V = \bigoplus_{j=0}^{\infty} \mathrm{Sym}^j V = K \oplus V \oplus \mathrm{Sym}^2 V \oplus \mathrm{Sym}^3 V \oplus \cdots$$

with the bilinear product on $\mathrm{Sym}^\bullet V$ induced from $\cdot : \mathrm{Sym}^s V \times \mathrm{Sym}^t V \to \mathrm{Sym}^{s+t}$ described in Definition 4.6.9 and extended by linearity.

3. The *alternating algebra* on V, denoted $\bigwedge V$, is the infinite direct sum

$$\bigwedge V = \bigoplus_{j=0}^{\infty} \bigwedge^j V = K \oplus V \oplus \bigwedge^2 V \oplus \bigwedge^3 V \oplus \cdots$$

with the bilinear product on $\bigwedge^\bullet V$ induced from the exterior product (Definition 4.6.20) $\wedge : \bigwedge^s V \times \bigwedge^t V \to \bigwedge^{s+t} V$ and extended by linearity.

As long as V is a nontrivial vector space, $T^\bullet V$ and $\mathrm{Sym}\, V$ are infinite dimensional. However, if V is finite dimensional with $\dim V = n$, then $\bigwedge^k V$ is trivial for $k > n$ and

$$\dim \bigwedge V = \sum_{j=0}^{n} \dim \bigwedge^j V = \sum_{j=0}^{n} \binom{n}{j} = 2^n.$$

The tensor, symmetric and alternating algebras on a vector space are associative. However, only the symmetric algebra is commutative.

Example 4.7.6. As an example of operations in $T^\bullet V$, suppose that $V = \mathbb{R}^3$ with basis $\{\vec{\imath}, \vec{\jmath}, \vec{k}\}$. Let $\alpha = 4 + 2\vec{\imath} - 3\vec{\imath} \otimes \vec{k}$ and let $\beta = 7 + 3\vec{\imath} - \vec{k}$. The addition of α and β on $T^\bullet V$, after collecting like terms, is

$$\alpha + \beta = 11 + 5\vec{\imath} - \vec{k} - 3\vec{\imath} \otimes \vec{k}.$$

The product follows from distributivity by

$$\alpha \otimes \beta = (4 + 2\vec{\imath} - 3\vec{\imath} \otimes \vec{k}) \otimes (7 + 3\vec{\imath} - \vec{k})$$
$$= 28 + 12\vec{\imath} - 4\vec{k} + 14\vec{\imath} + 6\vec{\imath} \otimes \vec{\imath} - 2\vec{\imath} \otimes \vec{k} - 21\vec{\imath} \otimes \vec{k}$$
$$- 9\vec{\imath} \otimes \vec{k} \otimes \vec{\imath} + 3\vec{\imath} \otimes \vec{k} \otimes \vec{k}$$
$$= 28 + 26\vec{\imath} - 4\vec{k} + 6\vec{\imath} \otimes \vec{\imath} - 23\vec{\imath} \otimes \vec{k} - 9\vec{\imath} \otimes \vec{k} \otimes \vec{\imath} + 3\vec{\imath} \otimes \vec{k} \otimes \vec{k}.$$

The tensor, symmetric and alternating algebras associated to a vector space are examples of *graded algebras*, graded by $\mathbb{N}$.

Definition 4.7.7. An $\mathbb{N}$-graded algebra is a vector space expressed as

$$V = \bigoplus_{j \in \mathbb{N}} V_j$$

where V_j is a vector space for all $j \in \mathbb{N}$ and in which, for and $j, k \in \mathbb{N}$, the product $\cdot$ has $V_j \cdot V_k \subseteq V_{j+k}$.

4.7.2   Generating Subsets

Recall that in linear algebra, if S is a nonempty subset of a vector space V, the span of S consists of all linear combinations of elements in S, namely,

$$\text{Span}(S) = \{c_1 u_1 + c_2 u_2 + \cdots + c_n u_n \mid n \in \mathbb{N}^*, c_1, \ldots, c_n \in K, \text{ and } u_1, \ldots, u_n \in S\}.$$

It is an easy exercise in linear algebra to show that for any nonempty set S, the set $\text{Span}(S)$ is a subspace of V. We say that S spans V is $\text{Span}(S) = V$.

In contrast for algebras, if S is a subset of an algebra A, we define the subset of A *generated by* S as the smallest subset T of A that contains S, is closed under multiplication by any scalar, closed under addition, and closed under the algebra product. By distributivity of scalar multiplication, associativity of addition, and distributivity of the product, the subset generated by S is a subalgebra of A. We say that A is generated by the subset S if the subalgebra generated by S is all of A.

Example 4.7.8. Consider the set $K[x]$ of all polynomials of scalars from a field K and consider the subset $\{1, x\}$. Using the product of x with itself produces the infinite set $\{x, x^2, x^3, \ldots\}$. By taking any finite sum of scalar multiples of elements in $\{1, x, x^2, \ldots\}$ gives every polynomial. Consequently, $\{1, x\}$ generates $K[x]$ as an algebra.

Example 4.7.9. Let V be a vector spaces with basis $\{u_1, u_2, \ldots, u_n\}$. It is not hard to see that, using their respective products, the tensor algebra, the symmetric algebra and the alternating algebra on V are all generated by $\{1, u_1, u_2, \ldots, u_n\}$.

4.7.3 Derivations

Among the examples of algebras that we presented above, consider the algebra $C^\infty(I, \mathbb{R})$ of differentiable real-valued functions over an interval I of $\mathbb{R}$. The derivative operator D on $C^\infty(I, \mathbb{R})$ is a linear transformation. The derivative of a product is not the product of derivatives but the derivative satisfies the product rule, also called Leibniz's law.

Definition 4.7.10. Let A be an algebra over a field K. A *derivation* on A is a linear transformation $D : A \to A$ that satisfies Leibniz's law,

$$D(ab) = D(a)b + aD(b), \qquad \text{for all } a, b \in A.$$

The set of all derivations on A is denoted by $\mathrm{Der}_K(A)$.

Example 4.7.11. Let $K[x]$ be the polynomial algebra. Define $D : K[x] \to K[x]$ by

$$D(a_n x^n + \cdots + a_1 x + a_0) = n a_n x^n + (n-1)a_{n-1}x^{n-1} + \cdots + a_1 x.$$

We recognize this as x times the derivative of the polynomial. We prove directly that this is a derivation.

It is easy to see that D is a linear transformation. We need to check the Leibniz rule. Let $a(x) = a_n x^n + \cdots + a_1 x + a_0$ and $b(x) = b_m x^m + \cdots + b_1 x + b_0$. Assuming $a_i = 0$ if $i < 0$ or $i > n$ and similarly for the coefficients of $b(x)$, then we can write the product as

$$a(x)b(x) = \sum_{k=0}^{m+n} \left(\sum_{i+j=k} a_i b_j \right) x^k.$$

So

$$D(a(x)b(x)) = \sum_{k=1}^{m+n} \left(\sum_{i+j=k} k a_i b_j \right) x^k = \sum_{k=1}^{m+n} \left(\sum_{i+j=k} (i+j) a_i b_j \right) x^k$$

$$= \sum_{k=1}^{m+n} \left(\sum_{i+j=k} i a_i b_j \right) x^k + \sum_{k=1}^{m+n} \left(\sum_{i+j=k} j a_i b_j \right) x^k$$

$$= D(a(x))b(x) + a(x)D(b(x)).$$

This shows that D is a derivation.

Proposition 4.7.12. *Suppose that an algebra A is generated by a subset S. Then two derivations D_1 and D_2 are equal if and only if they agree on all elements of S.*

Proof. By Problem 4.7.8, the set of derivations $\mathrm{Der}_K(A)$ is a subspace of $\mathrm{Hom}_K(A, A)$. In particular $D_1 - D_2$ is a derivation. So D_1 and D_2 are equal as derivations if and only if $D = D_1 - D_2$ is the trivial function. So it suffices to prove that a derivation is trivial if and only if it maps all elements of S to 0.

If D is trivial, it obviously maps all elements of S to 0. We need to prove the converse. Suppose that D maps all elements of the subset S to 0. For all $a \in A$, $0a = 0 = a0$, where 0 is the zero vector of A. So

$$\forall u \in S, \forall c \in K, \quad D(cv) = cD(v) = c0 = 0,$$
$$\forall u, v \in S, \quad D(u + v) = D(u) + D(v) = 0 + 0 = 0,$$
$$\forall u, v \in S, \quad D(uv) = D(u)v + uD(v) = 0v + u0 = 0.$$

Since D is trivial on all S and since having a 0 derivation is preserved with the three operations that define A recursively from S, then D is trivial on all of A. The proposition follows. $\qquad \square$

PROBLEMS

4.7.1. Let K be a field and let $M_n(K)$ be the set of $n \times n$ matrices with coefficients in K. Define the bracket operation on $M_n(K)$ by $[A, B] = AB - BA$.

    (a) Prove that $M_n(K)$ equipped with $[\,,]$ is an algebra.

    (b) Prove that this algebra is neither commutative nor associative.

4.7.2. Prove the set of bilinear products on V is a vector space and show a canonical isomorphism between the set of bilinear products on V and $V \otimes V^* \otimes V^*$.

4.7.3. Let V be a vector space over a field K. Prove that the direct sum of all tensor products of type (r, s),

$$\bigoplus_{r=0}^{\infty} \bigoplus_{s=0}^{\infty} V^{\otimes r} \otimes V^{*\otimes s}, \qquad \text{where } V^{\otimes 0} = K \text{ and } K \otimes K = K,$$

is an algebra with the usual tensor product $\otimes$ as the bilinear product.

4.7.4. Let V be a 2-dimensional vector space of K with basis $\{e_1, e_2\}$. Suppose that $S : V \to V$ is a linear transformation. Define $\bigwedge S : \bigwedge V \to \bigwedge V$ as

$$(\textstyle\bigwedge S)(c^1 + c^2 e_1 + c^3 e^2 + c^4 e_1 \wedge e_2) = c^1 + c^2 S(e_1) + c^3 S(e_2) + c^4 S(e_1) \wedge S(e_2).$$

    (a) Prove that $\wedge S$ is a linear transformation.

    (b) Suppose that with respect to the ordered basis (e_1, e_2) on V, the matrix of S is $\begin{pmatrix} a & b \\ c & d \end{pmatrix}$. Find the matrix of $\wedge S$ with respect to the ordered basis $(1, e_1, e_2, e_1 \wedge e_2)$ on $\bigwedge V$.

4.7.5. Repeat the previous exercise assuming that V is a vector space of dimension 3.

4.7.6. Let I be an interval of $\mathbb{R}$ and consider the vector space $C^\infty(I, \mathbb{R})$. Show that the second derivative D^2 is not a derivation.

4.7.7. Let I be an interval of $\mathbb{R}$. Consider the vector space $C^\infty(I, \mathbb{R})$ and let D be the derivative operator on $C^\infty(I, \mathbb{R})$. Let $g : I \to I$ be a differentiable function.

(a) Prove that $D_1 : C^\infty(I, \mathbb{R}) \to C^\infty(I, \mathbb{R})$ defined by $D_1(f) = g \cdot D(f)$ is a derivation.

(b) Explain why $D_2 : C^\infty(I, \mathbb{R}) \to C^\infty(I, \mathbb{R})$ defined by $D_2(f) = D(f \circ g)$ is not a derivation.

4.7.8. Prove that $\mathrm{Der}_K(A)$ is a vector subspace of $\mathrm{Hom}_K(A, A)$.

4.7.9. Let A be an algebra. Prove that $\mathrm{Der}_K(A)$ is an algebra when equipped with the bilinear transformation

$$[D_1, D_2] = D_1 \circ D_2 - D_2 \circ D_1.$$

[Hint: Use Problem 4.7.8.]

4.7.10. Let U be an open subset of $\mathbb{R}^n$ and consider the algebra $C^\infty(U, \mathbb{R})$ of smooth functions on U. Let $a_i(x_1, x_2, \ldots, x_n)$ be smooth functions over U for $i = 1, 2, \ldots, n$. Prove that

$$a_1(x_1, x_2, \ldots, x_n)\frac{\partial}{\partial x^1} + \cdots + a_n(x_1, x_2, \ldots, x_n)\frac{\partial}{\partial x^n}$$

is a derivation on $C^\infty(U, \mathbb{R})$.

CHAPTER 5

Analysis on Manifolds

In Chapter 3, we introduced the concept of a differentiable manifold as motivated by a search for topological spaces over which it is possible to do calculus and ultimately dynamics. The idea of having a topological space locally homeomorphic to $\mathbb{R}^n$ drove the definition of a differentiable manifold. Subsequent sections in that chapter discussed differentiable maps between manifolds and the differentials of such maps. We used these to introduce the important notions of immersions, submersions, and submanifolds as qualifiers of how manifolds may relate to one another.

The astute reader might observe that we have not so far made good on our promise to do physics on a manifold, no matter how amorphous that expression may be. As an illustrative example, consider Newton's second law of motion applied to, say, simple gravity, as follows:

$$m\vec{x}''(t) = m\vec{g}, \tag{5.1}$$

where m is constant and $\vec{g}$ is a constant vector. the parametrized curve $\vec{x}(t)$ in $\mathbb{R}^3$ is called the trajectory and its acceleration vector $\vec{x}''(t)$ is also a vector function in $\mathbb{R}^3$. In order for (5.1) to have meaning, it is essential that the quantities on both sides of the equation exist in the same Euclidean space. Applying this type of equation to the context of manifolds poses a variety of difficulties.

First, note that a curve in a manifold M is a submanifold $\gamma : I \to M$, where I is an open interval of $\mathbb{R}$, whereas the velocity vector of a curve at a point p is an element of the tangent plane to M at p. Second, the discussion of differentials in Chapter 3 does not readily extend to a concept of second derivatives for a curve in a manifold. It is not even obvious in what space a second derivative would exist. Consequently, it is not at all obvious how to transcribe equations of curves in $\mathbb{R}^3$ that involve $\vec{x}$, $\vec{x}'$, and $\vec{x}''$ to the context of manifolds.

Another difficulty arises when we try to express in the context of differentiable manifolds the classical local theory of surfaces in $\mathbb{R}^3$ (as presented in [5, Chapter 5]). It is not difficult to define the first fundamental form as a bilinear form on T_pM. However, since we do not view a given manifold M as a subset of any Euclidean (vector) space, the concept of normal vectors does not exist. Therefore, there is no equivalent of the second fundamental form, and all concepts of curvature become problematic to define (see Chapter 6 in [5]).

This chapter does not yet discuss how to do physics on a manifold, but it does begin to show how to do calculus. We study in greater detail the relationship between the tangent space to a manifold M at p. Also, in order to overcome the conceptual hurdles mentioned above, we introduce the formalism of vector bundles on a manifold, discuss vector (and tensor) fields on the manifold, develop the calculus of differential forms, and end by considering integration on manifolds.

In Chapter 4 we commented how geometers and physicists both use tensors but usually with very different notations (usually called coordinate-free or coordinate-dependent). This difference continues here as we use tensor fields on manifolds. If a reader is already familiar with one or the other habits of notation, it is very useful to recognize both as representing the same kind of object. However, we must begin by introducing the vector bundle formalism.

5.1   Vector Bundles on Manifolds

A vector bundle over a manifold is a particular case of a fiber bundle over a topological space. As we do not need the full generality of fiber bundles in this book, we refer the interested reader to [53] or [12] and present instead the specific formalism of vector bundles.

Chapter 3 discussed tangent spaces to manifolds. To each point $p \in M$, we associated a tangent space. The elements of the tangent space are differential operators of differentiable functions $f : M \to \mathbb{R}$. Despite their slightly more abstract definition, such differential operators properly model the role of tangent vectors. Since M is not a subset of some Euclidean space, the tangent spaces $T_p M$ are not subspaces of any ambient space either. A manifold equipped with tangent spaces at each point motivates the idea of "attaching" a vector space to each point p of a manifold M. Furthermore, from an intuitive perspective, we would like to attach these vector spaces, in some sense, continuously. The following definition formalizes this perspective.

Definition 5.1.1. Let M^n be a differentiable manifold with atlas $\mathcal{A} = \{(U_\alpha, \phi_\alpha)\}_{\alpha \in I}$, and let V be a finite-dimensional, real, vector space. A *vector bundle* over M of *fiber* V is a Hausdorff topological space E with a continuous surjection $\pi : E \to M$ (called a *bundle projection*) and a collection Ψ of homeomorphisms (called *trivializations*) $\psi_\alpha : U_\alpha \times V \to \pi^{-1}(U_\alpha)$, satisfying

1. $\pi \circ \psi_\alpha(p, v) = p$ for all $(p, v) \in U_\alpha \times V$ and $E_p \overset{\text{def}}{=} \pi^{-1}(p)$ is homeomorphic to V;

2. if $U_\alpha \cap U_\beta \neq \emptyset$, then $\psi_\beta^{-1} \circ \psi_\alpha : (U_\alpha \cap U_\beta) \times V \to (U_\alpha \cap U_\beta) \times V$ is of the form

$$\psi_\beta^{-1} \circ \psi_\alpha(p, v) = (p, \theta_{\beta\alpha}(p)v),$$

where $\theta_{\beta\alpha}(p) : U_\alpha \cap U_\beta \to \mathrm{GL}(V)$ is a continuous map into the general linear group (i.e., the set of invertible transformations from V to V).

The vector bundle is called differentiable (respectively, C^k or smooth) if M is differentiable (respectively, C^k or smooth) and if all the maps $\theta_{\beta\alpha}$ are differentiable (respectively, C^k or smooth) as maps between manifolds.

We point out that when V is a finite dimensional vector space of dimension n, we can identify $\mathrm{GL}(V)$ as an open subset of the set $\mathbb{R}^{n^2}$. Hence, $\mathrm{GL}(V)$ naturally carries the structure of a differentiable manifold. It is in this sense that the functions $\theta_{\beta\alpha}$ can be differentiable maps between manifolds.

A vector bundle whose fiber is one-dimensional is called a *line bundle*.

Vector bundles are often denoted by a single Greek letter ξ or η. The topological space E is called the *total space* and denoted by $E(\xi)$ while the manifold M is called the *base space* and denoted by $B(\xi)$.

Example 5.1.2 (The Trivial Bundle). Let M^n be a manifold with atlas $\mathcal{A} = \{\phi_\alpha\}$, and let V be a real vector space. The topological space $M \times V$ is a vector bundle over M. The trivialization maps ψ_α are all the identity maps on $U_\alpha \times V$ and the maps $\theta_{\beta\alpha}$ are the identity linear transformation.

Example 5.1.3 (Infinite Möbius Strip). Consider the circle $\mathbb{S}^1$ as a manifold with the atlas $\{(U_1, \phi_1), (U_2, \phi_2)\}$ defined by:

$$\phi_1 : U_1 = \mathbb{S}^1 - \{(1,0)\} \to (0, 2\pi), \quad \text{with } \phi_1(\cos u, \sin u) = u,$$
$$\phi_2 : U_2 = \mathbb{S}^1 - \{(-1,0)\} \to (\pi, 3\pi), \quad \text{with } \phi_2(\cos u, \sin u) = u.$$

So ϕ_1 uses as a coordinate the angle around $\mathbb{S}^1$ from $(1,0)$, while ϕ_2 also uses as a coordinate the angle around $\mathbb{S}^1$ starting $(1,0)$ but, with the value of the angle taken in $(\pi, 3\pi)$. The transition map between these two charts is

$$\phi_{21} = \phi_2 \circ \phi_1^{-1} : (0, \pi) \cup (\pi, 2\pi) \to (\pi, 2\pi) \cup (2\pi, 3\pi)$$

$$u \mapsto \begin{cases} u + 2\pi & \text{if } 0 < u < \pi, \\ u & \text{if } \pi < u < 2\pi. \end{cases}$$

Now define the vector bundle ξ of fiber $\mathbb{R}$ over $\mathbb{S}^1$ as a total space E with the surjective map $\pi : E \to \mathbb{S}^1$ defined by homeomorphisms $\psi_i : U_i \times \mathbb{R} \to \pi^{-1}(U_i)$ for $i = 1, 2$, such that $\psi_2^{-1} \circ \psi_1 : (U_1 \cap U_2) \times \mathbb{R} \to (U_1 \cap U_2) \times \mathbb{R}$ is given by

$$\psi_2^{-1} \circ \psi_1(p, v) = (p, \theta_{21}(p)(v))$$

where

$$\theta_{21}(p) = \begin{cases} -1, & \text{if } 0 < \phi_1(p) < \pi, \\ 1, & \text{if } \pi < \phi_1(p) < 2\pi. \end{cases} \tag{5.2}$$

Note that $\theta_{21} : U_1 \cap U_2 \to \mathrm{GL}(\mathbb{R}) = \mathbb{R} - \{0\}$ is constant on the connected components of $U_1 \cap U_2 = \mathbb{S}^1 - \{(1,0), (-1,0)\}$ and hence is continuous and also smooth. The function θ_{12} is the inverse function of θ_{21} so has the same properties.

We will show that the image M of the parametrized surface in $\mathbb{R}^4$ described by

$$Y(u,t) = \left(\cos u, \sin u, t \cos \left(\frac{u}{2} \right), t \sin \left(\frac{u}{2} \right) \right), \quad \text{with } (u,t) \in [0, 2\pi] \times \mathbb{R}$$

realizes this vector bundle ξ. The function $\pi : M \to \mathbb{S}^1$ defined by projection onto the first two coordinates in $\mathbb{R}^4$. Note that above each point $p \in \mathbb{S}^1$, i.e., the points of $\pi^{-1}(p)$ are lines in $\mathbb{R}^4$. Furthermore, it is not hard to see that $\pi^{-1}(U_i)$ is homeomorphic to $U_i \times \mathbb{R}$ for $i = 1, 2$, each of which we can visualize as an infinite strip. If we define $\psi_i : U_i \times \mathbb{R} \to \pi^{-1}(U_i)$ as

$$\psi_i(p, t) = Y(\phi_i(p), t),$$

then

$$\psi_2(p, t) = Y(\phi_2(p), t) = Y(\phi_{21}(\phi_1(p)), t).$$

Since $\phi_{\beta\alpha}(u)$ is always $u + 2\pi k$ for $k \in \{-1, 0, 1\}$ and $\cos x$ and $\sin x$ are periodic 2π, then

$$\pi(\psi_i(p, t)) = \pi(Y(\phi_i(p), t)) = p.$$

Furthermore, if $0 < \phi_1(p) < \pi$,

$$\begin{aligned}
\psi_2(p, t) &= Y(\phi_1(p) + 2\pi, t) \\
&= (\cos(\phi_1(p)), \sin(\phi_1(p)), t\cos(\phi_1(p) + \pi), t\sin(\phi_1(p) + \pi)) \\
&= (\cos(\phi_1(p)), \sin(\phi_1(p)), -t\cos(\phi_1(p)), -t\sin(\phi_1(p))) = \psi_1(p, -t),
\end{aligned}$$

while, if $\pi < \phi_1(p) < 2\pi$, we have $\psi_2(p, t) = Y(\phi_1(p), t) = \psi_1(p, t)$. Consequently,

$$\psi_2^{-1} \circ \psi_1(p, t) = (p, \theta_{21}(p)(t))$$

for the transition functions θ_{21} defined in (5.2).

The subset M is evidently not the cylinder $\mathbb{S}^1 \times \mathbb{R}$. Furthermore, one can get an intuition for this set as a Möbius band of infinite width.

The intuitive stance behind Definition 5.1.1 is that a vector bundle is not just a manifold with a vector space V associated to each point but that the vector spaces "vary continuously."

Consider now a differentiable manifold, and consider also the *disjoint* union of all the tangent planes to M at points $p \in M$, i.e.,

$$\coprod_{p \in M} T_p M = \{ (p, X) \mid p \in M \text{ and } X \in T_p M \}.$$

The identity map $i : M \to M$ is certainly differentiable, and we calculate its differential at a point $p \in U_\alpha \cap U_\beta$ in overlapping charts. Label the coordinate charts

$x = \phi_\alpha$ and $\bar{x} = \phi_\beta$. According to (3.13), the matrix of the differential of the identity map is

$$[di_p] = \left(\frac{\partial \bar{x}^j}{\partial x^i} \Big|_p \right),$$

and the reader should recall that the explicit meaning of this partial derivative is given in (3.14). Given any pair of overlapping coordinate charts, this differential is invertible so it is an element in $GL_n(\mathbb{R})$ and corresponds to the maps $\theta_{\beta\alpha}$.

We can arrive at this same result in another way. Consider the coordinate systems defined by $\bar{x}$ and x over $U_\alpha \cap U_\beta$. The chain rule gives, as operators,

$$\frac{\partial}{\partial \bar{x}^j} \Big|_p = \sum_{i=1}^n \frac{\partial x^i}{\partial \bar{x}^j} \Big|_p \frac{\partial}{\partial x^i} \Big|_p. \tag{5.3}$$

(The subscript $|_p$ becomes tedious and so in the remaining paragraphs, we understand the differential operators and the matrices as depending on $p \in M$.) Recall that by the chain rule $(\partial x^i / \partial \bar{x}^j)$ and $(\partial \bar{x}^j / \partial x^i)$ are inverse matrices to each other, so, in particular,

$$\sum_{i=1}^n \frac{\partial x^i}{\partial \bar{x}^j} \frac{\partial \bar{x}^k}{\partial x^i} = \delta_j^k, \tag{5.4}$$

where δ_j^k is the Kronecker delta. Note that (5.4) follows from (5.3) by applying $\partial / \partial \bar{x}^j$ to $\bar{x}^k$.

Let $X \in T_p M$ be a vector in the tangent space. Suppose that the vector X has coordinates a^j in the basis $(\partial / \partial x^j)$ and coordinates $\bar{a}^j$ in the basis $(\partial / \partial \bar{x}^j)$. Using Einstein summation convention, we have $X = \bar{a}^j \partial / \partial \bar{x}^j$. Then

$$X = \bar{a}^j \left(\frac{\partial x^i}{\partial \bar{x}^j} \frac{\partial}{\partial x^i} \right) = a^i \frac{\partial}{\partial x^i} \qquad \text{so} \qquad a^i = \bar{a}^j \frac{\partial x^i}{\partial \bar{x}^j}.$$

Multiplying by $\frac{\partial \bar{x}^k}{\partial x^i}$ and summing over i, we obtain

$$\frac{\partial \bar{x}^k}{\partial x^i} a^i = \bar{a}^j \frac{\partial x^i}{\partial \bar{x}^j} \frac{\partial \bar{x}^k}{\partial x^i} = \bar{a}^j \left(\frac{\partial x^i}{\partial \bar{x}^j} \frac{\partial \bar{x}^k}{\partial x^i} \right) = \frac{\partial \bar{x}^k}{\partial x^i} a^i = \bar{a}^j \delta_j^k = \bar{a}^k.$$

This leads to the change-of-coordinates formula

$$\bar{a}^k = \frac{\partial \bar{x}^k}{\partial x^i} a^i. \tag{5.5}$$

The above calculations are important in their own right, but in terms of vector bundles, they lead to the following proposition.

Proposition 5.1.4. *Let M^n be a differentiable manifold. The disjoint union of all the tangent planes to M*

$$\coprod_{p \in M} T_p M,$$

is a vector bundle with fiber $\mathbb{R}^n$ over M.

Proof. An element of TM is of the form (p, X_p), where $p \in M$ and $X_p \in T_p M$. We point out first that the bundle projection $\pi : TM \to M$ is simply the function $\pi(p, X_p) = p$.

We have already seen that for each $p \in M$, the matrix $\left(\partial \bar{x}^k / \partial x^i \big|_p \right)$ is invertible. It remains to be verified that this matrix varies continuously in $p \in M$ over $U_\alpha \cap U_\beta$, where x is the coordinate system over U_α and $\bar{x}$ is the coordinate system over U_β. However, $\left(\partial \bar{x}^k / \partial x^i \big|_p \right)$ is the matrix of the differential of $\bar{x} \circ x^{-1}$ and the fact that this is continuous is part of the definition of a differentiable manifold (see Definition 3.1.3). $\qquad \square$

Definition 5.1.5. The vector bundle in Proposition 5.1.4 is called the *tangent bundle* to M and is denoted by TM.

There is an inherent difficulty in visualizing the tangent bundle, and more generally any vector bundle, to a manifold. Consider the tangent bundle to a circle. The circle $\mathbb{S}^1$ is a one-dimensional manifold that we typically visualize as the unit circle as a subset of $\mathbb{R}^2$. Viewing the tangent spaces to the circle as subspaces of $\mathbb{R}^2$ or even as the geometric tangent lines to $\mathbb{S}^1$ at p, we should view the union $\bigcup_p T_p M$ as a subset of $\mathbb{R}^2$. This is not what is meant by the definition of TM. The spaces $T_p M$ and $T_q M$ do not intersect if $p \neq q$. At best, if M is an embedded submanifold of $\mathbb{R}^n$, then $T_p M$ may be viewed as a subspace of a *different* Euclidean space. Thus, for example, for the circle $\mathbb{S}^1$, the tangent bundle $T(\mathbb{S}^1)$ can be realized as an embedded submanifold of $\mathbb{R}^4$. In fact, we can parametrize $T(\mathbb{S}^1)$ by

$$Y(u, t) = (\cos u, \sin u, -t \sin u, t \cos u) \qquad \text{for} \qquad (u, t) \in [0, 2\pi] \times \mathbb{R}.$$

Therefore, even in this simple example, visualizing the tangent bundle requires more than three dimensions. Nonetheless, it is not uncommon to illustrate the tangent bundle over a manifold by a picture akin to Figure 5.1.

Proposition 5.1.6. *If M^m is a differentiable manifold of dimension m, and V is a real vector space of dimension n, then a differentiable vector bundle of fiber V over M is a differentiable manifold of dimension $m + n$.*

Proof. Let E be a vector bundle of fiber V over a differentiable manifold M with the data described in Definition 5.1.1. Since V is isomorphic to $\mathbb{R}^n$, without loss of generality, let us take $V = \mathbb{R}^n$. On each open set $\pi^{-1}(U_\alpha)$ in the vector bundle E, consider the function τ_α defined by the composition

$$\tau_\alpha : \pi^{-1}(U_\alpha) \xrightarrow{\psi_\alpha^{-1}} U_\alpha \times \mathbb{R}^n \xrightarrow{\phi_\alpha \times \text{id}} \mathbb{R}^m \times \mathbb{R}^n = \mathbb{R}^{m+n},$$

where by $\phi_\alpha \times \text{id}$ we mean the function $(\phi_\alpha \times \text{id})(p, v) = (\phi_\alpha(p), v)$. We prove that the collection of functions $\{(\pi^{-1}(U_\alpha), \tau_\alpha)\}$ is an atlas that equips E with the structure of a differentiable manifold.

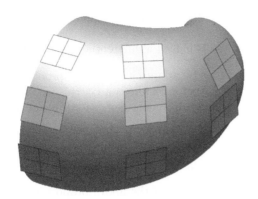

Figure 5.1: Intuitive picture for a tangent bundle.

Since π is continuous, $\pi^{-1}(U_\alpha)$ is open and, by construction, the collection of open sets $\pi^{-1}(U_\alpha)$ cover E. The function $\phi_\alpha : U_\alpha \to V_\alpha$ is a homeomorphism, where V_α is an open subset of $\mathbb{R}^m$. Therefore, it is easy to check that for each $\alpha \in I$,

$$\phi_\alpha \times \mathrm{id} : U_\alpha \times \mathbb{R}^n \to V_\alpha \times \mathbb{R}^n$$

is a homeomorphism. Thus, since ψ_α is a homeomorphism by definition, then $\tau_\alpha = (\phi_\alpha \times \mathrm{id}) \circ \psi_\alpha^{-1}$ is also a homeomorphism.

Let $(y, v) \in \phi_\beta(U_\alpha \cap U_\beta) \times \mathbb{R}^n$, and let $(p, v) = (\phi_\beta \times \mathrm{id})^{-1}(y, v)$ so that (p, v) is in the domain of the trivialization for ψ_β. Then we calculate that

$$(\tau_\alpha \circ \tau_\beta^{-1})(y, v) = \big((\phi_\alpha \times \mathrm{id}) \circ \psi_\alpha^{-1} \circ \psi_\beta \circ (\phi_\beta^{-1} \times \mathrm{id})\big)(y, v)$$
$$= \big(\phi_\alpha \circ \phi_\beta^{-1}(y), \theta_{\alpha\beta}(p)v\big)$$

because $\psi_\alpha^{-1} \circ \psi_\beta(p, v) = (p, \theta_{\alpha\beta}(p)v)$ by definition of a vector bundle.

At this stage, we must use the fact that $\theta_{\alpha\beta}$ is a differentiable map between the differentiable manifolds M and $\mathrm{GL}(\mathbb{R}^n)$. Since $\mathrm{GL}(\mathbb{R}^n)$ inherits its manifold structure as an embedded submanifold of $\mathbb{R}^{n^2}$, the following quantities exist as $n \times n$ matrices:

$$\frac{\partial(\theta_{\alpha\beta} \circ \phi_\beta^{-1})}{\partial y^i} \qquad \text{for } 1 \le i \le m.$$

To simplify notations, we set $F = \theta_{\alpha\beta} \circ \phi_\beta^{-1}$. Then a simple calculation for the function $\tau_\alpha \circ \tau_\beta^{-1}$ as a function from $\mathbb{R}^{m+n}$ into itself gives the following differential as a block matrix:

$$[d(\tau_\alpha \circ \tau_\beta^{-1})_{(y,v)}] = \left(\begin{array}{ccc|c} [d(\phi_\alpha \circ \phi_\beta^{-1})_y] & & & 0 \\ \hline \dfrac{\partial F}{\partial y^1}v & \cdots & \dfrac{\partial F}{\partial y^m}v & F(y) \end{array} \right).$$

Furthermore, each of the entries in the above matrix is continuous. This shows that all the transition functions $\tau_\alpha \circ \tau_\beta^{-1}$ are of class C^1, establishing that the differentiable vector bundle is indeed a differentiable manifold. $\square$

It is not hard to see that by adapting the above proof, we can also show that a C^k (respectively, smooth) vector bundle is a C^k (respectively, smooth) manifold. However, we point out the following consequence for tangent bundles to a manifold.

Corollary 5.1.7. *If M is a manifold of class C^k and dimension m, then TM is a manifold of class C^{k-1} and dimension $2m$. Furthermore, if M is a smooth manifold, then TM is a smooth as well.*

Proof. This follows from the proof of the above proposition and the fact that the linear transformation $\theta_{\alpha\beta}$ is $(\partial \bar{x}^j / \partial x^i)$, where $(\bar{x}^j)$ are the coordinates with respect to ϕ_β and (x^i) are the coordinates with respect to ϕ_α. Therefore, in order for the functions $\theta_{\alpha\beta}$ to be of class C^l, the transition functions $\phi_\beta \circ \phi_\alpha^{-1}$ must be of class C^{l+1}.

The second claim of the corollary follows immediately. $\square$

Example 5.1.8 (Tangent Bundle of $\mathbb{R}^n$). As we saw in Example 3.3.7, the tangent plane to any point p in $\mathbb{R}^n$ is again $\mathbb{R}^n$. However, we can now make the stronger claim that the tangent bundle of $\mathbb{R}^n$ is $T(\mathbb{R}^n) = \mathbb{R}^n \times \mathbb{R}^n$. We can see this from the fact that $\mathbb{R}^n$ is a manifold that can be equipped with an atlas of just one coordinate chart. Then, from Definition 5.1.1, there is only one trivialization map. Thus, the tangent bundle is a trivial bundle.

Chapter 4 introduced various constructions associated to a vector space V, namely the dual V^*, the space $V^{\otimes p} \otimes V^{*\otimes q}$, the symmetric product $\operatorname{Sym}^k V$, and the alternating product $\bigwedge^k V$. Also, if we are given a vector space W of dimension n, the direct product $V \oplus W$ and the tensor product $V \otimes W$ are new vector spaces. In each case, if V and W are equipped with bases, there exist natural bases on the new vector spaces.

Constructions on vector spaces carry over to vector bundles over a differentiable manifold M in the following way. Let ξ be a vector bundle over M with fiber V, and let η be a vector bundle over M with fiber W. It is possible to construct the following vector bundles over M in such a way that their bundle data are compatible with the data for ξ and η and the properties of the associated fiber:

- The dual bundle ξ^*. The fiber is the vector space V^*.

- The direct sum $\xi \oplus \eta$. The fiber is the vector space $V \oplus W$. The direct sum is also called the *Whitney sum* of two vector bundles.

- The tensor product $\xi \otimes \eta$. The fiber is the vector space $V \otimes W$.

- The symmetric product $\operatorname{Sym}^k \xi$ for some positive integer k. The fiber is the vector space $\operatorname{Sym}^k V$.

- The alternating product $\bigwedge^k \xi$ for some positive integer k. The fiber is the vector space $\bigwedge^k V$.

Each of the above situations requires careful construction and proof that they are in fact vector bundles over M. We omit the details here but refer the reader to Chapter 3 in [40] for a careful discussion of how to get new vector bundles from old ones.

One of the first useful bundles constructed from the tangent bundle is the *cotangent bundle*, TM^*, the dual to the tangent bundle. Recall that if $p \in M^m$, U is an open neighborhood of p in M, and $x : U \to \mathbb{R}^m$ is a coordinate chart for U, then the operators

$$\partial_1, \ldots, \partial_m \stackrel{\text{def}}{=} \frac{\partial}{\partial x^1}\Big|_p, \ldots, \frac{\partial}{\partial x^m}\Big|_p$$

form the associated basis of T_pM. The cobasis for the dual bundle T_pM^* is denoted by

$$dx^1, \ dx^2, \ldots, \ dx^m, \tag{5.6}$$

defined as the linear functions on $T_pM \to \mathbb{R}$ such that

$$dx^i(\partial_j) = dx^i\left(\frac{\partial}{\partial x^j}\Big|_p\right) = \delta^i_j = \begin{cases} 1 & \text{if } i = j, \\ 0 & \text{if } i \neq j. \end{cases} \tag{5.7}$$

The dependence on the point $p \in M$ is understood by context.

Example 5.1.9. Consider a regular surface M in $\mathbb{R}^3$. M is an embedded two-dimensional submanifold of $\mathbb{R}^3$. Consider the bundle $TM^* \otimes TM^*$ over M. Via a comment after Proposition 4.5.2, we identify $TM^* \otimes TM^*$ as the vector bundle over M such that each fiber at a point $p \in M$ corresponds to the vector space of all bilinear forms on T_pM.

The formalism of vector bundles over manifolds may initially appear unnecessarily pedantic. However, since in general a manifold need not be given as a subset of an ambient Euclidean space, it is only in the context of the tangent bundle on a manifold that we can make sense of tangent vectors to M at various points $p \in M$. We discussed how to obtain new bundles from old ones so that it would be possible to discuss other linear algebraic objects associated to the tangent bundle, such as bilinear forms on TM, as in Example 5.1.9.

The value for physics is that in order to study the motion of a particle or a system of particles that is not in $\mathbb{R}^n$, then the ambient space for this system would be a manifold. Without the structure of a differentiable manifold, we cannot talk about differentiability at all. However, on a differentiable manifold, any kind of differentiation will be given in reference to the tangent bundle. It is not hard to imagine the need to do physics on a sphere, say when studying global earth phenomenon but only looking at the surface of the earth. In some natural problems, the configuration space (the space in which the variables of interest exist) is not a Euclidean space, and in this context, the equations of dynamics must take into account the fact that the ambient space is a manifold. Perhaps the most blatant examples of the need for manifolds come from cosmology, in which it is now well understood that our universe is not flat. Therefore, doing cosmological calculations (calculations on large portions of the universe) requires the manifold formalism.

PROBLEMS

5.1.1. Consider the unit sphere $\mathbb{S}^2$ equipped with the oriented stereographic atlas $\{\pi_N, \bar{\pi}_S\}$ described in Examples 3.7.3 and 3.1.4. Explicitly describe an atlas for the tangent bundle $T(\mathbb{S}^2)$ as a manifold and write down the transition functions for this atlas.

5.1.2. *Normal Bundle.* Consider a regular surface S in $\mathbb{R}^3$. At each point $p \in S$, let $N(p)$ be the set of all normal vectors. Explicitly show that the points in S, along with its normal vectors at corresponding points, form a vector bundle (in fact a line bundle). Suppose that for each coordinate patch U_α parametrized by $\vec{X}(u, v)$ we define $\psi_\alpha : U\alpha \times \mathbb{R} \to \pi^{-1}(U_\alpha)$ as

$$\psi_\alpha(p, t) = p + t\vec{X}_u(u_0, v_0) \times \vec{X}_v(u_0, v_0),$$

where $p = \vec{X}(u_0, v_0)$. Determine the functions $\theta_{\beta\alpha}$ between different trivialization maps. (This vector bundle is called the *normal bundle*.)

5.1.3. *Normal Bundle.* Let M^m be a differentiable manifold embedded in $\mathbb{R}^n$ where $m < n$. For all $p \in M$, let N_p be the orthogonal complement to T_pM in $\mathbb{R}^n$. Prove that the disjoint union of all N_p subspaces is a vector bundle over M. (This vector bundle is called the *normal bundle* to M and generalizes the situation in the previous exercise.)

5.1.4. In the study of dynamics of a particle, one locates the position of a point in $\mathbb{R}^3$ using its three coordinates. Therefore, the variable space is $\mathbb{R}^3$. Explain why the variable space for a general solid object (or system of particles rigidly attached to each other) is $\mathbb{R}^3 \times \mathrm{SO}(3)$. In particular, explain why we require six variables to completely describe the position of a solid object in $\mathbb{R}^3$.

5.1.5. Provide appropriate details behind the construction of the Whitney sum of two vector bundles.

5.1.6. Consider the real projective space $M = \mathbb{RP}^n$. We view $\mathbb{RP}^n$ as the set of one-dimensional subspaces of $\mathbb{R}^{n+1}$. Consider the set $\{(V, \vec{u}) \in \mathbb{RP}^n \times \mathbb{R}^{n+1} \mid \vec{u} \in V\}$.

 (a) Show that this is a line bundle.

 (b) Show that this line bundle is not the trivial bundle.

(This bundle is called the canonical line bundle on $\mathbb{RP}^n$.)

5.1.7. Consider the following parametrization for a torus $\mathbb{S}^1 \times \mathbb{S}^1$ as a subset of $\mathbb{R}^3$

$$\vec{X}(u, v) = \big((2 + \cos u) \cos v, (2 + \cos u) \sin v, \sin u\big),$$

for $(u, v) \in [0, 2\pi] \times [0, 2\pi]$. Using $\vec{X}$, given the associated parametrization of the manifold $T(\mathbb{S}^1 \times \mathbb{S}^1)$ as a subset of $\mathbb{R}^6$.

5.2 Vector and Tensor Fields on Manifolds

5.2.1 Vector and Tensor Fields

Definition 5.2.1. Let ξ be a vector bundle over a manifold M with fiber V, with projection $\pi : E(\xi) \to M$. A *global section* of ξ is a continuous map $s : M \to E(\xi)$ such that $\pi \circ s = \mathrm{id}_M$, the identity function on M. The set of all global sections is denoted by $\Gamma(\xi)$. Given an open set $U \subseteq M$, we call a *local section* over U a continuous map $s : U \to E(\xi)$ such that $\pi \circ s = \mathrm{id}_U$. The set of all local sections on U is denoted by $\Gamma(U; \xi)$.

Note that sections of a vector bundle (whether local or global) can be added or multiplied by a scalar in the following sense. If $s_1, s_2 \in \Gamma(U; \xi)$, then for each $p \in U \subseteq M$, $s_1(p)$ and $s_2(p)$ are vectors in the same fiber $\pi^{-1}(p)$. Consequently, for any scalars $a, b \in \mathbb{R}$, the linear combination $as_1(p) + bs_2(p)$ is well defined as an element in $\pi^{-1}(p)$.

Definition 5.2.2. Let M be a differentiable manifold. A global section of TM is called a *vector field* on M. In other words, a vector field associates to each $p \in M$ a vector $X(p)$ (also denoted by X_p) in T_pM. The set of all vector fields on M is denoted by $\mathfrak{X}(M)$. A vector field X is said to be of class C^k if $X : M \to TM$ is a map of class C^k between manifolds.

We point out that if U is a open subset of a smooth manifold M, then a vector field $X \in \mathfrak{X}(U)$ is a derivation on $C^\infty(U, \mathbb{R})$.

Example 5.2.3 (Metric Tensor of a Surface). Let M be a regular surface in $\mathbb{R}^3$. In the local theory of regular surfaces in $\mathbb{R}^3$, the first fundamental form (alternatively called the metric tensor) is the bilinear product $g = I_p(\cdot, \cdot)$ on T_pM obtained as the restriction of the dot product in $\mathbb{R}^3$ to the tangent plane T_pM. Therefore, with the formalism of vector bundles and using Example 5.1.9, the first fundamental form is a section of $TM^* \otimes TM^*$. In fact, since $I_p(\cdot, \cdot)$ is symmetric and defined for all p, independent of any particular basis on $TM^* \otimes TM^*$, then the metric tensor is a global section of $\mathrm{Sym}^2 TM^*$.

Let p be a point of M, and let U be a coordinate neighborhood of p with coordinates (x^1, x^2). This coordinate system defines the basis

$$dx^1 \otimes dx^1, \ dx^1 \otimes dx^2, \ dx^2 \otimes dx^1, \ dx^2 \otimes dx^2$$

on $T_pM^* \otimes T_pM^*$. Furthermore, each basis vector is a local section in $\Gamma(U, TM^* \otimes TM^*)$. The coefficient functions g_{ij} of the metric tensor are functions such that, as an element of $\Gamma(U, TM^* \otimes TM^*)$, the metric tensor can be written as

$$g = g_{11}dx^1 \otimes dx^1 + g_{12}dx^1 \otimes dx^2 + g_{21}dx^2 \otimes dx^1 + g_{22}dx^2 \otimes dx^2.$$

Definition 5.2.4. A *tensor field* of type (r, s) is a global section of the vector bundle $TM^{\otimes r} \otimes TM^{*\otimes s}$. The index r is called the *contravariant* index, while the index s is called the *covariant* index.

Let A be a tensor field of type (r,s) on a manifold M^n. Over a coordinate patch U of M with coordinates $(x^1, x^2, \ldots, x^n)$, we write the components of A as $A^{i_1 i_2 \cdots i_r}_{j_1 j_2 \cdots j_s}$. This means that $A^{i_1 i_2 \cdots i_r}_{j_1 j_2 \cdots j_s}$ are n^{r+s} functions $U \to \mathbb{R}$ such that with respect to the basis on $T_p M^{\otimes r} \otimes T_p M^{*\otimes s}$,

$$A = A^{i_1 i_2 \cdots i_r}_{j_1 j_2 \cdots j_s} \frac{\partial}{\partial x^{i_1}} \otimes \frac{\partial}{\partial x^{i_2}} \otimes \cdots \otimes \frac{\partial}{\partial x^{i_r}} \otimes dx^{j_1} \otimes dx^{j_2} \otimes \cdots \otimes dx^{j_s}.$$

If U' is another coordinate patch on M with coordinates $(\bar{x}^1, \bar{x}^2, \ldots, \bar{x}^n)$, we label the components of A in reference to this system as $\bar{A}^{k_1 k_2 \cdots k_r}_{l_1 l_2 \cdots l_s}$. Again, these components are a collection of n^{r+s} functions $U' \to \mathbb{R}$. On the intersection $U \cap U'$, both sets of components describe the same tensor but in reference to different bases. By Proposition 4.5.2, the components of A change according to

$$\bar{A}^{k_1 k_2 \cdots k_r}_{l_1 l_2 \cdots l_s} = \frac{\partial \bar{x}^{k_1}}{\partial x^{i_1}} \frac{\partial \bar{x}^{k_2}}{\partial x^{i_2}} \cdots \frac{\partial \bar{x}^{k_r}}{\partial x^{i_r}} \frac{\partial x^{j_1}}{\partial \bar{x}^{l_1}} \frac{\partial x^{j_2}}{\partial \bar{x}^{l_2}} \cdots \frac{\partial x^{j_s}}{\partial \bar{x}^{l_s}} A^{i_1 i_2 \cdots i_r}_{j_1 j_2 \cdots j_s}. \tag{5.8}$$

As anticipated by the comments in Section 4.5.4, we have generalized the notion of tensors in $\mathbb{R}^n$ to tensor fields over a manifold M.

As a point of terminology, a vector field on a manifold is a tensor field of type $(1, 0)$. In contrast, a tensor field of type $(0, 1)$ is often called a covariant vector field, or shorter, a covector field. We call any tensor field of type $(r, 0)$ a contravariant tensor field, and any tensor field of type $(0, s)$ is called a covariant tensor field.

5.2.2   Operations of Tensor Fields

Referring to the multilinear algebra developed in Section 4.4, there exist a number of natural operations on tensor fields. Let M be a differentiable manifold. Let A be a tensor field of type (r, s) and let B be a tensor field of type (k, ℓ) on M. We define the *tensor product* of A and B as the tensor field of type $(r + k, s + \ell)$ defined by

$$(A \otimes B)_p = A_p \otimes B_p \qquad \text{for } p \in M.$$

In this sense, the $\otimes$ operator is a bilinear transformation

$$\otimes : \Gamma(TM^{\otimes r} \otimes TM^{*\otimes s}) \times \Gamma(TM^{\otimes k} \otimes TM^{*\otimes \ell}) \to \Gamma(TM^{\otimes(r+k)} \otimes TM^{*\otimes(s+\ell)}).$$

Let $X \in \mathcal{X}(M)$ be a differentiable vector field on a manifold M and let $A \in \Gamma(TM^{\otimes r} \otimes TM^{*\otimes s})$ be a tensor field on M. Then the contraction operation between X and A is the linear transformation

$$\mathcal{X}(M) \otimes \Gamma(TM^{\otimes r} \otimes TM^{*\otimes s}) \to \Gamma(TM^{\otimes r} \otimes TM^{*\otimes(s-1)})$$

defined by contraction on the first covariant index of A. This is also denoted by $i_X A$, where we view i_X as a linear transformation $\Gamma(TM^{\otimes r} \otimes TM^{*\otimes s}) \to \Gamma(TM^{\otimes r} \otimes TM^{*\otimes(s-1)})$.

5.2.3   Push-Forwards of Vector Fields

We remind the reader that a tangent vector field X on a manifold M is such that, at each point $p \in M$, we have a differential operator on real-valued functions $X_p : C^1(M, \mathbb{R}) \to \mathbb{R}$. So a vector field is a function of both $p \in M$ and $f \in C^1(M, \mathbb{R})$. If we apply X to the function f first, then we can think of a vector field as a mapping $X : C^1(M, \mathbb{R}) \to C^0(M, \mathbb{R})$ via the identification

$$Xf = (p \mapsto X_p(f)). \tag{5.9}$$

Over a coordinate chart U of M with coordinate system $(x^1, x^2, \ldots, x^n)$, we write

$$X_p = \sum_{i=1}^{n} X^i(p) \partial_i,$$

so the real-valued function Xf on M is defined by

$$(Xf)(p) = \sum_{i=1}^{n} X^i(p) \frac{\partial f}{\partial x^i}\Big|_p.$$

Recall from the definition of the differential, if $F : M \to N$ is a differentiable map and X is a vector field on M, then for each point $p \in M$ we define the vector $F_* X_p = dF_p(X_p) \in T_{F(p)} N$ as the *push-forward* of X by F. Unfortunately, this does not in general define a vector field on N. If F is not surjective, there is no natural way to define a vector field associated to X on $N - F(M)$. (Even proposing to define the push-forward vector field to 0 on $N - F(M)$ would not ensure a continuous vector field on N.) Furthermore, if F is not injective and if p_1 and p_2 are preimages of a point $q \in F(M)$, then nothing guarantees that $F_*(X_{p_1}) = F_*(X_{p_2})$. Thus, the push-forward is not well defined in this case either. However, we can make the following definition.

Definition 5.2.5. Let M and N be differentiable manifolds, let $F : M \to N$ be a differentiable map, let X be a vector field on M, and let Y be a vector field on N. We say that X and Y are *F-related* if $F_*(X_p) = Y_{F(p)}$ for all $p \in M$.

With this terminology, the above comments can be rephrased to say that if X is a vector field on M and $F : M \to N$ is a differentiable map, then there does not necessarily exist a vector field on N that is F-related to X.

Proposition 5.2.6. *Let $F : M \to N$ be a differentiable map between differentiable manifolds. Let $X \in \mathfrak{X}(M)$ and $Y \in \mathfrak{X}(N)$. The vector field X is F-related to Y if and only if for every open subset U of N and every function $f \in C^1(U, \mathbb{R})$ we have*

$$X(f \circ F) = (Yf) \circ F.$$

Proof. For any $p \in M$ and any $f \in C^1(U, \mathbb{R})$, where U is a neighborhood of $F(p)$, by Proposition 3.4.2 we have

$$X(f \circ F)(p) = X_p(f \circ F) = dF_p(X_p)(f) = F_*(X_p)(f).$$

On the other hand,

$$((Yf) \circ F)(p) = (Yf)(F(p)) = Y_{F(p)}f.$$

Thus, $X(f \circ F) = (Yf) \circ F$ is true for all f if and only if $F_*(X_p) = Y_{F(p)}$ for all $p \in M$. The proposition follows. □

Though in general vector fields cannot be pushed forward via a differentiable map, we show one particular case in which push-forwards for vector fields exist.

Proposition 5.2.7. *Let $X \in \mathfrak{X}(M)$ be a vector field, and let $F : M \to N$ be a diffeomorphism. There exists a unique vector field $Y \in \mathfrak{X}(N)$ that is F-related to X. Furthermore, if X is of class C^k and F is a diffeomorphism of class C^k, then so is Y.*

Proof. In order for X and Y to be F-related, we must have $F_*X_p = Y_{F(p)}$. Therefore, we define $Y_q = F_*(X_{F^{-1}(q)})$. Since F is a diffeomorphism, the association $q \mapsto Y_q$ is well defined. However, we must check this association is continuous before we can call it a vector field.

If (x^i) is a coordinate system on a neighborhood of $p = F^{-1}(q)$ and if (y^j) is a coordinate system on a neighborhood of q, then the coordinates of Y_q are

$$Y_q = \left.\frac{\partial F^j}{\partial x^i}\right|_{F^{-1}(q)} X^i_{F^{-1}(q)} \left.\frac{\partial}{\partial y^j}\right|_q.$$

Finally, if F^{-1} and X are of class C^k, then by composition and product rule, the global section $N \to TN$ defined by $q \mapsto (q, Y_q)$ is of class C^k. □

Definition 5.2.8. If $X \in \mathfrak{X}(M)$ and $F : M \to N$ is a diffeomorphism, then the vector field Y in Proposition 5.2.7 is called the push-forward of X by F and is denoted by F_*X.

5.2.4   Integral Curves and Flows

As we promised in the introduction to this chapter, vector bundles on a manifold allow for the possibility of doing physics on a manifold. We begin to see this through the existence of trajectories in what we might view as a velocity vector field.

Let $\delta > 0$, and let $\gamma : (-\delta, \delta) \to M$ be a differentiable curve on M^m. Recall that we must understand γ as a differentiable function between manifolds. Let X be a vector field on M so that for each $p \in M$, X_p is a tangent vector in T_pM. Referring to Example 3.4.3 for notation, the curve γ is called a *trajectory* of X through p if $\gamma(0) = p$ and

$$\gamma'(t) \stackrel{\text{def}}{=} \gamma_* \left(\frac{d}{dt}\Big|_t \right) = X_{\gamma(t)} \tag{5.10}$$

for all $t \in (-\delta, \delta)$. A trajectory is also called an *integral curve* of the vector field X because it solves the differential equation represented in (5.10).

If $x : U \to \mathbb{R}^n$ is a coordinate patch of M around p, it is by definition a diffeomorphism with $x(U)$. Hence, the push-forward $x_*(X)$ is a vector field on $x(U)$. Call this $\vec{G}$, so $x_*X = \vec{G} : x(U) \to \mathbb{R}^m$. Call $\vec{c} : (-\delta, \delta) \to x(U)$, the curve with $\vec{c}(t) = (x \circ \gamma)(t)$. Then applying x_* to (5.10) gives

$$x_*\gamma_* \left(\frac{d}{dt}\Big|_t \right) = (x_*X)_{x \circ \gamma(t)} \iff \frac{d\vec{c}}{dt} = \vec{G}(\vec{c}(t)).$$

Consequently, (5.10) is equivalent (locally) to an ordinary differential equation in $\mathbb{R}^m$. Theorems of existence, uniqueness, and continuous dependence on initial conditions for ordinary differential equations carry over to the context of differentiable manifolds. (See for example Sections 7 and 35 in [3] for the classic results in this area.) Instead of proving the difficult theorems behind the following application to differentiable manifolds, we restate [52, Theorem 5, Chapter 5].

Theorem 5.2.9. *Let M be a differentiable manifold of class C^k with $k \geq 2$, and let X be a vector field on M of class C^k. Let $p \in M$. There exists an open neighborhood $V \subset M$ of p and a positive real $\delta > 0$ such that there is a unique collection of diffeomorphisms $\varphi_t : V \to \varphi_t(V)$ for $|t| < \delta$ with the following properties:*

1. *$\varphi_0 = \text{id}_V$, i.e., $\varphi_0(q) = q$ for all $q \in V$.*
2. *$\varphi : (-\delta, \delta) \times V \to M$, defined by $\varphi(t, q) = \varphi_t(q)$, is C^k.*
3. *If $|s| < \delta$, $|t| < \delta$, and $|s + t| < \delta$, and both $q, \varphi_t(q) \in V$, then $\varphi_{s+t}(q) = \varphi_s \circ \varphi_t(q)$.*
4. *If $q \in V$ then*

$$\frac{\partial \varphi}{\partial t} = X_{\varphi(t,q)}; \tag{5.11}$$

in other words, for ε small enough, the curve $\gamma : (-\varepsilon, \varepsilon) \to M$ defined by $\gamma(t) = \varphi_t(q)$ is a trajectory of X, i.e., satisfies $\gamma'(0) = X_q$.

Definition 5.2.10. The function $\varphi : (-\delta, \delta) \times U \to M$ is called the *flow* of X on M near p.

Figure 5.2 depicts a vector field X on two-dimensional manifold M (embedded in $\mathbb{R}^3$). The black curve is a particular trajectory since every tangent vector to the curve at p (a point on the curve) is parallel to X_p. To be precise, the shown curve is only the locus of the trajectory since the trajectory itself is a curve parametrized in such a way that the velocity vector at each point p is exactly X_p.

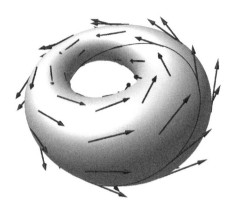

Figure 5.2: A vector field on a manifold.

According to Definition 3.3.1, the last condition of Theorem 5.2.9 means that $X_q = D_\gamma$. Thus, for all real-valued differentiable functions f on a neighborhood of q,

$$(Xf)(q) = X_q(f) = D_\gamma(f) = \frac{d}{dt}(f(\gamma(t)))\big|_{t=0} = \lim_{h \to 0} \frac{f(\varphi_h(q)) - f(q)}{h}. \qquad (5.12)$$

This equation simplifies many calculations, as we will soon see.

Example 5.2.11. As an example of a flow using this differential geometry notation, consider the Euclidean plane $M = \mathbb{R}^2$ and the vector field $X = -y\partial_x + x\partial_y$. Recall that the notation $\partial\varphi/\partial t$ means $\varphi_*(\partial/\partial t)$. Furthermore, $\varphi : (-\delta, \delta) \times \mathbb{R}^2 \to \mathbb{R}^2$ and, with respect to the standard basis on the tangent plane of $\mathbb{R}^2$ the matrix of φ_* is

$$[\varphi_*] = \begin{pmatrix} \dfrac{\partial\varphi^1}{\partial t} & \dfrac{\partial\varphi^1}{\partial x} & \dfrac{\partial\varphi^1}{\partial y} \\ \dfrac{\partial\varphi^2}{\partial t} & \dfrac{\partial\varphi^2}{\partial x} & \dfrac{\partial\varphi^2}{\partial y} \end{pmatrix}.$$

The vector $\partial/\partial t$ is the first basis vector of that tangent space to a point of $(-\delta, \delta) \times M$. Thus, (5.11) means in components with respect to the basis $\{\partial_x, \partial_y\}$,

$$\begin{pmatrix} \dfrac{\partial\varphi^1}{\partial t} \\ \dfrac{\partial\varphi^2}{\partial t} \end{pmatrix} = \begin{pmatrix} -y \circ \varphi \\ x \circ \varphi \end{pmatrix} = \begin{pmatrix} -\varphi^2 \\ \varphi^1 \end{pmatrix}.$$

This implies that

$$\frac{\partial^2\varphi^1}{\partial t^2} = -\frac{\partial\varphi^2}{\partial t} = -\varphi^1.$$

Using techniques to solve linear second order differential equations with constant coefficients, we see that $\varphi^1(t, x, y) = A\cos t + B\sin t$, where A and B are functions of x and y. Since $\varphi^2 = -\partial\varphi^1/\partial t$, we have $\varphi^2(t, x, y) = A\sin t - B\cos t$. However, the condition that $\varphi_0(q) = q$ for all $q \in M$ means for this function that $A = x$ and $B = -y$. Thus, we find that

$$\varphi(t, x, y) = \begin{pmatrix} x\cos t - y\sin t \\ x\sin t + y\cos t \end{pmatrix}.$$

It is not hard to check that the trajectories for the flow of this vector field X consists of circles centered at the origin.

Problems

5.2.1. Let $M = \mathbb{S}^2$ be the unit sphere and let U be the coordinate patch parametrized by

$$x^{-1}(u^1, u^2) = (\cos u^1 \sin u^2, \sin u^1 \sin u^2, \cos u^2)$$

with $(u^1, u^2) \in (0, 2\pi) \times (0, \pi)$. Let $X = \cos u^1 \sin u^2 \partial_1 + \sin u^1 \sin u^2 \partial_2$, $Y = \partial_1$, and $Z = \sin u^2 \partial_1$ be vector fields over U.

(a) Show that X and Z can be extended continuously to vector fields over all of M.

(b) Show that Y cannot be extended continuously to a vector field in $\mathfrak{X}(M)$.

5.2.2. Let S be a regular surface in $\mathbb{R}^3$, and let X be a vector field on $\mathbb{R}^3$. For every $p \in S$, define Y_p as the orthogonal projection of X_p onto T_pS. Show that Y is a vector field on S.

5.2.3. Suppose that M is the torus that has a dense coordinate patch parametrized by

$$x^{-1}(u, v) = \big((3 + \cos v)\cos u, (3 + \cos v)\sin u, \sin v\big).$$

Consider the vector field $X = -z\partial/\partial x + x\partial/\partial z \in \mathbb{R}^3$. In terms of the coordinates (u, v), calculate the vector field on M induced from X by orthogonal projection, as described in the previous exercise.

5.2.4. Let $M = \mathbb{S}^1 \times \mathbb{S}^1 \times \mathbb{S}^1$ be the 3-torus given as an embedded submanifold of $\mathbb{R}^4$ by the parametrization

$$x^{-1}(u, v, w) \mapsto \Big(\big(4 + (2 + \cos u)\cos v\big)\cos w, \big(4 + (2 + \cos u)\cos v\big)\sin w, (2 + \cos u)\sin v, \sin u\Big).$$

Consider the radial vector field in $\mathbb{R}^4$ given by $Z = x^1\partial_1 + x^2\partial_2 + x^3\partial_3 + x^4\partial_4$. In terms of the coordinates (u, v, w), calculate the vector field X on M induced from Z by orthogonal projection of Z_p onto T_pM for all $p \in M$.

5.2.5. Find a vector field on $\mathbb{S}^2$ that vanishes at one point. Write down a formula expression for this vector field in some coordinate patch of $\mathbb{S}^2$.

5.2.6. Referring to Problem 5.2.1, prove that the flow of Z is

$$\varphi_t(u_1, u_2) = \begin{pmatrix} u^1 + t \sin u^2 \\ u^2 \end{pmatrix}.$$

5.2.7. Prove that $T\mathbb{S}^1$ is diffeomorphic to $\mathbb{S}^1 \times \mathbb{R}$.

5.2.8. Let M be any differentiable manifold. Show that $\mathfrak{X}(M)$ is an infinite-dimensional vector space.

5.3   Lie Bracket and Lie Derivative

5.3.1   Lie Bracket

Consider a C^2-manifold M and let X be a differentiable vector field over M. Recall that to call a vector field differentiable it means that the function $X : M \to TM$ is a differentiable map, or equivalently over any coordinate chart, the corresponding components X^i of X are differentiable functions $M \to \mathbb{R}$. With the above interpretation of a vector field on M, we can talk about the functions $Y(Xf)$ or $X(Yf)$, where X and Y are two differentiable vector fields on M. However, neither of the composition operations XY or YX leads to another vector field.

Letting $X = X^i \partial_i$ and $Y = Y^j \partial_j$, for any function $f \in C^2(M, \mathbb{R})$, we have

$$X(Yf) = X(b^j \partial_j f) = X^i \partial_i (Y^j \partial_j f) = X^i \partial_i Y^j \partial_j f + X^i Y^j \partial_i (\partial_j f). \qquad (5.13)$$

Thus, we see that $f \mapsto X(Yf)(p)$ is not a tangent vector to M at p since it involves a repeated differentiation of f. Nonetheless, we do have the following proposition.

Proposition 5.3.1. *Let M be a C^2-manifold, and let X and Y be two vector fields of class C^1. Then the operation $f \mapsto (XY - YX)f$ is another vector field.*

Proof. Since the second derivatives of f are continuous, then the mixed partials with respect to the same variables, though ordered differently, are equal. By using Equation (5.13) twice, we find that

$$(XY - YX)f = \left(X^i \partial_i Y^j \partial_j f \right) - \left(Y^j \partial_j X^i \partial_i f \right) = \left(X^i \partial_i Y^j - Y^i \partial_i X^j \right) \frac{\partial f}{\partial x^j}.$$

Since for all $j = 1, \ldots, n$ the expressions in the above parentheses are continuous real-valued functions on M, then $(XY - YX)$ has the structure of a vector field.

We leave it as an exercise for the reader to show that the coordinates of $(XY - YX)$ change contravariantly under a basis change in T_pM. $\square$

Definition 5.3.2. The vector field defined in Proposition 5.3.1 is called the *Lie bracket* of X and Y and is denoted by $[X, Y] = XY - YX$. If X and Y are of class

C^n with $n \geq 1$, then $[X, Y]$ is of class C^{n-1}. Also, if X and Y are smooth vector fields, then so is $[X, Y]$. More precisely,

$$[X, Y]_p(f) = X_p(Yf) - Y_p(Xf) \tag{5.14}$$

for all $p \in M$ and all $f \in C^2(M, \mathbb{R})$.

The proof of Proposition 5.3.1 shows that, in a coordinate neighborhood, if $X = X^i \partial_i$ and $Y = Y^j \partial_j$, the Lie bracket is

$$[X, Y] = (X^i \partial_i Y^j - Y^i \partial_i X^j) \partial_j. \tag{5.15}$$

This formula gives a coordinate-dependent definition of the Lie bracket, while (5.14) is a coordinate-free definition.

Example 5.3.3. Consider the manifold $\mathbb{R}^3 - \{(x, y, z) | z = 0\}$, and consider the two vector fields

$$X = xy \frac{\partial}{\partial x} + \frac{1}{z} \frac{\partial}{\partial y} - 3yz^3 \frac{\partial}{\partial z},$$

$$Y = \frac{\partial}{\partial x} + (x + y) \frac{\partial}{\partial z}.$$

The one iterated derivation is

$$XYf = \left(xy \frac{\partial}{\partial x} + \frac{1}{z} \frac{\partial}{\partial y} - 3yz^3 \frac{\partial}{\partial z} \right) \left(\frac{\partial f}{\partial x} + (x + y) \frac{\partial f}{\partial z} \right)$$

$$= xy \frac{\partial^2 f}{\partial x^2} + xy \frac{\partial f}{\partial z} + xy(x + y) \frac{\partial^2 f}{\partial x \partial z} + \frac{1}{z} \frac{\partial^2 f}{\partial y \partial x} + \frac{1}{z} \frac{\partial f}{\partial z}$$

$$+ \frac{1}{z}(x + y) \frac{\partial^2 f}{\partial y \partial z} - 3yz^3 \frac{\partial^2 f}{\partial z \partial x} - 3yz^3(x + y) \frac{\partial^2 f}{\partial z^2}.$$

The expression YXf has exactly the same second derivative expressions for f, and upon subtracting, we find that

$$[X, Y] = (XY - YX) = -y \frac{\partial}{\partial x} + \frac{1}{z^2}(x + y) \frac{\partial}{\partial y} + \left(xy + \frac{1}{z} + 9yz^2(x + y) \right) \frac{\partial}{\partial z}.$$

The Lie bracket has the following algebraic properties.

Proposition 5.3.4. *Let X, Y, and Z be differentiable vector fields on a differentiable manifold M. Let $a, b \in \mathbb{R}$, and let f and g be differentiable functions $M \to \mathbb{R}$. Then the following hold:*

1. *Anticommutativity:* $[Y, X] = -[X, Y]$.
2. *Bilinearity:* $[aX + bY, Z] = a[X, Z] + b[Y, Z]$ *and similarly for the second input to the bracket.*
3. *Jacobi identity:* $[[X, Y], Z] + [[Y, Z], X] + [[Z, X], Y] = 0$.
4. $[fX, gY] = fg[X, Y] + fX(g)Y - gY(f)X$.

Proof. (The proofs of these facts are straightforward and are left as exercises for the reader.) □

5.3.2   Lie Derivative of Vector Fields

A key concept in analysis is the ability to take derivatives. We have studied differential operators of functions $f : M \to \mathbb{R}$. However, a central theme of analysis on manifolds pertains to defining operators on vector fields and tensor fields that behave like derivatives. Though we begin this theme here, we revisit it in Section 5.6 and in Section 6.2.

Suppose that f is a differentiable function on some open set U of $\mathbb{R}^n$, let $p \in U$ and let V be a vector in $\mathbb{R}^n$. In calculus, assuming that v is a unit vector, we define the directional derivative of f along v at p by

$$\frac{d}{dt}(f(p+tv))\big|_{t=0} = \lim_{h \to 0} \frac{1}{h}(f(p+hv) - f(p)),$$

We can extend this concept to vector fields in a natural way. If X is a differentiable vector field on U, then the directional derivative of X at p along v is

$$\frac{d}{dt}(X(p+tv))\big|_{t=0} = \lim_{h \to 0} \frac{1}{h}(X(p+hv) - X(p)),$$

For each vector V, this defines a new vector field on $\mathbb{R}^n$. If the Cartesian coordinate functions of X are X^i, then the ith component of the directional derivative is

$$\frac{d}{dt}(X^i(p^1 + tv^1, \ldots, p^n + tv^n))\big|_{t=0} = \frac{\partial X^i}{\partial x^j}\Big|_p v^j.$$

This notion does not easily generalize to manifolds because the Euclidean space is both a manifold and a vector space. Furthermore, identifying $T_p\mathbb{R}^n$ with $\mathbb{R}^n$ for all p allows the expression $X(p+tv) - X(p)$ to have meaning. In a general manifold, X_{p+tv} and X_p are in different tangent spaces, so taking their difference makes no sense. However, using the push-forward of a "backwards flow" we can propose an operation that makes sense.

Definition 5.3.5. Let M be a C^2-manifold and let $X \in \mathfrak{X}(M)$ be a C^2 vector field on M and let φ_t be the flow of X on M.

1. If $f \in C^1(M, \mathbb{R})$, we define the Lie derivative of f by X as the function $\mathcal{L}_X f \stackrel{\text{def}}{=} Xf$.

2. If $Y \in \mathfrak{X}(M)$ is another differentiable vector field, we define the Lie derivative of Y by X as the vector field $\mathcal{L}_X Y$ with

$$(\mathcal{L}_X Y)_p \stackrel{\text{def}}{=} \frac{d}{dt}(\varphi_{-t})_*(Y_{\varphi_t(p)})\big|_{t=0} = \lim_{h \to 0} \frac{1}{h}(((\varphi_{-h})_* Y)_p - Y_p). \qquad (5.16)$$

Figure 5.3 illustrates the definition of the Lie derivative by depicting the trajectory of X through p as well as the trajectories of Y through p and $\varphi_t(p)$.

By Theorem 5.2.9, the flow exists for small $t \neq 0$ and $(\varphi_{-t})_* Y_{\varphi_t(p)}$ is a vector in $T_p M$ so the difference of vectors in (5.16) is well-defined. To clarify notation,

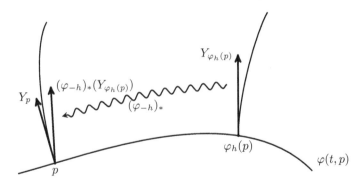

Figure 5.3: Illustration of (5.16).

$(\varphi_{-t})_*(Y_{\varphi_t(p)})$ is the differential of φ_{-t} applied to the tangent vector $Y_{\varphi_t(p)}$, while $((\varphi_{-t})_*Y)_p$ is the push-forward of the vector field Y by the diffeomorphism φ_{-t}, then evaluated at p. The above definition claimed that $\mathcal{L}_X Y$ is a vector field; we need to prove it.

Proposition 5.3.6. *Let M be a manifold of class C^k and let X be a vector field of class C^k on M and Y a vector field of class C^ℓ on M. Then $\mathcal{L}_X Y$ is a vector field on M of class C^r, where $r = \min(k-2, \ell-1)$. If the manifold and vector fields are smooth, then so is $\mathcal{L}_X Y$.*

Proof. Let φ be the flow of X on M. By Theorem 5.2.9(2), the flow $\varphi : (-\delta, \delta) \times M \to M$ is of class C^k. Let $p \in M$ and let (U, x) be a coordinate chart of a neighborhood of p. There exists a domain $(-\varepsilon, \varepsilon) \times U_0$ with $p \in U_0$ such that φ maps $(-\varepsilon, \varepsilon) \times U_0$ into U. Write $\varphi^i = x^i \circ \varphi$ as the component functions of the flow in U. The components of the matrix of the differential $(\varphi_{-t})_* : T_{\varphi_t(p)}M \to T_pM$ are

$$\frac{\partial \varphi^i(-t, \varphi(t, p))}{\partial x^j}. \tag{5.17}$$

Consequently, if $Y = Y^j \partial_j$ over U, then

$$(\varphi_{-t})_* Y_{\varphi_t(p)} = \frac{\partial \varphi^i(-t, \varphi(t, p))}{\partial x^j} Y^j(\varphi(t, p)) \frac{\partial}{\partial x^i}\Big|_p. \tag{5.18}$$

We obtain $\mathcal{L}_X Y$ by taking the derivative of the component functions in (5.18) with respect to t and setting $t = 0$. The functions Y^j are of class C^ℓ, $\varphi(t, p)$ of class C^k and $\partial \varphi^i / \partial x^j$ of class C^{k-1}, so taking the derivative with respect to t decreases the differentiability class by 1. The result follows. $\qquad\square$

Example 5.3.7. As a specific example of using (5.16), let us revisit Example 5.2.11, where $M = \mathbb{R}^2$ and $X = -y\partial_x + x\partial_y$. In standard coordinates with respect to these bases,

$$\varphi_{-t}(x,y) = \begin{pmatrix} x\cos t + y\sin t \\ -x\sin t + y\cos t \end{pmatrix} \quad \text{and} \quad (\varphi_{-t})_* = \begin{pmatrix} \cos t & \sin t \\ -\sin t & \cos t \end{pmatrix}.$$

This example is particularly simple because $(\varphi_{-t})_*$ is independent of (x,y). Hence, the matrix in (5.17) is the same as $(\varphi_{-t})_*$ expressed here. Then for a vector field Y, with component functions $Y^1(x,y)$ and $Y^2(x,y)$, using (5.18) we get

$$(\varphi_{-t})_*(Y_{\varphi_t(x,y)}) = \begin{pmatrix} \cos t & \sin t \\ -\sin t & \cos t \end{pmatrix} \begin{pmatrix} Y^1(x\cos t - y\sin t, x\sin t + y\cos t) \\ Y^2(x\cos t - y\sin t, x\sin t + y\cos t) \end{pmatrix}.$$

There are a few other natural ways we can bring two nearby vectors into the same tangent space in order to perform a limiting difference. However, they turn out to be equal to that given in (5.16). We leave it as an exercise to the reader to prove the following proposition.

Proposition 5.3.8. *If X and Y are differentiable vector fields on a C^2-manifold, then*

$$(\mathcal{L}_X Y)_p = \lim_{h\to 0} \frac{1}{h}\left(Y_p - (\varphi_h)_*(Y_{\varphi_{-h}(p)})\right) = \lim_{h\to 0} \frac{1}{h}\left(Y_{\varphi_h(p)} - (\varphi_h)_*(Y_p)\right).$$

We usually understand an operator on functions to be a differential operator if it is linear and satisfies an appropriate Leibniz rule (product rule). The following proposition shows this is the case for the Lie derivative.

Proposition 5.3.9. *Let X, Y, and Z be vector fields on a differentiable manifold M and let $f : M \to \mathbb{R}$ be a differentiable function. Then*

1. *$\mathcal{L}_X(Y + Z) = \mathcal{L}_X Y + \mathcal{L}_X Z$;*
2. *$\mathcal{L}_X(fY) = (\mathcal{L}_X f)Y + f(\mathcal{L}_X Y)$.*

Proof. Part (1) follows immediately from the linearity properties of $(\varphi_{-t})_*$ and of the derivative operator d/dt.

For Part (2), if φ_t is the flow of X on M, then

$$\mathcal{L}_X(fY) = \lim_{h\to 0} \frac{1}{h}((\varphi_{-h})_*((fY)_{\varphi_h(p)}) - (fY)_p)$$

$$= \lim_{h\to 0} \frac{1}{h}((\varphi_{-h})_*(f(\varphi_h(p))Y_{\varphi_h(p)}) - f(p)Y_p)$$

$$= \lim_{h\to 0} \frac{1}{h}(f(\varphi_h(p))(\varphi_{-h})_*(Y_{\varphi_h(p)}) - f(p)Y_p).$$

Then using the typical trick to prove the product rule gives

$$\mathcal{L}_X(fY) = \lim_{h\to 0} \frac{1}{h}\left(f(\varphi_h(p))(\varphi_{-h})_*(Y_{\varphi_h(p)}) - f(\varphi_h(p))(Y_p)\right) + \lim_{h\to 0} \frac{1}{h}\left(f(\varphi_h(p))(Y_p) - f(p)Y_p\right)$$

$$= \left(\lim_{h\to 0} f(\varphi_h(p))\right)\lim_{h\to 0} \frac{1}{h}((\varphi_{-h})_*(Y_{\varphi_h(p)}) - Y_p) + \left(\lim_{h\to 0} \frac{1}{h}(f(\varphi_h(p)) - f(p))\right)Y_p$$

$$= f(p)(\mathcal{L}_X Y)_p + (Xf)(p)Y_p.$$

The formula follows from the definition $\mathcal{L}_X f = Xf$. $\qquad\qquad\qquad\qquad\qquad\square$

With the identities of Proposition 5.3.9, we can calculate the coordinate-dependent formula for the Lie derivative, which leads to the following interesting result.

Theorem 5.3.10. *Let X and Y be C^2 vector fields on a C^2-manifold M. Then*

$$\mathcal{L}_X Y = [X, Y].$$

Proof. Let $p \in M$ and let (U, x) be a coordinate chart on a neighborhood of p. Suppose that $X = X^i \partial_i$ and $Y = Y^i \partial_i$ in coordinates over this chart and let $\varphi_t(x)$ be the flow of X over U.

We point out two facts about the flow of X. First, since $\lim_{h \to 0} \varphi_h(x) = x$ for all $x \in U$, then

$$\lim_{h \to 0} \frac{\partial \varphi_h^i}{\partial x^j} = \delta_j^i \tag{5.19}$$

for all i, j, so in particular,

$$\lim_{h \to 0} \frac{\partial \varphi_{-h}^i}{\partial x^j}(\varphi_h(x)) - \delta_j^i = 0.$$

This leads to our second observation: we claim that

$$\lim_{h \to 0} \frac{1}{h}\left(\frac{\partial \varphi_{-h}^i}{\partial x^j}(\varphi_h(x)) - \delta_j^i\right) = -\frac{\partial X^i}{\partial x^j}. \tag{5.20}$$

To see this, note that by definition of differentiability in t, the component functions $\varphi_t^i(x)$ satisfy

$$\varphi_t^i(x) = x^i + \left.\frac{d\varphi_t^i(x)}{dt}\right|_{t=0} t + tR^i(t, x) = x^i + X^i(x)t + tR^i(t, x)$$

for remainder functions $R^i : (-\varepsilon, \varepsilon) \times U \to \mathbb{R}$ such that $\lim_{t \to 0} R^i(t, x) = 0$. Then

$$\frac{\partial \varphi_{-h}^i}{\partial x^j}(x) = \delta_j^i - \frac{\partial X^i}{\partial x^j}(x)h - h\frac{\partial R^i}{\partial x^j}(-h, x).$$

This gives

$$\lim_{h \to 0} \frac{1}{h}\left(\frac{\partial \varphi_{-h}^i}{\partial x^j}(\varphi_h(x)) - \delta_j^i\right) = \lim_{h \to 0} \frac{1}{h}\left(-\frac{\partial X^i}{\partial x^j}(\varphi_h(x))h - h\frac{\partial R^i}{\partial x^j}(-h, \varphi_h(x))\right)$$

$$= \lim_{h \to 0}\left(-\frac{\partial X^i}{\partial x^j}(\varphi_h(x)) - \frac{\partial R^i}{\partial x^j}(-h, \varphi_h(x))\right)$$

$$= -\frac{\partial X^i}{\partial x^j}(\varphi_0(x)) - \frac{\partial}{\partial x^j}\left(\lim_{h \to 0} R^i(-h, \varphi_h(x))\right) = -\frac{\partial X^i}{\partial x^j}(x).$$

We can now calculate the Lie derivative in components. By definition of differentiability near $t = 0$, we can write the component functions of $Y_{\varphi_t(x)}$, which we express for now by $Y(\varphi_t(x))$, as

$$Y^j(\varphi_t(x)) = Y^j(x) + \frac{d}{dt}(Y^j(\varphi_t(x)))\big|_{t=0} t + tS^j(t, x),$$

for some remainder functions $S^j(t, x)$ that satisfy $\lim_{t \to 0} S^j(t, x) = 0$. Then

$$Y^j(\varphi_t(x)) = Y^j(x) + \left[\frac{\partial Y^j}{\partial x^k}(\varphi_t(x))\frac{d\varphi_t^k}{dt}(x)\right]_{t=0} t + tS^j(t, x)$$

$$= Y^j(x) + \frac{\partial Y^j}{\partial x^k}(x)X^k(x)t + tS^j(t, x).$$

By (5.18) the components of $(\varphi_{-t})_* Y_{\varphi_t(p)}$ are

$$\frac{\partial \varphi_{-t}^i}{\partial x^j}(\varphi_t(p))\left(Y^j(p) + \frac{\partial Y^j}{\partial x^k}(p)X^k(p)t + tS^j(t, p)\right),$$

so the components of $\mathcal{L}_X Y$ are

$$(\mathcal{L}_X Y)^i(p) = \lim_{h \to 0} \frac{1}{h}\left(\frac{\partial \varphi_{-h}^i}{\partial x^j}(\varphi_h(p))Y^j(p) + h\frac{\partial \varphi_{-h}^i}{\partial x^j}(\varphi_h(p))\frac{\partial Y^j}{\partial x^k}(p)X^k(p)\right.$$

$$\left. + h\frac{\partial \varphi_{-h}^i}{\partial x^j}(\varphi_h(p))S^j(h, p) - Y^i(p)\right)$$

$$= \lim_{h \to 0}\left(\frac{1}{h}\left(\frac{\partial \varphi_{-h}^i}{\partial x^j}(\varphi_h(p))Y^j(p) - Y^i(p)\right)\right.$$

$$\left. + \frac{\partial \varphi_{-h}^i}{\partial x^j}(\varphi_h(p))\frac{\partial Y^j}{\partial x^k}(p)X^k(p) + \frac{\partial \varphi_{-h}^i}{\partial x^j}(\varphi_h(p))S^j(h, p)\right)$$

so by (5.19),

$$(\mathcal{L}_X Y)^i(p) = \lim_{h \to 0}\left(\frac{1}{h}\left(\frac{\partial \varphi_{-h}^i}{\partial x^j}(\varphi_h(p)) - \delta_j^i\right)Y^j(p) + \delta_j^i\frac{\partial Y^j}{\partial x^k}(p)X^k(p) + \delta_j^i S^j(h, p)\right)$$

$$= -\frac{\partial X^i}{\partial x^j}(p)Y^j(p) + \frac{\partial Y^i}{\partial x^k}(p)X^k(p)$$

where the last equality holds by (5.20). After replacing the summation variable k with j, we recover the component description of the Lie bracket given in (5.15). $\square$

We mention this first corollary simply to reinterate the coordinate-dependent expression for the Lie derivative.

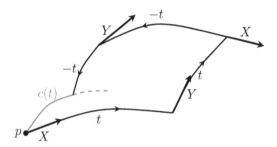

Figure 5.4: The curve paths defining $c(t)$.

Corollary 5.3.11. *Let X and Y be vector fields on a C^2-manifold M. Suppose that over a coordinate patch (U, x) of M we have $X = X^i \partial_i$ and $Y = Y_i \partial_i$ in components. Then the components of the Lie derivative are*

$$(\mathcal{L}_X Y)^i = X^j \frac{\partial Y^i}{\partial x^j} - \frac{\partial X^i}{\partial x^j} Y^j.$$

Theorem 5.3.10 immediately leads to the following interesting corollary.

Corollary 5.3.12. *Let X, Y, and Z be differentiable vector fields on M and let $f \in C^1(M, \mathbb{R})$. Then*

1. $\mathcal{L}_Y X = -\mathcal{L}_X Y$.

2. $\mathcal{L}_{(X+Y)} Z = \mathcal{L}_X Z + \mathcal{L}_Y Z$.

This is rather striking because of the following observation. The linearity rules of the Lie derivative $\mathcal{L}_X$ as described in Proposition 5.3.9 followed easily from the linearity of $(\varphi_{-t})_*$ and a product rule. However, proving that $\mathcal{L}_{(X+Y)} Z = \mathcal{L}_X Z + \mathcal{L}_Y Z$ from the definition would be intractable because there is no immediately obvious connection between the sum of two vector fields and their flows on the manifold.

Theorem 5.3.10 implies a number of nonobvious properties for the Lie derivative, the proofs of which we leave as exercises for the reader.

Proposition 5.3.13. *Let X, Y, and Z be differentiable vector fields on M, and let $F : M \to N$ be a diffeomorphism between manifolds. Then*

1. $\mathcal{L}_X [Y, Z] = [\mathcal{L}_X Y, \mathcal{L}_X Z]$.

2. $\mathcal{L}_{[X,Y]} Z = \mathcal{L}_X \mathcal{L}_Y Z - \mathcal{L}_Y \mathcal{L}_X Z$.

3. $F_* (\mathcal{L}_X Y) = \mathcal{L}_{F_* X} (F_* Y)$.

In Section 5.6, we will expand the definition of the Lie derivative to tensor fields of all types, and not just functions and tangent vector fields.

Besides the algebraic properties, the Lie bracket also carries a more geometric interpretation. The bracket $[X, Y]$ measures an instantaneous path dependence between the integral curves of X and Y. To be more precise, for sufficiently small $t \in (-\varepsilon, \varepsilon)$, consider the curve $c(t)$ that

- starts at a point $c(0) = p$;

- follows the integral curve of X starting at p for time t;

- starting from there, follows the integral curve of Y for time t;

- then follows the integral curve of X backwards by time $-t$;

- then follows the integral curve of Y backwards by time $-t$.

See Figure 5.4. If φ_t is the flow for X and ψ_t is the flow for Y, then this curve $c : (-\varepsilon, \varepsilon) \to M$ is

$$c(t) = \psi_{-t}(\varphi_{-t}(\psi_t(\varphi_t(p)))).$$

Two properties are obvious. If t approaches 0, then $c(t)$ approaches p. Also, if x is a system of coordinates on a patch U of M and if $X = \partial_1$ and $Y = \partial_2$, then the above steps for the description of $c(t)$ travel around a "square" with side t based at p, and thus $c(t)$ is constant. Other properties are not so obvious, and we refer the reader to [52, Proposition 5.15, Theorem 5.16] for proofs.

Proposition 5.3.14. *Defining the curve $c(t)$ as above,*

1. *$c'(0) = 0$;*

2. *if we define $c''(0)$ as the operator satisfying $c''(0)(f) = (f \circ c)''(0)$, then $c''(0)$ is a derivation and hence an element of $T_p M$;*

3. *$c''(0) = 2[X, Y]_p$.*

Consequently, from an intuitive perspective, the Lie bracket $[X, Y]$ is a vector field that at p measures the second-order derivation of $c(t)$ at p. Since the first derivative $c'(0)$ is 0, then $c''(0) = 2[X, Y]_p$ gives the direction of motion of $c(t)$ out of p as a second-order approximation.

PROBLEMS

5.3.1. Let $M = \mathbb{R}^2$. Calculate the Lie bracket $[X, Y]$ for each of the following pairs of vector fields:

(a) $X = x\dfrac{\partial}{\partial x} + y\dfrac{\partial}{\partial y}$ and $Y = -y\dfrac{\partial}{\partial x} + x\dfrac{\partial}{\partial y}$.

(b) $X = \sin(x+y)\dfrac{\partial}{\partial x} + \cos x\dfrac{\partial}{\partial y}$ and $Y = \cos x\dfrac{\partial}{\partial x} + \sin y\dfrac{\partial}{\partial y}$.

5.3.2. Let $M = \mathbb{R}^3$. Calculate the Lie bracket $[X, Y]$ for each of the following pairs of vector fields:

(a) $X = z^2 \dfrac{\partial}{\partial x} + xy \dfrac{\partial}{\partial z}$ and $Y = (x + y^3) \dfrac{\partial}{\partial y} + yz \dfrac{\partial}{\partial z}$.

(b) $X = yz \dfrac{\partial}{\partial x} + xz \dfrac{\partial}{\partial y} + xy \dfrac{\partial}{\partial z}$ and $Y = x \dfrac{\partial}{\partial x} + y \dfrac{\partial}{\partial y} + z \dfrac{\partial}{\partial z}$.

(c) $X = \ln(x^2 + 1) \dfrac{\partial}{\partial y} + \tan^{-1}(xy) \dfrac{\partial}{\partial z}$ and $Y = \dfrac{\partial}{\partial x}$.

5.3.3. Consider Exercise 5.2.1, calculate the components of $[X, Z]$ over U.

5.3.4. Prove that Equation 5.15 that comes out of Proposition 5.3.1 changes contravariantly, as a vector, under a change of coordinates on $T_p M$.

5.3.5. Referring to Example 5.3.7, use Definition 5.3.5 to calculate $\mathcal{L}_X Y$ for an arbitrary vector field on $\mathbb{R}^2$. Calculate the Lie bracket $[X, Y]$ directly. [Hint: They should be equal.]

5.3.6. Prove Proposition 5.3.8.

5.3.7. Prove Proposition 5.3.4.

5.3.8. Let $F : M \to N$ be a differentiable map. Let $X_1, X_2 \in \mathfrak{X}(M)$, and let $Y_1, Y_2 \in \mathfrak{X}(N)$. Suppose that X_i is F-related to Y_i. Prove that $[X_1, X_2]$ is F-related to $[Y_1, Y_2]$.

5.3.9. Prove Proposition 5.3.13.

5.3.10. Let X and Y be differentiable vector fields on a C^2 manifold M. Let φ be the flow of X on M. Prove that $\mathcal{L}_X Y = 0$ everywhere if and only if Y is invariant under the flow of X (i.e., $Y_{\varphi_t(p)} = (\varphi_t)_* Y_p$).

5.4 Differential Forms

We now consider a particular class of tensor fields called differential forms. As we will see, differential forms have many uses in geometry and in physics, in particular for integration on manifolds. We introduced the linear algebra necessary for differential forms in Section 4.6.

Though it would be possible to continue the discussion with manifolds and functions of class C^k, we will restrict our attention to smooth manifolds for simplicity.

5.4.1 Definitions

Definition 5.4.1. Let M^n be a smooth manifold. A *differential form* ω of type r on M (or more succinctly, r-form) is a smooth global section (tensor field) of $\bigwedge^r (TM^*)$.

Intuitively, for each $p \in M$, we associate $\omega_p \in \bigwedge^r (T_p M^*)$ in such a way that ω_p varies smoothly with p. The tensor ω_p is an alternating r-multilinear function $T_p M^{\otimes r} \to \mathbb{R}$. Hence, a differential form is a particular type of covariant tensor field and that 1-forms are simply covector fields. A differential form of type 0 is simply a smooth real-valued function on M.

Let U be a coordinate neighborhood of M with coordinates $x = (x^1, x^2, \ldots, x^n)$. Define $\mathcal{I}(r, n)$ as the set of all increasing sequences of length r with values in $\{1, 2, \ldots, n\}$. For example, $(2, 3, 7) \in \mathcal{I}(3, 7)$ because there are three elements in the sequence, they are listed in increasing order, and their values are in $\{1, 2, \ldots, 7\}$. By Proposition 4.6.18, over the coordinate patch U, an r-form ω can be written in a unique way as

$$\omega = \sum_{I \in \mathcal{I}(r,n)} a_I \, dx^I,$$

where each a_I is a smooth function, and where we denote $dx^I = dx^{i_1} \wedge \cdots \wedge dx^{i_r}$ when I is the r-tuple $I = (i_1, \ldots, i_r)$. Recall that the symbol dx^i is defined in Equations (5.6) and (5.7) and that this wedge product is defined as the alternation

$$dx^{i_1} \wedge \cdots \wedge dx^{i_r} = \mathsf{A}(dx^{i_1} \otimes dx^{i_2} \otimes \cdots \otimes dx^{i_r}). \tag{5.21}$$

Alternatively, in reference to a coordinate system, a differential form is a smooth covariant tensor field $\omega = \omega_{i_1, i_2, \ldots, i_r} dx^{i_1} \otimes dx^{i_2} \otimes \cdots \otimes dx^{i_r}$ such that component functions satisfy

$$\omega_{i_\sigma^{-1}(1), i_\sigma^{-1}(2), \ldots, i_\sigma^{-1}(r)} = \text{sign}(\sigma) \omega_{i_1, i_2, \ldots, i_r}$$

for any permutation $\sigma \in S_r$ of the indices.

Definition 5.4.2. If U is an open subset of M, we denote by $\Omega^r(U)$ the set of all differential forms of type r on U.

We remark that, similar to Problem 5.2.8, for each r, the set $\Omega^r(U)$ is an infinite-dimensional vector space. In particular, if $\omega, \eta \in \Omega^r(U)$ and $\lambda \in \mathbb{R}$, then $\omega + \eta \in \Omega^r(U)$ and $\lambda\omega \in \Omega^r(U)$, where by definition

$$(\omega + \eta)_p = \omega_p + \eta_p \quad \text{and} \quad (\lambda\omega)_p = \lambda\omega_p \quad \text{in} \bigwedge^r TM^*.$$

Not only is each $\Omega^r(U)$ closed under scalar multiplication, but it is closed under multiplication by a smooth function. More precisely, for all smooth functions $f : U \to \mathbb{R}$, we have $f\omega \in \Omega^r(U)$, where $(f\omega)_p = f(p)\omega_p$ for all $p \in U$.

Finally, similar to the alternating products of a fixed vector space, for $\omega \in \Omega^r(U)$ and $\eta \in \Omega^s(U)$, we define the *exterior product* $\omega \wedge \eta \in \Omega^{r+s}(U)$ as the differential form defined by $(\omega \wedge \eta)_p = \omega_p \wedge \eta_p$ for all $p \in M$.

Example 5.4.3. Consider the sphere $\mathbb{S}^2$, and let U be the coordinate neighborhood with a system of coordinates x defined by the parametrization $x^{-1}(u, v) = (\cos u \sin v, \sin u \sin v, \cos v)$ defined on $(0, 2\pi) \times (0, \pi)$. Let

$$\omega = (\sin^2 v) \, du + (\sin v \cos v) \, dv$$

$$\eta = \cos u \sin v \, du + (\sin u \cos v - \sin v) \, dv$$

be two 1-forms on $\mathbb{S}^2$. Remarking that $du \wedge du = dv \wedge dv = 0$, we calculate

$$\omega \wedge \eta = \sin^2 v (\sin u \cos v - \cos u \cos v - \sin v) du \wedge dv.$$

5.4.2 Exterior Differential

Let f be a smooth real-valued function on a smooth manifold M, and let $X \in \mathfrak{X}(M)$ be a vector field. Viewing $f : M \to \mathbb{R}$ as a differential map between manifolds, the differential df is such that, at each point $p \in M$, it evaluates $df_p(X_p)$ to a tangent vector in $T_{f(p)}(\mathbb{R})$. However, the tangent space $T_{f(p)}(\mathbb{R})$ is equal to $\mathbb{R}$, so $df_p(X_p)$ is just a real number. Hence, $df_p \in T_p M^*$, and since all of the operations vary smoothly with p, then $df \in \Omega^1(M)$. If $x = (x^1, \ldots, x^n)$ is a coordinate system on an open set $U \subseteq M$, then in coordinates we have

$$df = \sum_{i=1}^{n} \frac{\partial f}{\partial x^i} \, dx^i. \tag{5.22}$$

Since $C^\infty(U, \mathbb{R}) = \Omega^0(U)$, the differential d defines a linear transformation $d : \Omega^0(U) \to \Omega^1(U)$. We now generalize this remark by the following definition.

Definition 5.4.4. Let $\omega = \sum_I a_I \, dx^I$ be a smooth differential r-form over U. The *exterior differential* of ω is the $(r+1)$-form written as $d\omega$ and defined by

$$d\omega = \sum_{I \in \mathcal{I}(r,n)} (da_I) \wedge dx^I. \tag{5.23}$$

Example 5.4.5. Revisiting Example 5.4.3, we calculate $d\omega$ and $d\eta$. First, for $d\omega$ we have

$$
\begin{aligned}
d\omega &= (d(\sin^2 v)) \wedge du + (d(\sin v \cos v)) \wedge dv \\
&= (2 \sin v \cos v \, dv) \wedge du + ((\cos^2 v - \sin^2 v) \, dv) \wedge dv \\
&= (-2 \sin v \cos v) \, du \wedge dv.
\end{aligned}
$$

For $d\eta$, we calculate

$$
\begin{aligned}
d\eta &= (d(\cos u \sin v)) \wedge du + (d(\sin u \cos v - \sin v)) \wedge dv \\
&= ((-\sin u \sin v) \, du + (\cos u \cos v) \, dv) \wedge du \\
&\qquad + ((\cos u \cos v) \, du + (-\sin u \sin v - \cos v) \, dv) \wedge dv \\
&= (\cos u \cos v) \, dv \wedge du + (\cos u \cos v) \, du \wedge dv = 0.
\end{aligned}
$$

The differential form η has the unexpected property that $d\eta = 0$. We will say that η is a closed 1-form (see Definition 5.4.7).

Proposition 5.4.6. *Let M be a smooth manifold, and let U be an open subset of M. The exterior differential satisfies the following:*

1. *For each $0 \leq r \leq n - 1$, the operator $d : \Omega^r(U) \to \Omega^{r+1}(U)$ is a linear map.*

2. *If $\omega \in \Omega^r(U)$ and $\eta \in \Omega^s(U)$, then*

$$d(\omega \wedge \eta) = d\omega \wedge \eta + (-1)^r \omega \wedge d\eta.$$

3. For all $\omega \in \Omega^r(U)$, we have $d(d\omega) = 0$.

Proof. For Part 1, set $\omega = \sum_I a_I \, dx^I$, where the summation is over all $I \in \mathcal{I}(r,n)$ and a^I are smooth real-valued functions on M. Then from Equation (5.23), each da^I is a 1-form, so obviously the summation is over $(r+1)$-forms.

Now let $\eta = \sum_I b_I \, dx^I$ be another r-form, and let $\lambda, \mu \in \mathbb{R}$. Then

$$d(\lambda\omega + \mu\eta) = \sum_I d(\lambda a_I + \mu b_I) \wedge dx^I$$

$$= \sum_I \left(\sum_{j=1}^n \left(\lambda \frac{\partial a^I}{\partial x^j} + \mu \frac{\partial b^I}{\partial x^j} \right) dx^j \right) \wedge dx^I$$

$$= \lambda \sum_I \left(\sum_{j=1}^n \frac{\partial a^I}{\partial x^j} \, dx^j \right) \wedge dx^I + \mu \sum_I \left(\sum_{j=1}^n \frac{\partial b^I}{\partial x^j} \, dx^j \right) \wedge dx^I$$

$$= \lambda d\omega + \mu d\eta.$$

This proves linearity of d.

For Part 2, again let ω be as above, and let $\eta \in \Omega^s(U)$ expressed as $\eta = \sum_J b_J \, dx^J$, where the summation in J runs over $\mathcal{I}(s,n)$. By the linearity of the wedge product, we can write

$$\omega \wedge \eta = \sum_I \sum_J a_I b_J \, dx^I \wedge dx^J.$$

Note that for various combinations of I and J, the wedge products $dx^I \wedge dx^J$ will cancel if I and J share any common indices. Then

$$d(\omega \wedge \eta) = \sum_I \sum_J \left(\sum_{k=1}^n \frac{\partial a_I b_J}{\partial x^k} \, dx^k \right) \wedge dx^I \wedge dx^J$$

$$= \sum_I \sum_J \left(\sum_{k=1}^n \left(\frac{\partial a_I}{\partial x^k} b_J + a_I \frac{\partial b_J}{\partial x^k} \right) dx^k \right) \wedge dx^I \wedge dx^J$$

$$= \sum_I \sum_J \left(\sum_{k=1}^n \frac{\partial a_I}{\partial x^k} b_J \, dx^k \right) \wedge dx^I \wedge dx^J$$

$$+ \sum_I \sum_J \left(\sum_{k=1}^n a_I \frac{\partial b_J}{\partial x^k} \, dx^k \right) \wedge dx^I \wedge dx^J.$$

But by the properties of wedge products, $dx^k \wedge dx^I \wedge dx^J = (-1)^r dx^I \wedge dx^k \wedge dx^J$

(see Proposition 4.6.21). Thus,

$$d(\omega \wedge \eta) = \sum_I \sum_{k=1}^n \sum_J \frac{\partial a_I}{\partial x^k} b_J \, dx^k \wedge dx^I \wedge dx^J$$

$$+ (-1)^r \sum_I \sum_J \sum_{k=1}^n a_I \frac{\partial b_J}{\partial x^k} \, dx^I \wedge dx^k \wedge dx^J$$

$$= \left(\sum_I \sum_{k=1}^n \frac{\partial a_I}{\partial x^k} dx^k \wedge dx^I \right) \wedge \eta$$

$$+ (-1)^r \omega \wedge \left(\sum_J \sum_{k=1}^n \frac{\partial b_J}{\partial x^k} dx^k \wedge dx^J \right)$$

$$= d\omega \wedge \eta + (-1)^r \omega \wedge d\eta.$$

To prove Part 3, we first show that $d(df) = 0$ for a smooth function f on M. We have

$$d(df) = d\left(\sum_{i=1}^n \frac{\partial f}{\partial x^i} dx^i \right) = \sum_{i=1}^n \sum_{j=1}^n \frac{\partial^2 f}{\partial x^j \partial x^i} \, dx^j \wedge dx^i$$

$$= \sum_{I \in \mathcal{I}(2,n)} \left(\frac{\partial^2 f}{\partial x^{i_1} \partial x^{i_2}} - \frac{\partial^2 f}{\partial x^{i_2} \partial x^{i_1}} \right) dx^I,$$

where we assume $I = (i_1, i_2)$. However, since the function f is smooth, by Clairaut's Theorem on mixed partials each component function is 0. Thus $d(df) = 0$.

Now for any r-form $\omega = \sum_I a_I \, dx^I$ we have

$$d(d\omega) = d\left(\sum_I d(a_I) \wedge dx^I \right) = \sum_I d(d(a_I) \wedge dx^I) \text{(by linearity)}$$

$$= \sum_I \left(d(da^I) \wedge dx^I - da_I \wedge d(dx^I) \right) \qquad \text{(by part 2)}$$

$$= 0,$$

where the last line follows because $d(da_I) = 0$ and $d(dx^I) = 0$ for all I. $\qquad \square$

It is illuminating to compare the exterior differential to differential operators on vector fields in $\mathbb{R}^n$. We emphasize three particular cases.

First, let f be a smooth real-valued function on $\mathbb{R}^n$. Then

$$df = \sum_{i=1}^n \frac{\partial f}{\partial x^i} \, dx^i.$$

Thus, df has exactly the same components as the gradient, defined in multivariable calculus as

$$\text{grad } f = \vec{\nabla} f = (\partial_1 f, \partial_2 f, \ldots, \partial_n f).$$

Therefore, in our presentation, the gradient of a function f is in fact a covector field, i.e., a vector field in $TM^* = (\mathbb{R}^n)^*$.

In calculus courses, we do not distinguish between vectors and covectors, i.e., vectors in $\mathbb{R}^n$ or in $(\mathbb{R}^n)^*$, since these are isomorphic as vector spaces. However, as we saw in (4.2) and Proposition 4.1.6, vector fields and covector fields have different transformational properties under changes of coordinates. Example 4.5.8 showed that the gradient of a function transforms covariantly, but it is also instructive to see how this plays out in common formulas. For example, the chain rule for paths states that if $\vec{c}(t)$ is a differentiable curve in $\mathbb{R}^n$ and $f : \mathbb{R}^n \to \mathbb{R}$ is differentiable, then

$$\frac{d}{dt} f(\vec{c}(t)) = \vec{\nabla} f_{\vec{c}(t)} \cdot \vec{c}'(t).$$

However, from the perspective of multilinear algebra, we should understand the dot product in this context as the contraction map $V^* \otimes V \to \mathbb{R}$ defined by $\lambda \otimes \vec{v} \mapsto \lambda(\vec{v})$. Since by definition $\vec{c}'(t)$ is a tangent vector to $\mathbb{R}^n$ at $\vec{c}(t)$, then we should view the gradient $\vec{\nabla} f$ as a covector in $(\mathbb{R}^n)^*$.

As a second illustration, consider $(n-1)$-forms over $\mathbb{R}^n$. For each $1 \leq j \leq n$, define the $(n-1)$-forms η^j as

$$\eta^j = (-1)^{j-1} dx^1 \wedge \cdots \wedge dx^{j-1} \wedge dx^{j+1} \wedge \cdots \wedge dx^n. \tag{5.24}$$

For each $p \in M$, the set $\{\eta_p^j\}_{j=1}^n$ is a basis for $\bigwedge^{n-1}(T_p\mathbb{R}^n)^*$. So any $(n-1)$-form ω can be written as $\omega = \sum_{j=1}^n a_j \eta^j$ for functions $a_j : M \to \mathbb{R}$. Note that having the $(-1)^{j-1}$ factor in the definition of η^j leads to the identity

$$dx^i \wedge \eta^j = \begin{cases} 0, & \text{if } i \neq j, \\ dx^1 \wedge dx^2 \wedge \cdots \wedge dx^n, & \text{if } i = j. \end{cases} \tag{5.25}$$

Thus, for the differential of ω, we have

$$d\omega = \sum_{i=1}^n \sum_{j=1}^n \frac{\partial a_j}{\partial x^i} dx^i \wedge \eta^j = \left(\sum_{i=1}^n \frac{\partial a_i}{\partial x^i} \right) dx^1 \wedge dx^2 \wedge \cdots \wedge dx^n.$$

Hence, for the case of $(n-1)$-forms, the exterior differential d operates like the divergence operator $\text{div} = \vec{\nabla} \cdot$ on a vector field $(a_1, \ldots, a_n)$ in $\mathbb{R}^n$.

In the case of $\mathbb{R}^3$, the exterior differential carries another point of significance.

Let $\omega \in \Omega^1(\mathbb{R}^3)$, and write $\omega = \sum_{i=1}^n a_i \, dx^i$. Then

$$dw = \sum_{i=1}^n \sum_{j=1}^n \frac{\partial a_i}{\partial x^j} \, dx^j \wedge dx^i$$

$$= \left(\frac{\partial a_2}{\partial x^1} - \frac{\partial a_1}{\partial x^2} \right) dx^1 \wedge dx^2 + \left(\frac{\partial a_3}{\partial x^1} - \frac{\partial a_1}{\partial x^3} \right) dx^1 \wedge dx^3$$

$$+ \left(\frac{\partial a_3}{\partial x^2} - \frac{\partial a_2}{\partial x^3} \right) dx^2 \wedge dx^3$$

$$= \left(\frac{\partial a_3}{\partial x^2} - \frac{\partial a_2}{\partial x^3} \right) \eta^1 + \left(\frac{\partial a_1}{\partial x^3} - \frac{\partial a_3}{\partial x^1} \right) \eta^2 + \left(\frac{\partial a_2}{\partial x^1} - \frac{\partial a_1}{\partial x^2} \right) \eta^3,$$

which is precisely the curl of the vector field (a_1, a_2, a_3).

It is particularly interesting to note that the property $d(d\omega)$ in Proposition 5.4.6 summarizes simultaneously the following two standard theorems in multivariable calculus:

$$\text{curl grad } f = \vec{0} \qquad \text{([55, Theorem 17.3]),}$$
$$\text{div curl } \vec{F} = 0 \qquad \text{([55, Theorem 17.11]),}$$

where $f : \mathbb{R}^n \to \mathbb{R}$ is a function of class C^2 and $\vec{F} : \mathbb{R}^3 \to \mathbb{R}^3$ is a vector field of class C^2.

We point out that the forms η^j defined in (5.24) are instances of the Hodge star operator $\star$ which we discuss in Appendix C.3. The Hodge star operator exists in the general context of a vector space equipped with an inner product (a bilinear form that is symmetric and nondegenerate). In the above situation, we have $V = \mathbb{R}^n$ and the inner product $\langle , \rangle$ is the standard Euclidean dot product. Then according to Proposition C.3.3, we have

$$\eta^j = \star dx^j.$$

5.4.3 Closed and Exact Forms

Definition 5.4.7. Let M be a smooth manifold. A differential form $\omega \in \Omega^r(M)$ is called *closed* if $d\omega = 0$ and is called *exact* if there exist $\eta \in \Omega^{r-1}(M)$ such that $\omega = d\eta$.

Example 5.4.8. As an example, consider the explicit covector fields ω and η on $\mathbb{S}^2$ described in Examples 5.4.3 and 5.4.5. In Example 5.4.5 we showed that $d\eta = 0$, meaning that η is closed. If

$$\eta = \cos u \sin v \, du + (\sin u \cos v - \sin v) \, dv$$

is exact, then there exists a 0-form, i.e., differentiable function, $f : \mathbb{S}^2 \to \mathbb{R}$ such that $\eta = df$. Thus

$$\frac{\partial f}{\partial u} = \cos u \sin v \qquad \text{and} \qquad \frac{\partial f}{\partial v} = \sin u \cos v - \sin v.$$

By integrating with respect to u, we must have $f(u, v) = \sin u \sin v + g(v)$. Then differentiating with respect to v, gives $\partial f / \partial v = \sin u \cos v + g'(v) = \sin u \cos v - \sin v$. Thus, $g(v) = \cos v + C$ for some constant C. Thus $f(u, v) = \sin u \sin v + \cos v + C$. A priori, this is only defined for $(u, v) \in (0, 2\pi) \times (0, \pi)$. However, $f(u, 0) = 1$ and $f(u, \pi) = -1$, regardless of $u \in \mathbb{R}$ and for $v \in (0, \pi)$, we also have $f(u + 2\pi, v) = f(u, v)$. Hence, f extends continuously to a well-defined function on all of $\mathbb{S}^2$. (We also note that with respect to the typical embedding of $\mathbb{S}^2$ in $\mathbb{R}^3$, the function f is equal to $y + z + C$ restricted to $\mathbb{S}^2$.)

Our calculations show that $\eta = df$, so η is an exact.

This shows that not just any pair of smooth functions $a_1(u, v)$ and $a_2(u, v)$ allow the 1-form

$$\omega = a_1(u, v) \, du + a_2(u, v) \, dv$$

to extend over $\mathbb{S}^2$ to create a smooth 1-form on $\mathbb{S}^2$. For example, not even $u \, dv$, defined on the same coordinate chart described in the above example, extends continuously to a 1-form on all of $\mathbb{S}^2$. This restriction shows that $\Omega^1(\mathbb{S}^2)$ is affected by the global geometry of $\mathbb{S}^2$. The principle behind this example is true in general: the vector spaces $\Omega^r(M)$, though infinite-dimensional, depend on the global structure of M.

The identity $d(d\omega) = 0$ for any differential form means that every exact form is closed. The converse is not true in general, and it is precisely this fact that leads to profound results in topology. In the language of homology, the sequence of vector spaces and linear maps

$$\Omega^0(M) \xrightarrow{\ d\ } \Omega^1(M) \xrightarrow{\ d\ } \Omega^2(M) \xrightarrow{\ d\ } \cdots \xrightarrow{\ d\ } \Omega^n(M)$$

satisfying the identity $d \circ d = 0$ is called a *complex*. To distinguish between types, we often write d^r for the differential $d : \Omega^r(M) \to \Omega^{r+1}(M)$. The fact that every exact form is closed can be restated once more by saying that $\operatorname{Im} d^{r-1}$ is a vector subspace of $\ker d^r$. The quotient vector space

$$\ker d^r / \operatorname{Im} d^{r-1} = \ker(d : \Omega^r(M) \to \Omega^{r+1}(M)) / \operatorname{Im}(d : \Omega^{r-1}(M) \to \Omega^r(M))$$

is called the *r*th *de Rham cohomology group* of M, denoted $H^r_{dR}(M)$. The de Rham cohomology groups are in fact global properties of the manifold M and are related to profound topological invariants of M. This topic exceeds the scope of this book, but we wish to point out two ways in which one can glimpse why the groups $H^r_{dR}(M)$ are global properties of M.

In Example 5.4.8 we observed that defining a form on all of $\mathbb{S}^2$ carries some restrictions. Hence, the space of functions carries information about the global structure of M. Similarly, Problem 5.4.13 gives an example of a 1-form on $\mathbb{R}^2 - \{(0,0)\}$ that is closed but not exact.

As a second example, we determine $H^0_{dR}(M)$ for any manifold. Of course, d^{-1} does not exist explicitly so we set, by convention, $\Omega^{-1}(M) = 0$, i.e., the zero-dimensional vector space. Then $\operatorname{Im} d^{-1} = \{0\}$ is the trivial subspace in $\Omega^0(M)$.

Furthermore, since $\Omega^0(M)$ is the space of all smooth real-valued functions on M, the 0th cohomology group is

$$H^0_{dR}(M) = \ker(d : C^\infty(M) \to \Omega^1(M))/\{0\} = \ker(d : C^\infty(M) \to \Omega^1(M)),$$

namely, the subspace of all smooth functions on M whose differentials are 0. In other words, $H^0_{dR}(M)$ is the space of all functions that are constant on each connected component of M. Thus, $H^0_{dR}(M) = \mathbb{R}^\ell$, where ℓ is the number of connected components of M, a global property.

5.4.4 Algebra of Differential Forms

We conclude this section with a brief comment about the algebra of differential forms. Not unlike the tensor algebra or the alternating algebra over a vector space V defined in Section 4.7, we define the algebra of smooth differential forms over a smooth m-dimensional manifold M as

$$\Omega^\bullet(M) = \bigoplus_{k=0}^m \Omega^k(M) = C^\infty(M, \mathbb{R}) \oplus \Omega^1(M) \oplus \cdots \oplus \Omega^n(M),$$

equipped with the exterior product $\wedge$ as the bilinear product.

PROBLEMS

5.4.1. Let M be a smooth manifold. Let $\omega \in \Omega^r(M)$ be a nonzero r-form. Characterize the forms $\eta \in \Omega^s(M)$ such that $\omega \wedge \eta = 0$.

5.4.2. Let $M = \mathbb{R}^3$. Find the exterior differential of the following:

(a) $x\,dy \wedge dz + y\,dz \wedge dx + z\,dx \wedge dy$.

(b) $xy^2z^3\,dx + y\sin(xz)\,dz$.

(c) $\dfrac{dx \wedge dy + x\,dy \wedge dz}{x^2 + y^2 + z^2 + 1}$.

5.4.3. Let $M = \mathbb{R}^n$. Let $\omega = x^1\,dx^1 + \cdots + x^n\,dx^n$ and $\eta = x^2\,dx^1 + \cdots + x^n\,dx^{n-1} + x^1\,dx^n$.

(a) Calculate $d\omega$ and $d\eta$.

(b) Calculate $\omega \wedge \eta$ and $d(\omega \wedge \eta)$.

(c) Calculate the exterior differential of $x^1\eta^1 + x^2\eta^2 + \cdots + x^n\eta^n$, where the forms η^i are defined as in Equation (5.24).

5.4.4. Let $M = \mathbb{S}^1 \times \mathbb{S}^1$ be the torus in $\mathbb{R}^3$ that has a coordinate neighborhood (U, x) that can be parametrized by

$$x^{-1}(u, v) \mapsto \big((3 + \cos u)\cos v, (3 + \cos u)\sin v, \sin u \big) \text{ for } (u, v) \in (0, 2\pi)^2.$$

Consider the two differential forms ω and η, given over U by $\omega = \cos(u + v)\,du + 2\sin^2 u\,dv$ and $\eta = 3\sin^2 v\,du - 4\,dv$.

(a) Show why ω and η extend to differential forms over the whole torus.

(b) Calculate $\omega \wedge \omega$ and $\omega \wedge \eta$.

(c) Calculate $d\omega$ and $d\eta$.

5.4.5. Consider the manifold $\mathbb{RP}^3$ with the standard atlas described in Example 3.1.6. Consider also the 1-form that is described in coordinates over U_0 as $\omega = x^1 \, dx^1 + x^2(x^3)^3 \, dx^2 + x^1 x^2 \, dx^3$.

(a) Write down a coordinate expression for ω in U_1, U_2, and U_3.

(b) Calculate $d\omega$ and $\omega \wedge \omega$ in coordinates over U_0.

(c) Calculate $d\omega$ in coordinates over U_1 and show explicitly that the coordinates change as expected over $U_0 \cap U_1$.

5.4.6. Set $\omega = x^1 x^2 \, dx^2 + (x^2 + 3x^4 x^5) \, dx^3 + ((x^2)^2 + (x^3)^2) \, dx^5$ as a 1-form over $\mathbb{R}^5$. Calculate $d\omega$, $\omega \wedge d\omega$, $\omega \wedge \omega$, and $d\omega \wedge d\omega \wedge \omega$.

5.4.7. Consider the spacetime variables $(x^0, x^1, x^2, x^3) = (ct, x, y, z)$ in $\mathbb{R}^{1+3}$ and consider the two 2-forms α and β defined by

$$\alpha = -\sum_{i=1}^{3} E_i \, dx^0 \wedge dx^i + \sum_{j=1}^{3} B_j \eta^j \quad \text{and} \quad \beta = \sum_{i=1}^{3} B_i \, dx^0 \wedge dx^i + \sum_{j=1}^{3} E_j \eta^j,$$

where the forms η^j are the 2-forms defined in Equation (5.24) over the space variables, i.e., $\eta^1 = dx^2 \wedge dx^3$, $\eta^2 = -dx^1 \wedge dx^3$, and $\eta^3 = dx^1 \wedge dx^2$.

(a) Writing $\vec{E} = (E_1, E_2, E_3)$ and $\vec{B} = (B_1, B_2, B_3)$ as time-dependent vector fields in $\mathbb{R}^3$, show that the source-free Maxwell's equations

$$\nabla \times \vec{E} = -\frac{1}{c}\frac{\partial \vec{B}}{\partial t}, \qquad \nabla \cdot \vec{E} = 0,$$

$$\nabla \times \vec{B} = \frac{1}{c}\frac{\partial \vec{E}}{\partial t}, \qquad \nabla \cdot \vec{B} = 0,$$

can be expressed in the form

$$d\alpha = 0 \quad \text{and} \quad d\beta = 0.$$

(b) If we write the 1-form $\lambda = -\phi \, dx^0 + A_1 \, dx^1 + A_2 \, dx^2 + A_3 \, dx^3$, show that $d\lambda = \alpha$ if and only if

$$\vec{E} = -\nabla\phi - \frac{1}{c}\frac{\partial \vec{A}}{\partial t} \qquad \text{and} \qquad \vec{B} = \nabla \times \vec{A}.$$

5.4.8. In the theory of differential equations, if $A(x, y)$ and $B(x, y)$ are functions of x and y, an *integrating factor* for an expression of the form $M\frac{dy}{dx} + N$ is a function $I(x, y)$ such that

$$I(x, y)\left(A(x, y)\frac{dy}{dx} + B(x, y)\right) = \frac{d}{dx}F(x, y)$$

for some function $F(x, y)$. If M is a smooth manifold and $\omega \in \Omega^1(M)$, we call an integrating factor of ω a smooth function f that is nowhere 0 on M and such that $f\omega$ is exact. Prove that if such a function f exists, then $\omega \wedge d\omega = 0$.

5.4.9. Let $\omega = (1 + xy^2)e^{xy^2} dx + 2x^2ye^{xy^2} dy$ be a 1-form on $\mathbb{R}^2$. Show that $d\omega = 0$. Then find a function $f : \mathbb{R}^2 \to \mathbb{R}$ such that $\omega = df$.

5.4.10. Let $\omega = yz\, dx \wedge dz + (-y + xz)\, dy \wedge dz$ be a 2-form on $\mathbb{R}^3$. Show that $d\omega = 0$. Then find a 1-form λ such that $\omega = d\lambda$.

5.4.11. Suppose that $\omega \in \Omega^1(M)$ for some smooth manifold M. Suppose that over each coordinate chart of M, if we write ω in components as $\omega = \omega_i\, dx^i$ and if the component functions have the property that $\partial_j\omega_i = \partial_i\omega_j$, then ω is a closed form.

5.4.12. Let ω and η be forms on a smooth manifold M.

   (a) Show that if ω and η are closed, then so is $\omega \wedge \eta$.

   (b) Show that if ω and η are exact, then so is $\omega \wedge \eta$.

5.4.13. Consider the manifold $M = \mathbb{R}^2 - \{(0,0)\}$ with the structure inherited from $\mathbb{R}^2$ and let
$$\omega = \frac{y}{x^2 + y^2} dx - \frac{x}{x^2 + y^2} dy.$$
Prove that ω is closed but not exact. [Note: In this case, there does exist a differentiable function ψ such that $d\psi = \omega$ on $\{(x,y) \in \mathbb{R}^2 \mid x > 0\}$ but not on all of M. This particular form ω shows that $\dim H^1_{dR}(M) \geq 1$.]

5.4.14. Let M be a manifold of dimension $m \geq 4$. Let ω be a 2-form on M, and let $\{\alpha, \beta\}$ be a set of linearly independent 1-forms. Show that
$$\omega \wedge \alpha \wedge \beta = 0$$
if and only if there exist 1-forms λ and η such that
$$\omega = \lambda \wedge \alpha + \eta \wedge \beta.$$

5.4.15. Consider the manifold $GL_n(\mathbb{R})$ of invertible matrices, and consider the function $\det : GL_n(\mathbb{R}) \to \mathbb{R}$ as a function between manifolds.

   (a) Prove that for all $X \in GL_n(\mathbb{R})$, the tangent space is $T_X GL_n(\mathbb{R}) \cong \mathbb{R}^{n \times n}$, the space of $n \times n$ matrices.

   (b) Writing the entries of a matrix $X \in GL_n(\mathbb{R})$ as $X = (x^i_j)$, prove that
$$\frac{\partial \det}{\partial x^i_j}(X) = (\det X)(X^{-1})^i_j.$$

   (c) Prove that the differential of the determinant map can be written as
$$d(\det)_X(A) = (\det X)\operatorname{Tr}(X^{-1}A),$$
   where $\operatorname{Tr} M = \sum_i m^i_i$ is the trace of the matrix.

5.4.16. This exercise presents the *interior product* of k-forms on a smooth manifold M. The interior product of a k-form with $k \geq 1$ is defined as the contraction of the form with a vector field. More precisely, if X is vector field of M we define $i_X : \Omega^k(M) \to \Omega^{k-1}(M)$ such that for all $p \in M$
$$(i_X\omega)_p(v_1, v_2, \ldots, v_{k-1}) = \omega_p(X_p, v_1, \ldots, v_{k-1}),$$
for all $v_1, \ldots, v_{k-1} \in T_pM$. Prove the following properties of the interior product. Let X be a smooth vector field over M. Suppose that over a coordinate patch (U, x), the vector field is written in components as $X^i\partial_i$.

(a) Prove that $i_X(dx^1 \wedge dx^2 \wedge dx^3) = X^3 dx^2 \wedge dx^3 - X^2 dx^1 \wedge dx^3 + X^1 dx^2 \wedge dx^3$. [Hint: Refer to (5.21).]

(b) Suppose $I = (i_1, i_2, \ldots, i_r)$ with $i_1 < i_2 < \cdots < i_r$. Prove that

$$i_X(dx^I) = \sum_{j=1}^{r} (-1)^{j-1} X^{i_j} dx^{i_1} \wedge \cdots \wedge \widehat{dx^{i_j}} \wedge \cdots \wedge dx^{i_r},$$

where the $\widehat{dx^i}$ means to remove that term.

(c) If α is an r-form and β an s-form, then $i_X(\alpha \wedge \beta) = (i_X \alpha) \wedge \beta + (-1)^r \alpha \wedge (i_X \beta)$. [Hint: Using coordinates, first prove this result on $\alpha = dx^I$ and $\beta = dx^J$ with $I \in \mathcal{I}(r, n)$ and $J \in \mathcal{I}(s, n)$.]

(d) If Y is another vector field, then $i_X i_Y \omega = -i_Y i_X \omega$.

5.5 Pull-Backs of Covariant Tensor Fields

In this section we define the notion of a pull-back of a covariant tensor fields by a smooth function between manifolds. Though the construction of pull-backs is interesting in its own right, in subsequent sections we will see a few applications of the pull-back, including how to integrate differential forms over a manifold.

Definition 5.5.1. Let $f : M^m \to N^n$ be a smooth map between two smooth manifolds, and let $\alpha \in \Gamma(TN^{*\otimes s})$ be a covariant tensor field on N. Define the *pull-back* of α by f, written $f^*\alpha$, by the multilinear function on $T_p M$ that is defined by

$$(f^*\alpha)_p(v_1, v_2, \ldots, v_r) = \alpha_{f(p)}(df_p(v_1), df_p(v_2), \ldots, df_p(v_r)), \qquad (5.26)$$

where v_i are tangent vectors in $T_p M$.

According to this definition, $(f^*\alpha)_p \in T_p M^{*\otimes s}$, so $f^*\alpha$ is a global section from M into the vector bundle $TM^{*\otimes s}$. Furthermore, it is not hard to see that if ω is a differential form in $\Omega^s(N)$, then $(f^*\omega)_p$ is also an alternating multilinear function on $T_p M$, so $f^*\omega$ is a differential form in $\Omega^s(M)$.

The above definition is coordinate-free. We now work to express the pull-back of a covariant tensor field in terms of coordinates. Let x be a local coordinate system on M and y is a coordinate system on N. Suppose that over a coordinate neighborhood (V, y) of N, the covariant tensor field α is written as

$$\alpha = \alpha_{i_1 i_2 \cdots i_s} dy^{i_1} \otimes dy^{i_2} \otimes \cdots \otimes dy^{i_s},$$

where $\alpha_{i_1 i_2 \cdots i_s}$ is a smooth function of V for each s-tuple $(i_1, i_2, \ldots, i_s)$. Then locally, for every $v \in T_p M$, expressed in terms of coordinates we have $df_p(v) =$

$\sum_{i=1}^{m} \partial_i f^j v^i$ for $j = 1, \ldots, n$, where the functions f^j are the components $f^j = y^j \circ f : M \to \mathbb{R}$. Then

$$(f^*(dy^{i_1} \otimes dy^{i_2} \otimes \cdots \otimes dy^{i_s}))_p(v_1, \ldots, v_s)$$
$$= (dy^{i_1} \otimes dy^{i_2} \otimes \cdots \otimes dy^{i_s})(df_p(v_1), \ldots, df_p(v_s))$$
$$= dy^{i_1}(df_p(v_1)) \otimes dy^{i_2}(df_p(v_2)) \otimes \cdots \otimes dy^{i_s}(df_p(v_s))$$
$$= df^{i_1}(v_1) \otimes df^{i_2}(v_2) \otimes \cdots \otimes df^{i_s}(v_s)$$
$$= (df^{i_1} \otimes df^{i_2} \otimes \cdots \otimes df^{i_s})_p(v_1, v_2, \ldots, v_s).$$

We conclude that in coordinates, as a covariant tensor field over M,

$$f^*\alpha = (\alpha_{i_1 i_2 \cdots i_s} \circ f) \, df^{i_1} \otimes df^{i_2} \otimes \cdots df^{i_s} \tag{5.27}$$

$$= (\alpha_{i_1 i_2 \cdots i_s} \circ f) \frac{\partial f^{i_1}}{\partial x^{j_1}} \frac{\partial f^{i_2}}{\partial x^{j_2}} \cdots \frac{\partial f^{i_r}}{\partial x^{j_r}} \, dx^{i_1} \otimes dx^{i_2} \otimes \cdots \otimes dx^{i_r}. \tag{5.28}$$

If ω happens to be a differential form of type s, then in coordinates

$$f^*\omega = f^*\left(\sum_{I \in \mathcal{I}(r,m)} a_I \, dy^I\right) = \sum_{I \in \mathcal{I}(r,m)} (a_I \circ f) \, df^{i_1} \wedge \cdots \wedge df^{i_r}, \tag{5.29}$$

where we are writing $I = (i_1, i_2, \ldots, i_r)$.

Example 5.5.2. Let $M = \mathbb{R}$ and let N^n be a differentiable manifold. Consider an immersion $\gamma : \mathbb{R} \to N$, which we can understand as a regular curve on N. Let ω be a 1-form on N such that over a coordinate neighborhood of N with coordinate $y = (y_1, y_2, \ldots, y_n)$, we write

$$\omega = \omega_1 dy^1 + \omega_2 dy^2 + \cdots + \omega_n dy^n.$$

Using t as the coordinate of $\mathbb{R}$, we write in coordinates

$$(\gamma^*\omega)_t = \omega_1(\gamma(t))\frac{d\gamma^1}{dt}dt + \omega_2(\gamma(t))\frac{d\gamma^2}{dt}dt + \cdots + \omega_n(\gamma(t))\frac{d\gamma^n}{dt}dt.$$

As we will see in the section, this pull back is related to calculating line integrals.

Example 5.5.3. Consider the unit sphere $\mathbb{S}^2$ and let (θ, φ) be the usual longitude-latitude coordinate patch. The typical embedding of $f : \mathbb{S}^2 \to \mathbb{R}^3$ corresponds to the functions

$$f(\theta, \varphi) = (\cos \theta \sin \varphi, \sin \theta \sin \varphi, \cos \varphi).$$

The dot product on $\mathbb{R}^3$ corresponds to a covariant tensor field of type 2, expressed in usual coordinates (x, y, z) by

$$\omega = dx \otimes dx + dy \otimes dy + dz \otimes dz.$$

With respect to the given coordinate systems, the pull-back of ω is

$$
\begin{aligned}
f^*\omega &= (-\sin\theta\sin\varphi\,d\theta + \cos\theta\cos\varphi\,d\varphi) \otimes (-\sin\theta\sin\varphi\,d\theta + \cos\theta\cos\varphi\,d\varphi) \\
&\quad + (\cos\theta\sin\varphi\,d\theta + \sin\theta\cos\varphi\,d\varphi) \otimes (\cos\theta\sin\varphi\,d\theta + \sin\theta\cos\varphi\,d\varphi) \\
&\quad + (-\sin\varphi\,d\varphi) \otimes (-\sin\varphi\,d\varphi) \\
&= (\sin^2\theta\sin^2\varphi + \cos^2\theta\sin^2\varphi)d\theta \otimes d\theta \\
&\quad + (-\sin\theta\cos\theta\sin\varphi\cos\varphi + \sin\theta\cos\theta\sin\varphi\cos\varphi)d\theta \otimes d\varphi \\
&\quad + (-\sin\theta\cos\theta\sin\varphi\cos\varphi + \sin\theta\cos\theta\sin\varphi\cos\varphi)d\varphi \otimes d\theta \\
&\quad + (\cos^2\theta\cos^2\varphi + \sin^2\theta\cos^2\varphi + \sin^2\varphi)d\varphi \otimes d\varphi \\
&= \sin^2\varphi\,d\theta \otimes d\theta + d\varphi \otimes d\varphi.
\end{aligned}
$$

As we will see, this is the standard metric tensor on the sphere with longitude-latitude coordinate system.

Example 5.5.4. As an another example, whose details we leave as an exercise (Problem 5.5.5), we deduce the following fundamental formula. Let M and N be smooth manifolds of the same dimension n, and f a smooth map between them. In reference to a coordinate chart (U, x) on M and a chart (V, y) on N, for all $p \in U$,

$$
f^*(dy^1 \wedge dy^2 \wedge \cdots \wedge dy^n)_p = (\det df_p)dx^1 \wedge dx^2 \wedge \cdots \wedge dx^n. \tag{5.30}
$$

We notice that if $M = N = \mathbb{R}^n$, then $\det df_p$ is the Jacobian of the function f at the point p.

The pull-back of covariant tensor fields satisfies a few properties. The proofs are straightforward so we leave them as exercises.

Proposition 5.5.5. *Let $f : M^m \to N^n$ be a smooth map between smooth manifolds. Let α and β be covariant tensor fields on N.*

1. *The pull-back $f^* : \Gamma(TN^{*\otimes s}) \to \Gamma(TM^{*\otimes s})$ is a linear function.*
2. *If $a : N \to \mathbb{R}$ is a smooth function, then $f^*(a\alpha) = (a \circ f)f^*\alpha$.*
3. *$f^*(\alpha \otimes \beta) = f^*(\alpha) \otimes f^*(\beta)$.*
4. *$\mathrm{id}_N^*(\alpha) = \alpha$.*

Proof. (Left as an exercise for the reader. See Exercise 5.5.2.) □

The pull-back of r-forms satisfies a few more properties.

Proposition 5.5.6. *Let $f : M^m \to N^n$ be a smooth map between smooth manifolds. The following hold for all $r \leq \min(m, n)$:*

1. *Considering $\Omega^s(N)$ as a subspace of $\Gamma(TN^{*\otimes s})$, then $f^*(\Omega^s(N)) \subseteq \Omega^s(M)$.*
2. *For all $\omega \in \Omega^r(N)$ and $\eta \in \Omega^s(N)$, $f^*(\omega \wedge \eta) = (f^*\omega) \wedge (f^*\eta)$.*
3. *For all $\omega \in \Omega^r(N)$ with $r < \min(m, n)$, $f^*(d\omega) = d(f^*\omega)$.*

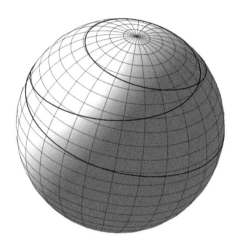

Figure 5.5: The curve on $\mathbb{S}^2$ in Example 5.5.8.

Proof. Part 1 follows immediately from the functional definition in Equation (5.26). Part 2 is an easy application of Equation (5.29). Finally, for part 3, note that $d(df^{i_1} \wedge \cdots \wedge df^{i_r}) = 0$ by a repeated use of Proposition 5.4.6(2) and the fact that $d(df^i) = 0$. Then if $\omega = \sum_I a_I dx^I$, Equation (5.29) gives

$$
\begin{aligned}
d(f^*\omega) &= \sum_{I \in \mathcal{I}(r,m)} d\big((a_I \circ f)\, df^{i_1} \wedge \cdots \wedge df^{i_r}\big) \\
&= \sum_{I \in \mathcal{I}(r,m)} d(a_I \circ f) \wedge df^{i_1} \wedge \cdots \wedge df^{i_r} + (a_I \circ f)d(df^{i_1} \wedge \cdots \wedge df^{i_r}) \\
&= \sum_{I \in \mathcal{I}(r,m)} d(a_I \circ f) \wedge df^{i_1} \wedge \cdots \wedge df^{i_r} = \sum_{I \in \mathcal{I}(r,m)} d(a_I \circ f) \wedge f^*(dx^I) \\
&= f^*(d\omega). \qquad \qquad \square
\end{aligned}
$$

Proposition 5.5.7. *Let $f : M \to N$ and $g : U \to M$ be smooth functions between smooth manifolds. Then $(f \circ g)^* = g^* \circ f^*$.*

Proof. (Left as an exercise for the reader.) $\qquad \square$

Example 5.5.8. As a slightly more involved example, we revisit Example 5.4.3, where $M = \mathbb{R}$ and $N = \mathbb{S}^2$. Let V be the coordinate neighborhood on $\mathbb{S}^2$ with a system of coordinates y defined by the parametrization $x^{-1}(u, v) = (\cos u \sin v, \sin u \sin v, \cos v)$ defined on $(0, 2\pi) \times (0, \pi)$. We consider the 1-form

$$
\omega = (\sin^2 v)\, du + (\sin v \cos v)\, dv.
$$

Consider the function $f : \mathbb{R} \to \mathbb{S}^2$ defined in coordinates by $(u, v) = f(t) = (3t, 1 + \frac{1}{2}\sin t)$. The image of this immersion is depicted in Figure 5.5. Defined in this way, we see that $f^1(t) = 3t$ and $f^2(t) = 1 + \frac{1}{2}\sin(t)$. Then

$$(f^*\omega)_t = \sin^2\left(1 + \frac{1}{2}\sin(t)\right) 3\,dt$$

$$+ \sin\left(1 + \frac{1}{2}\sin(t)\right)\cos\left(1 + \frac{1}{2}\sin(t)\right)\left(\frac{1}{2}\cos t\right)dt.$$

PROBLEMS

5.5.1. Prove that (5.29) follows from (5.27).

5.5.2. Prove Proposition 5.5.5.

5.5.3. let $f : M \to N$ be a differentiable map of manifolds. Prove that if α is a global section $\operatorname{Sym}^k(TN^*)$, then $f^*\alpha$ is a global section of $\operatorname{Sym}^k(TM^*)$.

5.5.4. Prove Proposition 5.5.7.

5.5.5. Prove the formula mentioned in Example 5.5.4.

5.5.6. Prove the product rule for the Lie derivative of the product between a function and a covariant tensor field: Let M be a smooth manifold, let $f \in C^1(M, \mathbb{R})$, let X be a differentiable vector field on M, and let α be a differentiable covariant tensor field on M. Prove that $\mathcal{L}_X(f\alpha) = (\mathcal{L}_X f)\alpha + f(\mathcal{L}_X\alpha)$. [Hint: Use a coordinate-dependent approach.]

5.5.7. This exercises generalizes Example 5.5.3. Let S be a parametrized surface in $\mathbb{R}^3$, which we can think of as an immersion of 2-manifold in $\mathbb{R}^3$. Let (u, v) be a coordinate patch of S and suppose that the immersion of S in $\mathbb{R}^3$ is given by a function $F(u, v)$. Let

$$\omega = dx \otimes dx + dy \otimes dy + dz \otimes dz$$

be the usual dot product on (the tangent spaces of) $\mathbb{R}^3$. Prove that

$$F^*\omega = (F_u \cdot F_u)du \otimes du + (F_u \cdot F_v)du \otimes dv + (F_v \cdot F_u)dv \otimes du + (F_v \cdot F_v)dv \otimes dv,$$

where by $F_u \cdot F_u$, we mean the dot product of the vector F_u with itself, and so on.

5.5.8. Let $M = \mathbb{RP}^2$ be the manifold of the real projective plane. (Recall Example 3.1.6.) We use the homogeneous coordinates $(x : y : z)$ with $(x, y, z) \neq 0$ to locate points in $\mathbb{RP}^2$. Define the three open sets $U_1 = \{(x : y : z) \in \mathbb{RP}^2 \mid x \neq 0\}$ and similarly U_2 and U_3 where $y \neq 0$ and $z \neq 0$ respectively. We define the coordinate maps $\phi_1 : U_1 \to \mathbb{R}^2$ as $\phi_1(x : y : z) = (y/x, z/x)$, and similarly for ϕ_2 and ϕ_3. Use coordinates (u, v) for the (U_3, ϕ_3) chart and (r, s) for the (U_2, ϕ_2) chart.

(a) Setting $(u, v) = \phi_{32}(r, s)$, show that $\phi_{32}(r, s) = (r/s, 1/s)$ and determine $d\phi_{32}$.

(b) Consider the function $f : \mathbb{RP}^2 \to \mathbb{R}$ defined by $f(x : y : z) = 3yz/(x^2 + 2y^2 + z^2)$. Show that this function is well-defined on all of $\mathbb{RP}^2$.

(c) Show that over U_3 we have

$$(\phi_3^{-1})^*(df) = \frac{6uv}{(u^2 + 2v^2 + 1)^2}du + \frac{3(u^2 - 2v^2 + 1)}{(u^2 + 2v^2 + 1)^2}dv.$$

(d) Determine the expression of df in the U_2 coordinate chart, namely determine $(\phi_2^{-1})^*(df)$, and then show directly that

$$(\phi_3^{-1})^*(df) = \phi_{32}^*((\phi_2^{-1})^*(df)).$$

5.6 Lie Derivative of Tensor Fields

As promised in earlier sections, we want to develop the notion of a type of derivative on tensor fields. Now that we have the notion of a pull-back of a covariant tensor field at our disposal, we can extend Definition 5.3.5 of the Lie derivative to tensors of any type. We start with Lie derivatives of covariant tensor fields.

Definition 5.6.1. Let M be a C^2-manifold and let $X \in \mathfrak{X}(M)$ be a differentiable vector field on M and let φ_t be the flow of X on M. If $\alpha \in \Gamma(TM^{*\otimes s})$ is a smooth covariant tensor field on M, we define the Lie derivative of α by X as the covector field given by

$$(\mathcal{L}_X \alpha)_p \overset{\text{def}}{=} \frac{d}{dt}(\varphi_t)^*(\alpha)_p\big|_{t=0} = \lim_{h \to 0} \frac{1}{h}((\varphi_h^*)_p(\alpha_{\varphi_h(p)}) - \alpha_p). \tag{5.31}$$

We emphasize that for all h near 0, the difference $(\varphi_h^*)_p(\alpha_{\varphi_h(p)}) - \alpha_p$ is a difference of elements in $T_p M^{*\otimes s}$ so it makes sense. This is a coordinate-free description of the Lie derivative.

Proposition 5.6.2. *Let (U, x) be a coordinate chart on a manifold M, let X be a differentiable vector field on M and let α be a covariant tensor field of type $(0, s)$. Suppose that with respect to the coordinate chart (U, x), the components of X are X^i and the components of α are $\alpha_{j_1 j_2 \cdots j_s}$. Then the components of the Lie derivative of α are given by*

$$(\mathcal{L}_X \alpha)_{j_1 j_2 \cdots j_s} = X^k \frac{\partial \alpha_{j_1 j_2 \cdots j_s}}{\partial x^k} + \frac{\partial X^k}{\partial x^{j_1}} \alpha_{k j_2 \cdots j_s}$$
$$+ \frac{\partial X^k}{\partial x^{j_2}} \alpha_{j_1 k \cdots j_s} + \cdots + \frac{\partial X^k}{\partial x^{j_s}} \alpha_{j_1 j_2 \cdots j_{s-1} k}.$$

Proof. Let $p \in U$ and let $v_1, v_2, \ldots, v_s$ be s arbitrary vectors in $T_p M$. Then

$$(\mathcal{L}_X \alpha)_p(v_1, v_2, \ldots, v_s) = \frac{d}{dt}(\varphi_t^* \alpha)_p(v_1, v_2, \ldots, v_s)\big|_{t=0}$$

$$= \frac{d}{dt} \alpha_{\varphi_t(p)}((d\varphi_t)_p(v_1), (d\varphi_t)_p(v_2), \ldots, (d\varphi_t)_p(v_s))\big|_{t=0}$$

$$= \frac{d}{dt}\left(\alpha_{j_1 j_2 \cdots j_s}(\varphi_t(p)) \frac{\partial \varphi_t^{j_1}}{\partial x^{\ell_1}} v_1^{\ell_1} \frac{\partial \varphi_t^{j_2}}{\partial x^{\ell_2}} v_1^{\ell_2} \cdots \frac{\partial \varphi_t^{j_s}}{\partial x^{\ell_s}} v_1^{\ell_s}\right)\bigg|_{t=0}.$$

This expression corresponds to s summations and on each term we involve a product rule with $s + 1$ functions in the parameter t. Using (5.19) and (5.20), the product rule gives

$$(\mathcal{L}_X \alpha)_p(v_1, v_2, \ldots, v_s)$$
$$= \left(\frac{\partial \alpha_{j_1 j_2 \cdots j_s}}{\partial x^k} \frac{d\varphi_t^k}{dt} \Big|_{t=0} \delta_{\ell_1}^{j_1} \cdots \delta_{\ell_s}^{j_s} + \alpha_{j_1 j_2 \cdots j_s}(\varphi_0(p)) \frac{\partial X^{j_1}}{\partial x^{\ell_1}} \delta_{\ell_2}^{j_2} \cdots \delta_{\ell_s}^{j_s} + \cdots \right.$$
$$\left. + \alpha_{j_1 j_2 \cdots j_s}(\varphi_0(p)) \delta_{\ell_1}^{j_1} \cdots \delta_{\ell_{s-1}}^{j_{s-1}} \frac{\partial X^{j_s}}{\partial x^{\ell_s}} \right) v_1^{\ell_1} v_2^{\ell_2} \cdots v_s^{\ell_s}.$$

After relabeling the indices of summation as necessary, we find that

$$(\mathcal{L}_X \alpha)_p(v_1, v_2, \ldots, v_s) = \left(\frac{\partial \alpha_{j_1 j_2 \cdots j_s}}{\partial x^k} X^k + \frac{\partial X^k}{\partial x^{j_1}} \alpha_{k j_2 \cdots j_s} + \cdots + \frac{\partial X^k}{\partial x^{j_s}} \alpha_{j_1 j_2 \cdots j_{s-1} k} \right) v_1^{j_1} v_2^{j_2} \cdots v_s^{j_s},$$

with all component functions evaluated at p. The proposition follows. $\qquad \square$

So far we have defined the Lie derivative on (a) functions on M, (b) vector fields on M, and (c) covariant tensor fields on M. The latter case includes covector fields and k-forms. Before we give a complete definition for the Lie derivative, we consider how the Lie derivative interacts with various operations on tensors or forms that we have introduced so far.

Problem 5.5.6 generalizes to contraction of a vector field with any covariant tensor field. This establishes the following proposition.

Proposition 5.6.3. *Let X be a vector field on M and let α be any covariant tensor field of rank $(0, s)$. Then if $Y_1, Y_2, \ldots, Y_s$ are s vector fields on M, then the Lie derivative of the contraction is*

$$\mathcal{L}_X(\alpha(Y_1, \ldots, Y_s)) = (\mathcal{L}_X \alpha)(Y_1, \ldots, Y_s) + \alpha(\mathcal{L}_X Y_1, Y_2, \ldots, Y_s) +$$
$$\cdots + \alpha(Y_1, Y_2, \ldots, \mathcal{L}_X Y_s).$$

Finally, let $f \in C^2(M, \mathbb{R})$ be a differentiable function on M. Then the differential df is a 1-form, i.e., a smooth covariant vector field.

Proposition 5.6.4. *For any differentiable vector field X on M, the operators $\mathcal{L}_X$ and d commute on $C^2(M, \mathbb{R})$. In other words*

$$\mathcal{L}_X(df) = d(\mathcal{L}_X f).$$

Proof. (Left as an exercise for the reader. See Problem 5.6.1.) $\qquad \square$

Before defining the Lie derivative for a general tensor field, we list a few results we have established so far. Let M be a differentiable manifold, let f be a differentiable function on M, let $X, Y, Z, Y_1, \ldots, Y_s$ be vector fields on M, and let α be a covariant tensor field of type $(0, s)$ on M.

1. Definition 5.3.5: $\mathcal{L}_X f = X(f)$.

2. Theorem 5.3.10: $\mathcal{L}_X Y = [X, Y]$.

3. Linearity. Proposition 5.3.9: $\mathcal{L}_X(Y + Z) = \mathcal{L}_X Y + \mathcal{L}_X Z$.

4. Product rule. Proposition 5.3.9: $\mathcal{L}_X(fY) = (\mathcal{L}_X f)Y + f(\mathcal{L}_X Y)$.

5. Contraction. Proposition 5.6.3:

$$\mathcal{L}_X(\alpha(Y_1, \ldots, Y_s)) = (\mathcal{L}_X \alpha)(Y_1, \ldots, Y_s)$$
$$+ \alpha(\mathcal{L}_X Y_1, \ldots, Y_s) + \cdots + \alpha(Y_1, \ldots, \mathcal{L}_X Y_s).$$

6. Proposition 5.6.4: $\mathcal{L}_X \circ d = d \circ \mathcal{L}_X$ on functions.

We can now present a definition of the Lie derivative on tensors of type (r, s) with $r \geq 2$ or with $r = 1$ and $s > 0$.

Definition 5.6.5. Let X be a differentiable vector field on a manifold M. For all pairs (r, s) of nonnegative integers, we define the Lie derivative $\mathcal{L}_X$ as the linear transformation on the vector space $\Gamma(TM^{\otimes r} \otimes TM^{*\otimes s})$ of tensor fields satisfying Definition 5.3.5 for vector fields, Defintion 5.31 for covector fields, as well as the product rule

$$\mathcal{L}_X(S \otimes T) = (\mathcal{L}_X S) \otimes T + S \otimes (\mathcal{L}_X T) \tag{5.32}$$

for any tensor fields S and T.

It is not hard to show that the full Definition 5.31 for any covariant tensor field satisfies the product rule (5.32) applied to tensor products of covariant tensor fields. By virtue of the properties already established, imposing this additional product rule allows us to define the Lie derivative on any tensor field.

We took a coordinate-free approach to defining the Lie derivative. This is essential to know that this construction has mathematical meaning. The following proposition gives the coordinate dependent description of the the Lie derivative. The proof of this proposition is left as an exercise. Furthermore, this proposition gives a coordinate dependent way to show that the Lie derivative of a tensor field of type (r, s) is again a tensor field of type (r, s). (See Problem 5.6.3.)

Proposition 5.6.6. *Let M be a smooth manifold and let $X \in \mathfrak{X}(M)$. Let $A \in \Gamma(TM^{\otimes r} \otimes TM^{*\otimes s})$ be a smooth tensor field of type (r, s). Suppose that over some coordinate chart (U, x) of M, the components of X are X^k and that the components of A are $A^{i_1, i_2, \ldots, i_r}_{j_1, j_2, \ldots, j_s}$. Then the components of $\mathcal{L}_X A$ are*

$$(\mathcal{L}_X A)^{i_1 i_2 \cdots i_r}_{j_1 j_2 \cdots j_s} = X^k \frac{\partial A^{i_1 i_2 \cdots i_r}_{j_1 j_2 \cdots j_s}}{\partial x^k} - \frac{\partial X^{i_1}}{\partial x^k} A^{k i_2 \cdots i_r}_{j_1 j_2 \cdots j_s} \cdots - \frac{\partial X^{i_r}}{\partial x^k} A^{i_1 i_2 \cdots i_{r-1} k}_{j_1 j_2 \cdots j_s}$$
$$+ \frac{\partial X^k}{\partial x^{j_1}} A^{i_1 i_2 \cdots i_r}_{k j_2 \cdots j_s} + \cdots + \frac{\partial X^k}{\partial x^{j_s}} A^{i_1 i_2 \cdots i_r}_{j_1 j_2 \cdots j_{s-1} k}. \tag{5.33}$$

Intuitively speaking, the Lie derivative of a tensor field generalizes the concept of a directional derivative in $\mathbb{R}^n$ to any manifold and applied to any tensor field.

We finish this section with the Cartan formula, also called the Cartan magic formula. The result is interesting in itself but the proof is interesting as well since it affords us the opportunity to use some of the more algebraic techniques presented in Section 4.7.

In Problem 5.4.16 we discussed the interior product of a vector field X with an r-form ω, written $i_X\omega$. This interior product is essentially the contraction of X with the first component of the r-form but we must remember that for an r-tuples of indices, $i_1 < i_2 < \cdots < i_r$,

$$dx^{i_1} \wedge \cdots \wedge dx^{i_r} = \mathsf{A}(dx^{i_1} \otimes \cdots \otimes dx^{i_r}).$$

Proposition 5.6.7 (Cartan Formula). *Let X be a differentiable vector field on a smooth manifold M. Then as operators $\Omega^\bullet(M) \to \Omega^\bullet(M)$,*

$$\mathcal{L}_X = d \circ i_X + i_X \circ d.$$

Before proving the Cartan formula, we point out one of the reasons this result might be surprising. The operators involved are shown in the following diagram.

$$
\begin{array}{ccc}
\Omega^s(M) & \xrightarrow{d} & \Omega^{s+1}(M) \\
{\scriptstyle i_X}\downarrow & & \downarrow{\scriptstyle i_X} \\
\Omega^{s-1}(M) & \xrightarrow[d]{} & \Omega^s(M)
\end{array}
$$

This diagram is not commutative, i.e., that generally $i_X \circ d$ is not equal to $d \circ i_X$. However, it is interesting to see the Lie derivative $\mathcal{L}_X$, decomposes into a part that goes through $\Omega^{s+1}(M)$ and another part that goes through $\Omega^{s-1}(M)$.

Proof of Cartan formula. By definition of the Lie product, since it obeys the Leibniz rule, $\mathcal{L}_X$ is a derivation on the algebra of differential forms $\Omega^\bullet(M)$.

Now let $\omega \in \Omega^r(M)$ and $\eta \in \Omega^s(M)$. Then using the result of Problem 5.4.16,

$$
\begin{aligned}
(d \circ & i_X + i_X \circ d)(\omega \wedge \eta) \\
&= d(i_X(\omega \wedge \eta)) + i_X(d(\omega \wedge \eta)) \\
&= d((i_X\omega) \wedge \eta + (-1)^r \omega \wedge (i_X\eta)) + i_X(d\omega \wedge \eta + (-1)^r \omega \wedge d\eta) \\
&= d(i_X\omega) \wedge \eta + (-1)^{r-1}(i_X\omega) \wedge (d\eta) + (-1)^r(d\omega) \wedge (i_X\eta) \\
&\quad + (-1)^{2r}\omega \wedge (d(i_X\eta)) + (i_X(d\omega)) \wedge \eta + (-1)^{r+1}(d\omega) \wedge (i_X\eta) \\
&\quad + (-1)^r(i_X\omega) \wedge (d\eta) + (-1)^{2r}\omega \wedge (i_X(d\eta)) \\
&= (d(i_X\omega) + i_X(d\omega)) \wedge \eta + \omega \wedge (d(i_X\eta) + i_X(d\eta)).
\end{aligned}
$$

Thus, the operation $d \circ i_X + i_X \circ d$ is a derivation on $\Omega^\bullet(M)$.

We prove the Cartan formula by using Proposition 4.7.12 and observing that over every coordinate chart (U, x) of M, as an algebra, $\Omega^\bullet(U)$ is generated by $\Omega^0(U) = C^\infty(U, \mathbb{R})$ and the 1-forms dx^i.

We first prove that $\mathcal{L}_X$ and $d \circ i_X + i_X \circ d$ are equal on $C^\infty(U, \mathbb{R})$. By definition, $\mathcal{L}_X f = X(f)$ for all $f \in C^\infty(M, \mathbb{R})$. With respect to a coordinate system, $X(f) = X^i \partial_i f$. On the other hand $i_X f = 0$ by definition so in coordinates

$$(i_X \circ d + d \circ i_X)(f) = i_X(df) = i_X\left(\partial_j f \, dx^j\right) = X^j \partial_j f.$$

This shows that $\mathcal{L}_X$ and $i_X \circ d + d \circ i_X$ agree on the set of differentiable functions. Considering the 1-forms dx^i, by Proposition 5.6.6, $\mathcal{L}_X(dx^i) = \partial X^i / \partial x^j \, dx^j$ and

$$(d \circ i_X + i_X \circ d)(dx^i) = d(i_X(dx^i)) = d(X^i) = \frac{\partial X^i}{\partial x^j} dx^j.$$

Thus, $\mathcal{L}_X$ and $i_X \circ d + d \circ i_X$ agree also on dx^i for all $i = 1, 2, \ldots, n$.

We have shown that for any coordinate chart U of M, the operations $\mathcal{L}_X$ and $i_X \circ d + d \circ i_X$ are derivations on $\Omega^\bullet(U)$ that agree on a generating set of $\Omega^\bullet(U)$. By Proposition 4.7.12, they are equal on $\Omega^\bullet(U)$. Since this is true for any coordinate chart, $\mathcal{L}_X = (i_X \circ d + d \circ i_X)$ on $\Omega^\bullet(M)$. $\qquad\square$

PROBLEMS

5.6.1. Prove Proposition 5.6.4. [Hint: Use a coordinate dependent approach.]

5.6.2. Prove Proposition 5.6.6.

5.6.3. Let X^i represent the components of a vector field on a manifold and let $A^{i_1 i_2 \cdots i_r}_{j_1 j_2 \cdots j_s}$ be the components of a tensor field of type (r, s). Following a similar coordinate-dependent approach as taken in Example 4.5.9, prove that the collection of functions defined on the right hand side of (5.33) form the components of a tensor field of type (r, s).

5.6.4. Let X and Y be vector fields and let T be any tensor field. Prove that $\mathcal{L}_{[X,Y]} T = \mathcal{L}_X \mathcal{L}_Y T - \mathcal{L}_Y \mathcal{L}_X T$. consequently, we can write as operators $\mathcal{L}_{[X,Y]} = \mathcal{L}_X \mathcal{L}_Y - \mathcal{L}_Y \mathcal{L}_X$.

5.7 Integration on Manifolds - Definition

The sections in the chapter so far discussed vector fields and tensor fields on manifolds, and two methods that provide a sort of derivative, namely the exterior differential on r-forms and the Lie derivative by a vector field. The remaining sections present the theory of integration on manifolds. This section develops the definition of integration, Section 5.8 presents calculations and applications with integration, and finally Section 5.9 discusses Stokes' Theorem.

The theory of integration on manifolds must generalize all types of integration introduced in the usual calculus sequence. This includes

- integration of a one-variable, real-valued function over an interval;

- integration of a multivariable, real-valued function over a domain in $\mathbb{R}^n$;

- line integrals of functions in $\mathbb{R}^n$;

- line integrals of vector fields in $\mathbb{R}^n$;

- surface integrals of a real-valued function defined over a closed and bounded region of a regular surface;

- surface integrals of vector fields in $\mathbb{R}^3$.

One of the beauties of differential forms is that they will allow for a single concise description that does generalize all of these types of integrals.

Readers may be aware of the difference between Riemannian integration, the theory introduced in the usual calculus sequence, and Lebesgue integration, which relies on measure theory. The theory developed here does not inherently depend on either of these theories of integration but could use either. The definitions for integration on a manifold use the fact that a manifold is locally diffeomorphic to an open subset in $\mathbb{R}^n$ and define an integral on a manifold in reference to integration on $\mathbb{R}^n$. Therefore, we can presuppose the use of either Riemannian integration, Lebesgue integration, or any other theory of integration of functions over $\mathbb{R}^n$.

5.7.1   Partitions of Unity

The basis for defining integration on a smooth manifold M^n relies on relating the integral on M to integration in $\mathbb{R}^n$. However, since a manifold is only locally homeomorphic to an open set in $\mathbb{R}^n$, one can only define directly integration on a manifold over a coordinate patch.

We begin this section by introducing a technical construction that makes it possible, even from just a theoretical perspective, to piece together the integrals of a function over the different coordinate patches of the manifold's atlas.

Definition 5.7.1. Let M be a manifold, and let $\mathcal{V} = \{V_\alpha\}_{\alpha \in I}$ be a collection of open sets that covers M. A *partition of unity subordinate to* $\mathcal{V}$ is a collection of continuous functions $\{\psi_\alpha : M \to \mathbb{R}\}_{\alpha \in I}$ that satisfy the following properties:

1. $0 \leq \psi_\alpha(x) \leq 1$ for all $\alpha \in I$ and all $x \in M$.

2. $\psi_\alpha(x)$ vanishes outside a compact subset of V_α.

3. For all $x \in M$, there exists only a finite number of $\alpha \in I$ such that $\psi_\alpha(x) \neq 0$.

4. $\sum_{\alpha \in I} \psi_\alpha(x) = 1$ for all $x \in M$.

The summation in the fourth condition always exists since, for all $x \in M$, it is only a finite sum by the third criterion. Therefore, we do not worry about issues of convergence in this definition. The terminology "partition of unity" comes from the fact that the collection of functions $\{\psi_\alpha\}$ add up to the constant function 1 on M.

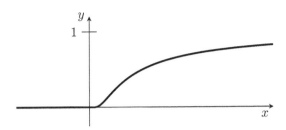

Figure 5.6: $f(x) = e^{-1/x}$ if $x > 0$ and 0 if $x \leq 0$.

Theorem 5.7.2 (Existence of Partitions of Unity). *Let M be a smooth manifold with atlas $\mathcal{A} = \{(U_\alpha, \phi_\alpha)\}_{\alpha \in I}$. There exists a smooth partition of unity of M subordinate to $\mathcal{A}$.*

For the sake of space, we forgo a complete proof of this theorem and refer the reader to [33, pp. 54–55], [49, Theorem 10.8], or [15, Section 14.1]. The proof relies on the existence of smooth real-valued functions that are nonzero in an open set $U \subset \mathbb{R}^n$ but identically 0 outside of U. Many of the common examples of partitions of unity depend on the following lemma.

Lemma 5.7.3. *The function $f : \mathbb{R} \to \mathbb{R}$ defined by*

$$f(x) = \begin{cases} 0, & \text{if } x \leq 0, \\ e^{-1/x}, & \text{if } x > 0, \end{cases}$$

is a smooth function. (See Figure 5.6.)

The proof for this lemma is an exercise in calculating higher derivatives and evaluating limits. Interestingly enough, this function at $x = 0$ is an example of a function that is smooth, i.e., has all its higher derivatives, but is not analytic, i.e., equal to its Taylor series over a neighborhood of $x = 0$.

The function $f(x)$ in Lemma 5.7.3 is useful because it passes smoothly from constant behavior to nonconstant behavior. This function $f(x)$ also leads immediately to functions with other desirable properties. For example, $f(x - a) + b$ is a smooth function that is constant and equal to b for $x \leq a$ and then nonconstant for $x > a$. In contrast, $f(a - x) + b$ is a smooth function that is constant and equal to b for $x \geq a$ and then nonconstant for $x < a$. More useful still for our purposes, if $a < b$, the function $g(x) = f(x - a)f(b - x)$ is smooth, identically equal to 0 for $x \notin (a, b)$, and is nonzero for $x \in (a, b)$. We can call this a *bump function* over (a, b) (see Figure 5.7). Also, the function

$$h(x) = \frac{f(b - x)}{f(x - a) + f(b - x)} \tag{5.34}$$

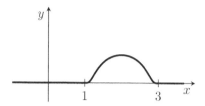

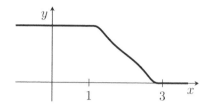

Figure 5.7: Bump function.   Figure 5.8: Cut-off function.

is smooth, is identically equal to 1 for $x \leq a$, identically equal to 0 for $x \geq b$, and strictly decreasing over (a, b). The function $h(x)$ is sometimes called a *cut-off function* (see Figure 5.8).

We will illustrate how to construct partitions of unity over a manifold with the following two simple examples.

Example 5.7.4. Consider the real line $\mathbb{R}$ as a 1-manifold, and consider the open cover $\mathcal{U} = \{U_i\}$, where $U_i = (i - 1, i + 1)$. In this open cover, we note that if n is an integer, then n is only contained in one set, U_n, and if t is not an integer, then t is contained in both $U_{\lfloor t \rfloor}$ and $U_{\lfloor t \rfloor + 1}$. Consider first the bump functions $g_i(x) = f(x - (i - 0.9))f((i + 0.9) - x)$ which has

$$g_i(x) = \begin{cases} 0, & \text{if } x \leq i - 0.9, \\ e^{1.8/(x-i+0.9)(x-i-0.9)}, & \text{if } i - 0.9 < x < i + 0.9, \\ 0, & \text{if } x \geq i + 0.9, \end{cases}$$

where we use the function f as defined in Lemma 5.7.3. It is not hard to show that these functions are smooth. Furthermore, by definition, $g_i(x) = 0$ for $x \notin [i - 0.9, i + 0.9] = K_i$, which is a compact subset of U_i. For any $i \in \mathbb{Z}$, the only functions that are not identically 0 on U_i are g_{i-1}, g_i, and g_{i+1}. Now define

$$\psi_i(x) = \frac{g_i(x)}{g_{i-1}(x) + g_i(x) + g_{i+1}(x)}.$$

We claim that the collection $\{\psi_i\}_{i \in \mathbb{Z}}$ forms a smooth partition of unity subordinate to $\mathcal{U}$. Again, $\psi_i(x) \neq 0$ for $x \in K_i$ and $\psi_i(x) = 0$ for $x \notin K_i$. Furthermore, the only functions ψ_k that are not identically 0 on U_i are ψ_{i-1}, ψ_i, and ψ_{i+1}. If $x = n$ is an integer, then

$$\sum_{i \in \mathbb{Z}} \psi_i(x) = \psi_n(n) = \frac{g_n(n)}{g_{n-1}(n) + g_n(n) + g_{n+1}(n)} = \frac{g_n(n)}{g_n(n)} = 1.$$

Figure 5.9: Example 5.7.5.

If instead x is not an integer, then when we set $n = \lfloor x \rfloor$, we have

$$\sum_{i \in \mathbb{Z}} \psi_i(x) = \psi_n(x) + \psi_{n+1}(x)$$

$$= \frac{g_n(x)}{g_{n-1}(x) + g_n(x) + g_{n+1}(x)} + \frac{g_{n+1}(x)}{g_n(x) + g_{n+1}(x) + g_{n+2}(x)}$$

$$= \frac{g_n(x)}{g_n(x) + g_{n+1}(x)} + \frac{g_{n+1}(x)}{g_n(x) + g_{n+1}(x)} = 1$$

since $g_{n-1}(x) = g_{n+2}(x) = 0$ for $x \in U_n \cap U_{n+1}$.

Example 5.7.5. Consider the unit sphere $\mathbb{S}^2$ given as a subset of $\mathbb{R}^3$. Cover $\mathbb{S}^2$ with two coordinate patches (U_1, x) and $(U_2, \bar{x})$, where the coordinate functions have the following inverses:

$$x^{-1}(u, v) = (\cos u \sin v, \sin u \sin v, \cos v)$$
$$\bar{x}^{-1}(\bar{u}, \bar{v}) = (- \cos \bar{u} \sin \bar{v}, - \cos \bar{v}, - \sin \bar{u} \sin \bar{v})$$

for $(u, v) \in (0, 2\pi) \times (0, \pi)$.

Define now the bump functions

$$g_1(u, v) = f(u - 0.1)f(6 - u)f(v - 0.1)f(3 - v),$$
$$g_2(\bar{u}, \bar{v}) = f(\bar{u} - 0.1)f(6 - \bar{u})f(\bar{v} - 0.1)f(3 - \bar{v}),$$

where f is the function in Lemma 5.7.3. These functions are smooth and vanish outside $[0.1, 6] \times [0.1, 3] = K$, which is a compact subset of $(0, 2\pi) \times (0, \pi)$. Define also the bump functions $h_i : \mathbb{S}^2 \to \mathbb{R}$ by

$$h_1(p) = \begin{cases} g_1 \circ x(p), & \text{if } p \in U_1, \\ 0, & \text{if } p \notin U_1, \end{cases} \quad \text{and} \quad h_2(p) = \begin{cases} g_2 \circ \bar{x}(p), & \text{if } p \in U_2, \\ 0, & \text{if } p \notin U_2. \end{cases}$$

By construction, these functions are smooth on $\mathbb{S}^2$ and vanish outside a compact subset of U_1 and U_2, namely, $x^{-1}(K)$ and $\bar{x}^{-1}(K)$ respectively. In Figure 5.9, the half-circles depict the complements of U_1 and U_2 on $\mathbb{S}^2$, and the piecewise-smooth curves that surround the semicircles show the boundary of $x^{-1}(K)$ and $\bar{x}^{-1}(K)$.

Finally, define the functions $\psi_i : \mathbb{S}^2 \to \mathbb{R}$ by

$$\psi_i(p) = \frac{h_i(p)}{h_1(p) + h_2(p)}.$$

These functions are well defined since h_1 and h_2 are nonzero on the interior of $x^{-1}(K)$ and $\bar{x}^{-1}(K)$, respectively, and these interiors cover $\mathbb{S}^2$. The pair of functions $\{\psi_1, \psi_2\}$ is a smooth partition of unity that is subordinate to the atlas that we defined on $\mathbb{S}^2$.

An object that recurs when dealing with partitions of unity is the set over which the function is nonzero. We make the following definition.

Definition 5.7.6. Let $f : M \to \mathbb{R}$ be a real-valued function from a manifold M. The *support* of f, written $\mathrm{Supp}\, f$, is defined as the closure of the non-zero set, i.e.

$$\mathrm{Supp}\, f = \overline{\{p \in M \mid f(p) \neq 0\}}.$$

A function is said to have *compact support* if $\mathrm{Supp}\, f$ is a compact set.

With this terminology, the second criterion concerning functions in a partition of unity $\{\psi_\alpha\}_{\alpha \in \mathcal{I}}$ subordinate to a given atlas is that each function has a support that is compact and in an open set of the atlas.

5.7.2   Integrating Differential Forms

We are now in a position to define integration of n-forms on a smooth n-dimensional manifold. We must begin by connecting integration of forms in $\mathbb{R}^n$ to usual integration.

Definition 5.7.7. Let ω be a differential n-form over $\mathbb{R}^n$. Let K be a compact subset of $\mathbb{R}^n$. If

$$\omega = f(x^1, \ldots, x^n)\, dx^1 \wedge \cdots \wedge dx^n,$$

then we define the integral

$$\int_K \omega \overset{\text{def}}{=} \int_K f(x^1, \ldots, x^n)\, dx^1\, dx^2 \cdots dx^n = \int_K f\, dV,$$

where the right-hand side represents the usual Riemann integral.

(As pointed out at the beginning of this section, we can also use the Lebesgue integral instead of the Riemann integral.) Also, if ω is a form that vanishes outside a compact set K, which is a subset of an open set U, then we define $\int_U \omega = \int_K \omega$.

In order to connect the integration on a manifold M^n to integration in $\mathbb{R}^n$, we must first show that this can be done independent of the coordinate system.

Lemma 5.7.8. *Let M^n be a smooth, oriented manifold with atlas $\mathcal{A} = \{(U_i, \phi_i)\}_{i \in I}$. Let K be a compact set with $K \subseteq U_1 \cap U_2$, and let ω be an n-form that vanishes outside of K. Setting $V_i = \phi_i(U_i)$ for $i = 1, 2$, then the following integrals are equal*

$$\int_{V_1} (\phi_1^{-1})^*(\omega) = \int_{V_2} (\phi_2^{-1})^*(\omega).$$

Proof. Using the standard notation for transition functions, write $V_{\alpha\beta} = \phi_\alpha(U_\alpha \cap U_\beta)$ and $\phi_{21} = \phi_2 \circ \phi_1^{-1}$, a homeomorphism from V_{12} to V_{21}. We use coordinates $(x^1, \ldots, x^n)$ on the chart (U_1, ϕ_1) and $(y^1, \ldots, y^n)$ on the patch (U_2, ϕ_2). We write

$$(\phi_1^{-1})^*(\omega) = f(x^1, \ldots, x^n) \, dx^1 \wedge \cdots \wedge dx^n$$
$$(\phi_2^{-1})^*(\omega) = \bar{f}(y^1, \ldots, y^n) \, dy^1 \wedge \cdots \wedge dy^n$$

as n-forms in $\mathbb{R}^n$, which by hypothesis are zero outside of V_{12} and V_{21} respectively. According to Definition 5.7.7,

$$\int_{V_{12}} (\phi_1^{-1})^*(\omega) = \int f(x^1, \ldots, x^n) \, dx^1 \cdots dx^n$$

$$\int_{V_{21}} (\phi_2^{-1})^*(\omega) = \int \bar{f}(y^1, \ldots, y^n) \, dy^1 \cdots dy^n$$

We note that $\phi_2^{-1} = \phi_1^{-1} \circ \phi_{12}$ so $(\phi_2^{-1})^* = \phi_{12}^* \circ (\phi_1^{-1})^*$. Hence

$$\bar{f}(y_1, \ldots, y^n) \, dy^1 \wedge \cdots \wedge dy^n = (\phi_2^{-1})^*(\omega) = \phi_{12}^* \left((\phi_1^{-1})^* \omega \right)$$
$$= \phi_{12}^*(f(x^1, \ldots, x^n) \, dx^1 \wedge \cdots \wedge dx^n)$$
$$= (f \circ \phi_{12})(y^1, \ldots, y^n)(\det d\phi_{12}) dy^1 \wedge \cdots \wedge dy^n,$$

where the last equality follows from (5.30). Since the manifold is oriented, $\det d\phi_{12} > 0$. Consequently, we have

$$\int_{V_{21}} (\phi_2^{-1})^*(\omega) = \int_{V_{21}} (f \circ \phi_{12})(y^1, \ldots, y^n)(\det d\phi_{12}) \, dy^1 \cdots dy^n$$
$$= \int_{V_{21}} (f \circ \phi_{12})(y^1, \ldots, y^n) |\det d\phi_{12}| \, dy^1 \cdots dy^n$$
$$= \int_{V_{12}} f(x^1, \ldots, x^n) \, dx^1 \cdots dx^n = \int_{V_{12}} (\phi_1^{-1})^*(\omega),$$

where the second to last equality holds by the usual substitution-of-variables formula for integration. $\qquad\square$

This lemma justifies the following definition in that it is independent of the choice of coordinate system.

Definition 5.7.9. Let M be an oriented, smooth n-dimensional manifold. Let ω be an n-form that vanishes outside of a compact subset K of M, and suppose that K is also a subset of a coordinate neighborhood (U, ϕ). Then we define the integral as

$$\int_M \omega = \int_{\phi(U)} (\phi^{-1})^*(\omega),$$

where the right-hand side is an integral over $\mathbb{R}^n$ given by Definition 5.7.7.

This definition explains how to integrate an n-form when it vanishes outside a compact subset of a coordinate patch. If this latter criterion does not hold, we use partitions of unity to piece together calculations that fall under Definition 5.7.9.

If a manifold M is not orientable, then for any atlas there will exist two coordinate charts ϕ_α and ϕ_β such that $\det(d(\phi_\beta \circ \phi_\alpha^{-1})) < 0$. From the proof of Lemma 5.7.8 integrating a form over the intersection of these two coordinate charts, with respect to one chart versus the other, will give a difference of signs. Then Definition 5.7.9 is not well-defined. Consequently, it is impossible to define integration over a non-orientable manifold. On the other hand, if M is non-orientable and U is an open subset of M, it may be possible that U is orientable. In this case, Definition 5.7.9 applies.

Definition 5.7.10. Let M^n be an oriented, smooth manifold, and let ω be an n-form that vanishes outside a compact set. Let $\{\psi_i\}_{i \in I}$ be a partition of unity subordinate to the atlas on M. Define

$$\int_M \omega = \sum_{i \in I} \int_M \psi_i \omega$$

where we calculate each summand on the right using Definition 5.7.9.

The summation only involves a finite number of nonzero terms since ω vanishes outside a compact set. The reader may wonder why we only consider forms that vanish outside of a compact subset of the manifold. This is similar to restricting one's attention to definite integrals in standard calculus courses. Otherwise, we face improper integrals and must discuss limits. As it is, many manifolds we consider are themselves compact; in the context of compact manifolds, the requirement that ω vanish outside a compact subset is superfluous.

The next proposition outlines some properties of integration of n-forms on n-dimensional manifolds that easily follow from properties of integration of functions in $\mathbb{R}^n$ as seen in ordinary calculus. However, we first give a lemma that restates the change-of-variables rule in integration over $\mathbb{R}^n$.

Lemma 5.7.11. *Let A and B be compact subsets of $\mathbb{R}^n$. Let $f : A \to B$ be a smooth map whose restriction to the interior A° is a diffeomorphism with the interior B°. Then on A°, f is either orientation-preserving or orientation-reversing on each connected component. Furthermore,*

$$\int_B \omega = \pm 1 \int_A f^* \omega,$$

where the sign is +1 (respectively, −1) if f is orientation-preserving (respectively, orientation-reversing) over A.

Proof. By the Inverse Function Theorem, f^{-1} is differentiable at a point $f(p)$ if and only if df_p is invertible and if and only if $\det df_p \neq 0$. Since each component function in the matrix of df_p is continuous, then $\det df_p$ is a continuous function from A to $\mathbb{R}$. By the Intermediate Value Theorem, $\det df_p$ does not change signs over any connected component of A°. Thus, f is orientation-preserving or orientation-reversing on each connected component of A°.

Let $(x^1, x^2, \cdots, x^n)$ be coordinates on $A \subset \mathbb{R}^n$ and $(y^1, y^2, \cdots, y^n)$ coordinates on $B \subset \mathbb{R}^n$. Then we can write $\omega = \alpha \, dy^1 \wedge \cdots \wedge dy^n$ for a smooth function $\alpha : \mathbb{R}^n \to \mathbb{R}$. By Problem 5.5.5,

$$f^*\omega = \alpha \circ f (\det df) \, dx^1 \wedge \cdots \wedge dx^n.$$

Furthermore, according to the change-of-variables formula for integration in $\mathbb{R}^n$ (see [55, Section 16.9, Equations (9) and (13)]) in the usual calculus notation, we have

$$\int_B \alpha \, dy^1 \, dy^2 \cdots dy^n = \int_A \alpha \circ f \, |\det df| \, dx^1 \, dx^2 \cdots dx^n.$$

Therefore, if f is orientation-preserving on A,

$$\int_B \omega = \int_B \alpha \, dy^1 \, dy^2 \cdots dy^n = \int_A \alpha \circ f \, |\det df| \, dx^1 \, dx^2 \cdots dx^n$$
$$= \int_A \alpha \circ f (\det df) \, dx^1 \, dx^2 \cdots dx^n = \int_A f^*\omega.$$

If f is orientation-reversing, the above reasoning simply changes by $|\det df| = -\det df$ and a -1 factors out of the integral. $\qquad\square$

Proposition 5.7.12 (Properties of Integration). *Let M and N be oriented, smooth manifolds with or without boundaries. Let ω and η be smooth forms that vanish outside of a compact set on M.*

1. *Linearity: For all $a, b \in \mathbb{R}$, $\displaystyle\int_M (a\omega + b\eta) = a \int_M \omega + b \int_M \eta$.*

2. *Orientation change: If we denote by $(-M)$ the manifold M but with the opposite orientation, then*

$$\int_{(-M)} \omega = -\int_M \omega.$$

3. *Substitution rule: If $g : N \to M$ is an orientation-preserving diffeomorphism,*

$$\int_M \omega = \int_N g^*\omega.$$

Proof. Part 1 is left as an exercise for the reader.

If M is an oriented manifold with atlas $\{(U_\alpha, \phi_\alpha)\}_{\alpha \in I}$, then equipping M with an opposite orientation means giving a different atlas $\{(V_\beta, \tilde{\phi}_\beta)\}_{\beta \in J}$ such that $\det d(\tilde{\phi}_\beta \circ \phi_\alpha^{-1}) < 0$ whenever $\tilde{\phi}_\beta \circ \phi_\alpha^{-1}$ is defined. Following the proof of Lemma 5.7.8, one can show from the reversal in orientation that

$$\int_{(-K)} \omega = -\int_K \omega$$

for any compact set K in any intersection $U_\alpha \cap V_\beta$. Hence, by using appropriate partitions of unity and piecing together the integral according to Definition 5.7.10, we deduce part 2 of the proposition.

To prove part 3, assume again that ω is compactly supported in just one coordinate chart (U, ϕ) of M. Otherwise, using a partition of unity, we can write ω as a finite sum of n-forms, each compactly supported in just one coordinate neighborhood. Without loss of generality, suppose that $g^{-1}(U)$ is a subset of a coordinate chart (V, ψ) on N. Saying that g is orientation-preserving means that $\det(\phi \circ g \circ \psi^{-1}) > 0$. Since $g^{-1}(U) \subset V$, then V contains the support of $g^*\omega$. Now, by applying Lemma 5.7.11 to the diffeomorphism $\phi \circ g \circ \psi^{-1}$, we have

$$\int_M \omega = \int_{\phi(U)} (\phi^{-1})^*\omega = \int_{\psi(V)} (\phi \circ g \circ \psi^{-1})^*(\phi^{-1})^*\omega = \int_{\psi(V)} (\phi^{-1} \circ \phi \circ g \circ \psi^{-1})^*\omega$$

$$= \int_{\psi(V)} (g \circ \psi^{-1})^*\omega = \int_{\psi(V)} (\psi^{-1})^*(g^*\omega) = \int_N g^*\omega. \qquad \square$$

In calculus we defined line integrals along piecewise-smooth curves or surface integrals on piecewise-smooth surfaces. Though we have not, to this point, defined piecewise-smooth manifolds, we can do so in a way that allows us to give a definition of the integral over a piecewise-smooth manifold.

Definition 5.7.13. A *piecewise-smooth* manifold M is a topological manifold that is the finite union of smooth manifolds $M_1, M_2, \ldots, M_k$ that intersect only on their boundaries. A piecewise-smooth manifold is *oriented* if each manifold M_i is oriented in such a way that if M_i and M_j intersect along a boundary component C, then the orientation induced on C from M_i is opposite the orientation induced from M_j.

Definition 5.7.14. Let M^n be a piecewise-smooth manifold as in Definition 5.7.13. Let ω be an n-form that is smooth on each piece M_i. Then we defined the integral

$$\int_M \omega = \int_{M_1} \omega + \cdots + \int_{M_k} \omega.$$

PROBLEMS

5.7.1. Prove that a function $f : \mathbb{R} \to \mathbb{R}$ that is identically 0 for $x \leq 0$ and positive for $x > 0$ cannot be analytic.

5.7.2. The manifold $\mathbb{RP}^3$ is orientable. Let $U_i = \{(x^1 : x^2 : x^3 : x^4) \in \mathbb{RP}^3 \,|\, x_i \neq 0\}$ be the coordinate open set as described in Example 3.1.6. Use $\mathcal{A} = \{(U_i, \phi_i)\}_{i=1}^4$ as the atlas for $\mathbb{RP}^3$. Define

$$f(x : y : z : w) = \begin{cases} e^{-w^2/(4w^2 - x^2 - y^2 - z^2)} & \text{if } x^2 + y^2 + z^2 < 4w^2 \\ 0 & \text{otherwise.} \end{cases}$$

Define also $h_1(x^1 : x^2 : x^3 : x^4) = f(x^2 : x^3 : x^4 : x^1)$, $h_2(x^1 : x^2 : x^3 : x^4) = (x^3 : x^4 : x^1 : x^2)$ and similarly for h_3 and h_4. Finally define

$$\psi_i(x^1 : x^2 : x^3 : x^4) = \frac{h_i(x^1 : x^2 : x^3 : x^4)}{\sum_{j=1}^n h_j(x^1 : x^2 : x^3 : x^4)}.$$

(a) Prove that f (and hence ψ_i for $i = 1, 2, 3, 4$) is a well-defined function on $\mathbb{RP}^3$.

(b) Prove that f is smooth.

(c) Prove that $\{\psi_1, \psi_2, \psi_3, \psi_4\}$ is a smooth partition of unity of $\mathbb{RP}^3$ subordinate to $\mathcal{A}$.

5.7.3. Prove that Proposition 5.7.12 holds for oriented piecewise-smooth manifolds.

5.8   Integration on Manifolds - Applications

The reader might have noticed the impracticality of n-forms on an n-dimensional manifold from the definition. By virtue of the structure of a manifold, simply to provide a consistent definition, we are compelled to use a formula similar to that presented in Definition 5.7.10. On the other hand, integrals involving terms such as $e^{-1/x}$ or bump functions as described in Example 5.7.5 are intractable to compute by hand.

The following useful proposition gives a method to calculate integrals of forms on a manifold using parametrizations while avoiding the use of an explicit partition of unity. The proposition breaks the calculation into integrals over compact subsets of $\mathbb{R}^n$, but we need to first comment on what types of compact sets we can allow. We will consider compact sets $C \subset \mathbb{R}^n$ whose boundary ∂C has "measure 0." By "measure 0," we mean $\int_{\partial C} 1 \, dV = 0$. More intuitively, we do not want C to be strange enough that its boundary ∂C has any n-volume.

Proposition 5.8.1. *Let M^m be a smooth, oriented manifold with or without boundary. Suppose that there exists a finite collection $\{C_i\}_{i=1}^k$ of compact subsets of $\mathbb{R}^m$, each with boundary ∂C_i of measure 0, along with a collection of smooth functions $F_i : C_i \to M$ such that: (1) each F_i is a diffeomorphism from the interior C_i° onto the interior $F_i(C_i)^\circ$ and (2) any pair $F_i(C_i)$ and $F_j(C_j)$ intersect only along their boundary. Then for any n-form ω on M, which has a compact support that is contained in $F_1(C_1) \cup \cdots \cup F_k(C_k)$,*

$$\int_M \omega = \sum_{i=1}^k \int_{C_i} F_i^* \omega. \tag{5.35}$$

Proof. We need the following remarks from set theory and topology. Recall that for any function $f : X \to Y$ and any subsets A, B of Y, we have $f^{-1}(A \cup B) = f^{-1}(A) \cup f^{-1}(B)$ and $f^{-1}(A \cap B) = f^{-1}(A) \cap f^{-1}(B)$. For general functions, the same equalities do not hold when one replaces f with f^{-1}. However, if f is bijective, the equality does hold in both directions.

Let $\mathcal{A} = \{(U_\alpha, \phi_\alpha)\}_{\alpha \in I}$ be the atlas given on M. Let K be the support of ω. Note that since each F_i is continuous, then $F_i(C_i)$ is compact.

Suppose first that K is a subset of a single coordinate chart (U_0, ϕ). Since $K \subset F_1(C_1) \cup \cdots \cup F_k(C_k)$,

$$K = K \cap \big(F_1(C_1) \cup \cdots \cup F_k(C_k) \big) = (F_1(C_1) \cap K) \cup \cdots \cup (F_k(C_k) \cap K),$$

and, again, because ϕ is a bijection,

$$\phi(K) = \big(\phi \circ F_1(C_1) \cap \phi(K) \big) \cup \cdots \cup \big(\phi \circ F_k(C_k) \cap K \big). \tag{5.36}$$

Since ϕ is a homeomorphism and since any pair $F_i(C_i)$ and $F_j(C_j)$ intersect only along their boundaries, then the same holds for any pair $K \cap F_i(C_i)$ and $K \cap F_j(C_j)$ and also for any pair $\phi(K) \cap (\phi \circ F_i)(C_i)$ and $\phi(K) \cap (\phi \circ F_j)(C_j)$.

By definition of integration of n-forms over a coordinate chart, i.e., Definition 5.7.9,

$$\int_M \omega = \int_{\phi(U)} (\phi^{-1})^* \omega = \int_{\phi(K)} (\phi^{-1})^* \omega.$$

By Equation (5.36) and the theorem on subdividing an integral by nonoverlapping regions in $\mathbb{R}^n$ (see [55, Section 16.3, Equation (9)] for the statement for integrals over $\mathbb{R}^2$),

$$\int_M \omega = \sum_{i=1}^k \int_{\phi(K) \cap (\phi \circ F_i)(C_i)} (\phi^{-1})^* \omega.$$

Note that this is precisely where we need to require that the C_i have boundaries of measure 0.

The setup for the proposition was specifically designed to apply Lemma 5.7.11 to the function $\phi \circ F_i : C_i \to (\phi \circ F_i)(C_i)$ for each $i \in \{1, \ldots, k\}$. We have

$$\int_{\phi(K) \cap \phi \circ F_i(C_i)} (\phi^{-1})^* \omega = \int_{F_i^{-1}(K) \cap C_i} (\phi \circ F_i)^* (\phi^{-1})^* \omega$$

$$= \int_{F_i^{-1}(K) \cap C_i} F_i^* \omega = \int_{C_i} F_i^* \omega,$$

and the proposition follows for when K is a subset of a single coordinate chart.

If K is not a subset of a single coordinate chart, we use a partition of unity subordinate to the atlas of M. In this case, the proposition again follows, using Proposition 5.5.6(2), so that for each partition-of-unity function ψ_j, we have

$$F_i^*(\psi_j \omega) = (\psi_j \circ F_i) F_i^* \omega.$$

$\square$

With this proposition at our disposal, we are finally in a position to present some examples of integration of n-forms on a smooth n-manifold.

Example 5.8.2. Consider the 2-torus $M = \mathbb{T}^2 = \mathbb{S}^1 \times \mathbb{S}^1$. We choose an atlas on M so that one of the coordinate functions is $\phi : M \to (0, 2\pi)^2$ corresponding to pairs of angles going around each $\mathbb{S}^1$. It should be clear by now that in order to express an n-form explicitly, we need a coordinate-depend description. Let ω be a 2-form on M such that

$$(\phi^{-1})^*(\omega) = (3 + \cos v)^2 \cos v \, du \wedge dv.$$

Then from Proposition 5.8.1 we calculate that

$$\int_M \omega = \int_0^{2\pi} \int_0^{2\pi} (3 + \cos v)^2 \cos v \, du \, dv$$

$$= 2\pi \int_0^{2\pi} 3 \cos v + 6 \cos^2 v + 9 \cos^3 v \, dv = 12\pi^2.$$

This example illustrates a special case of a particular situation. We often think of the torus as an embedded submanifold of $\mathbb{R}^3$. This motivates the following definition.

Definition 5.8.3 (Integration on Submanifolds). If M is an immersed submanifold of dimension m with the immersion $f : M^m \to N^n$ and if $\omega \in \Omega^m(N)$, then we define

$$\int_{f(M)} \omega = \int_M f^* \omega. \tag{5.37}$$

This definition applies in particular to embedded submanifolds.

Example 5.8.4. We revisit Example 5.8.2 to show how it relates to Definition 5.8.3. Suppose that we embed the torus in $\mathbb{R}^3$ using the parametrization

$$F(u, v) = \big((3 + \cos v) \cos u, (3 + \cos v) \sin u, \sin v\big) \qquad \text{for } (u, v) \in [0, 2\pi]^2.$$

Notice that this parametrization is described by $F = f \circ \phi^{-1}$, where ϕ is the coordinate chart described in Example 5.8.2 and $f : M \to \mathbb{R}^3$ is the actual embedding function of the torus into $\mathbb{R}^3$.

Consider the 2-form on $\mathbb{R}^3$ defined by $\eta = -y \, dx \wedge dz + x \, dy \wedge dz$. We calculate that

$$F^*(dx \wedge dz) = d((3 + \cos v) \cos u) \wedge d(\sin v)$$
$$= -\sin u \cos v (3 + \cos v) \, du \wedge dv,$$
$$F^*(dy \wedge dz) = d((3 + \cos v) \sin u) \wedge d(\sin v)$$
$$= \cos u \cos v (3 + \cos v) \, du \wedge dv.$$

Thus,

$$F^*\eta = -(3 + \cos v)\sin u \, F^*(dx \wedge dz) + (3 + \cos v)\cos u \, F^*(dy \wedge dz)$$
$$= (3 + \cos v)^2 \cos v \, du \wedge dv.$$

So $F^*\eta = (\phi^{-1})^*(\omega)$ from the previous example. Since $F^* = (\phi^{-1})^* \circ f^*$, we see that the form ω chosen in Example 5.8.2 is $f^*\eta$. So by Definition 5.8.3, we connect these integrations by

$$\int_{f(M)} \eta = \int_M f^*\eta = \int_M \omega = \int_{[0,2\pi]^2} (\phi^{-1})^*(\omega)$$
$$= \int_{[0,2\pi]^2} (3 + \cos v)^2 \cos v \, du \, dv = 12\pi^2,$$

which we calculated in Example 5.8.2.

Example 5.8.5. As another example, consider the unit sphere $\mathbb{S}^2$ in $\mathbb{R}^3$ covered by the six coordinate patches described in Example 3.1.5. Adjusting notation to $F_1 = \vec{X}_{(1)}$ and $F_2 = \vec{X}_{(2)}$, we observe that if we use the compact set $C_1 = C_2$ as the closed unit disk $\{(u, v) \mid u^2 + v^2 \leq 1\}$, then the sphere can be covered by $F_1(C_1)$ and $F_2(C_2)$. Thus, we have $k = 2$ in the setup of Proposition 5.8.1.

Consider the 2-form $\omega = xz^3 \, dy \wedge dz$ on $\mathbb{S}^2$, with the x, y, z representing the coordinates in $\mathbb{R}^3$. We have

$$F_1(u, v) = (u, v, \sqrt{1 - u^2 - v^2}) \quad \text{and} \quad F_2(u, v) = (u, v, -\sqrt{1 - u^2 - v^2}),$$

so we calculate that

$$F_1^*\omega = u(1 - u^2 - v^2)^{3/2} \, dv \wedge \left(-\frac{u}{\sqrt{1 - u^2 - v^2}} \, du - \frac{v}{\sqrt{1 - u^2 - v^2}} \, dv\right)$$
$$= u^2(1 - u^2 - v^2) \, du \wedge dv,$$

and similarly, $F_2^*\omega = u^2(1 - u^2 - v^2) \, du \wedge dv$. Then by Proposition 5.8.1,

$$\int_{\mathbb{S}^2} \omega = \int_{C_1} F_1^*\omega + \int_{C_2} F_2^*\omega = 2 \int_{C_1} u^2(1 - u^2 - v^2) \, du \, dv.$$

Putting this in polar coordinates, we get

$$\int_{\mathbb{S}^2} \omega = 2 \int_0^{2\pi} \int_0^1 r^2 \cos^2\theta (1 - r^2) r \, dr \, d\theta$$
$$2 \left(\int_0^{2\pi} \cos^2\theta \, d\theta\right) \left(\int_0^1 r^3 - r^5 \, dr\right) = \frac{\pi}{6}.$$

An important case of integration on submanifolds is the line integral over a curve.

Definition 5.8.6. Let $\gamma : [a, b] \to M$ be a smooth curve, and let ω be a 1-form on M. We define the *line integral* of ω over γ as

$$\int_\gamma \omega = \int_{[a,b]} \gamma^* \omega.$$

In addition, if γ is a piecewise-smooth curve, we define

$$\int_\gamma \omega = \sum_{i=1}^k \int_{[c_{i-1}, c_i]} \gamma^* \omega,$$

where $[c_{i-1}, c_i]$, with $i = 1, \ldots, k$, are the smooth arcs of γ.

At the beginning of this section, we proposed to find a definition of integration that generalizes many common notions from standard calculus. We explain now how the above two definitions generalize the concepts of line integrals in $\mathbb{R}^n$ and integrals of vector fields over surfaces.

Consider first the situation of line integrals in $\mathbb{R}^3$. (The case for $\mathbb{R}^n$ is identical in form.) In vector calculus (see [55, Definition 17.2.13]), one considers a continuous vector field $\vec{F} : \mathbb{R}^3 \to \mathbb{R}^3$ defined over a smooth curve $\vec{\gamma} : [a, b] \to \mathbb{R}^3$. Then the line integral is defined as

$$\int_{\vec{\gamma}} \vec{F} \cdot d\vec{r} = \int_a^b \vec{F}(\vec{\gamma}(t)) \cdot \vec{\gamma}'(t) \, dt.$$

To connect the classical line integral to the line integral in our present formulation, set $\omega = F_1 \, dx + F_2 \, dy + F_3 \, dz$, where $\vec{F} = (F_1, F_2, F_3)$. If we write $\gamma(t) = \vec{\gamma}(t) = (\gamma^1(t), \gamma^2(t), \gamma^3(t))$, then

$$\gamma^* \omega = F_1(\gamma(t)) d(\gamma^1) + F_2(\gamma(t)) d(\gamma^2) + F_3(\gamma(t)) d(\gamma^3)$$
$$= \Big(F_1(\gamma(t))(\gamma^1)'(t) + F_2(\gamma(t))(\gamma^2)'(t) + F_3(\gamma(t))(\gamma^3)'(t) \Big) dt$$
$$= \vec{F}(\vec{\gamma}(t)) \cdot \vec{\gamma}'(t) \, dt.$$

Thus, we have shown that the classical and modern line integrals are equal via

$$\int_\gamma \omega = \int_a^b \gamma^* \omega = \int_{\vec{\gamma}} \vec{F} \cdot d\vec{r}.$$

Second, consider the situation for surface integrals. In vector calculus (see [55, Definitions 17.7.8 and 17.7.9]), one considers a continuous vector field $\vec{F} : \mathbb{R}^3 \to \mathbb{R}^3$ defined over an oriented surface S parametrized by $\vec{r} : D \to \mathbb{R}^3$, where D is a compact region in $\mathbb{R}^2$. If (u, v) are the variables used in D, then

$$\iint_S \vec{F} \cdot d\vec{S} = \iint_D \vec{F}(\vec{r}(u, v)) \cdot (\vec{r}_u \times \vec{r}_v) \, dA.$$

To demonstrate the connection with the modern formulation, if we write $\vec{F} = (F_1, F_2, F_3)$, then set

$$\omega = F_1 \eta^1 + F_2 \eta^2 + F_3 \eta^3,$$

where η^j are the 2-forms described in Equation (5.24). Set also $f(u, v) = \vec{r}(u, v)$, and write $f = (f^1, f^2, f^3)$ as component functions in $\mathbb{R}^3$. Then

$$f^*\omega = F_1(f(u, v))f^*\eta^1 + F_2(f(u, v))f^*\eta^2 + F_3(f(u, v))f^*\eta^3. \tag{5.38}$$

We calculate $f^*\eta^1$ as

$$f^*(dx^2 \wedge dx^3) = df^2 \wedge df^3 = \left(\frac{\partial f^2}{\partial u} du + \frac{\partial f^2}{\partial v} dv\right) \wedge \left(\frac{\partial f^3}{\partial u} du + \frac{\partial f^3}{\partial v} dv\right)$$

$$= \left(\frac{\partial f^2}{\partial u}\frac{\partial f^3}{\partial v} - \frac{\partial f^2}{\partial v}\frac{\partial f^3}{\partial u}\right) du \wedge dv.$$

Repeating similar calculations for $f^*\eta^2$ and $f^*\eta^3$ and putting the results in Equation (5.38), we arrive at

$$f^*\omega = \vec{F}(\vec{r}(u, v)) \cdot (\vec{r}_u \times \vec{r}_v)\, du \wedge dv.$$

Using Definition 5.8.3 for the integration on a submanifold, we conclude that

$$\int_S \omega = \int_D f^*\omega = \iint_S \vec{F} \cdot d\vec{S},$$

thereby showing how integration of 2-forms on a submanifold gives the classical surface integral.

It is interesting to observe how the integration of forms on manifolds and on submanifolds of a manifold generalizes simultaneously many of the integrals that are studied in classic calculus, which are in turn studied for their applicability to science. However, the reader who has been checking off the list at the beginning of this section of types of integration we proposed to generalize might notice that until now we have not provided generalizations for path integrals $\int_C f\, ds$ or integrals of scalar functions over a surface $\int_S f\, dA$. The reason for this is that these integrals involve an arclength element ds or a surface area element dA. However, given a smooth manifold M without any additional structure, there is no way to discuss distances, areas, or n-volumes on M. Riemannian manifolds, which we introduce in the next chapter, provide a structure that allows us to make geometric calculations of length and volume. In that context, one can easily define generalizations of path integrals and integrals of scalar functions over a surface.

Before moving on to applications to physics, we mention a special case where the line integral is easy to compute.

Theorem 5.8.7 (Fundamental Theorem for Line Integrals). *Let M be a smooth manifold, let $f : M \to \mathbb{R}$ be a smooth function, and let $\gamma : [a, b] \to M$ be a piecewise-smooth curve on M. Then*

$$\int_\gamma df = f(\gamma(b)) - f(\gamma(a)).$$

Proof. By Proposition 5.5.6(3), $\gamma^*(df) = d(\gamma^* f)$, so we have

$$\gamma^*(df) = d(\gamma^* f) = d(f \circ \gamma) = (f \circ \gamma)'(t)\,dt,$$

where t is the variable on the manifold $[a, b]$. Thus,

$$\int_\gamma df = \int_{[a,b]} \gamma^*(df) = \int_a^b (f \circ \gamma)'(t)\,dt = f(\gamma(b)) - f(\gamma(a)).$$

$\square$

5.8.1 Conservative Vector Fields

We now wish to look at a central topic from elementary physics through the lens of our theory of integration on a manifold.

In elementary physics, one of the first areas studied is the dynamics of a particle under the action of a force. We remind the reader of some basic facts from physics. Suppose a particle of constant mass m is acted upon by a force $\vec{F}$ (which may depend on time and space) and follows a trajectory parametrized by $\vec{r}(t)$. Writing $\vec{v} = \vec{r}'$ for the velocity and $v = \|\vec{v}\|$ for the speed of the particle, we define the kinetic energy by $T = \frac{1}{2}mv^2$. Furthermore, since m is constant for a particle, according to Newton's law of motion, $\vec{F} = m\vec{v}'$. Finally, as the particle travels for $t_1 \le t \le t_2$, we define the *work* done by $\vec{F}$ as the line integral

$$W = \int_{t_1}^{t_2} \vec{F} \cdot d\vec{r}.$$

The kinetic energy depends on time and we have

$$\frac{dT}{dt} = \frac{d}{dt}\left(\frac{1}{2}m\vec{v} \cdot \vec{v}\right) = m\frac{d\vec{v}}{dt} \cdot \vec{v} = \vec{F} \cdot \vec{v}. \tag{5.39}$$

Thus, as a particle moves along $\vec{r}(t)$ for $t_1 \le t \le t_2$, the change in kinetic energy is

$$T_2 - T_1 = T(t_2) - T(t_1) = \int_{t_1}^{t_2} \vec{F} \cdot \vec{v}\,dt = \int_\gamma \vec{F} \cdot d\vec{r} = W, \tag{5.40}$$

where the last integral is a line integral over γ, the curve traced out by the trajectory $\vec{r}(t)$ of the particle. Thus, the change in kinetic energy is equal to the work done by the external forces. This result is often called the *Energy Theorem*. A force is called *conservative* if it does not depend on time and, if, as a particle travels over any closed, piecewise, smooth curve the kinetic energy does not change.

Though in physics one simply speaks of vectors or vector fields, from the perspective of manifold theory, certain objects may be vectors or covectors, depending on their use or their transformational properties under coordinate systems. Some objects viewed as vector fields in classical physics should even be understood as a

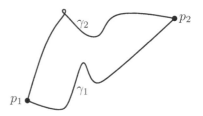

Figure 5.10: Two paths between p_1 and p_2.

2-form; this possible confusion arises from the fact that over a smooth manifold 3-manifold, for each $p \in M$, $\bigwedge^1 T_p M^*$, $\bigwedge^2 T_p M^*$, and $T_p M$ are all isomorphic as vector spaces.

Because of how it appears in the Energy Theorem (5.40), a force field should be viewed as a 1-form ω defined in $\mathbb{R}^3$. Then the Energy Theorem for the trajectory of a particle can be written as

$$\int_\gamma \omega = T(\gamma(t_2)) - T(\gamma(t_1)).$$

At any given point p along the trajectory of the particle, the velocity of the particle is a tangent vector $v \in T_p \mathbb{R}^3$. Then the instantaneous change of energy in (5.39) is simply the contraction $\omega(v)_p$.

Definition 5.8.8. A 1-form (covector field) on a smooth manifold M is called *conservative* if

$$\int_\gamma \omega = 0$$

for all closed, piecewise-smooth curves γ on M.

This definition has a different and perhaps more useful characterization. If γ_1 and γ_2 are two piecewise-smooth paths from points p_1 to p_2, then the path $\gamma_1 \bullet (-\gamma_2)$ defined by first traveling from p_1 to p_2 along γ_1 and then traveling backwards from p_2 to p_1 along γ_2 is a closed, piecewise-smooth curve (see Figure 5.10). It is not hard to show that for any 1-form ω,

$$\int_{\gamma_1 \bullet (-\gamma_2)} \omega = \int_{\gamma_1} \omega + \int_{(-\gamma_2)} \omega = \int_{\gamma_1} \omega - \int_{\gamma_2} \omega.$$

Hence, a covector field ω is conservative if and only if the integral of ω between any two points p_1 and p_2 is independent of the path between them.

A smooth 1-form has another alternative characterization, whose proof we leave as an exercise for the reader.

Theorem 5.8.9. *Let M be a smooth, oriented manifold. A 1-form $\omega \in \Omega^1(M)$ is conservative if and only if ω is exact.*

Returning to physics in Euclidean $\mathbb{R}^3$, according to the Energy Theorem from Equation (5.40), a force is conservative if and only if the work done over a piecewise-smooth path between any two points p_1 and p_2 is independent of the path chosen. Thus, if $\vec{F}$ is conservative, one defines the *potential energy* by

$$V(x, y, z) = -\int_{(x_0, y_0, z_0)}^{(x, y, z)} \vec{F} \cdot d\vec{r}.$$

where (x_0, y_0, z_0) is any fixed point. Obviously, the potential energy of $\vec{F}$ is a function that is well-defined only up to a constant that corresponds to the selected origin point (x_0, y_0, z_0). It is easy to check that

$$\vec{F} = -\nabla V.$$

For a conservative force $\vec{F}$ with potential energy V, the work of $\vec{F}$ as the particle travels along $\vec{r}(t)$ for $t \in [t_1, t_2]$ is

$$W = \int_{t_1}^{t_2} \vec{F} \cdot d\vec{r} = -\int_{t_1}^{t_2} \nabla V \cdot d\vec{r} = -\big(V(\vec{r}(t_2)) - V(\vec{r}(t_1))\big) = -(V_2 - V_1).$$

Hence, the Energy Theorem can be rewritten as

$$T_1 + V_1 = T_2 + V_2.$$

The sum $T + V$ of kinetic and potential energy is often referred to simply as the energy or total energy of a particle. This justifies the terminology "conservative": the total energy of a particle moving under the action of a conservative force is conserved along any path.

As further examples of applications of manifold theory to physics, Problems 5.8.9 through 5.8.11 discuss conservative properties and calculations of flux across surfaces for inverse square forces.

PROBLEMS

5.8.1. Let γ be the curve in $\mathbb{R}^4$ parametrized by $\gamma(t) = (1 + t^2, 2t - 1, t^3 - 4t, \frac{1}{t})$ for $t \in [1, 3]$. Let $\omega = x\, dy + (y^2 + z)\, dz + xw\, dw$. Calculate the line integral $\int_\gamma \omega$.

5.8.2. Calculate the line integral $\int_\gamma \omega$, where γ is the triangle in $\mathbb{R}^3$ with vertices $(0, 1, 2)$, $(1, 2, 4)$ and $(-3, 4, -2)$ and where ω is the 1-form given in $\mathbb{R}^3$ by

$$\omega = (2xy + 1)\, dx + 3x\, dy + yz\, dz.$$

5.8.3. Evaluate $\int_M \omega$, where M is the portion of paraboloid in $\mathbb{R}^3$ given by $z = 9 - x^2 - y^2$ above the xy-plane and where ω is the 2-form $\omega = y^2\, dx \wedge dy + z^2\, dx \wedge dz + 2\, dy \wedge dz$.

5.8.4. Let $\mathbb{T}^2$ be the torus embedded in $\mathbb{R}^4$ that is given by the equations $x^2 + y^2 = z^2 + w^2 = 1$. Note that the flat torus can be parametrized by

$$F(u, v) = (\cos u, \sin u, \cos v, \sin v)$$

for appropriate u and v. Compute the integral $\int_{\mathbb{T}^2} \omega$, where ω is the 2-form in $\mathbb{R}^4$ given by

(a) $\omega = x^3 \, dy \wedge dw$;

(b) $\omega = x^3 z \, dy \wedge dw$;

(c) $\omega = (x^2 yz + 1) \, dx \wedge dz + e^x yz \, dy \wedge dw$.

5.8.5. Consider the unit sphere $M = \mathbb{S}^2$ embedded in $\mathbb{R}^3$. Let

$$\omega = \frac{z^2 \, dx \wedge dy + x \, dx \wedge dz + xy \, dy \wedge dz}{x^2 + y^2 + z^2}$$

be a 2-form pulled back to $\mathbb{S}^2$. Calculate directly the integral $\int_M \omega$ using:

(a) the latitude-longitude parametrization;

(b) the stereographic parametrizations $\{\pi_N, \pi_S\}$ defined in Problem 3.2.5 and Example 3.7.3. [Hint: Use two coordinate patches.]

5.8.6. Consider the 3-torus described in Problem 5.2.4. Calculate $\int_M \omega$, where

(a) $\omega = \left(\cos^2 v + \dfrac{1 + \sin w}{2 + \cos u} \right) du \wedge dv \wedge dw$ given in local coordinates;

(b) $\omega = x^1 \, dx^1 \wedge dx^2 \wedge dx^3 + x^2 \, dx^2 \wedge dx^3 \wedge dx^4$ in coordinates in $\mathbb{R}^4$.

5.8.7. Let $\mathbb{T}^2 = \mathbb{S}^1 \times \mathbb{S}^1$ be the 2-torus where we use a pair of angles $(\theta, \varphi) \in (0, 2\pi)^2$ as one of the coordinate charts and complete it in the natural manner to cover the whole torus. Show that ω expressed as $3 \cos^2 \theta \sin \varphi \, d\theta + (2 + \cos^2 \varphi) \, d\varphi$ over the given coordinate chart extends to a 1-form over the whole torus. Consider the curve C on $\mathbb{T}^2$ given as a submanifold $\gamma : [0, 2\pi] \to \mathbb{T}^2$ expressed over this coordinate chart as $(\theta, \varphi) = \gamma(t) = (2t, 3t)$. Calculate the integral $\int_\gamma \omega$.

5.8.8. Prove part 1 of Proposition 5.7.12.

5.8.9. The force exerted by an electric charge placed at the origin on a charged particle is given by the force field $\vec{F}(\vec{r}) = K\vec{r}/\|\vec{r}\|^3$, where K is a constant and $\vec{r} = (x, y, z)$ is the position vector of the charged particle. Write this force field as the covector

$$\omega = \frac{Kx}{(x^2 + y^2 + z^2)^{3/2}} \, dx + \frac{Ky}{(x^2 + y^2 + z^2)^{3/2}} \, dy + \frac{Kz}{(x^2 + y^2 + z^2)^{3/2}} \, dz$$

over the manifold $\mathbb{R}^3$.

(a) Calculate the work exerted by the force on a charged particle that travels along the straight line from $(3, -1, 2)$ to $(4, 5, -1)$.

(b) Prove that $\vec{F}$ is a conservative force, i.e., that ω is a conservative covector field.

(c) Prove that $\omega = df$ where $f(x, y, z) = -K/r = -K/(x^2 + y^2 + z^2)^{1/2}$.

5.8.10. This exercise continues Problem 5.8.9. Consider the sphere of radius R and center 0 as an embedded submanifold $f : \mathbb{S}^2 \to \mathbb{R}^3$. Prove that

$$\int_{f(\mathbb{S}^2)} \star\omega = -4\pi K.$$

[Hint: Use the longitude-latitude coordinate system with $(u, v) \in [0, 2\pi] \times [0, \pi]$.]

5.8.11. Let $\mathbb{T}^2$ be the 2-torus embedded in $\mathbb{R}^3$, using the function $f : \mathbb{T}^2 \to \mathbb{R}^3$ given in Example 5.8.4. Show by direct calculation that

$$\int_{f(\mathbb{T}^2)} \star\omega = 0$$

where ω is the 2-form described in Problem 5.8.9.

5.8.12. Let M be a smooth, oriented manifold. Referring to Example 5.8.9, prove that a smooth (co)vector field ω on M is conservative if and only if ω is exact.

5.9 Stokes' Theorem

In the last section of this chapter, we present Stokes' Theorem, a central result in the theory of integration on manifolds.

In multivariable calculus, one encounters a theorem by the same name. What is called Stokes' Theorem for vector fields in $\mathbb{R}^3$ states that if S is an oriented, piecewise-smooth surface that is bounded by a simple, closed, piecewise-smooth curve C, then for any C^1 vector field $\vec{F}$ defined over an open region that contains S,

$$\int_C \vec{F} \cdot d\vec{r} = \int_S (\nabla \times \vec{F}) \cdot d\vec{S}.$$

(See [55, Section 17.8].)

It is a striking result that the generalization of this theorem to the context of manifolds simultaneously subsumes the Fundamental Theorem of Integral Calculus, Green's Theorem, the classical Stokes' Theorem, and the Divergence Theorem.

Before giving the theorem, we state a convention for what it means to integrate a 0-form on an oriented, zero-dimensional manifold. If N is an oriented zero-dimensional manifold, then $N = \{p_1, p_2, \ldots, p_c\}$ is a discrete set of points equipped with an association of signs $s_i = \pm 1$ for each $i = 1, \ldots, c$. Then by convention for any 0-form f (i.e., a function on N),

$$\int_N f = \sum_{i=1}^{c} s_i f(p_i). \tag{5.41}$$

Theorem 5.9.1 (Stokes' Theorem). *Let M^n be a piecewise-smooth, oriented manifold with or without boundary, and let ω be an $(n-1)$-form that is compactly supported on M. Equipping ∂M with the induced orientation, the following integrals are equal*

$$\int_M d\omega = \int_{\partial M} \omega, \tag{5.42}$$

where on the right side we take ω to mean the restriction of ω to ∂M. If $\partial M = \emptyset$, we understand the right side as 0.

Proof. We first treat the case where $n > 1$. Furthermore, we first prove Stokes' Theorem when M is a smooth manifold.

Suppose first that ω is compactly supported in a single coordinate chart (U, ϕ). Then by the definition of integration and by Proposition 5.5.6,

$$\int_M d\omega = \int_{\mathbb{R}^n_+} (\phi^{-1})^* d\omega = \int_{\mathbb{R}^n_+} d\big((\phi^{-1})^* \omega\big).$$

Using the $(n-1)$-forms η^j defined in Equation (5.24) as a basis for $\Omega^n(\mathbb{R}^n_+)$, write $(\phi^{-1})^* \omega = \sum_{j=1}^n \omega_j \eta^j$. Then, for the exterior differential, we have

$$d\big((\phi^{-1})^* \omega\big) = \sum_{j=1}^n \left(\sum_{i=1}^n \frac{\partial \omega_j}{\partial x^i} dx^i \right) \wedge \eta^j = \left(\sum_{i=1}^n \frac{\partial \omega_i}{\partial x^i} \right) dx^1 \wedge \cdots \wedge dx^n,$$

where the second equality follows from Equation (5.25).

Since ω is compactly supported in U, then for large enough R, the component functions $\omega_i(x^1, \ldots, x^n)$ vanish identically outside the parallelepiped

$$D_R = \underbrace{[-R, R] \times \cdots \times [-R, R]}_{n-1} \times [0, R].$$

Therefore, we remark that for all $i = 1, \ldots, n-1$, we have

$$\int_{-R}^R \frac{\partial \omega_i}{\partial x^i} dx^i = \big[\omega_i(x) \big]_{x^i = -R}^{x^i = R} = 0. \tag{5.43}$$

Consequently, we deduce that

$$\int_M d\omega = \int_0^R \int_{-R}^R \cdots \int_{-R}^R \left(\sum_{i=1}^n \frac{\partial \omega_i}{\partial x^i} \right) dx^1 \cdots dx^{n-1} dx^n$$

$$= \sum_{i=1}^n \int_0^R \int_{-R}^R \cdots \int_{-R}^R \frac{\partial \omega_i}{\partial x^i} dx^1 \cdots dx^n$$

$$= \sum_{i=1}^{n-1} \int_0^R \int_{-R}^R \cdots \int_{-R}^R \left(\int_{-R}^R \frac{\partial \omega_i}{\partial x^i} dx^i \right) dx^1 \cdots \widehat{dx^i} \cdots dx^n$$

$$+ \int_{-R}^R \cdots \int_{-R}^R \left(\int_0^R \frac{\partial \omega_n}{\partial x^n} dx^n \right) dx^1 \cdots dx^{n-1}$$

$$= \int_{-R}^R \cdots \int_{-R}^R [\omega_n(x)]_{x^n=0}^{x^n=R} dx^1 \cdots dx^{n-1} \qquad \text{by Equation (5.43)}$$

$$= -\int_{-R}^R \cdots \int_{-R}^R \omega_n(x^1, \ldots, x^{n-1}, 0) dx^1 \cdots dx^{n-1}, \qquad (5.44)$$

where the last equality holds because $\omega_n(x^1, \ldots, x^{n-1}, R) = 0$. Note that if the support of ω does not meet the boundary ∂M, then $\omega_n(x^1, \ldots, x^{n-1}, 0)$ is also identically 0 and $\int_M \omega = 0$.

To understand the right-hand side of Equation (5.42), let $i : \partial M \to M$ be the embedding of the boundary into M. The restriction of ω to ∂M is $i^*(\omega)$. Furthermore, in coordinates in (U, ϕ), $i^*(dx^k) = dx^k$ if $k = 1, \ldots, n-1$ and $i^*(dx^n) = 0$. Hence, $i^*(\eta^j) = 0$ for all $j \neq n$. Thus, in coordinates,

$$i^*(\omega) = \omega_n(x^1, \cdots, x^{n-1}, 0)\eta^n$$
$$= (-1)^{n-1}\omega_n(x^1, \cdots, x^{n-1}, 0) dx^1 \wedge \cdots \wedge dx^{n-1}.$$

However, by Definition 3.7.8 for the orientation induced on the boundary of a manifold, we have

$$\int_{\partial M} a(x) dx^1 \wedge \cdots \wedge dx^{n-1} = (-1)^n \int_{\mathbb{R}^{n-1}} a(x) dx^1 \cdots dx^{n-1}$$

for the $(n-1)$-form $a\, dx^1 \wedge \cdots \wedge dx^{n-1}$. Thus,

$$\int_{\partial M} \omega = \int_{\partial M} i^*(\omega)$$

$$= \int_{D_R \cap \{x^n=0\}} (-1)^{n-1}\omega_n(x^1, \cdots, x^{n-1}, 0) dx^1 \wedge \cdots \wedge dx^{n-1}$$

$$= -\int_{-R}^R \cdots \int_{-R}^R \omega_n(x^1, \ldots, x^{n-1}, 0) dx^1 \cdots dx^{n-1},$$

which, by (5.44), is equal to $\int_M d\omega$. This proves (5.42) for the case when ω is supported in a compact subset of a single coordinate patch.

Suppose now that ω is supported over a compact subset K of M that is not necessarily a subset of any particular coordinate patch in the atlas $\mathcal{A} = \{(U_\alpha, \phi_\alpha)\}$ for M. Then we use a partition of unity $\{\psi_\alpha\}$ that is subordinate to $\mathcal{A}$. Since K is compact, we can cover it with a finite collection of coordinate patches $\{(U_i, \phi_i)\}_{i=1}^k$. Then

$$\int_{\partial M} \omega = \sum_{i=1}^k \int_{\partial M} \psi_i \omega = \sum_{i=1}^k \int_M d(\psi_i \omega)$$

by application of Stokes' Theorem for each form $\psi_i \omega$ that is supported over a compact set in the coordinate patch U_i. But $d(\psi_i \omega) = d\psi_i \wedge \omega + \psi_i d\omega$, so

$$\int_{\partial M} \omega = \sum_{i=1}^k \int_M \left(d\psi_i \wedge \omega + \psi_i d\omega\right) = \sum_{i=1}^k \int_M d\psi_i \wedge \omega + \sum_{i=1}^k \int_M \psi_i d\omega$$

$$= \int_M d\left(\sum_{i=1}^k \psi_i\right) \wedge \omega + \sum_{i=1}^k \int_M \psi_i d\omega$$

$$= \int_M d(1) \wedge \omega + \sum_{i=1}^k \int_M \psi_i d\omega = 0 + \sum_{i=1}^k \int_M \psi_i d\omega = \int_M d\omega.$$

This establishes Stokes' Theorem for $n > 1$ and M a smooth manifold.

Let $n > 1$ and suppose now that M is a piecewise-smooth, oriented manifold, consisting of smooth submanifolds $M_1, M_2, \ldots, M_\ell$. By definition of integration on piecewise-smooth manifolds, if ω is a compactly supported $(n-1)$-form, then

$$\int_M d\omega = \int_{M_1} d\omega + \cdots + \int_{M_\ell} d\omega = \int_{\partial M_1} \omega + \cdots + \int_{\partial M_2} \omega,$$

where the second equality follows by Stokes' Theorem on smooth, oriented manifolds. By the definition of an orientation on an oriented, piecewise-smooth manifold, if M_i and M_j share a boundary component, then these components have induced opposite orientation. Consequently, the boundary components in the set $\{\partial M_1, \ldots, \partial M_\ell\}$ which do not cancel out precisely form the components of the boundary ∂M. Hence, we again recover

$$\int_M d\omega = \int_{\partial M} \omega.$$

Now consider the case of a 1-manifold M. The boundary ∂M is a 0-dimensional manifold. The 0-form ω is simply a real-valued function on M. For a compact set K contained in a coordinate system $\phi : U \to \mathbb{R}_+$ on M, the intersection $K \cap \partial M$ is either empty or consists of a single point $\{p_i\}$. Thus, with the assumption that ω

is supported over a compact set contained in a coordinate patch of M, we conclude that

$$\int_M df = f(p_i)s_i$$

by the usual Fundamental Theorem of Integral Calculus. By the convention in (5.41) for integration on a zero-dimensional manifold, we also have $\int_{\partial M} f = f(p_i)s_i$. Utilizing a partition of unity when ω is not assumed to be supported in a single coordinate patch, one also immediately recovers Stokes' Theorem.   □

Two cases of Stokes' Theorem occur frequently enough to warrant special emphasis. The proofs are implicit in the above proof of Stokes' Theorem.

Corollary 5.9.2. *If M is a smooth manifold without boundary and ω is a smooth $(n-1)$-form, then*

$$\int_M d\omega = 0.$$

Corollary 5.9.3. *If M is a smooth manifold with or without boundary and ω is a smooth $(n-1)$-form that is closed (i.e., $d\omega = 0$), then*

$$\int_{\partial M} \omega = 0.$$

The convention for integrating 0-forms on a zero-dimensional manifold allows Stokes' Theorem to directly generalize the Fundamental Theorem of Calculus in the following way. Consider the interval $[a, b]$ as a one-dimensional manifold M with boundary with orientation of displacement from a to b. Then $\partial M = \{a, b\}$ with an orientation of -1 for a and $+1$ for b. A 0-form on M is a smooth function $f : [a, b] \to \mathbb{R}$. Then Theorem 5.9.1 simply states that

$$\int_{[a,b]} df = \int_a^b f'(x)\, dx = f(b) - f(a),$$

which is precisely the Fundamental Theorem of Calculus.

The reader might remark that, as stated, Stokes' Theorem on manifolds only generalizes the Fundamental Theorem of Calculus (FTC) when f is a smooth function, whereas most calculus texts only presuppose that f is C^1 over $[a, b]$. The history behind the FTC is long and we encourage the reader to consult [13] for an excellent historical account of the work on defining integrals properly. Since we restricted our attention to smooth manifolds and smooth functions, these technical details are moot.

PROBLEMS

5.9.1. Explicitly show how Stokes' Theorem on manifolds generalizes Stokes' Theorem and the Divergence Theorem from standard multivariable calculus.

5.9.2. Use Stokes' Theorem to evaluate $\int_S d\omega$, where S is the image in $\mathbb{R}^4$ of the parametrization
$$r(u,v) = (1-v)(\cos u, \sin u, \sin 2u, 0) + v(2, \cos u, \sin u, \sin 2u)$$
and where $\omega = x^2\,dx^1 + x^3\,dx^2 + x^4\,dx^3 - x^1\,dx^4$.

5.9.3. Let M be the hypercube in $\mathbb{R}^4$ consisting of the 16 vertices $(\pm 1, \pm 1, \pm 1, \pm 1)$. This is a manifold with boundary embedded in $\mathbb{R}^4$. Let $\omega = x\,dy \wedge dz \wedge dw + (3\sin(y+z) + e^{x^2})\,dx \wedge dz \wedge dw$ be a 3-form in $\mathbb{R}^3$, which we consider as a 3-form on the surface of the hypercube ∂M. Use Stokes' Theorem to calculate $\int_{\partial M} \omega$.

5.9.4. Let $\mathbb{B}^4 = \{x \in \mathbb{R}^4 \mid \|x\| \le 1\}$ be the unit ball in $\mathbb{R}^4$, and note that $\partial \mathbb{B}^4 = \mathbb{S}^3$. We use the coordinates (x,y,z,w) in $\mathbb{R}^4$ and hence in $\mathbb{B}^4$. Use Stokes' Theorem to evaluate
$$\int_{\mathbb{S}^3} (e^{xy}\cos w\,dx \wedge dy \wedge dz + x^2 z\,dx \wedge dy \wedge dw).$$
[Hint: After applying Stokes' Theorem, consider symmetry across the $w = 0$ plane, then use a combination of Cartesian and spherical coordinates integration.]

5.9.5. Let M be a compact, oriented, n-manifold, and let $\omega \in \Omega^j(M)$ and $\eta \in \Omega^k(M)$, where $j+k = n-1$. Suppose that η vanishes on the boundary ∂M or that $\partial M = \emptyset$. Show that
$$\int_M \omega \wedge d\eta = (-1)^{j-1} \int_M d\omega \wedge \eta.$$

5.9.6. Let M be a compact, oriented n-manifold. Let ω and η be forms of type j and k respectively, such that $j+k = n-2$. Show that
$$\int_M d\omega \wedge d\eta = \int_{\partial M} \omega \wedge d\eta.$$
Explain how this generalizes the well-known result in multivariable calculus that
$$\int_C (f\nabla g) \cdot d\vec{r} = \iint_S (\nabla f \times \nabla g) \cdot d\vec{S},$$
where S is a regular surface in $\mathbb{R}^3$ with boundary C and where f adn g are real-valued functions that are defined and have continuous second derivatives over an open set containing S.

5.9.7. *Integration by Parts on a Curve.* Let M be a compact and connected one-dimensional smooth manifold. Let $f, g : M \to \mathbb{R}$ be two smooth functions on the curve M. Show that ∂M consists of two discrete points $\{p, q\}$. Suppose that M is oriented so that the orientation induced on ∂M is -1 for p and $+1$ for q. Show that
$$\int_M f\,dg = f(q)g(q) - f(p)g(p) - \int_M g\,df.$$

5.9.8. Let M be an embedded submanifold of $\mathbb{R}^n$ of dimension $n-1$. Suppose that M encloses a compact region $\mathcal{R}$. Setting $\omega = \frac{1}{n}(\sum_{i=1}^n x^i \eta^i)$, where the η^j are defined by Equation (5.24), show that the n-volume of $\mathcal{R}$ is $\int_M \omega$.

5.9.9. Consider $M = \mathbb{R}^n - \{(0,\ldots,0)\}$ as a submanifold of $\mathbb{R}^n$, and let $\mathbb{S}^{n-1}$ be the unit sphere in $\mathbb{R}^n$ centered at the origin.

(a) Show that if $\omega \in \Omega^{n-1}(M)$ is exact, then $\int_{\mathbb{S}^{n-1}} \omega = 0$.

(b) Find an example of a closed form $\omega \in \Omega^{n-1}(M)$ such that $\int_{\mathbb{S}^{n-1}} \omega \ne 0$.

CHAPTER 6

Introduction to Riemannian Geometry

To recapitulate what we have done in the past two chapters, manifolds are topological spaces that are locally homeomorphic to a Euclidean space $\mathbb{R}^n$ in which one could do calculus. Chapter 5 introduced the analysis on manifolds by connecting it to analysis on $\mathbb{R}^n$ via coordinate charts. However, the astute reader might have noticed that our presentation of analysis on manifolds so far has not recovered one of the foundational aspects of Euclidean calculus: the concept of distance. And related to the concept of distance are angles, areas, volumes, curvature...

In the local theory of regular surfaces S in $\mathbb{R}^3$, the first fundamental form (see Example 5.2.3) allows one to calculate the length of curves on a surface, the angle between two intersecting curves, and the area of a compact set on S (see [5, Section 6.1] for details). This should not be surprising: we defined the first fundamental form on S as the restriction of the usual Euclidean dot product in $\mathbb{R}^3$ to the tangent space $T_p(S)$ for any given point $p \in S$, and the dot product is the basis for measures of distances and angles in $\mathbb{R}^3$.

In general, manifolds are not given as topological subspaces of $\mathbb{R}^n$ so one does not immediately have a first fundamental form as we defined in Example 5.2.3. Furthermore, from the definition of a differentiable manifold, it is not at all obvious that it has a metric (though we will see in Proposition 6.1.8 that every smooth manifold has a metric structure). Consequently, one must equip a manifold with a metric structure, which we will call a "Riemannian structure." Applications of manifolds to geometry and curved space in physics will require this additional metric structure.

As in many mathematics texts, our treatment of manifolds and Riemannian metrics does not emphasize how long it took these ideas to develop nor have the previous two chapters followed the historical trajectory of the subject. After the discovery of non-Euclidean geometries (see [11] for a good historical discussion), by using only an intuitive notion of a manifold, it was Riemann [47, Section II] in 1854 who first proposed the idea of a metric that varied at each point of a manifold. During the following 50 or more years, many mathematicians (Codazzi, Beltrami, Ricci-Curbastro, Levi-Civita, Klein to name a few) developed the theories of curvature and of geodesy for Riemann spaces. However, the concept of a differential manifold as presented in Chapter 3 did not appear until 1913 in the work of H. Weyl [59, I.§4]. According to Steenrod [53, p. v], general definitions for fiber bundles and vector bundles, which we introduced in part in Chapter 5, did not appear until the work of Whitney in 1935–1940.

Turning to physics, general relativity, one of the landmark achievements in science of the early 20th century, stands as the most visible application of Riemann manifolds to science. Starting from the principle that the speed of light in a vacuum is constant regardless of reference frame [20, p. 42], Einstein developed the theory of special relativity, defined in the absence of gravity. The "interpretation" of the law that "the gravitational mass of a body is equal to its inertial mass" [20, p. 65] and the intention to preserve the principle of the constancy of the speed of light led Einstein to understand spacetime as a curved space where "the geometrical properties of space are not independent, but [...] determined by matter" [20, p. 113]. Riemannian metrics, curvature, and the associated theorems for geodesics gave Einstein precisely the mathematical tools he needed to express his conception of a curvilinear spacetime.

The reader should be aware that other applications of manifolds to science do not (and should not) always require a metric structure. Applications of manifolds to either geometry or physics may require a different structure from or additional structure to a Riemann metric. For example, in its properly generalized context, Hamiltonian mechanics require the structure of what is called a symplectic manifold.

6.1 Riemannian Metrics

6.1.1 Definitions and Examples

Definition 6.1.1. Let M be a smooth manifold. A *Riemannian metric* on M is a tensor field g in $\mathrm{Sym}^2 TM^*$ that is positive definite. In more detail, at each point $p \in M$, a Riemannian metric determines an inner product on T_pM. A smooth manifold M together with a Riemannian metric g is called a *Riemannian manifold* and is denoted by the pair (M, g).

Over a coordinate patch of M with coordinate system $(x^1, \ldots, x^n)$, as a section of $TM^{*\otimes 2}$, one writes the metric g as

$$g_{ij}\, dx^i \otimes dx^j,$$

where g_{ij} are smooth functions on M. (In this chapter, we regularly use Einstein's summation convention.) Since at each point, g is a symmetric tensor, $g_{ij} = g_{ji}$ identically. Furthermore, using the notation from Section 4.6, since g is a section of $\mathrm{Sym}^2 TM^*$, we write

$$g_{ij}\, dx^i\, dx^j. \tag{6.1}$$

The square root of the expression in Equation (6.1) is called the *line element ds* associated to this metric. Many texts, in particular, physics texts, give the metric in reference to the line element by writing $ds^2 = g_{ij}\, dx^i\, dx^j$.

For vectors $X, Y \in T_pM$, we sometimes use the same notation as the first fundamental form ([6, Section 6.1]) and write $\langle X, Y \rangle_p$ for $g_p(X, Y)$, and it is also common to drop the subscript p whenever the point p is implied by context. By analogy with the dot product, the Riemannian metric allows one to define many common notions in geometry.

Definition 6.1.2. Let (M, g) be a Riemannian manifold. Suppose that X, Y are vectors in T_pM.

1. The *length* of X, denoted $\|X\|$, is defined by $\|X\| = \sqrt{g(X, X)}$.

2. The *angle* θ between X and Y is defined by

$$\cos\theta = \frac{g(X, Y)}{\|X\|\, \|Y\|}.$$

3. X and Y are called orthogonal if $g(X, Y) = 0$.

Whenever one introduces a new mathematical structure, one must discuss functions between any two instances of them and when two structures are considered equivalent. In the context of Riemannian manifolds, one still studies any smooth functions between two manifolds. However, two Riemannian manifolds are considered the same if they have the same metric. The following definition makes this precise.

Definition 6.1.3. Let M and N be two Riemannian manifolds. A diffeomorphism $f : M \to N$ is called an *isometry* if for all $p \in M$,

$$\langle X, Y \rangle_p = \langle df_p(X), df_p(Y) \rangle_{f(p)} \qquad \text{for all } X, Y \in T_pM.$$

Two Riemannian manifolds are called *isometric* if there exists an isometry between them.

From an intuitive perspective, an isometry is a transformation that bends (which also includes rigid motions) one manifold into another without stretching or cutting. Problem 6.1.6 asks the reader to show that the catenoid and the helicoid are isometric. Figure 6.1 shows intermediate stages of bending the catenoid into the helicoid. Though one might think this transformation incorporates some stretching because the longitudinal lines straighten out, the twist created in the helicoid strip "balances out" the flattening of the lines in just the right way so that one only needs to bend the surface.

Many examples of Riemannian metrics arise naturally as submanifolds of Riemannian manifolds.

Definition 6.1.4. Let $(N, \tilde{g})$ be a Riemannian manifold and M any smooth manifold. Let $f : M \to N$ be an immersion of M into N, i.e., f is differentiable and df_p

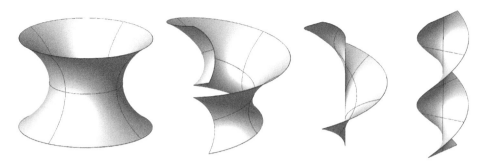

Figure 6.1: Bending the catenoid into the helicoid.

is injective for all p. The metric g on M *induced* by f (or "from N") is defined as the pull-back $g = f^* \tilde{g}$. In other words,

$$\langle X, Y \rangle_p = \langle df_p(X), df_p(Y) \rangle_{f(p)} \qquad \text{for all } p \in M \text{ and } X, Y \in T_p M.$$

The property that df_p is injective ensures that $\langle \, , \, \rangle_p$ remains positive definite when induced on M.

Example 6.1.5 (Euclidean Spaces). Consider the manifold $M = \mathbb{R}^n$, where $T_p(\mathbb{R}^n) = \mathbb{R}^n$ is naturally equipped with a Riemannian metric: the usual dot product. In particular,

$$g(\partial_i, \partial_j) = \delta_{ij} = \begin{cases} 1, & \text{if } i = j, \\ 0, & \text{if } i \neq j. \end{cases}$$

This metric is called the *Euclidean metric*.

Example 6.1.6 (First Fundamental Form). A regular surface S is a 2-manifold embedded in $\mathbb{R}^3$, where the embedding map is simply the injection $i : S \to \mathbb{R}^3$. The first fundamental form (see Example 5.2.3) is precisely the metric on S induced by i from the Euclidean metric on $\mathbb{R}^3$. This connection gives us immediately a whole host of examples of Riemannian 2-manifolds that we take from the local theory of regular surfaces.

Proposition 6.1.7. *Let M be an m-dimensional manifold embedded in $\mathbb{R}^n$. If $\vec{F}(u_1, \ldots, u_m)$ is a parametrization of a coordinate patch of M, then over this coordinate patch, the coefficients of the metric g on M induced from $\mathbb{R}^n$ are*

$$g_{ij} = \frac{\partial \vec{F}}{\partial u^i} \cdot \frac{\partial \vec{F}}{\partial u^j}.$$

Proof. Let $(x^1, \ldots, x^n)$ be the coordinates on $\mathbb{R}^n$. Suppose that a coordinate patch (U, ϕ) of M has coordinates $(u_1, \ldots, u_m)$ and that a parametrization of this coordinate patch is $\vec{F}(u_1, \ldots, u_m) = \phi^{-1}(u_1, \ldots, u_m)$. By Equation (3.14) the matrix of

$d\vec{F}$ in the given coordinate systems is

$$\left(\frac{\partial F^i}{\partial u^j}\right),$$

where $\vec{F} = (F^1, \ldots, F^n)$.

Set $\langle\,,\,\rangle$ as the usual dot product in $\mathbb{R}^n$. Then at each point in U, the coefficients $g_{k\ell}$ of the metric g satisfy

$$g_{k\ell} = g\left(\frac{\partial}{\partial u^k}, \frac{\partial}{\partial u^\ell}\right) = \left\langle d\vec{F}\left(\frac{\partial}{\partial u^k}\right), d\vec{F}\left(\frac{\partial}{\partial u^\ell}\right)\right\rangle = \frac{\partial F^i}{\partial u^k}\frac{\partial F^j}{\partial u^\ell}\left\langle\frac{\partial}{\partial x^i}, \frac{\partial}{\partial x^j}\right\rangle$$

$$= \frac{\partial F^i}{\partial u^k}\frac{\partial F^j}{\partial u^\ell}\delta_{ij} = \frac{\partial \vec{F}}{\partial u^k}\cdot\frac{\partial \vec{F}}{\partial u^\ell}. \qquad\qquad \square$$

We should emphasize at this point that a given manifold can be equipped with nonisometric Riemannian metrics. Problem 6.1.2 presents two different metrics on the 3-torus, each depending on a different embedding into some Euclidean space. In both cases, the 3-torus can be equipped with the same atlas, and so in both situations, the 3-torus is the same as a smooth manifold.

As another example, already in his seminal dissertation [47], Riemann introduced the following metric on the open unit ball in $\mathbb{R}^n$:

$$g_{ii} = \frac{4}{(1 - \|x\|^2)^2} \qquad \text{and} \qquad g_{ij} = 0 \text{ if } i \neq j. \qquad (6.2)$$

As we will see, this is not isometric with the open unit ball equipped with the Euclidean metric.

Example 6.1.5 could be misleading in its simplicity. The reader might consider the possibility of defining a metric on any smooth manifold M by taking $\langle\,,\,\rangle_p$ as the usual dot product in each T_pM with respect to the coordinate basis associated to a particular coordinate system. The problem with this idea is that it does not define a smooth section in $\mathrm{Sym}^2 TM^*$ over the whole manifold. Nonetheless, as the proof of the following proposition shows, we can use a partition of unity and stitch these bilinear forms together.

Proposition 6.1.8. *Every smooth manifold M has a Riemannian metric.*

Proof. Let M be a smooth manifold with atlas $\mathcal{A} = \{(U_\alpha, \phi_\alpha)\}_{\alpha \in I}$. For each $\alpha \in I$, label $\langle\,,\,\rangle^\alpha$ as the usual dot product with respect to the coordinate basis over U_α. Let $\{\psi_\alpha\}$ be a partition of unity that is subordinate to $\mathcal{A}$. For each $p \in M$, define the bilinear form $\langle\,,\,\rangle_p$ on T_pM by

$$\langle X, Y \rangle_p \stackrel{\mathrm{def}}{=} \sum_{\alpha \in I} \psi_\alpha(p)\langle X, Y \rangle_p^\alpha \qquad (6.3)$$

for any $X, Y \in T_pM$.

Since for each $p \in M$, only a finite number of $\alpha \in I$ have $\psi_\alpha(p) \neq 0$, then the sum in Equation (6.3) is finite. It is obvious by construction that $\langle X, Y \rangle_p$ is symmetric. To prove that $\langle \, , \, \rangle_p$ is positive definite, note that each $\langle X, Y \rangle_p^\alpha$ is. Let I' be the set of all indices $\alpha \in I$ such that $\psi_\alpha(p) \neq 0$. By definition of a partition of unity, $0 < \psi_\alpha(p) \leq 1$. Thus, for all $X \in T_pM$, clearly $\langle X, X \rangle_p \geq 0$. Furthermore, if $\langle X, X \rangle_p = 0$, then at least one summand in

$$\sum_{\alpha \in I'} \psi_\alpha(p) \langle X, X \rangle_p^\alpha$$

is 0. (In fact, all summands are 0.) Thus, there exists an $\alpha \in I$ with $\langle X, X \rangle_p^\alpha = 0$. Since $\langle \, , \, \rangle_p^\alpha$ is positive definite, then $X = 0$. Hence, $\langle X, X \rangle_p$ itself is positive definite. $\qquad \square$

Though Equation (6.3) presents a Riemannian metric on any smooth manifold M, this is not in general easy to work with for specific calculations since it uses a partition of unity, which involves functions that are usually complicated. At any given point $p \in M$, Equation (6.3) does not involve one coordinate system around p but all of the atlas's coordinate neighborhoods of p.

More importantly, the Riemannian metric constructed in the above proof might not have any natural meaning.

Example 6.1.9 (Projective Space). There is a natural metric on projective space $\mathbb{RP}^n$ that is induced from the Euclidean metric on $\mathbb{R}^{n+1}$. First, let g be the metric on $\mathbb{S}^n$ induced from Euclidean $\mathbb{R}^n$ as an embedded submanifold. Recall that the projection map $\pi : \mathbb{S}^n \to \mathbb{RP}^n$ as presented for $n = 2$ in Example 3.2.3 is a smooth function between manifolds. Define the metric $\tilde{g}$ on $\mathbb{RP}^n$ by

$$\tilde{g}_{\pi(p)}(v, w) = g_p \left((d\pi_p)^{-1}(v), (d\pi_p)^{-1}(w) \right), \tag{6.4}$$

for all $v, w \in T_{\pi(p)}\mathbb{RP}^n$. Note first that for all $p \in \mathbb{S}^n$, the linear transformation $d\pi_p$ is surjective between spaces of the same dimension, so is invertible. More importantly, $\tilde{g}$ is well-defined: If $A : \mathbb{S}^n \to \mathbb{S}^n$ is the antipodal map $A(p) = -p$, then $\pi \circ A = \pi$. Hence, $d\pi_p = d(\pi \circ A)_p = d\pi_{-p} \circ dA_p$. Hence

$$g_p \left((d\pi_p)^{-1}(v), (d\pi_p)^{-1}(w) \right)$$
$$= g_p \left((dA_p)^{-1}((d\pi_{-p})^{-1}(v)), (dA_p)^{-1}((d\pi_{-p})^{-1}(w)) \right)$$
$$= g_{-p}((d\pi_{-p})^{-1}(v), (d\pi_{-p})^{-1}(w)),$$

where the second equality follows because $A : \mathbb{S}^n \to \mathbb{S}^n$ is an isometry. Consequently, in (6.4), the choice of p or $-p$ for the pre-image of $\pi(p)$ in $\mathbb{RP}^n$ is irrelevant.

6.1.2 Arclength and Volume

Using integration, the Riemannian metric allows for formulas that measure nonlocal properties, such as length of a curve and volume of a region on a manifold.

For example, consider a C^1 curve $\gamma : [a, b] \to M$ on a Riemannian manifold (M, g). At each point $\gamma(t)$ in M, the vector $\gamma'(t) = \frac{d\gamma}{dt}$ is a tangent vector, called the *velocity vector*. The Riemannian metric $g = \langle\,,\,\rangle$ allows one to calculate the length $\|\gamma'(t)\|$, which we call the *speed*. This motivates the following definition.

Definition 6.1.10. Let $\gamma : [a, b] \to M$ be a curve on a Riemannian manifold M of class C^1. The arclength of the curve γ is

$$\ell(\gamma) = \int_a^b \sqrt{\left\langle \frac{d\gamma}{dt}, \frac{d\gamma}{dt} \right\rangle_{\gamma(t)}}\, dt.$$

Proposition 6.1.11. *Let (M, g) be an oriented Riemannian manifold of dimension n. There exists a unique n-form, denoted dV, such that at all $p \in M$, it satisfies $dV_p(e_1, \ldots, e_n) = 1$ for all bases $(e_1, \ldots, e_n)$ in T_pM that are orthonormal with respect to $g_p(\,,\,)$. Furthermore, over any coordinate patch U with coordinates $x = (x^1, \ldots, x^n)$,*

$$dV = \sqrt{\det(g_{ij})}\, dx^1 \wedge \cdots \wedge dx^n, \tag{6.5}$$

where $g_{ij} = g(\partial_i, \partial_j) = \langle \partial/\partial x^i, \partial/\partial x^j \rangle$.

Proof. The content of this proposition is primarily linear algebra. By Proposition C.2.1, on each coordinate patch U_α, the form $dV|_{U_\alpha} = \omega_\alpha$ exists on each T_pM and is given by Equation (6.5). The existence of this n-form ω with the desired property explicitly requires that g_p be an inner product on T_pM. In order to define the form dV on the whole manifold, we refer to a partition of unity $\{\psi_\alpha\}$ subordinate to the atlas on M and define

$$dV = \sum_\alpha \psi_\alpha \omega_\alpha. \qquad \square$$

Definition 6.1.12. The form dV described in Proposition 6.1.11 is called the *volume form* of (M, g) and is denoted dV_M if there is a chance of confusion about the manifold. If M^m is a compact manifold, then the m-volume of M is the integral

$$\mathrm{Vol}(M) = \int_M dV. \tag{6.6}$$

If $i : M^m \to N^n$ is an embedded submanifold of a Riemannian manifold $(N, \tilde{g})$ then we can also calculate the m-volume of the submanifold M by equipping M with the metric $g = i^*\tilde{g}$, i.e., the metric induced from N. The reader should be aware that the volume form on M is not necessarily $i^*(dV_N)$. In particular, if $m < n$, then $i^*(dV_N)$ would be an n-form on M, but there are no n-forms on M. We illustrate this with a few examples.

Example 6.1.13 (Arclength). Definition 6.1.10 should actually be a corollary of Proposition 6.1.11 and Definition 6.1.12. Let $(M, \tilde{g})$ be a Riemannian manifold and let $\gamma : [a, b] \to M$ be a curve of class C^1 on M. The induced metric g on γ is defined by

$$g_t(v, w) = \tilde{g}_{\gamma(t)}\left(\frac{d\gamma}{dt}v, \frac{d\gamma}{dt}w\right).$$

However, since a curve is one dimensional, both tangent vectors v and w are scalars. So

$$g_t(v, w) = \tilde{g}_{\gamma(t)}((d\gamma/dt), (d\gamma/dt))vw.$$

In the coordinate t on $[a, b]$, the domain of γ, the metric tensor can be represented by a 1×1 matrix $(\tilde{g}_{\gamma(t)}((d\gamma/dt), (d\gamma/dt)))$. Thus, the volume form on γ from the metric induced from M, is

$$dV = \sqrt{\tilde{g}_{\gamma(t)}((d\gamma/dt), (d\gamma/dt))}\, dt.$$

Definition 6.1.10 follows as a corollary.

Example 6.1.14 (Volume form on $\mathbb{S}^n$). Consider the unit n-sphere $\mathbb{S}^n$ as a Riemannian manifold, equipped with the metric induced from its usual embedding in $\mathbb{R}^{n+1}$.

Consider the usual longitude-latitude parametrization of the sphere $\mathbb{S}^2$:

$$\vec{X}(u, v) = (\cos u \sin v, \sin u \sin v, \cos v) \quad \text{for} \quad (u, v) \in [0, 2\pi] \times [0, \pi].$$

Note that if we restrict the domain to $(0, 2\pi) \times (0, \pi)$, we obtain a dense open subset of $\mathbb{S}^2$. By Proposition 6.1.7, with respect to this coordinate system, the coefficients of the metric tensor are

$$g_{ij} = \begin{pmatrix} \sin^2 v & 0 \\ 0 & 1 \end{pmatrix}.$$

Since $\sin v \geq 0$ for $v \in [0, \pi]$, the volume form on $\mathbb{S}^2$ with respect to this coordinate system is $dV = \sin v\, du \wedge dv$. By Proposition 5.8.1, the volume of the sphere is calculated by

$$V = \int_{\mathbb{S}^2} dV = \int_0^\pi \int_0^{2\pi} \sin v\, du\, dv = 2\pi\left[-\cos v\right]_0^\pi = 4\pi.$$

We now calculate the volume form on $\mathbb{S}^n$ using an alternate approach. By Example 4.6.24, we see that the volume form on $\mathbb{R}^{n+1}$ is

$$e^{*1} \wedge \cdots \wedge e^{*(n+1)},$$

where $\{e_1, \ldots, e_{n+1}\}$ is the standard basis on $\mathbb{R}^{n+1}$. Furthermore, recall that as an alternating function,

$$e^{*1} \wedge \cdots \wedge e^{*(n+1)}(v_1, \ldots, v_{n+1}) = \det\begin{pmatrix} v_1 & \cdots & v_{n+1} \end{pmatrix}$$

for any $(n + 1)$-tuple of vectors $(v_1, \ldots, v_{n+1})$.

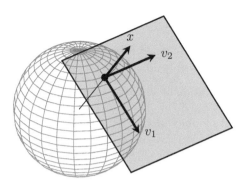

Figure 6.2: Volume form on the sphere.

Define a form $\omega \in \Omega^n(\mathbb{R}^{n+1})$, where for each $x \in \mathbb{R}^{n+1}$ and for any vectors u_i

$$\omega_x(u_1, \ldots, u_n) = \det \begin{pmatrix} x & u_1 & \cdots & u_n \end{pmatrix}.$$

This is in fact, the same construction as the inner product on forms $i_x\omega$, but where we are taking advantage of the identification of $\mathbb{R}^{n+1}$ with its own tangent space. By the properties of the determinant, for each x, ω_x is an alternating n-multilinear function on $\mathbb{R}^{n+1}$, so ω is indeed an n-form. Using the Laplace expansion of the determinant, it is easy to show that

$$\omega = \sum_{i=1}^{n+1} (-1)^{i-1} x^i \, dx^1 \wedge \cdots \wedge \widehat{dx^i} \wedge \cdots \wedge dx^{n+1}, \tag{6.7}$$

where the $\widehat{}$ notation means to exclude the bracketed term. Using the forms η^j introduced in Equation (5.24), we can write $\omega = \sum_j x^j \eta^j$.

Now if $x \in \mathbb{S}^n$, then from the geometry of the sphere, x is perpendicular to $T_x\mathbb{S}^n$ as a subspace of $\mathbb{R}^{n+1}$ (equipped with the Euclidean metric). Thus, if $\{v_1, \ldots, v_n\}$ forms a basis of $T_x\mathbb{S}^n$, then $\{x, v_1, \ldots, v_n\}$ forms a basis of $\mathbb{R}^{n+1}$. See Figure 6.2. Furthermore, if the n-tuple $(v_1, \ldots, v_n)$ is an orthonormal, positively-oriented basis of $T_x\mathbb{S}^n$, then $(x, v_1, \ldots, v_n)$ is an orthonormal, positively-oriented basis of $\mathbb{R}^{n+1}$. But then the restriction of ω to $\mathbb{S}^n$ has the properties described in Proposition 6.1.11. Hence, if $f : \mathbb{S}^n \to \mathbb{R}^{n+1}$ is the usual embedding of the sphere in Euclidean space, then we obtain the volume form of $\mathbb{S}^n$ as

$$dV_{\mathbb{S}^n} = \omega\big|_{\mathbb{S}^n} = f^*(\omega).$$

In Section 5.7, we attempted to generalize with the single technique of integration on manifolds all the types of integration introduced in a standard calculus sequence. However, there were two types of integrals in the list at the beginning of the section that did not fit in the formalism we had developed for the integration of n-forms on n-dimensional smooth manifolds, namely:

- Line integrals of functions in $\mathbb{R}^n$.

- Surface integrals of a real-valued function defined over a closed and bounded region of a regular surface in $\mathbb{R}^3$.

Both of these types of integrals fit into the theory of integration volume forms on manifolds in the following ways.

For the line integral of functions in $\mathbb{R}^n$ over a piecewise-smooth curve C, let $\gamma : [a, b] \rightarrow \mathbb{R}^n$ be a parametrization of C. Let $\langle\,,\,\rangle$ be the Euclidean form on $\mathbb{R}^n$ (i.e., the dot product). Then each smooth piece of γ is a one-dimensional submanifold of $\mathbb{R}^n$, equipped with the metric induced from $\mathbb{R}^n$. The volume form on γ is dV_γ so that for any smooth function f defined on a neighborhood of C,

$$\int_\gamma f \, dV_\gamma = \int_a^b f(t) \sqrt{\left\langle \frac{d\gamma}{dt}, \frac{d\gamma}{dt} \right\rangle_{\gamma(t)}} \, dt = \int_C f \, ds. \tag{6.8}$$

For surface integrals of a function f on a compact regular surface $\mathcal{S} \subset \mathbb{R}^3$, it is not hard to show (see Problem 6.1.1) that $dS = dV_\mathcal{S}$, where $dV_\mathcal{S}$ is the volume form on $\mathcal{S}$ equipped with the metric induced from the Euclidean metric. Thus, connecting the classical notation with the notation introduced in this section,

$$\int_\mathcal{S} f \, dS = \int_\mathcal{S} f \, dV_\mathcal{S}. \tag{6.9}$$

6.1.3   Raising and Lowering Indices

One interesting property of metrics on manifolds is that they give us a natural way to go back and forth between vectors and covectors.

Recall from Section 4.1 that for any real vector space V, the dual vector space V^* consists of all linear transformations $V \rightarrow \mathbb{R}$. If V is a vector space equipped with any bilinear form $\langle\,,\,\rangle$, then this form defines a linear transformation into the dual V^* by

$$i : V \longrightarrow V^*,$$

$$v \longmapsto i_v \;=\; (w \mapsto \langle v, w \rangle).$$

If V is finite-dimensional and the bilinear form $\langle\,,\,\rangle$ is nondegenerate, then the mapping $(v \mapsto i_v)$ is an isomorphism. We will assume from now on that $\langle\,,\,\rangle$ is an inner product. The positive-definite implies the bilinear form is nondegenerate so the above mapping is an isomorphism. This isomorphism allows us to define a natural bilinear form $\langle\,,\,\rangle^*$ on V^* by

$$\langle \eta, \tau \rangle^* = \langle i^{-1}(\eta), i^{-1}(\tau) \rangle.$$

Then $\langle\,,\,\rangle^*$ is an element of $V \otimes V$, so is a tensor of type of $(2, 0)$. Furthermore, if the components of $\langle\,,\,\rangle$ are c_{jk}, then the components of $\langle\,,\,\rangle^*$ with respect to the associated basis are denoted by c^{jk}, where these are the entries of the inverse matrix C^{-1}, where $C = (c_{jk})$. (See Exercises 4.2.4 and 4.2.5 for where we prove these results.)

In coordinates, let $\mathcal{B} = \{u_1, \ldots, u_n\}$ be a basis of V and let $\mathcal{B}^* = \{u^{*1}, \ldots, u^{*n}\}$ be the associated cobasis. Let $C = (c_{jk})$ be the matrix of $\langle\,,\,\rangle$ with respect to $\mathcal{B}$, viz., $c_{jk} = \langle u_j, u_k \rangle$. Then

$$\langle v, w \rangle = [v]_{\mathcal{B}}^T C [w]_{\mathcal{B}} = c_{jk} v^j w^k. \tag{6.10}$$

This gives the coordinates of i_v with respect to $\mathcal{B}^*$ as $c_{jk}v^j$. Note that the indices for the components of i_v arise naturally as subscripts, consistent with our notation. Similarly, if $\lambda \in V^*$ is a functional on V, then $\lambda = \lambda_i u^{*i}$. If $\lambda = i_v$ for some $v \in V$, then

$$c^{k\ell} \lambda_k = c^{k\ell} a_{jk} v^j = c_{jk} a^{k\ell} v^j = \delta_j^\ell v^j = v^\ell. \tag{6.11}$$

We say that the process of mapping v to i_v "lowers the indices," while mapping λ to $i^{-1}(\lambda)$ "raises the indices."

Now consider a Riemannian manifold (M, g). If $X \in \mathfrak{X}(M)$ is a vector field on M, we define the covector field $X^\flat$ by

$$X^\flat(Y) \overset{\text{def}}{=} g(X, Y)$$

for all vector fields $Y \in \mathfrak{X}(M)$. On a coordinate patch, X has coordinates X^i. By the process described in the previous paragraphs, $X^\flat$ has components $g_{ij}X^i$. Mimicking musical notation, the function $\Gamma(TM) \to \Gamma(TM^*)$ that sends $X \mapsto X^\flat$ is call the *flat*, since it lowers the indices of the vector field X.

As we saw, a metric g^* on M also induces an inner product in $\text{Sym}^2 T_p M$ defined by $g^*(\eta, \tau) = g(i^{-1}(\eta), i^{-1}(\tau))$ for any $\eta, \tau \in T_p M^*$. So if $\omega \in \Omega^1(M)$ is a covector field on M, we define the vector field $\omega^\sharp$ by

$$\eta(\omega^\sharp) \overset{\text{def}}{=} g^*(\eta, \omega)$$

for all covector fields $\eta \in \Omega^1(M)$. On a given coordinate patch, ω has component functions ω_i and the components of $\omega^\sharp$ are $g^{ij}\omega_i$. Keeping the musical analogy, we denote call this vector field the *sharp* of ω since the process raises the indices.

More generally, if T is any tensor field of type (p, q) on M, then

$$g_{i_\alpha k} T^{i_1 \cdots i_p}_{j_1 \cdots j_q} \qquad \text{and} \qquad g^{j_\beta l} T^{i_1 \cdots i_p}_{j_1 \cdots j_q}$$

define tensors fields of type $(p - 1, q + 1)$ and type $(p + 1, q - 1)$, respectively. It is common to still use the $\flat$ and $\sharp$ notation, here $T^\flat$ and $T^\sharp$, but one must indicate upon which index one performs the lowering or raising operations.

Recall that the trace of a matrix A is defined as the sum of the diagonal elements. If A has components A_j^i, then the trace is just A_i^i, using the Einstein summation convention of summing along i. Now A corresponds to a linear transformation

$T(\vec{v}) = A\vec{v}$ on a vector space V. Since the trace $\operatorname{Tr} A$ is also the sum of the eigenvalues, the trace remains unchanged under a change in basis in V.

Now, if A is a symmetric $(0,2)$-tensor, then $A^{\sharp}$ is a $(1,1)$-tensor, and the trace $\operatorname{Tr} A$ is defined in its usual linear algebraic sense. This process is common enough that we define the *trace with respect to g* of A to be

$$\operatorname{Tr}_g A \overset{\text{def}}{=} \operatorname{Tr} A^{\sharp}. \tag{6.12}$$

In coordinates, $\operatorname{Tr}_g A = g^{ij} A_{ij}$.

PROBLEMS

6.1.1. Recall the following formula from calculus. Let K be a compact set on a regular surface S in $\mathbb{R}^3$. Suppose that S is parametrized by $\vec{r}(u,v)$ and that under this parametrization, $K = \vec{R}(D)$ for some compact region D. Then the surface area of K on S is

$$\iint_K dS = \iint_D \|\vec{r}_u \times \vec{r}_v\| \, du \, dv.$$

Consider the regular surface S as a 2-manifold embedded in the Riemannian manifold of $\mathbb{R}^3$, equipped with the dot product as its usual metric. The parametrization $\vec{r}(u,v)$ describes the embedding of S in $\mathbb{R}^3$ in reference to the coordinates in a chart of S. Prove that the volume form on S of metric on S induced from the dot product in $\mathbb{R}^3$ is

$$dV_S = \|\vec{r}_u \times \vec{r}_v\| \, du \wedge dv.$$

This establishes the familiar surface area integral from Definition 6.1.12.

6.1.2. Consider the 3-torus $\mathbb{T}^3 = \mathbb{S}^1 \times \mathbb{S}^1 \times \mathbb{S}^1$. Calculate the induced metrics for the following two embeddings:

(a) Into $\mathbb{R}^6$ as the image of the parametrization

$$\vec{F}(u^1, u^2, u^3) = \left(\cos u^1, \sin u^1, \cos u^2, \sin u^2, \cos u^3, \sin u^3\right).$$

(b) Into $\mathbb{R}^4$ as the image of the parametrization $\vec{F}(u^1, u^2, u^3)$ given by

$$\big((c + (b + a\cos u^1)\cos u^2)\cos u^3, (c + (b + a\cos u^1)\cos u^2)\sin u^3,$$
$$(b + a\cos u^1)\sin u^2, a\sin u^1\big),$$

where $b > a > 0$ and $c > a + b$.

(c) Prove that these two Riemannian manifolds are not isometric.

[Hint: This gives two different metrics on the 3-torus that can be equipped with the same atlas in each case.]

6.1.3. We consider an embedding of $\mathbb{S}^1 \times \mathbb{S}^2$ in $\mathbb{R}^4$ by analogy with the embedding of the torus $\mathbb{S}^1 \times \mathbb{S}^1$ in $\mathbb{R}^3$. Place the sphere $\mathbb{S}^2$ with radius a at $(0,0,b,0)$ (where $b > a$) as a subset of the $x^1 x^2 x^3$-subspace, and rotate this sphere about the origin with a motion parallel to the $x^3 x^4$-axis. Call this submanifold M and equip it with the metric induced from $\mathbb{R}^4$.

(a) Show that the described manifold M is an embedding as claimed.

(b) Find a parametrization $F : D \to \mathbb{R}^4$ where $D \subset \mathbb{R}^3$ such that, as sets of points $F(D) = M$, and $F(D^\circ)$ is an open subset of M that is homeomorphic to D°. (D° is the interior of the set D.)

(c) Calculate the coefficients g_{ij} of the metric on M in the coordinate patch defined by the above parametrization.

6.1.4. Let (M_1, g_1) and (M_2, g_2) be two Riemannian manifolds, and consider the product manifold $M_1 \times M_2$ with a $(0, 2)$-tensor field defined by

$$g(X_1 + X_2, Y_1 + Y_2) = g_1(X_1, Y_1) + g_2(X_2, Y_2).$$

(a) Show that g defines a metric on $M_1 \times M_2$.

(b) Let $(x^1, \ldots, x^n)$ be local coordinates on M_1 and $(x^{n+1}, \ldots, x^{n+m})$ be local coordinates on M_2 so that $(x^1, \ldots, x^{m+n})$ are local coordinates on $M_1 \times M_2$. Determine the components of the metric g on $M_1 \times M_2$ in terms of g_1 and g_2.

6.1.5. Repeat Problem 6.1.3 with $\mathbb{S}^1 \times S$, where S is a regular surface in $\mathbb{R}^3$ that does not intersect the plane $z = x^3 = 0$.

6.1.6. Consider the following two regular surfaces in $\mathbb{R}^3$. The catenoid parametrized by

$$F(\bar{u}^1, \bar{u}^2) = (\bar{u}^2 \cos \bar{u}^1, \bar{u}^2 \sin \bar{u}^1, \cosh^{-1} \bar{u}^2) \text{ for } (\bar{u}^1, \bar{u}^2) \in [0, 2\pi) \times [1, +\infty)$$

and the helicoid parametrized by

$$F(u^1, u^2) = (u^2 \cos u^1, u^2 \sin u^1, u^1) \text{ for } (u^1, u^2) \in [0, 2\pi) \times \mathbb{R}.$$

Prove that the helicoid and catenoid are isometric, and find an isometry between them.

6.1.7. Let M be a hypersurface of $\mathbb{R}^n$ (submanifold of dimension $n-1$), and equip M with Riemannian structure with the metric induced from $\mathbb{R}^n$. Suppose that an open set U of M is a graph of an $(n-1)$-variable function f, i.e., the parametrization of U is

$$x^1 = u^1, \ldots, x^{n-1} = u^{n-1}, x^n = f(u^1, \ldots, u^{n-1}),.$$

for $(u^1, \ldots, u^{n-1}) \in D$.

(a) Find the coefficients of the metric tensor g on M, and conclude that a formula for the $(n-1)$-volume of U is

$$\int_D \sqrt{1 + \|\text{grad } f\|^2} \, dV.$$

(b) Use this result to calculate the 3-volume of the surface in $\mathbb{R}^4$ given by $w = x^2 + y^2 + z^2$ for $x^2 + y^2 + z^2 \leq 4$.

6.1.8. Let M, N, and S be Riemannian manifolds, and let $f : M \to N$ and $h : N \to S$ be isometries.

(a) Show that f^{-1} is an isometry.

(b) Show that $h \circ f$ is an isometry.

(c) If you have seen some group theory, show that the set of isometries on a Riemannian manifold M forms a group.

6.1.9. Two metrics g and $\tilde{g}$ on a smooth manifold M are called *conformal* if there exists a smooth function $f \in C^{\infty}(M, \mathbb{R})$ such that $\tilde{g} = fg$. Prove that for all $p \in M$ and for all $X, Y \in T_p M$, the angle between X and Y with respect to $\tilde{g}$ is the same as the angle with respect to g.

6.1.10. Let (M, g) and $(N, \tilde{g})$ be Reimannian manifolds. A diffeomorphism $f : M \to N$ is called a conformal mapping if $f^* \tilde{g}$ is conformal to g. Repeat Problem 6.1.8 but replacing "isometry" with "conformal mapping." (See Problem 6.1.9.)

6.1.11. Let γ be a curve on a Riemannian manifold (M, g). Show precisely how the induced metric on γ generalizes Definition 6.1.10.

6.1.12. *Poincaré ball.* The Poincaré Ball is the open ball $\mathbb{B}_R^n$ in n dimensions of radius R equipped with the metric

$$\frac{4R^4}{(R^2 - \|x\|^2)^2} \left((dx^1)^2 + \cdots + (dx^n)^2 \right).$$

Note that this metric is conformal with the metric (see Problem 6.1.9) induced from the Euclidean metric in $\mathbb{R}^{n+1}$.

(a) Set $n = 2$ and $R = 1$. (This choice of parameters is called the unit Poincaré disk.) Calculate the area of the region $\mathcal{R}$ defined by $\|x\| \leq \frac{1}{2}$ and $0 \leq \theta \leq \pi/2$.

(b) Set $n = 3$ and $R = 2$. Calculate the length of the curve $\gamma(t) = (\cos t, \sin t, t/10)$ in the Poincaré ball for $0 \leq t \leq 4\pi$.

6.1.13. *Divergence Theorem.* Let (M, g) be an oriented, compact, Riemannian manifold with boundary. Given any vector field $X \in \mathfrak{X}(M)$ and any tensor field T of type (p, q), with $q \geq 1$, we define the contraction of X with T, denoted $i_X T$, as the tensor field of type $(p, q - 1)$ that over any coordinate chart has components

$$X^l T^{i_1 \cdots i_p}_{lj_2 \cdots j_q}.$$

We define the *divergence operator* $\operatorname{div} : \mathfrak{X}(M) \to C^{\infty}(M)$ implicitly by

$$d(i_X \, dV) = (\operatorname{div} X) \, dV.$$

Prove the Divergence Theorem, which states that for any $X \in \mathfrak{X}(M)$

$$\int_M \operatorname{div} X \, dV = \int_{\partial M} g(X, N) \, d\tilde{V},$$

where N is the outward unit normal to ∂M and $d\tilde{V}$ is the volume form associated to the metric on ∂M induced from M.

6.1.14. We consider the sphere $\mathbb{S}^3$ of radius R as a submanifold in $\mathbb{R}^4$ with the induced Euclidean metric.

(a) Show that

$$F(u^1, u^2, u^3) = (R \cos u^1 \sin u^2 \sin u^3, R \sin u^1 \sin u^2 \sin u^3, R \cos u^2 \sin u^3, R \cos u^3),$$

where $(u^1, u^2, u^3) \in [0, 2\pi] \times [0, \pi]^2$ gives a parametrization for $\mathbb{S}^3$ that is homeomorphic to its image when restricted to the open set $V = (0, 2\pi) \times (0, \pi)^2$.

(b) Calculate the components of the metric tensor on the coordinate patch $F(V) = U$.

(c) Use part (b) to calculate the volume of a 3-sphere of radius R.

(d) Leaving R unspecified, consider the function $f(x^1, x^2, x^3, x^4) = (x^1)^2 + (x^2)^2 + (x^3)^2$ and calculate the volume integral $\int_{\mathbb{S}^3} f \, dV$. (Note that this integral would give the radius of gyration of the spherical shell of radius R about a principal axis – if such a thing existed in $\mathbb{R}^4$!)

6.1.15. Calculate the 5-volume of the 5-sphere $\mathbb{S}^5$ of radius R as a submanifold of $\mathbb{R}^6$.

6.1.16. (**ODE**) A loxodrome on the unit sphere $\mathbb{S}^2$ is a curve that makes a constant angle with all meridian lines. We propose to study analogues of loxodromes on $\mathbb{S}^3$. Consider the unit 3-sphere $\mathbb{S}^3$ with the parametrization from Problem 6.1.14. Set $R = 1$. We will call a loxodrome on $\mathbb{S}^3$ any curve γ such that γ' makes a constant angle of α_2 with $\dfrac{\partial}{\partial u^2}$ and a constant angle of α_3 with $\dfrac{\partial}{\partial u^3}$.

(a) Find equations that the components of γ must satisfy.

(b) Solve the differential equations we get in part (a). [Hint: Obtain u^1 and u^2 as functions of u^3. You might only be able to obtain one of these functions implicitly.]

6.1.17. Consider the function $r : \mathbb{R}^{n+1} - \{0\} \to \mathbb{S}^n$ given by $r(x) = x/\|x\|$.

(a) Using Example 6.1.14, prove that

$$\tilde{\omega} = r^*(dV_{\mathbb{S}^n}) = \frac{1}{\|x\|^{n+1}} \sum_{j=1}^{n+1} x^j \eta^j.$$

(b) Show that $\tilde{\omega}$ is closed but not exact in $\mathbb{R}^{n+1} - \{0\}$.

(c) Use $dV_{\mathbb{S}^n}$ to show that

$$\text{Vol}(\mathbb{S}^n) = (n + 1)\text{Vol}(\mathbb{B}^{n+1}),$$

where $\mathbb{B}^{n+1}$ is the unit ball in $\mathbb{R}^{n+1}$.

6.1.18. Let M be a Riemannian manifold, and let $f : M \to M$ be an isometry on M. Prove that $f^*(dV_M) = \pm dV_M$. (The isometry f is called *orientation-preserving* if $f^*(dV_M) = dV_M$ and *orientation-reversing* if $f^*(dV_M) = -dV_M$.)

6.1.19. Suppose that J and K are disjoint, compact, oriented, connected, smooth submanifolds of $\mathbb{R}^{n+1}$ whose dimensions are greater than 0 and such that $\dim J + \dim K = n$. Define the function Ψ by

$$\Psi : J \times K \longrightarrow \mathbb{S}^n$$

$$(x, y) \longmapsto \frac{y - x}{\|y - x\|}.$$

The *linking number* between J and K is defined as

$$\mathrm{link}(J, K) = \frac{1}{\mathrm{Vol}(\mathbb{S}^n)} \int_{J \times K} \Psi^*(dV_{\mathbb{S}^n}).$$

Prove Gauss's Linking Formula for the linking number of two closed space curves:

$$\mathrm{link}(C_1, C_2) = \frac{1}{4\pi} \int_I \int_J \frac{\det(\vec{\alpha}(u) - \vec{\beta}(v), \vec{\alpha}'(u), \vec{\beta}'(v))}{\|\vec{\alpha}(u) - \vec{\beta}(v)\|^3} \, du \, dv. \tag{6.13}$$

[Note that a closed curve is homeomorphic to a circle $\mathbb{S}^1$ so $J \times K$ is homeomorphic to a torus. Hence, we can view Ψ as a function $\mathbb{T}^2 = \mathbb{S}^1 \times \mathbb{S}^1 \to \mathbb{S}^2$.]

The following exercises involve the Hodge star operation, which is introduced in Appendix C.3.

6.1.20. Let (M, g) be an oriented Riemannian manifold. Section C.3 defines the Hodge star operator on inner product spaces. Given a form $\eta \in \Omega^k(M)$, at each $p \in M$, the Hodge star operator defines an isomorphism $\star : \bigwedge^k T_p M^* \to \bigwedge^{n-k} T_p M^*$.

  (a) Show that the Hodge star operator $\star$ is a vector bundle map $\bigwedge^k TM^* \to \bigwedge^{n-k} TM^*$ that leaves every base point fixed and that varies smoothly.

  (b) Show that for all functions $f : C^\infty(M)$, the Hodge star operator is given by $\star f = f \, dV_g$.

6.1.21. Consider $\mathbb{R}^n$ as a manifold with the standard Euclidean metric.

  (a) Calculate $\star dx^i$ for any $i = 1, \ldots, n$.

  (b) Set $n = 4$, and calculate $\star(dx^i \wedge dx^j)$.

6.1.22. Let (M, g) be an oriented Riemannian manifold. Prove the following identities for any vector field $X \in \mathfrak{X}(M)$.

  (a) $\mathrm{div}\, X = \star d \star X^\flat$, where div is the divergence operator defined in Problem 6.1.13.

  (b) $i_X dV_g = \star X^\flat$.

6.1.23. Let (M, g) be a Riemannian manifold. Consider the operation that consists of $\star d \star d$.

  (a) Show that $\star d \star d$ is an $\mathbb{R}$-linear operator $\Omega^k(M) \to \Omega^k(M)$ for $k < \dim M$.

  (b) Let $M \mathbb{R}^n$ be a standard Euclidean space. Recalling that $\Omega^0(M) = C^\infty(M)$, show that for any smooth function f,

$$\star d \star df = \nabla^2 f,$$

where ∇^2 is the usual Laplacian $\nabla^2 = \frac{\partial^2}{\partial(x^1)^2} + \cdots + \frac{\partial^2}{\partial(x^n)^2}$.

(c) Find an expression in coordinates of $\star d \star df$ for smooth functions $f : \mathbb{S}^2 \to \mathbb{R}$, express in (u, v) longitude-latitude coordinates as used in Example 6.1.14.

6.2 Connections and Covariant Differentiation

6.2.1 Motivation

Despite all the "heavy machinery" we have developed in order to create a theory of analysis on manifolds, we are still unable to calculate or even define certain things that are simple in $\mathbb{R}^n$.

For example, if $\gamma : [a, b] \to M$ is a smooth curve on a smooth manifold, we have no way at present to talk about the acceleration of γ. Let $p \in M$, with $\gamma(t_0) = p$. In Definition 3.3.1, we presented the tangent vector $\gamma'(t_0)$ at p as the operator $D_\gamma : C^1(M) \to \mathbb{R}$ that evaluates

$$D_\gamma(f) = \frac{d}{dt} f(\gamma(t)) \Big|_{t_0}.$$

In Section 3.3, we developed the linear algebra of expressions D_γ for curves γ through p. The vector space of such operators is what we called the tangent space $T_p M$.

Mimicking what one does in standard calculus, one could try to define the acceleration vector $\gamma''(t_0)$ at p as a limiting ratio as $t \to t_0$ of $\gamma'(t) - \gamma'(t_0)$ with $t - t_0$. However, what we just wrote does not make sense in the context of manifolds because $\gamma'(t)$ and $\gamma'(t_0)$ are not even in the same vector spaces and so their difference is not defined.

Another attempt to define the acceleration might follow Definition 3.3.1 and try to define $\gamma''(t_0)$ at p as the operator $D_\gamma^{(2)} : C^2(M) \to \mathbb{R}$, where

$$D_\gamma^{(2)}(f) = \frac{d^2}{dt^2} f(\gamma(t)) \Big|_{t_0}.$$

This operator is well defined and linear. However, $D_\gamma^{(2)}$ does not satisfy Leibniz's rule, and therefore, there does not exist another curve $\tilde{\gamma}$, with $\tilde{\gamma}(t_0) = p$ such that $D_\gamma^{(2)} = D_{\tilde{\gamma}}$. Hence, $D_\gamma^{(2)} \notin T_p M$. We could study properties for operators of the form $D_\gamma^{(2)}$ but, since the operators do not exist in any $TM^{\otimes p} \otimes TM^{*\otimes q}$, this is not the direction the theory of manifolds developed.

Another lack in our current theory is the ability to take partial derivatives or, more generally, directional derivatives of a vector field. Over $\mathbb{R}^n$, it is easy to define $\partial \vec{F}/\partial x^j$, where $\vec{F}$ is a vector field, and, under suitable differentiability conditions, $\partial \vec{F}/\partial x^j$ is again another vector field. In contrast, if X is a vector field over a smooth manifold M and U is a coordinate neighborhood of $p \in M$, one encounters the same problem with defining a vector $\partial_i X_p$ as one does in defining the acceleration of a curve.

A more subtle attempt to define partial derivatives of a vector field X on M in a coordinate chart would be to imitate the exterior differential of forms (see Definition 5.4.4) and set as a differential for X the quantity

$$\left(\frac{\partial X^i}{\partial x^j}\, dx^j\right) \otimes \partial_i = \frac{\partial X^i}{\partial x^j}\, \partial_i \otimes dx^j.$$

However, this *does not* define a tensor field of type $(1,1)$ on M. It is easiest to see this by showing how the components violate the transformational properties of a tensor field. Let $\bar{x} = (\bar{x}^1, \ldots, \bar{x}^n)$ be another system of coordinates that overlaps with the coordinate patch for $x = (x^1, \ldots, x^n)$. Call $\bar{X}$ the components of the vector field X in the $\bar{x}$ system. We know that

$$\bar{X}^j(\bar{x}) = \frac{\partial \bar{x}^j}{\partial x^i} X^i(x).$$

Taking a derivative with respect to $\bar{x}^k$ and inserting appropriate chain rules, we have

$$\frac{\partial \bar{X}^j}{\partial \bar{x}^k} = \frac{\partial}{\partial \bar{x}^k}\left(\frac{\partial \bar{x}^j}{\partial x^i}\right) X^i + \frac{\partial \bar{x}^j}{\partial x^i}\frac{\partial X^i}{\partial \bar{x}^k} = \frac{\partial^2 \bar{x}^j}{\partial x^l \partial x^i}\frac{\partial x^l}{\partial \bar{x}^k} X^i + \frac{\partial \bar{x}^j}{\partial x^i}\frac{\partial x^l}{\partial \bar{x}^k}\frac{\partial X^i}{\partial x^l}. \qquad (6.14)$$

The presence of the first term on the right-hand side shows that the collection of functions $\partial_j X^i$ do not satisfy the coordinate change properties of tensors given in (5.8).

6.2.2   Connection on a Vector Bundle

To solve the above conceptual problems, we need some coordinate-invariant way to compare vectors in tangent spaces at nearby points. This is the role of a connection. A *connection* on a smooth manifold is an additional structure that, though we introduce it in this chapter, is entirely independent of any Riemannian structure. We can in fact define a connection on any vector bundle over M. Since we require this generality for our applications, we introduce connections in this manner.

Definition 6.2.1. Let M be a smooth manifold, and let ξ be a vector bundle over M. Let $\mathcal{E}(\xi)$ denote the subspace of $\Gamma(\xi)$ of smooth global sections of ξ. A *connection* on ξ is a map

$$\nabla : \mathfrak{X}(M) \times \mathcal{E}(\xi) \to \mathcal{E}(\xi),$$

written $\nabla_X Y$ instead of $\nabla(X, Y)$, that satisfies the following:

1. For all vector fields $Y \in \mathcal{E}(\xi)$, $\nabla(\_, Y)$ is linear over $C^\infty(M)$, i.e., for all $f, g \in C^\infty(M)$,
$$\nabla_{fX + g\tilde{X}} Y = f\nabla_X Y + g\nabla_{\tilde{X}} Y.$$

2. For all vector fields $X \in \mathfrak{X}(M)$, $\nabla(X, \_)$ is linear over $\mathbb{R}$, i.e., for all $a, b \in \mathbb{R}$,

$$\nabla_X(aY + b\tilde{Y}) = a\nabla_X Y + b\nabla_X \tilde{Y}.$$

3. For all vector fields $X \in \mathfrak{X}(M)$, $\nabla(X, \_)$ satisfies the product rule

$$\nabla_X(fY) = (Xf)Y + f\nabla_X Y$$

for all $f \in C^\infty(M)$.

The vector field $\nabla_X Y$ in $\mathcal{E}(\xi)$ is called the *covariant derivative of Y in the direction of X.*

The symbol ∇ is pronounced "del." The defining properties of the covariant derivative are modeled after the properties of directional derivatives of vector fields on $\mathbb{R}^n$ (see Problem 6.2.1). Intuitively, the connection explicitly defines how to take a partial derivative in $\mathcal{E}(\xi)$ with respect to vector fields in TM. In fact, Problems 6.2.3 and 6.2.4 show that $\nabla_X Y$ depends only on the values of X_p in $T_p M$ and the values of Y in a neighborhood of p on M. Therefore $\nabla_X Y\big|_p$ is truly a directional derivative of Y at p in the direction X_p. Hence, we often write $\nabla_{X_p} Y$ instead of $\nabla_X Y\big|_p$.

For the applications in differential geometry, we will usually be interested in using connections on vector bundles of the form $\xi = TM^{\otimes r} \otimes TM^{*\otimes s}$. As it will turn out, connections on these vector bundles are closely related to possible connections on TM. Therefore, we temporarily restrict our attention to connections

$$\nabla : \mathfrak{X}(M) \times \mathfrak{X}(M) \to \mathfrak{X}(M).$$

Over a coordinate patch U of M, the defining properties are such that ∇ is completely determined once one knows its values for $X = \partial_i$ and $Y = \partial_j$. Since $\nabla_{\partial_i}\partial_j$ is another vector field in M, we write

$$\nabla_{\partial_i}\partial_j = \Gamma_{ij}^k \partial_k. \tag{6.15}$$

The components Γ_{ij}^k are smooth functions $M \to \mathbb{R}$.

Definition 6.2.2. The functions Γ_{ij}^k in Equation (6.15) are called the *Christoffel symbols* of the connection ∇.

As it turns out, there are no restrictions besides smoothness on the functions Γ_{ij}^k.

Proposition 6.2.3. *Let M^n be a smooth manifold, and let U be a coordinate patch on M. There is a bijective correspondence between connections on $\mathfrak{X}(U)$ and collections of n^3 smooth functions Γ_{ij}^k defined on U. The bijection is given by the formula*

$$\nabla_X Y = (X^i \partial_i Y^k + \Gamma_{ij}^k X^i Y^j)\partial_k. \tag{6.16}$$

Proof. First, suppose that ∇ is a connection on $\mathfrak{X}(U)$ and let $X, Y \in \mathfrak{X}(U)$. Then by the relations in Definition 6.2.1,

$$\nabla_X Y = \nabla_{(X^i \partial_i)}(Y^j \partial_j) = X^i \nabla_{\partial_i}(Y^j \partial_j)$$
$$= X^i(\partial_i Y^j)\partial_j + X^i Y^j \nabla_{\partial_i} \partial_j = X^i(\partial_i Y^j)\partial_j + X^i Y^j \Gamma_{ij}^k \partial_k.$$

Equation (6.16) holds by changing the variable of summation from j to k in the first term of the last expression. Conversely, if Γ_{ij}^k are any smooth functions on U and if we define an operator $\mathfrak{X}(U) \times \mathfrak{X}(U) \to \mathfrak{X}(U)$ by Equation (6.16), it is quick to check that the three criteria of Definition 6.2.1 hold. Thus, Equation (6.16) defines a connection on $\mathfrak{X}(U)$. $\square$

At a first pass, the definition of a connection may seem rather burdensome and unintuitive. However, the component description given in (6.16) has the same format of something we have already seen. We encountered the same formula in (2.11) in the context of calculating partial derivatives of the components of a vector field expressed in reference to a variable frame in $\mathbb{R}^n$. In (2.11), the component functions Γ_{ij}^k precisely play the role described in (6.15). Consequently, in developing the concept of connections on the tangent bundle to a manifold, we could have started from (6.15) and worked back to the properties listed in Definition 6.2.1. Definition 6.2.1 is therefore simply a coordinate-free description of (6.16), which arose from a relationship that first appears in the analysis of moving frames in $\mathbb{R}^n$.

It is important to point out that the Lie derivative is not a connection because it violates the first criterion in Definition 6.2.1, namely the Lie derivative $\mathcal{L}_X Y$ is only $\mathbb{R}$-linear in X as opposed to $C^\infty(M)$-linear.

Example 6.2.4 (The Flat Connection on $\mathbb{R}^n$). In $\mathbb{R}^n$, the vector fields ∂_i are constant, and we identify them with the standard basis vector $\vec{e}_i$. According to Proposition 6.2.3, a connection exists for any collection of n^3 functions. However, if $Y = Y^j \partial_j$ is a vector field in $\mathbb{R}^n$, our usual way of taking partial derivatives of vector fields is

$$\nabla_{\partial_i} Y = \frac{\partial Y^j}{\partial x^i} \partial_j,$$

which takes partial derivatives componentwise on Y. By (6.16), we see that $\Gamma_{jk}^i = 0$ for all choices of the indices. A connection with this property is called a flat connection over the coordinate patch.

Even though the symbols Γ_{jk}^i resemble our notation for the components of a $(1,2)$-tensor, a connection is not a tensor field. The reason derives from the fact that $\partial_j X^i$ is not a $(1,1)$-tensor field. In fact, from (6.14) and the transformational properties of a vector field between overlapping coordinate systems on M, we can deduce the transformational properties of the component functions of a connection.

Proposition 6.2.5. *Let ∇ be a connection on $\mathfrak{X}(M)$. Suppose that U and $\bar{U}$ are overlapping coordinate patches, and denote by Γ_{jk}^i and $\bar{\Gamma}_{mn}^l$ the component functions* *of ∇ over these patches, respectively. Then over $U \cap \bar{U}$, the component functions are related to each other by*

$$\bar{\Gamma}^l_{mn} = \frac{\partial x^j}{\partial \bar{x}^m} \frac{\partial x^k}{\partial \bar{x}^n} \frac{\partial \bar{x}^l}{\partial x^i} \Gamma^i_{jk} - \frac{\partial x^j}{\partial \bar{x}^m} \frac{\partial x^k}{\partial \bar{x}^n} \frac{\partial^2 \bar{x}^l}{\partial x^j \partial x^k}.$$

Proof. (Left as an exercise for the reader.) $\qquad \square$

The astute reader might have noticed already from Definition 6.2.1 that a connection is not a tensor field of type $(1, 2)$. If an operator $F : \mathfrak{X}(M) \times \mathfrak{X}(M) \to \mathfrak{X}(M)$ were a tensor field in $TM \otimes TM^{*\otimes 2}$, then $F(X, \_)$ would be linear in $C^\infty(M)$ and would not satisfy the third property in Definition 6.2.1.

Example 6.2.6 (Polar Coordinates). We consider the connection ∇ on $\mathbb{R}^2$ that is flat over the Cartesian coordinate system. We calculate the components of ∇ with respect to polar coordinates. We could calculate the Christoffel symbols from Proposition 6.2.3, but instead, we use Proposition 6.2.5. Set $x^1 = x$, $x^2 = y$, $\bar{x}^1 = r$, and $\bar{x}^2 = \theta$, and denote by $\Gamma^i_{jk} = 0$ the Christoffel symbols for the flat connection on $\mathbb{R}^2$ and let $\bar{\Gamma}^l_{mn}$ denote the Christoffel symbols for ∇ in polar coordinates.

By direct calculation,

$$\bar{\Gamma}^2_{12} = -\sum_{j,k=1}^2 \frac{\partial x^j}{\partial \bar{x}^1} \frac{\partial x^k}{\partial \bar{x}^2} \frac{\partial^2 \bar{x}^2}{\partial x^j \partial x^k}$$

$$= -\left(-r \cos\theta \sin\theta \frac{\partial^2 \theta}{\partial x^2} + r^2 \cos^2\theta \frac{\partial^2 \theta}{\partial x \partial y} - r \sin^2\theta \frac{\partial^2 \theta}{\partial y \partial x} + r \sin\theta \cos\theta \frac{\partial^2 \theta}{\partial y^2} \right)$$

$$= -\left(-r \sin\theta \cos\theta \frac{2xy}{(x^2+y^2)^2} + r(\cos^2\theta - \sin^2\theta) \frac{y^2 - x^2}{(x^2+y^2)^2} - r \sin\theta \cos\theta \frac{2xy}{(x^2+y^2)^2} \right)$$

$$= \frac{1}{r} \left(2\sin^2\theta \cos^2\theta + (\cos^2\theta - \sin^2\theta)^2 + 2\sin^2\theta \cos^2\theta \right) = \frac{1}{r}.$$

It is not hard (though perhaps a little tedious) to show that

$$\bar{\Gamma}^1_{11} = 0, \qquad \bar{\Gamma}^1_{12} = \bar{\Gamma}^1_{21} = 0, \qquad \bar{\Gamma}^1_{22} = -r,$$

$$\bar{\Gamma}^2_{11} = 0, \qquad \bar{\Gamma}^2_{12} = \bar{\Gamma}^2_{21} = \frac{1}{r}, \qquad \bar{\Gamma}^2_{22} = 0.$$

We now wish to extend our discussion of connections on TM to connections on any tensor bundle $TM^{\otimes r} \otimes TM^{*\otimes s}$ in a natural manner for any pair (r, s). Two situations are settled: (1) if $f \in TM^0 = C^\infty(M)$, then we want $\nabla_X f = X(f)$, the expected directional derivative; and (2) if $X \in TM$, then the connection should follow the properties described in Definition 6.2.1 and Proposition 6.2.3.

Lemma 6.2.7. *Let M be a smooth manifold, and let ∇ be a connection on TM. For each pair $(r, s) \in \mathbb{N}^2$, there exists a unique connection on the tensor bundle*

$TM^{\otimes r} \otimes TM^{*\otimes s}$, also denoted ∇, given by the following conditions for any vector field X:

1. Consistency: ∇ is equal to the connection given on TM.

2. Directional derivative: $\nabla_X f = X(f)$ for all $f \in C^\infty(M) = TM^0$.

3. Contraction product rule: for all covector fields ω and vector fields Y

$$\nabla_X(\omega(Y)) = (\nabla_X \omega)(Y) + \omega(\nabla_X Y).$$

4. Tensor product rule: for all tensor fields A and B of any type,

$$\nabla_X(A \otimes B) = (\nabla_X A) \otimes B + A \otimes (\nabla_X B).$$

We omit the proof of this lemma since it is merely constructive. Property 3 determines uniquely how to define $\nabla_X \omega$ for any covector field and then Property 4 extends the connection to all other types of tensors.

Definition 6.2.8. Let M be a smooth manifold. We call ∇ an *affine* connection on $TM^{\otimes r} \otimes TM^{*\otimes s}$ if it satisfies the conditions of Lemma 6.2.7.

6.2.3   Covariant Derivative

Let ∇ be an affine connection on a smooth manifold M. Let F be a tensor field of type (r, s). Then the mapping ∇F that maps a vector field X to $\nabla_X F$ is a $C^\infty(M)$-linear transformation from $\mathfrak{X}(M)$ to the space of tensor fields of type (r, s). Thus, for each $p \in M$, $\nabla F\big|_p$ is a linear transformation $T_p M \to T_p M^{\otimes r} \otimes T_p M^{*\otimes s}$, so by Proposition 4.4.10,

$$\nabla F\big|_p \in \mathrm{Hom}(T_p M, T_p M^{\otimes r} \otimes T_p M^{*\otimes s}) = T_p M^{\otimes r} \otimes T_p M^{*\otimes(s+1)}.$$

Furthermore, since $\nabla F\big|_p$ varies smoothly with p, then ∇F is a smooth section of the tensor bundle $T_p M^{\otimes r} \otimes T_p M^{*\otimes s+1}$, and hence, it is a tensor field of type $(r, s+1)$.

Definition 6.2.9. Let M be a smooth manifold equipped with an affine connection ∇. If F is a tensor field of type (r, s), then the tensor field ∇F of type $(r, s+1)$ is called the *covariant derivative* of F.

Proposition 6.2.10. *Let F be a tensor field of type (r, s) over a manifold M. Suppose that F has components $F^{i_1 \cdots i_r}_{j_1 \cdots j_s}$ over a coordinate chart U. Then the components of the covariant derivative ∇F are*

$$F^{i_1 \cdots i_r}_{j_1 \cdots j_s; k} \overset{\text{def}}{=} \frac{\partial F^{i_1 \cdots i_r}_{j_1 \cdots j_s}}{\partial x^k} + \sum_{\alpha=1}^{r} \Gamma^{i_\alpha}_{k\mu} F^{i_1 \cdots i_{\alpha-1} \mu i_{\alpha+1} \cdots i_r}_{j_1 \cdots j_s}$$

$$- \sum_{\beta=1}^{s} \Gamma^{\mu}_{k j_\beta} F^{i_1 \cdots i_r}_{j_1 \cdots j_{\beta-1} \mu j_{\beta+1} \cdots j_s}. \tag{6.17}$$

(Some authors use the notation $F^{i_1\cdots i_r}_{j_1\cdots j_s|k}$ for the components of the covariant derivative.) The notation in Equation (6.17) is a little heavy, but it should become clear with a few examples. If ω is a 1-form, then $\nabla\omega$ is a 2-form with local components given by

$$\nabla\omega = \omega_{j;k}dx^j \otimes dx^k, \quad \text{where} \quad \omega_{j;k} = \partial_k\omega_j - \Gamma^\mu_{kj}\omega_\mu.$$

Similarly, if A^{ij}_k are the components of a $(2,1)$-tensor field A, then ∇A is a $(2,2)$-tensor field with local components given by

$$\nabla A = A^{ij}_{k;l}\partial_i \otimes \partial_j \otimes dx^k \otimes dx^l, \quad \text{where} \quad A^{ij}_{k;l} = \frac{\partial A^{ij}_k}{\partial x^l} + \Gamma^i_{l\mu}A^{\mu j}_k + \Gamma^j_{l\mu}A^{i\mu}_k - \Gamma^\mu_{lk}A^{ij}_\mu.$$

6.2.4 Levi-Civita Connection

Proposition 6.2.3 gives considerable freedom in choosing the components of a connection. In the context of Riemannian geometry, it is natural to wish for a connection that is in some sense "nice" with respect to the metric on the manifold. The following theorem is motivated by results in classical differential geometry of surfaces discussed in [5, Section 7.2] but is so central to Riemannian geometry that it is sometimes called the "miracle" of Riemannian geometry [45].

Theorem 6.2.11 (Levi-Civita Theorem). *Let (M,g) be a Riemannian manifold. There exists a unique affine connection ∇ that satisfies the following two conditions:*

1. Compatibility: *∇g is identically 0.*

2. Symmetry: *for all $X, Y \in \mathfrak{X}(M)$, $[X,Y] = \nabla_X Y - \nabla_Y X$.*

A few comments are in order before we prove this theorem. The condition that $\nabla g = 0$ intuitively says that ∇ is flat with respect to the metric. We say that ∇ is compatible with the metric. We leave it as an exercise for the reader (Problem 6.2.12) to show that if we write $g = \langle\,,\,\rangle$, then ∇g is identically 0 (i.e., $g_{ij;k} = 0$ in local coordinates) if and only if

$$\nabla_X(\langle Y, Z\rangle) = \langle\nabla_X Y, Z\rangle + \langle Y, \nabla_X Z\rangle. \tag{6.18}$$

Hence, if ∇ is compatible with the metric g, then it satisfies a product rule with respect to the metric.

By Problem 6.2.14, condition 2 implies that over any coordinate patch of the manifold, the Christoffel symbols Γ^i_{jk} of the connection ∇ satisfy $\Gamma^i_{jk} = \Gamma^i_{kj}$, which justifies the terminology of a symmetric connection.

Definition 6.2.12. The connection ∇ described in Theorem 6.2.11 is called the *Levi-Civita connection* or the *Riemannian connection* with respect to the metric g on M.

Proof of Theorem 6.2.11. Let $X, Y, Z \in \mathfrak{X}(M)$, and denote $g = \langle\,,\,\rangle$. Since $\langle X, Y \rangle$ is a smooth function on M, then we write $\nabla_Z(\langle X, Y \rangle) = Z\langle X, Y \rangle$.

Now suppose that such a connection ∇ exists. Then

$$X\langle Y, Z \rangle = \langle \nabla_X Y, Z \rangle + \langle Y, \nabla_X Z \rangle, \tag{6.19}$$

$$Y\langle Z, X \rangle = \langle \nabla_Y Z, X \rangle + \langle Z, \nabla_Y X \rangle, \tag{6.20}$$

$$Z\langle X, Y \rangle = \langle \nabla_Z X, Y \rangle + \langle X, \nabla_Z Y \rangle. \tag{6.21}$$

Adding Equations (6.19) and (6.20) and subtracting Equation (6.21), using the symmetry of the metric, we get

$$X\langle Y, Z \rangle + Y\langle Z, X \rangle - Z\langle X, Y \rangle$$
$$= \langle \nabla_X Y - \nabla_Y X, Z \rangle + \langle \nabla_X Z - \nabla_Z X, Y \rangle + \langle \nabla_Y Z - \nabla_Z Y, X \rangle + 2\langle Z, \nabla_Y X \rangle$$

Using the fact that ∇ is symmetric, we have

$$X\langle Y, Z \rangle + Y\langle Z, X \rangle - Z\langle X, Y \rangle$$
$$= \langle [X, Y], Z \rangle + \langle [X, Z], Y \rangle + \langle [Y, Z], X \rangle + 2\langle Z, \nabla_X Y \rangle,$$

and thus

$$\langle Z, \nabla_X Y \rangle = \frac{1}{2}\left(X\langle Y, Z \rangle + Y\langle Z, X \rangle - Z\langle X, Y \rangle \right.$$
$$\left. - \langle [X, Y], Z \rangle - \langle [X, Z], Y \rangle - \langle [Y, Z], X \rangle \right). \tag{6.22}$$

Now a connection on any coordinate patch is uniquely determined by its Christoffel symbols. However, setting $X = \partial_i$, $Y = \partial_j$ and $Z = \partial_k$, (6.22) gives a method to obtain the Christoffel symbols of ∇ strictly in terms of the metric. Hence, if a connection as described in the theorem exists, then it is unique.

To show that such a connection exists, simply start by defining ∇ using the identity in (6.22). Then it is not hard to show that the connection is both symmetric and compatible with g. $\qquad\square$

Proposition 6.2.13. *Let (M^n, g) be a smooth Riemannian manifold. Then over a coordinate patch of M with coordinates $(x^1, \ldots, x^n)$, the Christoffel symbols of the Levi-Civita connection are given by*

$$\Gamma^i_{jk} = \sum_{l=1}^n \frac{1}{2} g^{il} \left(\frac{\partial g_{kl}}{\partial x^j} + \frac{\partial g_{lj}}{\partial x^k} - \frac{\partial g_{jk}}{\partial x^l} \right), \tag{6.23}$$

where g^{ij} are the entries to the inverse matrix of (g_{kl}).

Proof. Set $g = \langle\,,\,\rangle$, and let $X = \partial_i$, $Y = \partial_j$, and $Z = \partial_k$. By the Levi-Civita connection defined in (6.22), we have

$$\left\langle \partial_k, \sum_{l=1}^n \Gamma^l_{ij} \partial_l \right\rangle = \frac{1}{2}(\partial_i\langle \partial_j, \partial_k \rangle + \partial_j\langle \partial_k, \partial_i \rangle - \partial_k\langle \partial_i, \partial_j \rangle$$

$$- \langle [\partial_i, \partial_j], \partial_k \rangle - \langle [\partial_i, \partial_k], \partial_j \rangle - \langle [\partial_j, \partial_k], \partial_i \rangle).$$

However, the smoothness condition implies that $[\partial_i, \partial_j] = 0$ for any indices i, j. Furthermore, by definition, $g_{ij} = \langle \partial_i, \partial_j \rangle$, so by the linearity of the metric on the left-hand side,

$$\sum_{l=1}^{n} g_{kl} \Gamma_{ij}^l = \frac{1}{2} \left(\partial_i g_{jk} + \partial_j g_{ki} - \partial_k g_{ij} \right).$$

The proposition follows by multiplying (and contracting) by g^{kl}, the components of the inverse of (g_{kl}). $\qquad \square$

The reader who is familiar with the differential geometry of surfaces has already seen Proposition 6.2.13 but in a more limited context. In Section 7.2 of [5], the authors talk about Gauss's equations for a regular surface over a parametrization $\vec{X}$. In that section, one sees that even though the normal vector to a surface is not an intrinsic property, $\vec{X}_{ij} \cdot \vec{X}_k$ is intrinsic and in fact is given by the Christoffel symbols of the first kind, which are precisely those in Equation (6.23), though with $n = 2$. This is not a mere coincidence. In defining the Levi-Civita connection, that we might want ∇ to be compatible with g made intuitive sense. However, the stipulation that we would want ∇ to be symmetric may have seemed somewhat artificial at the time. It is very interesting that the two conditions in Theorem 6.2.11 lead to Christoffel symbols that match those defined for surfaces in classical differential geometry.

It is possible to develop a theory of embedded submanifolds M^m of $\mathbb{R}^n$ following the theory of regular surfaces in $\mathbb{R}^3$. Mimicking the presentation in [5, Section 7.2], if $\vec{X}$ is a parametrization of a coordinate patch of M, then, by setting

$$\frac{\partial^2 \vec{X}}{\partial x^i \partial x^j} = \sum_{k=1}^{m} \Gamma_{ij}^k \frac{\partial \vec{X}}{\partial x^k} + (\text{Normal component}),$$

the components Γ_{ij}^k are again the Christoffel symbols of the second kind, given by the same formula in (6.23). This shows that for submanifolds of a Euclidean space, the Levi-Civita connection on a Riemannian manifold is essentially the flat connection on $\mathbb{R}^n$ restricted to the manifold.

One of the beauties of the condition that $\nabla g = 0$ is that the process of raising and lowering indices commutes with taking the covariant derivative associated to the Levi-Civita connection. In components, this means for example that if $A_j^i = g^{il} A_{jl}$, then

$$A_{j;k}^i = g^{il} A_{jl;k}.$$

This follows because in components $\nabla g = 0$ identically means that $g_{ij;k} = 0$ for the Levi-Civita covariant derivative. Then since $g_{ij} g^{jl} = \delta_i^l$ is a numerical tensor, we have

$$0 = \delta_{i;k}^l = g_{ij;k} g^{jl} + g_{ij} g_{;k}^{jl} = g_{ij} g_{;k}^{jl},$$

which implies that $g_{;k}^{jl}$ since g_{ij} is invertible. Thus, in our specific example,

$$A_{j;k}^i = g_{;k}^{ij} A_{jl} + g^{ij} A_{jl;k} = g^{ij} A_{jl;k}.$$

In this example we raised the index, but it is clear from the product rule and the fact that $\nabla g = 0$ and $\nabla g^{-1} = 0$, that this property holds for all tensors.

6.2.5   Divergence Operator

We finish this section with a comment on the divergence operator on tensors introduced in Problem 6.1.13. We will show in Problem 6.2.16 that, using the Levi-Civita connection, the divergence operator on a vector field $X \in \mathfrak{X}(M)$ can be written as

$$\operatorname{div} X = X^i_{;i}. \tag{6.24}$$

This motivates, first, the definition of the divergence of any tensor T of type (r,s), with $r \geq 1$, on a Riemannian manifold. If T has components $T^{i_1 \cdots i_r}_{j_1 \cdots j_s}$ in a coordinate system, then the divergence of T, written $\operatorname{div} T$ or $\nabla \cdot T$, is the tensor field of type $(r-1, s)$ with component functions

$$T^{\alpha i_2 \cdots i_r}_{j_1 j_2 \cdots j_s ;\alpha} = \nabla_{\partial_\alpha} T^{\alpha i_2 \cdots i_r}_{j_1 \cdots j_s}.$$

Similarly, we can take the divergence with respect to any contravariant index but we must specify which index. If the index is not specified, we assume the divergence is taken with respect to the first index.

We can also define the divergence of a covariant index by raising that index first. Thus, for example, if ω is a 1-form, then

$$\operatorname{div} \omega = (g^{ij} \omega_j)_{;i}. \tag{6.25}$$

Problem 6.2.16 shows that whether one raises the index before or after the covariant derivative is irrelevant.

PROBLEMS

6.2.1. Consider the special case of the manifold $M = \mathbb{R}^3$. Let X be the constant vector field $\vec{v}$, and let $\mathfrak{X}(\mathbb{R}^3)$ be the space of vector fields $\mathbb{R}^3 \to \mathbb{R}^3$. Show that the usual partial derivative $D_{\vec{v}}$ applied to $\mathfrak{X}(\mathbb{R}^3)$ satisfies conditions 2 and 3 of Definition 6.2.1.

6.2.2. Recall the permutation symbol defined in (4.35). Let M be a three-dimensional manifold equipped with a symmetric affine connection. Let A and B be vector fields on M. Show that

$$\varepsilon^{ijk} A_{j;i} = \varepsilon^{ijk} \frac{\partial A_j}{\partial x^i}$$

and that

$$(\varepsilon^{ijk} A_j B_k)_{;i} = \varepsilon^{ijk} A_{j;i} B_k - \varepsilon^{ijk} A_k B_{j;i}.$$

If $M = \mathbb{R}^3$, explain how the latter formula is equivalent to $\vec{\nabla} \cdot (\vec{A} \times \vec{B}) = (\vec{\nabla} \times \vec{A}) \cdot \vec{B} - \vec{A} \cdot (\vec{\nabla} \times \vec{B})$.

6.2.3. Let ∇ be a connection on a vector bundle ξ over a smooth manifold M. Prove that if $X = \tilde{X}$ and $Y = \tilde{Y}$ over a neighborhood of p, then

$$\nabla_X Y \big|_p = \nabla_{\tilde{X}} \tilde{Y} \big|_p.$$

6.2.4. Let ∇ be a connection on a vector bundle ξ over a smooth manifold M. Use the result of Problem 6.2.3 to show that $\nabla_X Y\big|_p$ depends only on X_p and the values of Y in a neighborhood of p.

6.2.5. Prove Proposition 6.2.5.

6.2.6. Prove that the Levi-Civita connection for the Euclidean space $\mathbb{R}^n$ is such that $\nabla_X Y = X(Y^k)\partial_k$.

6.2.7. Consider the open first quadrant $U = \{(u,v) \in \mathbb{R}^2 \mid u > 0,\ v > 0\}$, and equip U with the metric

$$(g_{ij}) = \begin{pmatrix} 1 & \frac{1}{\sqrt{u^2+v^2}} \\ \frac{1}{\sqrt{u^2+v^2}} & \frac{1}{u^2} \end{pmatrix}.$$

Calculate the Christoffel symbols for the associated Levi-Civita connection.

6.2.8. Let M be a two-dimensional manifold, and suppose that on a coordinate patch (x^1, x^2), the metric is of the form

$$g = \begin{pmatrix} f(r) & 0 \\ 0 & f(r) \end{pmatrix}, \qquad \text{where } r^2 = (x^1)^2 + (x^2)^2.$$

Find the function $f(r)$ that gives a flat connection.

6.2.9. Consider the cylinder in $\mathbb{S}^2 \times \mathbb{R}$ in $\mathbb{R}^4$ given by the parametrization

$$F(u^1, u^2, u^3) = (\cos u^1 \sin u^2, \sin u^1 \sin u^2, \cos u^2, u^3)$$

and equip it with the metric induced from $\mathbb{R}^4$. Over the open coordinate patch $U = (0, 2\pi) \times (0, \pi) \times \mathbb{R}$, calculate the metric coefficients and the Christoffel symbols for the Levi-Civita connection.

6.2.10. Consider the unit sphere $\mathbb{S}^3$ as a submanifold of $\mathbb{R}^4$ with the induced metric. Consider the coordinate patch on $\mathbb{S}^3$ given by the parametrization in 6.1.14(a). Calculate one nonzero Christoffel symbol Γ^i_{jk}. (It would be quite tedious to calculate all of the symbols since there could be as many as 27 of them.) [Hint: Show that the conditions of Problem 6.2.13 apply to this coordinate patch and use the result.]

6.2.11. Finish calculating directly the Christoffel symbols in Example 6.2.6.

6.2.12. Let (M, g) be a Riemannian manifold. Prove that a connection ∇ satisfies $\nabla g = 0$ identically if and only if (6.18) holds where $g = \langle\,,\,\rangle$.

6.2.13. Let (M, g) be a Riemannian manifold and let U be an orthogonal coordinate patch, i.e., $g_{ij} = 0$ if $i \neq j$ over U. Let ∇ be the Levi-Civita connection on M.

(a) Prove that on U the Christoffel symbols $\Gamma^k_{ij} = 0$ unless $k = i$, $i = j$, or $k = j$.

(b) Show that ∇ can be specified on U by $2n^2 - n$ smooth functions, i.e., there are at most that many distinct nonzero Christoffel symbols.

(c) Show that

$$\Gamma^k_{ii} = \pm\frac{1}{2}g^{kk}\frac{\partial g_{ii}}{\partial x^k}, \qquad \text{and} \qquad \Gamma^k_{ik} = \frac{1}{2}g^{kk}\frac{\partial g_{kk}}{\partial x^i}$$

where there is no summation in either of these formulas and where the sign of $\pm$ is $+1$ if $i = k$ and -1 if $i \neq k$.

6.2.14. Let M be a smooth manifold, and let ∇ be a connection on TM. Define a map $\tau : \mathfrak{X}(M) \times \mathfrak{X}(M) \to \mathfrak{X}(M)$ by

$$\tau(X,Y) = \nabla_X Y - \nabla_Y X - [X,Y].$$

(a) Show that τ is a tensor field of type $(1,2)$. This is called the *torsion tensor* associated to the connection ∇.

(b) Prove that the components of τ with respect to a basis are $\tau_{ij}^k = \Gamma_{ij}^k - \Gamma_{ji}^k$.

(c) The connection ∇ is called symmetric if its torsion vanishes identically. Deduce that ∇ is symmetric if and only if over every coordinate patch U, the component functions satisfy $\Gamma_{ij}^k = \Gamma_{ji}^k$.

6.2.15. Let ∇ be an affine connection on M. Prove that $\nabla + A$ is an affine connection, where A is a $(1,2)$-tensor field. Conversely, prove that every affine connection is of the form $\nabla + A$ for some $(1,2)$-tensor field A.

6.2.16. Consider the divergence operator introduced in Problem 6.1.13 and discussed at the end of this section.

(a) Show from the definition in Problem 6.1.13 that

$$\operatorname{div} X = X_{;i}^i,$$

where we've used the Levi-Civita connection to take the covariant derivative.

(b) Consider the definition in (6.25) for the divergence on a 1-form. Show that

$$\operatorname{div} \omega = (g^{ij}\omega_j)_{;i} = g^{ij}\omega_{j;i}.$$

6.2.17. Let $f \in C^\infty(M)$ be a smooth function on a manifold M equipped with any affine connection ∇. Show that

$$f_{;j;i} - f_{;i;j} = -\tau_{ij}^k f_{;k},$$

where τ is the torsion tensor from Exercise 6.2.14. Conclude that if ∇ is symmetric, then $f_{;i;j} = f_{;j;i}$.

6.2.18. Let M be a smooth manifold, let $\eta \in \Omega^2(M)$ be a 2-form, and let ∇ be any symmetric connection on M. Show that in any coordinate system,

$$c_{\alpha\beta\gamma} = \eta_{\alpha\beta;\gamma} + \eta_{\beta\gamma;\alpha} + \eta_{\gamma\alpha;\beta} = \partial_\gamma \eta_{\alpha\beta} + \partial_\alpha \eta_{\beta\gamma} + \partial_\beta \eta_{\gamma\alpha}. \tag{6.26}$$

Show that if we write $\eta = \frac{1}{2}\eta_{\alpha\beta}dx^\alpha \wedge dx^\beta$, then the left-hand side of (6.26) is the component of $d\eta$ in the basis $dx^\alpha \wedge dx^\beta \wedge dx^\gamma$ in the sense that

$$d\omega = c_{\alpha\beta\gamma}dx^\alpha \wedge dx^\beta \wedge dx^\gamma,$$

where we sum over all $\alpha, \beta, \gamma = 1, \ldots, n$.

6.2.19. Let (M,g) be a Riemannian metric with Levi-Civita connection ∇. Show that over every coordinate patch,

$$\frac{\partial(\ln\sqrt{\det g})}{\partial x^k} = \Gamma_{jk}^j,$$

where one sums over j on the right-hand side. [Hint: Use a result in Problem 2.3.12.]

6.2.20. Let ∇ be an affine connection on M, and let U be a coordinate patch on M.

    (a) Show that there exists a unique matrix of 1-forms ω_i^j defined on U such that

$$\nabla_X \partial_i = \omega_i^j(X)\partial_j$$

    for all $X \in \mathfrak{X}(M)$. (The matrix ω_i^j is called the *connection 1-forms* for this coordinate system.)

    (b) Suppose that (M, g) is a Riemannian manifold. Show that ∇ is compatible with the metric g if, over any coordinate system U,

$$g_{jk}\omega_i^k + g_{ik}\omega_j^k = dg_{ij}.$$

6.3   Vector Fields along Curves; Geodesics

Suppose we think of the trajectory of a particle on a manifold M. One would describe it as curve $\gamma(t)$ on M. Furthermore, in order to develop a theory of dynamics on manifolds, one would need to be able to make sense of the acceleration of the curve or of higher derivatives of the curve. In this section, we define vector fields on curves on manifolds. Once we define a covariant derivative of a vector field on a curve, we can then discuss parallel vector fields on the curve and the acceleration field along the curve. We then show that defining a geodesic as a curve whose acceleration is identically 0 leads to the classical understanding of a geodesic as a path of minimum length in some sense.

6.3.1   Vector Fields along Curves

Definition 6.3.1. Let M be a smooth manifold, and let $\gamma : I \to M$ be a smooth curve in M, where I is an interval in $\mathbb{R}$. We call V a *vector field along* γ if for each $t \in I$, $V(t)$ is a tangent vector in $T_{\gamma(t)}M$ and if V defines a smooth map $I \to TM$. We denote by $\mathfrak{X}_\gamma(M)$ the set of all smooth vector fields on M along γ.

    A vector field along a curve is not necessarily the restriction of a vector field on M to $\gamma(I)$. For example, whenever a curve self-intersects, $\gamma(t_0) = \gamma(t_1)$, with $t_0 \neq t_1$, but since $V(t_0) \neq V(t_1)$ there exists no vector field Y on M such that $V(t) = Y_{\gamma(t)}$ for all $t \in I$ (see Figure 6.3). If V is the restriction of a vector field Y, then we say that V is *induced* from Y or that V *extends* to Y.

Proposition 6.3.2. *Let M be a smooth manifold with an affine connection ∇, and let $\gamma : I \to M$ be a smooth curve on M. There exists a unique operator $D_t : \mathfrak{X}_\gamma(M) \to \mathfrak{X}_\gamma(M)$ (also denoted by $\frac{d}{dt}$) such that:*

    1. $D_t(V + W) = D_t V + D_t W$ for all $V, W \in \mathfrak{X}_\gamma(M)$.

    2. $D_t(fV) = \frac{df}{dt}V + fD_t V$ for all $V \in \mathfrak{X}_\gamma(M)$ and all $f \in C^\infty(I)$.

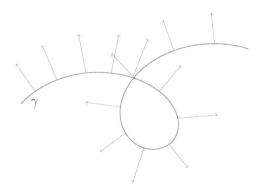

Figure 6.3: A nonextendable vector field on a curve.

3. *If V extends to a vector field $Y \in \mathfrak{X}(M)$, then $D_t V = \nabla_{\gamma'(t)} Y$.*

Note that the last condition makes sense by the fact that $\nabla_X Y|_p$ only depends on the values of Y in a neighborhood of p and on the value of X_p (see Exercise 6.2.4).

Before proving Proposition 6.3.2, we introduce the dot notation for derivatives. The only purpose is to slightly simplify our equations' notation. If $x(t)$ is a real-valued function of a real variable, we write

$$\dot{x}(t) \overset{\text{def}}{=} x'(t) = \frac{dx}{dt} \qquad \text{and} \qquad \ddot{x}(t) \overset{\text{def}}{=} x''(t) = \frac{d^2x}{dt^2}.$$

The dot notation is common in physics in the context of taking derivatives with respect to time. Therefore, $\dot{x}$ is usually used when one uses the letter t as the only independent variable for the function x.

Proof of Proposition 6.3.2. Let us first suppose that an operator D_t with Properties 1–3 exists. Let U be a coordinate patch of M with coordinates $x = (x^1, \ldots, x^n)$. For any $V \in \mathfrak{X}_\gamma(M)$, write $V = v^i \partial_i$ where $v^i \in C^\infty(I)$ are smooth functions over I. By Conditions 1 and 2 we have

$$D_t V = \dot{v}^j \partial_j + v^j D_t(\partial_j).$$

Now if we write $\gamma(t) = (\gamma^1(t), \ldots, \gamma^n(t))$ for the coordinate functions of γ over U, then $\gamma'(t) = \sum_{i=1}^n \dot{\gamma}^i \partial_i$. Thus, by Condition 3,

$$D_t(\partial_j) = \nabla_{\gamma'(t)} \partial_j = \sum_{i=1}^n \dot{\gamma}^i \nabla_{\partial_i} \partial_j = \dot{\gamma}^i \Gamma_{ij}^k \partial_k.$$

Hence, we deduce the following formula for $D_t V$ in coordinates over U:

$$D_t V = \left(\frac{dv^j}{dt} \partial_j \right) + \left(\Gamma_{ij}^k \frac{d\gamma^i}{dt} v^j \partial_k \right) = \left(\dot{v}^k + \Gamma_{ij}^k \dot{\gamma}^i v^j \right) \partial_k. \qquad (6.27)$$

Equation (6.27) shows that if there does exist an operator satisfying Conditions 1–3, then the operator is unique. To prove existence over all of M, we define D_t^α by (6.27) on each coordinate chart U_α. However, since D_t^α is unique on each coordinate chart, then $D_t^\alpha = D_t^\beta$ over $U_\alpha \cap U_\beta$ if U_α and U_β are overlapping coordinate charts. Hence, as α ranges over all coordinate charts in the atlas, the collection of operators D_t^α extends to a single operator D_t over all of M. $\qquad\square$

Note that Equation (6.27) in the above proof gives the formula for D_t over a coordinate patch of M. In particular, the expression in the parentheses on the right gives the component functions (in the index k) for $D_t V$.

Definition 6.3.3. The operator $D_t : \mathfrak{X}_\gamma(M) \to \mathfrak{X}_\gamma(M)$ defined in Proposition 6.3.2 is called the *covariant derivative along* γ.

In the context of Riemannian manifolds, the covariant derivative along a curve has the following interesting property.

Proposition 6.3.4. *Let γ be a smooth curve on a Riemannian manifold (M, g) equipped with the Levi-Civita connection. Write $g = \langle\,,\,\rangle$. Let V and W be vector fields along γ. Then*

$$\frac{d}{dt}\langle V, W \rangle = \langle D_t V, W \rangle + \langle V, D_t W \rangle.$$

Proof. (Left as an exercise for the reader. See Problem 6.3.9.) $\qquad\square$

The notion of a vector field along a curve (in a manifold M) leads us immediately to two useful notions: parallel transport and acceleration.

Definition 6.3.5. Let M be a smooth manifold with an affine connection ∇, and let $\gamma : I \to M$ be a smooth curve on M. A vector field V along γ is called *parallel* if $D_t V = 0$ identically.

The existence of parallel vector fields on a curve amounts to the solvability of a system of differential equations.

Proposition 6.3.6 (Parallel Transport)**.** *Let M be a smooth manifold with an affine connection ∇, and let $\gamma : I \to M$ be a smooth curve on M where I, is a compact interval of $\mathbb{R}$. Let $t_0 \in I$, set $p = \gamma(t_0)$, and let V_0 be any vector in $T_p M$. There exists a unique vector field of M along γ that is parallel and has $V(t_0) = V_0$.*

Proof. Suppose first that M is a manifold that is covered with a single coordinate system $x = (x^1, \ldots, x^n)$. By (6.27), the condition $D_t V = 0$ means that

$$\dot{v}^k + \Gamma_{ij}^k \dot{\gamma}^i v^j = 0 \qquad \text{for all } k = 1, \ldots, n. \tag{6.28}$$

The values Γ_{ij}^k depend on the position of $\gamma(t)$ as do the derivatives $\dot{\gamma}^i(t)$, but neither of these depend on the functions $v^i(t)$. Hence, (6.28) is a system of linear,

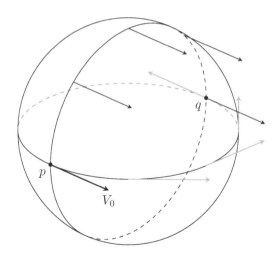

Figure 6.4: Path dependence of parallel transport.

homogeneous, ordinary, differential equations in the n functions $v^i(t)$. By a standard result of ordinary differential equations (see [18, Appendix A] or [3, Section 8]), given an initial value $t = t_0$ and initial conditions $v^i(t_0) = v_0^i$, there exists a unique solution to the system of equations satisfying these initial conditions. (The particular form of the nonautonomous system from (6.28) and the hypothesis that I is compact imply that the system satisfies the Lipschitz condition, which establishes the uniqueness of the solutions.) Hence V exists and is unique.

Now suppose that M cannot be covered by a single coordinate chart. We only need to consider coordinate charts that cover $\gamma(I)$. But since $\gamma(I)$ is compact, we can cover it with only a finite number of coordinate charts. However, on each of these charts, we have seen that there is a unique parallel vector field, as described. By identifying the vector fields over each coordinate chart, we obtain a single vector field over all of γ that is parallel to V_0. $\square$

Definition 6.3.7. The vector field V in Proposition 6.3.6 is called the *parallel transport* of V_0 along γ.

It is important to note that the parallel transport of V_0 from a point p to a point q along two different paths generally results in different vectors in $T_q M$. In Figure 6.4, the tangent vector V_0 at p produces different tangent vectors at q when transported along the black curve versus along the gray curve. One says that parallel transport is nonintegrable. However, it is not hard to see, either geometrically or by solving Equation (6.28), that in $\mathbb{R}^n$ parallel transport does not depend on the path. Therefore, this nonintegrability of parallel transport characterizes the notion of curvature, as we will see in the following section.

As a second application of the covariant derivative along a curve, we finally introduce the notion of acceleration of a curve on a manifold.

Definition 6.3.8. Let M be a smooth manifold with an affine connection and let $\gamma : I \to M$ be a smooth curve on M. For all $t \in I$, we define the *acceleration* of γ on M as the covariant derivative $D_t \gamma'(t)$ of $\gamma'(t)$ along γ.

Example 6.3.9. With the definition of the acceleration, we are in a position to be able to phrase Newton's second law of motion on a manifold. In $\mathbb{R}^3$, Newton's law states that if a particle has constant mass m and is influenced by the exterior forces $\vec{F}_i$, then the particle follows a path $\vec{x}(t)$ that satisfies $\sum_i \vec{F}_i = m\vec{x}''$. Translated into the theory of manifolds, if a force (or collection of forces) makes a particle move along some curve γ, then writing F as the vector field along γ that describes the force, γ must satisfy

$$m D_t \gamma'(t) = F(t).$$

The acceleration is itself a vector field along the curve γ so the notions of all the higher derivatives are defined as well.

6.3.2 Geodesics

Intuitively speaking, a geodesic on a manifold is a curve that generalizes the notion of a straight line in $\mathbb{R}^n$. This seemingly simple task is surprisingly difficult. Only now do we possess the necessary background to do so. Though everyone has an intuitive sense of what a straight line is, even Euclid's original definitions for a straight line do not satisfy today's standards of precision. We introduce geodesics using two different approaches, each taking a property of straight lines in $\mathbb{R}^n$ and translating it into the context of manifolds.

Definition 6.3.10. Let M be a smooth manifold with an affine connection ∇. A curve $\gamma : I \to M$ is called a *geodesic* if its acceleration is identically 0, i.e., $D_t \gamma'(t) = 0$.

Note that this definition does not require a metric structure on M, simply an affine connection. We should also observe that this definition relies on a specific parametrization of γ. The definition is modeled after the fact that the natural parametrization of a straight line in $\mathbb{R}^n$ by $\vec{\gamma}(t) = \vec{p} + t\vec{v}$ for constant vectors $\vec{p}$ and $\vec{v}$ satisfies $\vec{\gamma}''(t) = \vec{0}$. However, the curve $\vec{x}(t) = \vec{p} + t^3\vec{v}$ traces out the same set of points but $\vec{x}''(t) = 3t^2\vec{v}$, which is not identically 0. Despite this, we can leave Definition 6.3.10 as it is and keep in mind the role of the parametrization.

Proposition 6.3.11 (Geodesic Equations). *Let M be a smooth manifold equipped with an affine connection, and let $x = (x^1, \ldots, x^n)$ be a system of coordinates on a chart U. A curve γ is a geodesic on U if and only if the coordinate functions $\gamma(t) = (\gamma^1(t), \ldots, \gamma^n(t))$ satisfy*

$$\frac{d^2\gamma^i}{dt^2} + \Gamma^i_{jk}(\gamma(t))\frac{d\gamma^j}{dt}\frac{d\gamma^k}{dt} = 0 \qquad \text{for all } i. \tag{6.29}$$

Proof. This follows immediately from (6.27).   □

Equation (6.29) for a geodesic is a second-order system of ordinary differential equations in the functions $\gamma^i(t)$. Setting $v^i(t) = \dot{\gamma}^i$, we can write (6.29) as a first-order system in the $2n$ functions γ^i and v^i by

$$\begin{cases} \dot{\gamma}^i = v^i, \\ \dot{v}^i = -\Gamma^i_{jk}(\gamma(t))v^j v^k. \end{cases} \tag{6.30}$$

This system is now first-order and non-linear but autonomous (does not depend explicitly on t). Standard theorems in differential equations [3, Theorems 7.3, 7.4] imply the following foundational result.

Theorem 6.3.12. *Let M be a manifold with an affine connection. For any $p \in M$, for any $V \in T_pM$, and for any $t_0 \in \mathbb{R}$, there exists an open interval I containing t_0 and a unique geodesic $\gamma : I \to M$ satisfying $\gamma(t_0) = p$ and $\gamma'(t_0) = V$.*

This theorem shows the existence of the curve γ by solving (6.29) over a coordinate neighborhood. In this case, the interval I may be limited by virtue of the fact that $\gamma(I) \subset U$. It may be possible to extend γ over other coordinate patches. If $\gamma(t_1)$ for some $t_1 \in I$ is in another coordinate patch $\bar{U}$, then we can uniquely extend the geodesic over $\bar{U}$ as going through the point $\gamma(t_1)$ with velocity $\gamma'(t_1)$. We define a *maximal* geodesic as a geodesic $\gamma : I \to M$ whose domain interval cannot be extended. If γ is a maximal geodesic with $\gamma(t_0) = p$ and $\gamma'(t_0) = V$ for some $t_0 \in I$, we call γ the geodesic with initial point p and initial velocity $V \in T_pM$, and we denote it by γ_V.

Another defining property of a straight line in $\mathbb{R}^n$ is that the shortest path between two points is a straight line segment. If we use the concept of distance, we need a metric. Let (M, g) be a Riemannian metric equipped with the Levi-Civita connection, and let γ be a geodesic on M. By Proposition 6.3.4,

$$\frac{d}{dt}\langle \gamma'(t), \gamma'(t)\rangle = 2\langle D_t\gamma'(t), \gamma'(t)\rangle = 0,$$

so we can conclude the following initial result.

Proposition 6.3.13. *A geodesic on a Riemannian manifold has constant speed.*

Now on a Riemannian manifold, an alternate approach to defining geodesics is to call a geodesic a path of shortest length between two points. However, this definition is not quite good enough, as Figure 6.5 indicates. Both curves connecting p and q are geodesics, but one is shorter than the other. To be more precise, we call γ a geodesic connecting p_1 and p_2 if there is an interval $[t_1, t_2]$ such that $\gamma(t_1) = p_1$, $\gamma(t_2) = p_2$, and γ minimizes the arclength integral

$$L = \int_{t_1}^{t_2} \sqrt{g_{ij}(\gamma(t))\dot{\gamma}^i(t)\dot{\gamma}^j(t)}\, dt. \tag{6.31}$$

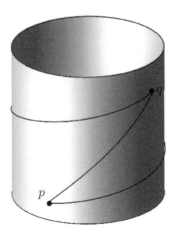

Figure 6.5: Two geodesics on a cylinder.

Techniques of calculus of variations discussed in Appendix B produce the differential equations for the curve γ that minimizes the arclength. However, similar to optimization methods in regular calculus, the solutions we obtain are local minima, which means in our case that there are no small deviations of γ that produce a shorter path between p and q. It is tedious to show, but Theorem B.3.1 implies that a curve γ that minimizes the integral in (6.31) must satisfy

$$\frac{d^2\gamma^i}{ds^2} + \Gamma^i_{jk}\frac{d\gamma^j}{ds}\frac{d\gamma^k}{ds} = 0, \tag{6.32}$$

where s is the arclength of γ. Proposition 6.3.11 and Proposition 6.3.13 show that defining a geodesic as having no acceleration is equivalent to defining it as minimizing length in the above sense.

Example 6.3.14 (Sphere). Consider the parametrization of the sphere given by

$$\vec{X}(x^1, x^2) = (R\cos x^1 \sin x^2, R\sin x^1 \sin x^2, R\cos x^2),$$

where x^1 is the longitude θ in spherical coordinates and x^2 is the angle φ down from the positive z-axis. In Example 6.1.14, we determined the coefficients of the metric tensor. Then it is easy to calculate the Christoffel symbols Γ^i_{jk} for the sphere. Equations (6.32) for geodesics on the sphere become

$$\frac{d^2x^1}{ds^2} + 2\cot(x^2)\frac{dx^1}{ds}\frac{dx^2}{ds} = 0,$$

$$\frac{d^2x^2}{ds^2} - \sin(x^2)\cos(x^2)\left(\frac{dx^1}{ds}\right)^2 = 0. \tag{6.33}$$

A geodesic on the sphere is now just a curve $\vec{\gamma}(s) = \vec{X}(x^1(s), x^2(s))$ where $x^1(s)$ and $x^2(s)$ satisfy the system of differential equations in (6.33). Taking a first derivative of $\vec{\gamma}(s)$ gives

$$\vec{\gamma}'(s) = R\left(-\sin x^1 \sin x^2 \frac{dx^1}{ds} + \cos x^1 \cos x^2 \frac{dx^2}{ds},\right.$$
$$\left. \cos x^1 \sin x^2 \frac{dx^1}{ds} + \sin x^1 \cos x^2 \frac{dx^2}{ds}, -\sin x^2 \frac{dx^2}{ds}\right),$$

and the second derivative, after simplification using (6.33), is

$$\frac{d^2\vec{\gamma}}{ds^2} = -\left[\sin^2(x^2)\left(\frac{dx^1}{ds}\right)^2 + \left(\frac{dx^2}{ds}\right)^2\right]\vec{\gamma}(s).$$

However, the term $R^2\left(\sin^2(x^2)\left(\frac{dx^1}{ds}\right)^2 + \left(\frac{dx^2}{ds}\right)^2\right)$ is the sphere metric applied to

$$((x^1)'(s), (x^2)'(s)),$$

which is precisely the square of the speed of $\vec{\gamma}(s)$. However, since the geodesic is parametrized by arclength its speed is identically 1. Thus, (6.33) leads to the differential equation

$$\vec{\gamma}''(s) + \frac{1}{R^2}\vec{\gamma}(s) = 0.$$

Standard techniques with differential equations allow one to show that all solutions to this differential equation are of the form

$$\vec{\gamma}(s) = \vec{a}\cos\left(\frac{s}{R}\right) + \vec{b}\sin\left(\frac{s}{R}\right),$$

where $\vec{a}$ and $\vec{b}$ are constant vectors. Note that $\vec{\gamma}(0) = \vec{a}$ and that $\vec{\gamma}'(0) = \frac{1}{R}\vec{b}$. Furthermore, to satisfy the conditions that $\vec{\gamma}(s)$ lie on the sphere of radius R and be parametrized by arclength, we deduce that $\vec{a}$ and $\vec{b}$ satisfy

$$\|\vec{a}\| = R, \qquad \|\vec{b}\| = R, \text{ and } \qquad \vec{a}\cdot\vec{b} = 0.$$

Therefore, we find that $\vec{\gamma}(s)$ traces out a great arc on the sphere that is the intersection of the sphere and the plane through the center of the sphere spanned by $\vec{\gamma}(0)$ and $\vec{\gamma}'(0)$.

There are many properties of lines that no longer hold for geodesics on manifolds. For example, lines in $\mathbb{R}^n$ are ("obviously") simple curves, i.e., they do not intersect themselves. In Example 6.3.14, we showed that the geodesics on a sphere are arcs of great circles (equators). In this case, a maximal geodesic is a whole circle that, as a closed curve, is still simple. In contrast, Figure 6.6 of a distorted sphere shows

Figure 6.6: A nonclosed geodesic on a manifold.

only a portion of a geodesic that is not closed and intersects itself many times. The problem of finding closed geodesics on surfaces illustrates how central the study of geodesics is in current research: in 1917, Birkhoff used techniques from dynamical systems to show that every deformed sphere has at least one closed geodesic [10]; in 1929, Lusternik and Schnirelmann improved upon this and proved that there always exist three closed geodesics on a deformed sphere [37]; and in 1992 and 1993, Franks and Bangert ([23] and [7]) proved that there exist an infinite number of closed geodesics on a deformed sphere. However, a proof of the existence of a closed geodesic would not necessarily help us construct one for any given surface.

We end this section by presenting the so-called exponential map. Theorem 6.3.12 allows us to define a map, for each $p \in M$, from the tangent plane T_pM to M by mapping V to a fixed distance along the unique geodesic γ_V.

Definition 6.3.15. Let p be a point on a Riemannian manifold (M, g). Let $\mathcal{D}_p$ be the set of tangent vectors $V \in T_pM$ such that the geodesic γ_V, with $\gamma_V(0) = p$, is defined over the interval $[0, 1]$. The *exponential map*, written $\exp_p$, is the function

$$\exp_p : \mathcal{D}_p \longrightarrow M,$$
$$V \longmapsto \gamma_V(1).$$

Lemma 6.3.16 (Scaling Lemma). *Let $V \in T_pM$, and let $c \in \mathbb{R}^{>0}$. Suppose that $\gamma_V(t)$ is defined over $(-\delta, \delta)$, with $\gamma_V(0) = p$. Then $\gamma_{cV}(t)$ is defined over the interval $(-\delta/c, \delta/c)$, and*

$$\gamma_{cV}(t) = \gamma_V(ct).$$

Proof. (Left as an exercise for the reader. See Problem 6.3.10.) $\qquad\square$

By virtue of the scaling lemma, we can write for the geodesic through p along V,

$$\gamma_V(t) = \exp_p(tV). \tag{6.34}$$

Proposition 6.3.17. *For all $p \in M$, there exists a neighborhood U of p on M and a neighborhood $\mathcal{D}$ of the origin in T_pM such that $\exp_p : \mathcal{D} \to U$ is a diffeomorphism.*

Proof. The differential of $\exp_p$ at 0 is a linear transformation $d(\exp_p)_0 : T_0(T_pM) \to T_pM$. However, since T_pM is a vector space, then the tangent space $T_0(T_pM)$ is naturally identified with T_pM. Thus, $d(\exp_p)_0$ is a linear transformation on the vector space T_pM. The proposition follows from the Inverse Function Theorem (Theorem 1.4.5) once we show that $d(\exp_p)_0 = (\exp_p)_*$ is invertible.

We show this indirectly using the chain rule. Let V be a tangent vector in $V \in T_pM$, and let $f : (-\delta, \delta) \to T_pM$ be the curve $f(t) = tV$. The function $\exp_p \circ f$ is a curve on M. Then

$$d(\exp_p \circ f)_0 = \left. \frac{d\exp_p(tV)}{dt} \right|_{t=0} = \left. \frac{d\gamma_V(t)}{dt} \right|_{t=0} = V.$$

However, by the chain rule, we also have

$$d(\exp_p \circ f)_0 = d(\exp_p)_0 df_0 = (\exp_p)_* V.$$

Hence, for all $V \in T_pM$, we have $(\exp_p)_* V = V$. Hence, $(\exp_p)_*$ is in fact the identity transformation so it is invertible, and the proposition follows. $\qquad\square$

Now if $\{e_\mu\}$ is any basis of T_pM, the exponential map sets up a coordinate system on a neighborhood of p on M defined by

$$\exp_p(X^\mu e_\mu).$$

We call this the *normal coordinate system* at p with respect to $\{e_\mu\}$. If q is a point in the neighborhood U, as in Proposition 6.3.17, then q is the image of a unique tangent vector X_q under $\exp_p$. The coordinates of q are X_q^μ.

Interestingly enough, the coefficients of the Levi-Civita connection vanish at p in the normal coordinate system X^μ at p. Consider a geodesic on M from p to q given by $c(t) = \exp_p(tX_q^\mu e_\mu)$, which in coordinates is just $X^\mu(t) = tX_q^\mu$. From the geodesic equation,

$$\frac{d^2 X^\mu}{dt^2} + \Gamma^\mu_{\lambda\nu} \frac{dX^\lambda}{dt} \frac{dX^\nu}{dt} = \Gamma^\mu_{\lambda\nu}(tX_q^i) X_q^\lambda X_q^\nu.$$

Setting $t = 0$, we find that $\Gamma^\mu_{\lambda\nu}(0) X_q^\lambda X_q^\nu = 0$ for any q. Thus, by appropriate choices of q, we determine that $\Gamma^\mu_{\lambda\nu}(0) = 0$, which are the components of the Levi-Civita connection at p in the normal coordinate system.

The exponential map allows us to redefine some common geometric objects in $\mathbb{R}^n$ in the context of Riemannian manifolds. Notice first that by Proposition 6.3.13, the arclength from p to $\exp_p(V)$ along $\gamma_V(t)$ is $\|V\|_p$. Now, let $r > 0$ be a positive real number and $B_r(0)$ be the open ball of radius r centered at the origin in T_pM. If r is small enough that $B_r(0)$ is contained in the neighborhood $\mathcal{U}$ from Proposition

6.3.17, then we call $\exp_p(B_r(0))$ the *geodesic ball* of radius r centered at p. If the sphere $S_r(0)$ of radius r centered at 0 in T_pM is contained in $\mathcal{U}$, then we call $\exp_p(S_r(0))$ the *geodesic sphere* of radius r centered at p.

PROBLEMS

6.3.1. Let S be a regular surface in $\mathbb{R}^3$, and let $\vec{X}$ be a parametrization of a coordinate chart U of S. Let ∇ be the Levi-Civita connection on S with respect to the first fundamental form metric. Let $\vec{\gamma}(t) = \gamma(t)$ be a curve on S. Prove that the acceleration $D_t\gamma'(t)$ is the orthogonal projection of $\vec{\gamma}''(t)$ onto the tangent plane to S at $\gamma(t)$.

6.3.2. Consider the torus parametrized by

$$\vec{X}(u,v) = ((a + b\cos v)\cos u, (a + b\cos v)\sin u, b\sin v),$$

where $a > b$. Show that the geodesics on a torus satisfy the differential equation

$$\frac{dr}{du} = \frac{1}{Cb}r\sqrt{r^2 - C^2}\sqrt{b^2 - (r-a)^2},$$

where C is a constant and $r = a + b\cos v$.

6.3.3. Find the differential equations that determine geodesics on a function graph $z = f(x,y)$.

6.3.4. If $\vec{X} : U \to \mathbb{R}^3$ is a parametrization of a coordinate patch on a regular surface S such that $g_{11} = E(u)$, $g_{12} = 0$, and $g_{22} = G(u)$, show that

(a) the u-parameter curves (i.e., over which v is a constant) are geodesics;

(b) the v-parameter curve $u = u_0$ is a geodesic if and only if $G_u(u_0) = 0$;

(c) the curve $\vec{x}(u, v(u))$ is a geodesic if and only if

$$v = \pm \int \frac{C\sqrt{E(u)}}{\sqrt{G(u)}\sqrt{G(u) - C^2}}\,du,$$

where C is a constant.

6.3.5. *Pseudosphere.* Consider a surface with a set of coordinates (u, v) defined over the upper half of the uv-plane, i.e., on $H = \{(u, v) \in \mathbb{R}^2 \mid v > 0\}$, such that the metric tensor is

$$(g_{ij}) = \begin{pmatrix} 1 & 0 \\ 0 & e^{2v} \end{pmatrix}.$$

Prove in this coordinate system that all the geodesics appear in the H as vertical lines or semicircles with center on the u-axis.

6.3.6. Let (M, g) be a two-dimensional Riemannian manifold. Suppose that on a coordinate patch U with coordinates $x = (x^1, x^2)$, the metric is given by $g_{11} = 1$, $g_{22} = (x^2)^2$, and $g_{12} = g_{21} = 0$. Show that the geodesics of M on U satisfy the existence and uniqueness of

$$x^1 = a\sec(x^2 + b).$$

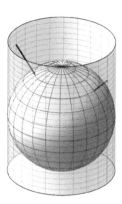

Figure 6.7: Mercator projection.

6.3.7. The Mercator projection used in cartography maps the globe (except the north and south poles, $\mathbb{S}^2 - \{(0,0,1),(0,0,-1)\}$) onto a cylinder, which is then unrolled into a flat map of the earth. However, one does not necessarily use the radial projection as shown in Figure 6.7. Consider a map f from $(x,y) \in (0,2\pi) \times \mathbb{R}$ to the spherical coordinates $(\theta,\phi) \in \mathbb{S}^2$ of the form $(\theta,\phi) = f(x,y) = (x,h(y))$.

(a) Recall that the usual Euclidean metric on $\mathbb{S}^2$ is

$$g = \begin{pmatrix} \sin^2\phi & 0 \\ 0 & 1 \end{pmatrix}.$$

The Mercator projection involves the above function $f(x,y)$, such that $h(y)$ gives a pull-back $f^*(g)$ that is a metric with a line element of the form $ds^2 = G(y)dx^2 + G(y)dy^2$. Prove that $h(y) = 2\cot^{-1}(e^y)$ works, and determine the corresponding function $G(y)$.

(b) Show that the geodesics on $\mathbb{R}^2$ equipped with the metric obtained from this $h(y)$ are of the form

$$\sinh y = \alpha\sin(x+\beta)$$

for some constants α and β.

6.3.8. Consider the Poincaré ball $\mathbb{B}_R^n$ from Problem 6.1.12. Prove that the geodesics in the Poincaré ball are either straight lines through the origin or circles that intersect the boundary $\partial\mathbb{B}_R^n$ perpendicularly. (The Poincaré ball is an example of a hyperbolic geometry. In this geometry, given a "straight line" (geodesic) L and a point p not on L, there exists a nonempty continuous set of lines (geodesics) through p that do not intersect L.)

6.3.9. Prove Proposition 6.3.4. [Hint: Use (6.27) and the fact that since the Levi-Civita connection is compatible with g, then $g_{ij;k} = 0$.]

6.3.10. Prove Lemma 6.3.16.

6.3.11. Consider the usual sphere $\mathbb{S}^2$ of radius R in $\mathbb{R}^3$. In the coordinate patch where $(\theta,\phi) \in (0,2\pi) \times (0,\pi)$, the Christoffel symbols are given in (6.33) of Example

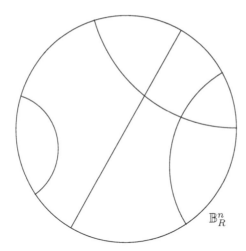

Figure 6.8: A few geodesics in the Poincaré disk.

6.3.14, where we use the coordinates $(\theta, \phi) = (x^1, x^2)$. Consider a point p on the sphere given by $P = (\theta_0, \phi_0)$. Let V_0 be a vector in $T_p\mathbb{S}^2$ with coordinates (V_0^1, V_0^2).

(a) Show that the stated Christoffel symbols used in (6.33) are correct.

(b) Calculate the coordinates of the parallel transport $V(t)$ of V_0 along the curve $\gamma(t) = (\theta_0, t)$, using the initial condition $t_0 = \phi_0$. Show that the length of the tangent vectors $V(t)$ does not change.

(c) Calculate the coordinates of the parallel transport $V(t)$ of V_0 along the curve $\gamma(t) = (t, \phi_0)$, using the initial condition $t_0 = \theta_0$. Show that $\|V(t)\|^2$ is constant.

6.3.12. Show that the locus of a geodesic on the n-sphere $\mathbb{S}^n$ (as a submanifold of $\mathbb{R}^{n+1}$) is the intersection of $\mathbb{S}^n$ with a 2-plane that passes through the sphere's center.

6.4 Curvature Tensor

In the study of curves and surfaces in classical differential geometry, the shape operator and the curvature tensor play a central role. We approach the notion of curvature on Riemannian manifolds in two different but equivalent ways.

6.4.1 Coordinate-Dependent

The first approach to curvature involves investigating mixed, partial, covariant derivatives. For smooth functions in $\mathbb{R}^n$, mixed, second-order partial derivatives are independent of the order of differentiation. Problem 6.2.17 showed that if a connection ∇ on M is not symmetric, the same result is no longer true for the mixed, covariant, partial derivatives of smooth functions on a manifold. We found that if $f : M \to \mathbb{R}$, then over a given coordinate path U, one has

$$f_{;j;i} - f_{;i;j} = -\tau_{ij}^k f_{;k},$$

where τ is the torsion tensor associated to ∇ (see Problem 6.2.14), and the coordinate components are

$$\tau_{ij}^k = \Gamma_{ij}^k - \Gamma_{ji}^k. \tag{6.35}$$

If we repeat the exercise with a vector field instead of a smooth function, a new phenomenon appears.

Proposition 6.4.1. *Let M be a smooth manifold equipped with an affine connection ∇. Let U be a coordinate patch on M and let X be a vector field defined over U. Then, in components, the mixed covariant derivatives satisfy*

$$X_{;j;i}^l - X_{;i;j}^l = K_{ijk}^l X^k - \tau_{ij}^k X_{;k}^l$$

where

$$K_{ijk}^l = \frac{\partial \Gamma_{jk}^l}{\partial x^i} - \frac{\partial \Gamma_{ik}^l}{\partial x^j} + \Gamma_{jk}^h \Gamma_{ih}^l - \Gamma_{ik}^h \Gamma_{jh}^l. \tag{6.36}$$

Proof. This is a simple matter of calculation. Starting from $X_{;i}^l = \partial X^l / \partial x^i + \Gamma_{ih}^l X^h$, we obtain

$$X_{;i;j}^l = \frac{\partial X_{;i}^l}{\partial x^j} + \Gamma_{jk}^l X_{;i}^k - \Gamma_{ji}^k X_{;k}^l$$

$$= \frac{\partial}{\partial x^j} \left(\frac{\partial X^l}{\partial x^i} + \Gamma_{ik}^l X^k \right) + \Gamma_{jk}^l \left(\frac{\partial X^k}{\partial x^i} + \Gamma_{im}^k X^m \right) - \Gamma_{ji}^k X_{;k}^l$$

$$= \frac{\partial^2 X^i}{\partial x^k \partial x^j} + \frac{\partial \Gamma_{ik}^l}{\partial x^j} X^k + \Gamma_{jk}^l \frac{\partial X^k}{\partial x^i} + \Gamma_{jm}^l \Gamma_{ik}^m X^k - \Gamma_{ji}^k X_{;k}^l$$

After collecting and canceling like terms, we find that

$$X_{;j;i}^l - X_{;i;j}^l = \left(\frac{\partial \Gamma_{jk}^l}{\partial x^i} - \frac{\partial \Gamma_{ik}^l}{\partial x^j} + \Gamma_{jk}^h \Gamma_{ih}^l - \Gamma_{ik}^h \Gamma_{jh}^l \right) X^k - (\Gamma_{ij}^k - \Gamma_{ji}^k) X_{;k}^l.$$

The proposition follows. $\square$

This result is particularly interesting because of the following proposition.

Proposition 6.4.2. *The collection of functions K_{jkl}^i defined in Equation (6.36) form the components of a tensor of type $(1,3)$.*

Proof. (This proposition relies on the coordinate-transformation properties of the component functions Γ_{jk}^i given in Proposition 6.2.5. The proof is left as an exercise for the reader.) $\square$

The functions K^i_{jkl} are the components of the so-called *curvature tensor* associated to the connection ∇.

The components of the curvature tensor came into play when we considered the mixed, covariant, partial derivatives of a vector field instead of just a smooth function. It is natural to ask whether some new quantity appears when one considers the mixed covariant partials of other tensors. Surprisingly, the answer is no.

Theorem 6.4.3 (Ricci's Identities). *Let $T^{i_1 \cdots i_r}_{j_1 \cdots j_s}$ be the components of a tensor field of type (r, s) over a coordinate patch of a manifold equipped with a connection ∇. Then the mixed, covariant, partial derivatives differ by*

$$T^{i_1 \cdots i_r}_{j_1 \cdots j_s;k;h} - T^{i_1 \cdots i_r}_{j_1 \cdots j_s;h;k} = \sum_{\alpha=1}^{r} K^{i_\alpha}_{hkm} T^{i_1 \cdots i_{\alpha-1} m i_{\alpha+1} \cdots i_r}_{j_1 \cdots j_s}$$

$$- \sum_{\beta=1}^{s} K^m_{hkj_\beta} T^{i_1 \cdots i_r}_{j_1 \cdots j_{\beta-1} m j_{\beta+1} \cdots j_s} - \tau^m_{hk} T^{i_1 \cdots i_r}_{j_1 \cdots j_s;m}.$$

Over the coordinate patch U, the components of the curvature tensor satisfy the *Bianchi identities*.

Proposition 6.4.4 (Bianchi Identities). *With K^i_{jkl} defined as in (6.36) and τ^i_{jk} defined as in Equation (6.35), then*

$$K^l_{ijk} + K^l_{jki} + K^l_{kij} = -\tau^l_{ij;k} - \tau^l_{jk;i} - \tau^l_{ki;j} - \tau^l_{im} \tau^m_{jk} - \tau^l_{jm} \tau^m_{ki} - \tau^l_{km} \tau^m_{ij}$$

and

$$K^l_{ijk;h} + K^l_{ikh;j} + K^l_{ihj;k} = -\tau^m_{jk} K^l_{imh} - \tau^m_{kh} K^l_{imj} - \tau^m_{hj} K^l_{imk}.$$

The second Bianchi identity is also called the differential Bianchi identity.

Proof. (Left as an exercise for the reader.) $\square$

In particular, if ∇ is a symmetric connection, the Bianchi identities reduce to

$$\text{first identity:} \quad K^l_{ijk} + K^l_{jki} + K^l_{kij} = 0, \tag{6.37}$$

$$\text{second identity:} \quad K^l_{ijk;h} + K^l_{ikh;j} + K^l_{ihj;k} = 0 \tag{6.38}$$

for any values of any of the indices.

(We need to mention at this point that some texts vary in how they assign meaning to the various indices of the Riemann curvature tensor and tensors associated to it. Because of the antisymmetry properties of the curvature tensor, the variances only lead to a possible difference in sign between component functions alternately defined. Fortunately, the coordinate-free definition for the curvature tensors seems to be uniformly accepted across the literature.)

6.4.2   Coordinate-Free

A second and more modern approach to curvature on a Riemannian manifold (M, g) defines the curvature tensor in a coordinate-free way, though still from a perspective of analyzing repeated covariant differentiation. If X, Y, and Z are vector fields on M, the difference in repeated covariant derivatives is

$$\nabla_X \nabla_Y Z - \nabla_Y \nabla_X Z. \tag{6.39}$$

Even with general vector fields in $\mathbb{R}^n$, (6.39) does not necessarily cancel out. However, by Problem 6.2.6, $\nabla_X \nabla_Y Z = X(Y(Z^k))\partial_k$, so

$$\nabla_X \nabla_Y Z - \nabla_Y \nabla_X Z = \nabla_{[X,Y]} Z. \tag{6.40}$$

This equality might not hold for all vector fields X, Y, Z on a manifold equipped with a connection ∇. This fact motivates defining the quantity

$$R(X, Y)Z \overset{\text{def}}{=} \nabla_X \nabla_Y Z - \nabla_Y \nabla_X Z - \nabla_{[X,Y]} Z. \tag{6.41}$$

The notation $R(X, Y)Z$ emphasizes the understanding that for each vector field X and Y, $R(X, Y)$ is an operator acting on Z. At first glance, $R(X, Y)Z$ is just a smooth mapping $\mathfrak{X}(M) \times \mathfrak{X}(M) \times \mathfrak{X}(M) \to \mathfrak{X}(M)$, smooth because the resulting vector field is smooth. However, more is true.

Proposition 6.4.5. *The function $R(X, Y)Z$ defined in Equation (6.41) is a tensor field of type $(1, 3)$, which is antisymmetric in X and Y.*

Proof. The antisymmetry property follows immediately from $[Y, X] = -[X, Y]$ and Definition 6.2.1. To prove the tensorial property, we need only to show that $R(X, Y)Z$ is multilinear over $C^\infty(M)$ in each of the three vector fields. We show linearity for the X variable, from which linearity immediately follows for the Y variable. We leave it as an exercise for the reader to prove linearity in Z.

Let $f_1, f_2 \in C^\infty(M)$. Then

$$R(f_1 X_1 + f_2 X_2, Y)Z = (f_1 \nabla_{X_1} + f_2 \nabla_{X_2})\nabla_Y Z$$
$$- \nabla_Y (f_1 \nabla_{X_1} + f_2 \nabla_{X_2})Z - \nabla_{[f_1 X_1, Y] + [f_2 X_2, Y]} Z.$$

By Proposition 5.3.4(4), $[f_i X_i, Y] = f_i[X_i, Y] - Y(f_i)X_i$. Thus,

$$R(f_1 X_1 + f_2 X_2, Y)Z$$
$$= f_1 \nabla_{X_1} \nabla_Y Z + f_2 \nabla_{X_2} \nabla_Y Z - f_1 \nabla_Y \nabla_{X_1} Z - Y(f_1)\nabla_{X_1} Z$$
$$- f_2 \nabla_Y \nabla_{X_2} Z - Y(f_2)\nabla_{X_2} Z - \nabla_{f_1[X_1,Y]-Y(f_1)X_1} Z - \nabla_{f_2[X_2,Y]-Y(f_2)X_2} Z$$
$$= f_1 \nabla_{X_1} \nabla_Y Z - f_1 \nabla_Y \nabla_{X_1} Z - f_1 \nabla_{[X_1,Y]} Z + f_2 \nabla_{X_2} \nabla_Y Z - f_2 \nabla_Y \nabla_{X_2} Z - f_2 \nabla_{[X_2,Y]}$$
$$- Y(f_1)\nabla_{X_1} Z - Y(f_2)\nabla_{X_2} Z + Y(f_1)\nabla_{X_1} Z + Y(f_2)\nabla_{X_2} Z$$
$$= f_1 R(X_1, Y)Z + f_2 R(X_2, Y)Z. \qquad \square$$

Definition 6.4.6. The tensor field R of type $(1,3)$ satisfying

$$R(X,Y)(Z) = \nabla_X \nabla_Y Z - \nabla_Y \nabla_X Z - \nabla_{[X,Y]} Z$$

is called the curvature tensor associated to the connection ∇. Occasionally, this is denoted R^∇ to explicitly indicate which connection.

We connect this approach to the coordinate-dependent Definition 6.36 as follows. Let x be a coordinate system on a coordinate patch of M. By the $C^\infty(M)$-linearity,

$$R(X,Y)Z = X^i Y^j Z^k \, R(\partial_i, \partial_j)\partial_k$$

where $X = X^i \partial_i$ and similarly for Y and Z. The components of R in local coordinates are R^l_{ijk}, where

$$R\left(\frac{\partial}{\partial x^i}, \frac{\partial}{\partial x^j}\right)\frac{\partial}{\partial x^k} = R^l_{ijk}\frac{\partial}{\partial x^l}.$$

Now since $[\partial_i, \partial_j] = 0$,

$$
\begin{aligned}
R(\partial_i, \partial_j)\partial_k &= \nabla_{\partial_i}\nabla_{\partial_j}\partial_k - \nabla_{\partial_j}\nabla_{\partial_i}\partial_k \\
&= \nabla_{\partial_i}\left(\Gamma^h_{jk}\partial_h\right) - \nabla_{\partial_j}\left(\Gamma^h_{ik}\partial_h\right) \\
&= \Gamma^h_{jk}\nabla_{\partial_i}\partial_h + \frac{\partial\Gamma^h_{jk}}{\partial x^i}\partial_h - \Gamma^h_{ik}\nabla_{\partial_j}\partial_h - \frac{\partial\Gamma^h_{ik}}{\partial x^j}\partial_h \\
&= \left(\Gamma^h_{jk}\Gamma^l_{ih} - \Gamma^h_{ik}\Gamma^l_{jh} + \frac{\partial\Gamma^l_{jk}}{\partial x^i} - \frac{\partial\Gamma^l_{ik}}{\partial x^j}\right)\partial_l,
\end{aligned}
$$

from which we obtain

$$R^l_{ijk} = \frac{\partial\Gamma^l_{jk}}{\partial x^i} - \frac{\partial\Gamma^l_{ik}}{\partial x^j} + \Gamma^h_{jk}\Gamma^l_{ih} - \Gamma^h_{ik}\Gamma^l_{jh},$$

which recovers exactly the coordinate-dependent Definition 6.36.

6.4.3 Riemannian Curvature

Our presentation of the curvature tensor so far applies to any affine connection. We turn to the specific example of a Riemannian manifold (M, g).

Definition 6.4.7. The curvature tensor associated to the Levi-Civita connection associated to the metric g is called the Riemann curvature tensor, denoted R.

As above, we denote the components of the curvature tensor by R^l_{ijk}. Since ∇ is symmetric, the torsion tensor τ associated to the Levi-Civita connection is identically 0. In coordinate-free expression, the Bianchi identities for the Riemann tensor are

$$R(X,Y)Z + R(Y,Z)X + R(Z,X)Y = 0 \qquad (6.42)$$
$$\nabla_W R(X,Y)Z + \nabla_Z R(X,W)Y + \nabla_Y R(X,Z)W = 0 \qquad (6.43)$$

By contracting with the metric tensor g, we obtain a tensor field of type $(0,4)$, which in components is

$$R_{ijkl} = g_{ml} R^m_{ijk}. \qquad (6.44)$$

We define this tensor also in a coordinate-free way.

Definition 6.4.8. If R is the Riemann curvature tensor on a Riemannian manifold (M,g), then $R^\flat$, which is commonly denoted Rm, is the Riemann *covariant curvature tensor*. In other words, for all vector fields, X, Y, Z, W on M,

$$Rm(X,Y,Z,W) = g(R(X,Y)Z, W).$$

We write the components of Rm with respect to a basis as R_{ijkl}.

Not all the component functions of R^l_{ijk} or of R_{ijkl} are independent. We now wish to determine the number of independent component functions in R_{ijkl}, which will be the same number of independent component functions of R^l_{ijk}.

By the definition from (6.36), we see that $R^l_{ijk} = -R^l_{jik}$ and, therefore, that

$$R_{ijkl} = -R_{jikl}. \qquad (6.45)$$

Furthermore, the first Bianchi identity gives

$$R_{ijkl} + R_{jkil} + R_{kijl} = 0. \qquad (6.46)$$

The compatibility condition of the Levi-Civita connection is expressed in coordinates as $g_{ij;k} = 0$ as functions for all indices i, j, k. This leads to another relation. Theorem 6.4.3 and the compatibility condition imply that

$$0 = g_{kl;j;i} - g_{kl;i;j} = -R^m_{ijk} g_{ml} - R^m_{ijl} g_{km} \qquad (6.47)$$

which is tantamount to

$$R_{ijkl} = -R_{ijlk}. \qquad (6.48)$$

Equations (6.45) and (6.48) show that the covariant curvature tensor is skew-symmetric in the first two indices and also in the last two indices. Furthermore, this skew-symmetry relation combined with the identity in (6.46) leads to

$$R_{ijkl} = R_{klij}. \qquad (6.49)$$

We can see from the following calculation, starting with the Bianchi identity, and using again in the middle:

$$
\begin{aligned}
0 &= R_{ikjl} + R_{kjil} + R_{jikl} = -R_{iklj} - R_{kjli} + R_{jikl} \\
&= R_{klij} + R_{likj} + R_{jlki} + R_{lkji} + R_{jikl} \\
&= 2R_{klij} - R_{lijk} - R_{jlik} + R_{jikl} \qquad \text{from (6.48) and (6.45)} \\
&= 2R_{klij} + R_{ijlk} + R_{jikl} = 2R_{klij} - 2R_{ijkl}
\end{aligned}
$$

Equation (6.49) follows.

We can now count the number of independent functions given the relations in (6.45), (6.46) and (6.48). There are five separate cases depending on how many indices are distinct. By virtue of (6.45), the cases when all indices are equal or when three of the indices are equal lead to identically 0 functions for the components of the covariant tensor. If there are two pairs of equal indices, then we must have $R_{iijj} = 0$ while the quantities R_{ijij} could be nonzero. In this case, the identities in (6.45), (6.46) and (6.49) explicitly determine all other possibilities with two pairs of equation indices from R_{ijij}. There are $\binom{n}{2}$ ways to select the pair $\{i, j\}$ to define R_{ijij}. If the indices have one pair of equal indices and the other two indices are different, then by (6.45) and (6.48), the only nonzero possibilities can be determined by R_{ijik} (where i, j, and k are all distinct). Hence, there are $n\binom{n-1}{2}$ choices of independent functions here. Lastly, suppose that all four indices are distinct. All the functions for combinations of indices can be obtained from the relations, given the functions for R_{ijkl} and R_{iljk}. Thus, there are $2\binom{n}{4}$ independent functions in this case. In total, the covariant curvature tensor is determined by

$$
\binom{n}{2} + n\binom{n-1}{2} + 2\binom{n}{4} = \frac{1}{12}n^2(n^2 - 1)
$$

independent functions.

Example 6.4.9. It is interesting to note that for manifolds of dimension $n = 2$, there is only one independent function in the curvature tensor, namely, R_{1212}. Equation (7.47) in [5] shows that the Gaussian curvature of the surface at any point is equal to

$$
K = -R_{1212}/\det(g_{ij}). \tag{6.50}
$$

Because of cancellations for repeated indices, an elegant way to rewrite this gives us the components of the Riemann covariant curvature tensor:

$$
R_{ijkl} = K(g_{il}g_{jk} - g_{ik}g_{jl}). \tag{6.51}
$$

Properties of the Riemann covariant curvature tensor presented in a coordinate-dependent manner have equivalent expressions in a coordinate-free formulation.

Proposition 6.4.10. *The covariant curvature tensor Rm satisfies the following symmetry properties for vector fields X, Y, Z, W, and T:*

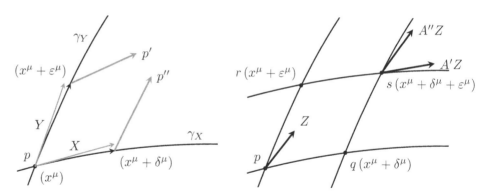

Figure 6.9: Geometric interpretation of the torsion tensor.   Figure 6.10: Geometric interpretation of the curvature tensor.

1. $Rm(X,Y,Z,W) = -Rm(Y,X,Z,W)$.

2. $Rm(X,Y,Z,W) = -Rm(X,Y,W,Z)$.

3. $Rm(X,Y,Z,W) = Rm(Z,W,X,Y)$.

4. *Bianchi's first identity:*

$$Rm(X,Y,Z,W) + Rm(Y,Z,X,W) + Rm(Z,X,Y,W) = 0.$$

5. *Bianchi's differential identity:*

$$\nabla Rm(X,Y,Z,W,T) + \nabla Rm(X,Y,W,T,Z) + \nabla Rm(X,Y,T,Z,W) = 0.$$

6.4.4 Geometric Interpretation

Until now, we have not given an interpretation for the geometric meaning of the curvature or torsion tensors.

Consider first the torsion tensor. (Of course, by definition, the Levi-Civita connection is symmetric and so the torsion is 0, but we give an interpretation for any affine connection.) We will use a first-order approximation discussion, following the presentation in [44, Section 7.3.2]. This reasoning differs slightly from a rigorous mathematical explanation, but we include it for the sake of familiarity with physics-style reasoning.

Let $p \in M$, with coordinates x^μ in a coordinate system on M. Let $X = \delta^\mu \partial_\mu$ and $Y = \varepsilon^\mu \partial_\mu$ be two vectors in $T_p M$. Let $\gamma_X(t)$ be the curve with coordinate functions $\delta^\mu t$ and let $\gamma_Y(t)$ be the curve with coordinate functions $\varepsilon^\mu t$. Consider the parallel transport of the vector X along $\gamma_Y(t)$. The coordinates of the resulting vector are $\delta^\mu + \Gamma^\mu_{\lambda\nu}\delta_\lambda\varepsilon_\nu$. The coordinates of p', the tip of the parallel transport of X, are

$$p' : \quad \delta^\mu + \varepsilon^\mu + \Gamma^\mu_{\lambda\nu}\delta_\lambda\varepsilon_\nu$$

If we take the parallel transport of Y along $\gamma_X(t)$, the coordinates of the resulting vector are $\varepsilon^\mu + \Gamma^\mu_{\lambda\nu}\varepsilon_\lambda\delta_\nu$. The coordinates of p'', the tip of the parallel transport of Y, are

$$p'': \quad \delta^\mu + \varepsilon^\mu + \Gamma^\mu_{\lambda\nu}\varepsilon_\lambda\delta_\nu.$$

The difference between these two parallel transports is $(\Gamma^\mu_{\lambda\nu} - \Gamma^\mu_{\nu\lambda})\delta_\lambda\varepsilon_\nu$, which is $T^\mu_{\lambda\nu}\delta_\lambda\varepsilon_\nu$. Therefore, intuitively speaking, the torsion tensor gives a local measure of how much the parallel transport of two noncollinear directions with respect to each other fails to close a parallelogram (see Figure 6.9).

The curvature tensor, on the other hand, measures the path dependence of parallel transport. In the coordinate-free definition of the curvature tensor from (6.41), the expression $\nabla_X\nabla_Y Z$ is a vector field that measures the rate of change of parallel transport of the vector field Z along an integral curve of Y and then a rate of change of parallel transport of this $\nabla_Y Z$ along an integral curve of X. The expression $\nabla_Y\nabla_X Z$ reverses the process.

As discussed in Section 5.2 in the subsection on Lie brackets (see also Figure 5.4), the successive flows of a distance h along the integral curves of Y and then along the integral curves of X do not in general lead one to the same point if one follows the integral curves of X and then of Y. Proposition 5.3.14 shows that $[X, Y]$ is a sort of measure for this nonclosure of integral paths in vector fields. Subtracting $\nabla_{[X,Y]}Z$ from $\nabla_X\nabla_Y Z - \nabla_Y\nabla_X Z$ eliminates the quantity of path dependence of parallel transport on a manifold that is naturally caused by the nonclosure of "square" paths of integral curves in vector fields.

Another perspective is to consider a vector Z based at p with coordinates x^μ and look at the path dependence of the parallel transport along two sides of a "parallelogram" based at p and spanned by directions δ^μ and ε^μ. Locally, i.e., when δ^μ and ε^μ are small, the parallel transport of Z from p to $q = (x^\mu + \delta^\mu)$ to $s = (x^\mu + \delta^\mu + \varepsilon^\mu)$ produces a vector $A'Z$. Similarly, the parallel transport of Z from p to $r = (x^\mu + \varepsilon^\mu)$ to $s = (x^\mu + \delta^\mu + \varepsilon^\mu)$ produces a vector $A''Z$ (see Figure 6.10). The difference $Z \mapsto A''Z - A'Z$ is a linear transformation defined locally at p that depends on the directions δ^μ and ε^μ. In fact, it is not hard to show that, in coordinates, the first order approximation in the variables δ and ε is

$$(A''Z - A'Z)^i = R^i_{jkl}\delta^j\varepsilon^k Z^l.$$

PROBLEMS

6.4.1. Calculate the 16 component functions of the curvature tensor for the sphere $\mathbb{S}^2$ in the standard (θ, ϕ) coordinate system.

6.4.2. Prove Proposition 6.4.2.

6.4.3. (a) Prove the first Bianchi identity in Proposition 6.4.4 using a coordinate-dependent approach.

(b) Prove the first Bianchi identity in Proposition 6.4.10 using a coordinate-free approach.

6.4.4. (a) Prove the second Bianchi identity in Proposition 6.4.4 using a coordinate-dependent approach.

(b) Prove the second (differential) Bianchi identity in Proposition 6.4.10 using a coordinate-free approach. [Hint: This can be long and tedious if done directly. Instead, since ∇Rm is $C^\infty(M)$-multilinear, choose X, Y, Z, W, T to be coordinate basis vector fields. Also, to make the computations even easier, use the normal coordinate system.]

6.4.5. Prove that the quantity $R(X,Y)Z$ defined in (6.41) is $C^\infty(M)$-linear in the Z variable.

6.4.6. A smooth family of smooth curves is a function $c : (-\varepsilon, \varepsilon) \times [a, b] \to M$ such that $c_s(t) = c(s, t)$ is a smooth curve in M for each $s \in (-\varepsilon, \varepsilon)$. Note that by symmetry, $c_t(s)$ is also a smooth curve for each $t \in [a, b]$. A vector field along c is a smooth map $V : (-\varepsilon, \varepsilon) \times [a, b] \to TM$ such that $V(s, t) \in T_{c(s,t)}M$ for each (s, t). Define the vector fields S and T on c by $S = \partial_s c$ and $\partial_t c$, i.e., the tangent vectors to c in the indicated direction. Show that for any vector field V on c,

$$D_s D_t V - D_t D_s = R(S, T)V.$$

(This gives another geometric interpretation of the curvature tensor.)

6.4.7. *The Jacobi Equation.* This exercise considers variations along a geodesic γ. A *variation through geodesics* along γ is a smooth family of smooth curves c (defined in Problem 6.4.6) such that for each s, the curve $c_s(t) = c(s, t)$ is a geodesic and $c(0, t) = \gamma(t)$. The *variation field* V of a variation through geodesics along γ is the vector field along γ defined by $V(t) = (\partial_s c)(0, t)$. Show that V satisfies the Jacobi equation

$$D_t^2 V + R(V, \dot{\gamma})\dot{\gamma} = 0. \tag{6.52}$$

6.4.8. Consider the 3-sphere $\mathbb{S}^3$, and consider the coordinate patch given by the parametrization described in Problem 6.1.14. Calculate the curvature tensor, the Ricci curvature tensor, and the scalar curvature.

6.4.9. Calculate the curvature tensor, the Ricci curvature tensor, and the scalar curvature for the Poincaré ball. (See Problem 6.1.12.)

6.4.10. Consider the 3-torus described in Problem 6.1.2(b) with the metric induced from $\mathbb{R}^4$. Calculate all the components of the curvature tensor, the Ricci tensor, and the scalar curvature, given in the coordinates defined by the parametrization given in Problem 6.1.2(b).

6.4.11. Consider the metric associated to spherical coordinates in $\mathbb{R}^3$, given by

$$g = dr^2 + r^2 \sin^2 \phi \, d\theta^2 + r^2 d\phi^2.$$

(Note, we have used the mathematics labeling of the longitude and latitude angles. Physics texts usually have the θ and ϕ reversed.) Prove that all the components of the curvature tensor are identically 0.

6.4.12. Consider the Riemannian manifold of dimension 2 equipped with the metric $g = f(u + v)(du^2 + dv^2)$ for some function f. Solve for which f lead to $R_{jklm} = 0$.

6.4.13. Let R be the Riemann curvature tensor (defined with respect to the Levi-Civita connection). Prove that

$$R_{ijkl} = \frac{1}{2}\left(\frac{\partial^2 g_{ik}}{\partial x^j \partial x^l} - \frac{\partial^2 g_{jk}}{\partial x^i \partial x^l} - \frac{\partial^2 g_{il}}{\partial x^j \partial x^k} + \frac{\partial^2 g_{jl}}{\partial x^i \partial x^k}\right) + g_{\lambda\mu}\left(\Gamma^\lambda_{ik}\Gamma^\mu_{jl} - \Gamma^\lambda_{il}\Gamma^\mu_{jk}\right).$$

Conclude that in normal coordinates centered at p, the following holds at p:

$$R_{ijkl} = \frac{1}{2}(\partial_j \partial_l g_{ik} - \partial_i \partial_l g_{jk} - \partial_j \partial_k g_{il} + \partial_i \partial_k g_{jl}).$$

6.4.14. *The Killing Equation.* Let (M,g) be a Riemannian manifold and let $X \in \mathfrak{X}(M)$. Consider the function $f_\varepsilon : M \to M$ defined by

$$f_\varepsilon(p) = \gamma(\varepsilon)$$

where γ is the integral curve of X through p. Thus, the linear approximation of f_ε for small ε maps $p = (x^i)$ to the point with coordinates $x^i + \varepsilon X^i(p)$. Suppose that f_ε is an isometry for infinitesimal ε.

(a) Use a linear approximation in ε on the change-of-coordinates formula for the metric g to show that g and X satisfy the Killing equation:

$$\frac{\partial g_{ij}}{\partial x^k} X^k + \frac{\partial X^l}{\partial x^i} g_{lj} + \frac{\partial X^l}{\partial x^j} g_{il} = 0. \tag{6.53}$$

(b) Let ∇ be the Levi-Civita connection. Show that the Killing equation is equivalent to the condition that $(\nabla X)^\flat$ is antisymmetric. In components related to a coordinate system, this means that

$$X_{i;j} + X_{j;i} = 0, \tag{6.54}$$

where $X_i = g_{ik}X^k$.

6.4.15. Consider a covector field ω on a Riemannian manifold (M,g). Suppose that ω satisfies the covariant Killing equation (see (6.54)), i.e,. $\omega_{i;j} + \omega_{j;i} = 0$. Show that along any geodesic $\gamma(s)$ of M, $\omega(\dot\gamma)$ is a nonzero constant.

6.4.16. Show that if $R_{ijkl} + R_{ljki} = 0$, then the covariant curvature tensor is identically 0.

6.5   Ricci Curvature and Einstein Tensor

We finish this chapter with a brief section on various tensors associated to the Riemann curvature tensor.

Since tensors of type $(1,3)$ or $(0,4)$ are so unwieldy, there are a few common ways to summarize some of the information contained in the curvature tensor.

One of the most common constructions is the *Ricci curvature tensor*, denoted by Rc or Ric. We tend to write R_{ij} instead of Rc_{ij} for the components of this tensor with respect to a coordinate system. The Ricci curvature tensor is $\mathrm{Tr}\,R$, or the trace (with respect to the first indices) of the Riemann curvature tensor. In coordinates, the components are defined by

$$R_{ij} = R^k_{kij} = g^{km} R_{kijm}.$$

By the symmetries of the curvature tensor, R_{ij} can be expressed equivalently as

$$R_{ij} = R^k_{kij} = -R^k_{ikj} = -g^{km} R_{ikjm} = -g^{km} R_{jmki}.$$

Proposition 6.5.1. *The Ricci tensor Rc is symmetric.*

Proof. We prove this within the context of a coordinate system. Since $R_{ij} = R^k_{kij}$, then

$$R_{ij} = \frac{\partial \Gamma^k_{ij}}{\partial x^k} - \frac{\partial \Gamma^k_{kj}}{\partial x^i} + \Gamma^h_{ij}\Gamma^k_{kh} - \Gamma^h_{kj}\Gamma^k_{ih}. \qquad (6.55)$$

In this expression, since the connection is symmetric, the first and third terms of the right-hand side are obviously symmetric in i and j. The fourth term $\Gamma^h_{kj}\Gamma^k_{ih}$ is also symmetric in i and j by a relabeling of the summation variables h and k. Surprisingly, the second term in (6.55) is also symmetric.

By Problem 6.2.19,

$$\frac{\partial(\ln \sqrt{\det g})}{\partial x^j} = \Gamma^k_{jk}.$$

Thus,

$$\frac{\partial \Gamma^k_{jk}}{\partial x^i} = \frac{\partial^2}{\partial x^i \partial x^j}(\ln \sqrt{\det g}) = \frac{\partial^2}{\partial x^j \partial x^i}(\ln \sqrt{\det g}) = \frac{\partial \Gamma^k_{ik}}{\partial x^j}.$$

Hence, all the terms in (6.55) are symmetric in i and j, so $R_{ij} = R_{ji}$ and the result follows. $\qquad \square$

Definition 6.5.2. The *scalar curvature* function R is defined as the trace of the Ricci tensor with respect to g, i.e.,

$$S = \mathrm{Tr}_g \, Rc = g^{ij} R_{ij}. \qquad (6.56)$$

Sometimes, texts use the letter R to denote the scalar curvature, but we have opted for the other common notation of S so as not to be confused with the curvature tensor symbol.

Example 6.5.3 (Ricci Tensor of a Surface). We observed in Example 6.4.9 that the covariant Riemann curvature Rm tensor for a 2-manifold depends on only one function, R_{1212}. Symmetry and antisymmetry properties of the tensor determine all the component functions from this one. Furthermore, we observed that $R_{1212} = -K \det(g_{ij})$, where K is the Gaussian curvature of the surface, that arises in classical differential geometry. We can write this as

$$R_{ijkl} = K(g_{il}g_{jk} - g_{ik}g_{jl}).$$

By definition of the Ricci tensor,

$$R_{ij} = g^{km}R_{kijm} = Kg^{km}(g_{km}g_{ij} - g_{kj}g_{im})$$
$$= K(g^{mk}g_{km}g_{ij} - g^{mk}g_{kj}g_{im}) = K(\delta^m_m g_{ij} - \delta^m_j g_{im})$$
$$= K(2g_{ij} - g_{ij}) = Kg_{ij}.$$

Hence, the Ricci curvature tensor is (locally) proportional to the metric tensor, by the factor of the Gaussian curvature. Furthermore, this implies that

$$S = Kg^{ij}g_{ij} = 2K.$$

So for all surfaces, whether embedded in $\mathbb{R}^3$ or not, the scalar curvature function is twice the Gaussian curvature.

The scalar curvature function allows us to define the Einstein tensor, which is of fundamental importance.

Definition 6.5.4. On any Riemannian manifold (M, g) the *Einstein tensor* G is the tensor of type $(0, 2)$ described in coordinates by

$$G_{\mu\nu} = R_{\mu\nu} - \frac{1}{2}g_{\mu\nu}S, \qquad (6.57)$$

where S is the scalar curvature.

Since the Ricci curvature tensor and the metric tensor are symmetric, i.e., in $\mathrm{Sym}^2(TM^*)$, then the Einstein tensor field is also symmetric. As we will see in Section 7.5, the Einstein tensor is of central importance in general relativity. From a purely geometric perspective, the Einstein tensor has the following important property.

Proposition 6.5.5. *Let G be the Einstein tensor on a Riemannian manifold. Then, using (6.25),*

$$\mathrm{div}\, G = 0.$$

In coordinates, this reads $G^\alpha_{\mu;\alpha} = (g^{\alpha\nu}G_{\mu\nu})_{;\alpha} = g^{\nu\alpha}G_{\mu\nu;\alpha} = 0.$

Proof. The proof of this proposition follows from the differential Bianchi identity. For the Riemann curvature tensor, by Proposition 6.4.10(5), we have

$$R_{ijkl;m} + R_{ijlm;k} + R_{ijmk;l} = 0.$$

Taking the trace with respect to g over the variable pair (i, l),

$$R_{jk;m} - R_{jm;k} + g^{il}R_{ijmk;l} = 0,$$

where the trace operator commutes with the covariant derivative because of the compatibility condition of the Levi-Civita connection. Multiplying by g^{jk} and contracting in both indices gives

$$S_{;m} - g^{jk}R_{jm;k} - g^{il}R_{im;l} = 0.$$

Relabeling summation indices and using the symmetry of g and Rc, we deduce that

$$S_{;m} - 2g^{jk}R_{mj;k} = 0. \tag{6.58}$$

But $G^{\alpha}_{\mu} = g^{\alpha\nu}G_{\mu\nu} = g^{\alpha\nu}R_{\mu\nu} - \frac{1}{2}\delta^{\alpha}_{\mu}S$, so

$$G^{\alpha}_{\mu;\alpha} = g^{\alpha\nu}R_{\mu\nu;\alpha} - \frac{1}{2}\delta^{\alpha}_{\mu}S_{;\alpha} = g^{\alpha\nu}R_{\mu\nu;\alpha} - \frac{1}{2}S_{;\mu},$$

and the vanishing divergence follows from (6.58). The last claim in the proposition follows from Problem 6.2.16. $\qquad\square$

Of particular interest in Riemannian geometry and in general relativity are manifolds in which the Ricci curvature is proportional to the metric tensor. The corresponding metric is called an *Einstein metric* and the manifold is called an *Einstein manifold*. More precisely, a Riemannian manifold (M, g) has an Einstein metric if

$$Rc = kg \tag{6.59}$$

for some constant $k \in \mathbb{R}$. Taking the trace with respect to g of (6.59) and noting that $\mathrm{Tr}_g\, g = \dim M$, we find that k must satisfy

$$k = \frac{S}{\dim M}. \tag{6.60}$$

This leads to the following interesting property.

Proposition 6.5.6. *If (M, g) is an Einstein manifold, then the scalar curvature is constant on each connected component of M.*

In part because of Proposition 6.5.6, Einstein metrics continue to remain an active area of research not only because of their applications to physics but more so because of their application to possible classification theorems for diffeomorphic manifolds. The Uniformization Theorem, a fundamental result in the theory of surfaces, establishes that every connected 2-manifold admits a Riemannian metric with constant Gaussian curvature. This in turn leads to a classification of diffeomorphism classes for surfaces.

One could hope that, in parallel with surfaces, all connected higher-dimensional manifolds ($\dim M > 2$) would possess an Einstein metric that would in turn lead to a classification theorem of diffeomorphism classes of manifolds. This turns out not to be the case. There do exist higher-dimensional compact manifolds that admit no Einstein metric ([9]). Nevertheless, in attempts to reach a generalization to the Uniformization Theorem for higher-dimensional manifolds, Einstein metrics play a vital role.

PROBLEMS

6.5.1. Suppose that on a Riemannian manifold (M, g), the curvature tensor satisfies $\operatorname{div} R = 0$, or in coordinates $R^i_{jkl;i} = 0$ for all j, k, l. Show that the following also hold:

    (a) $R_{ij;h} = R_{ih;j}$;

    (b) $S_{;j} = 0$;

    (c) $g^{ml} R^i_{jkl} R_{mh} + g^{ml} R^i_{khl} R_{mj} + g^{ml} R^i_{hjl} R_{mk} = 0$.

6.5.2. Some authors define an Einstein manifold to be Riemannian manifold such that the Ricci curvature tensor is proportional to the metric tensor in the sense that $Rc = \lambda g$, where $\lambda : C^\infty(M, \mathbb{R})$ is a smooth function on M. (The definition given in the text requires the λ be a constant.) Prove that if the manifold has dimension $n \geq 3$, then this alternate definition of an Einstein manifold also implies that the scalar curvature is constant on all connected components of the manifold. [Hint: Show that $S_{;h} = 0$.]

6.5.3. Let $G^i_j = g^{ik} G_{jk}$, where G_{jk} are the components of the Einstein tensor, and define $R^{il}_{jk} = g^{lm} R^i_{jkm}$. Prove that

$$G^i_j = -\frac{1}{4} \delta^{i\nu\kappa}_{j\lambda\mu} R^{\lambda\mu}_{\nu\kappa},$$

where we have used the generalized Kronecker symbol defined in (4.34).

CHAPTER 7

Applications of Manifolds to Physics

In the previous chapters, we set forth the goal of doing calculus on curved spaces as the motivating force behind the development of the theory of manifolds. Occasionally, we showcased applications to physics either in examples or exercise problems. Having developed a theory of manifolds, we now present five applications to physics that utilize this theory. Consequently, throughout this chapter, the motivation for topics is inverted as compared to the rest of the book: instead of starting from a mathematical structure and looking for applications to physics, we begin with concepts from physics and see how the theory of manifolds can provide a framework for the idea. Each section shows just the tip of the iceberg on very broad areas of active investigation.

Section 7.1 explores how Hamiltonian's equations of motion motivate the notion of symplectic manifolds. Because of these applications and fascinating properties, symplectic geometry has become a significant field. Historically, it was the Hamiltonian formulation of dynamics that lent itself best to quantization and hence to Schrödinger's equation in quantum mechanics.

In special relativity, Einstein's perspective of viewing spacetime as a single unit, equipped with a modified notion of metric, is properly modeled by Minkowski spaces. Section 7.2 discusses this, along with its natural generalization to pseudo-Riemannian manifolds.

A few exercises in this text have dealt with the theory of electromagnetism. In Section 7.3, we gather together some of the results we have seen in the theory of electromagnetism and rephrase them into the formalism of a Lorentzian spacetime.

We also discuss a few geometric concepts underlying string theory. Between 1900 and 1940, physics took two large steps in opposite directions of the size scale, with quantum mechanics describing the dynamics of the very small scale and general relativity describing the very large scale. These theories involve very different types of mathematics, which led physicists to look for reformulations or generalizations that could subsume both theories. However, despite extensive work to find a unifying theory, the task has proven exceedingly difficult, even on mathematical grounds. String theory is a model for the structure of elementary particles that currently holds promise to provide such a unification. We wish to mention string theory in this book because, at its core, the relativistic dynamics of a string involve a two-dimensional submanifold of a Minkowski space.

Finally, Einstein's theory of general relativity, introduced in Section 7.5, stands as a direct application of Riemannian manifolds. In fact, general relativity motivated some of the development and helped proliferate the notions of Riemannian (and pseudo-Riemannian) geometry beyond the confines of pure mathematics. Many of the "strange" (non-Newtonian) phenomena that fill the pages of popular books on cosmology occur as consequences of the mathematics of this geometry.

This chapter assumes that the reader has some experience in physics but no more than a first college course (calculus-based) in mechanics. All the other material will be introduced as needed. We do not discuss issues of quantization as those exceed the scope of this book.

7.1   Hamiltonian Mechanics

7.1.1   Equations of Motion

The classical study of dynamics relies almost exclusively on Newton's laws of motion, in particular, his second law. This law states that the sum of exterior forces on a particle or object is equal to the rate of change of momentum, i.e.,

$$\sum \vec{F}_{ext} = \frac{d\vec{p}}{dt}, \tag{7.1}$$

where $\vec{p} = m\,d\vec{x}/dt$ and $\vec{x}(t)$ is the position of the particle at time t. If m is constant, (7.1) reduces to

$$\sum \vec{F}_{ext} = m\frac{d^2\vec{x}}{dt^2}. \tag{7.2}$$

Furthermore, by a simple calculation, (7.1) directly implies the following law of motion for angular momentum about an origin O:

$$\frac{d\vec{L}}{dt} = \sum \vec{\tau}_{ext}, \tag{7.3}$$

where $\vec{L} = \vec{r} \times \vec{p}$ is the angular momentum of a particle or solid, where $\vec{r}$ is the position vector of the particle or center of mass of the solid, and where $\sum \vec{\tau}_{ext}$ is the sum of the torques about O. (Recall that the torque about the origin of a force $\vec{F}$ is $\tau = \vec{r} \times \vec{F}$.)

Though (7.1) undergirds all of classical dynamics, the value of ancillary equations, such as (7.3), arises from the fact that these other equations may elucidate conserved quantities or produce more tractable equations when using different variables besides the Cartesian coordinates. For example, when describing the orbits of planets around the sun, polar (cylindrical) coordinates are far better suited than

Cartesian coordinates. In particular, as shown in Example 2.2.3, the angular momentum is a conserved quantity for a particle under the influence of forces that are radial about some origin.

It turns out that in many cases (in particular when the forces are conservative), either (7.2) or (7.3) follows from a specific variational principle that has extensive consequences. Suppose that the state of a physical system is described by a system of coordinates q_k, with $k = 1, 2, \ldots, n$. Hamilton's principle states that the motion of a system evolves according to a path $\mathcal{P}$ parametrized by $(q_1(t), \ldots, q_n(t))$ between times t_1 and t_2 so as to minimize the integral

$$S = \int_{\mathcal{P}} L \, dt = \int_{t_1}^{t_2} L \, dt \tag{7.4}$$

where L is the *Lagrangian function*. The integral S is called the *action* of the system. When the system is under the influence of only conservative forces, the Lagrangian is $L = T - V$, where T is the kinetic energy and V is the potential energy. Recall that for a conservative force $\vec{F}$, its potential energy V, which is a function of the position variables alone, satisfies

$$\vec{F} = -\vec{\nabla} V = -\operatorname{grad} V.$$

Intuitively speaking, in the case of conservative forces, Hamilton's principle states that a system evolves in such a way as to minimize the total variation between kinetic and potential energy. However, even if a force is not conservative, it may still possess an associated Lagrangian that produces the appropriate equation of motion. (See Problem 7.1.8 for such an example.)

We consider the Lagrangian L as an explicit function of t, the coordinates q_k, and their time derivatives $\dot{q}_k = dq_k/dt$. According to Theorem B.3.1 in Appendix B, the Lagrangian must satisfy the Euler-Lagrange equation in each coordinate q_k, namely,

$$\frac{\partial L}{\partial q_k} - \frac{d}{dt}\left(\frac{\partial L}{\partial \dot{q}_k}\right) = 0. \tag{7.5}$$

This is called *Lagrange's equations of motion*. Though this system of equations moves away from the nice vector expression of (7.2), it has the distinct advantage of expressing equations of motion in a consistent way for any choice of coordinates.

Example 7.1.1. Consider a ball (or cylinder) of radius R rolling down a plane inclined with angle α, as depicted in Figure 7.1. Because the object rolls instead of sliding, the rotation about its center leads to an additional kinetic energy amount of $\frac{1}{2}I\dot{\theta}^2$, where $\dot{\theta}$ is the rate of rotation about its center. However, because there is no slipping, we deduce that $\dot{x} = R\dot{\theta}$, where x is the coordinate of the distance of the center of mass of the object up the incline. Thus, the Lagrangian of this system is

$$L = \frac{1}{2}I\dot{\theta}^2 + \frac{1}{2}mv^2 - mgh,$$

$$L(x, \dot{x}) = \frac{I}{2R^2}I\dot{x}^2 + \frac{1}{2}m\dot{x}^2 - mgx \sin \alpha.$$

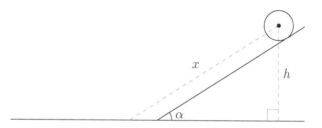

Figure 7.1: A round object rolling downhill.

The Euler-Lagrange Equation (7.5) gives

$$\frac{\partial L}{\partial x} - \frac{d}{dt}\left(\frac{\partial L}{\partial \dot{x}}\right) = -mg\sin\alpha - (I/R^2 + m)\ddot{x} = 0,$$

which leads to the equation of motion

$$\frac{d^2 x}{dt^2} = -\frac{g\sin\alpha}{\frac{I}{mR^2} + 1},$$

a well-known result from classical mechanics.

Though Example 7.1.1 involves a variable x that is essentially taken from $\mathbb{R}$, physical systems in general may typically be described by other types of variables. When studying the motion of a simple pendulum (Figure 7.2), we use as a variable the angle θ of deviation of the pendulum from the vertical. A system that is a double pendulum (see Figure 7.3) involves two angles.

If a physical system can be described by using n locally independent variables, then we say the system has n degrees of freedom. The set of all possible states of a physical system is a real manifold Q of dimension n, called the *configuration space* of the system. The variables (q_k) that locate a point on (a coordinate chart of) the manifold Q are called the *position variables*. (Note, we will use the subscript indices for the position variables to conform to physics texts and literature on symplectic manifolds, though one should remember at this stage that they are contravariant quantities.) For example, the configuration space of the system in Example 7.1.1 is simply $Q = \mathbb{R}$, while the configuration space for the simple pendulum is $Q = \mathbb{S}^1$ and the configuration space of the double pendulum is the torus $Q = \mathbb{S}^1 \times \mathbb{S}^1$.

The time development of a system corresponds to a curve $\gamma : t \mapsto (q_k(t))$ on the manifold, and the functions $\dot{q}_1, \ldots, \dot{q}_n$ are the coordinates of a tangent vector along γ in the tangent space TQ.

Now the Euler-Lagrange Equation (7.5) is a system of second-order, ordinary, differential equations. We would like to change this into a system of first-order differential equations for two reasons: (1) many theorems on differential equations

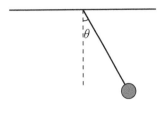

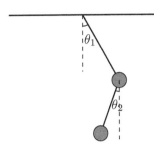

Figure 7.2: Simple Pendulum.   Figure 7.3: Compound Pendulum.

are stated for systems of first-order equations and (2) it is easier to discuss first-order equations in the context of manifolds. We do this in the following way.

Define the *generalized momenta* functionally by

$$p_k = \frac{\partial L}{\partial \dot{q}_k}. \tag{7.6}$$

The quantities p_k are the components of the momentum vector, which is in fact an element of $T_{\gamma(t)}Q^*$. We can see this as follows. Let W be an n-dimensional vector space, and let $f : W \to \mathbb{R}$ be any differentiable function. Then the differential $df_{\vec{v}}$ at a point $\vec{v} \in W$ is a linear transformation $df_{\vec{v}} : W \to \mathbb{R}$. Thus, by definition of the dual space, $df_{\vec{v}} \in W^*$. Consequently, the differential df gives a correspondence $df : W \to W^*$ via

$$\vec{v} \longmapsto df_{\vec{v}} = \sum_{i=1}^{n} \frac{\partial f}{\partial x^i}\bigg|_{\vec{v}} dx^i.$$

Taking W as the vector space $T_{\gamma(t)}Q$, the momentum at the point $\gamma(t)$ is the vector $dL_{(\dot{q}_k)} \in T_{\gamma(t)}Q^*$. Hence, we can think of the momentum vector p as a covector field along the curve γ given at each point by $dL_{(\dot{q}_k)}$.

Consider now the *Hamiltonian function H* defined by

$$H = \sum_{k=1}^{n} p_k \dot{q}_k - L(q_1, \ldots, q_n, \dot{q}_1, \ldots, \dot{q}_n, t).$$

Since we can write the quantity $\dot{q}_k$ in terms of the components p_k, we can view the Hamiltonian as a time-dependent function on TQ^*. Given any configuration space Q, we define the cotangent bundle TQ^* as the *phase space* of the system. If Q is an n-dimensional manifold, then TQ^* is a manifold of dimension $2n$.

The variables $\dot{q}_i$ are now functions of the independent variables

$$(t, q_1, \ldots, q_n, p_1, \ldots, p_n).$$

Taking derivatives of H, we find that

$$\frac{\partial H}{\partial p_i} = \dot{q}_i + \sum_{k=1}^{n} p_k \frac{\partial \dot{q}_k}{\partial p_i} - \sum_{k=1}^{n} \frac{\partial L}{\partial \dot{q}_k} \frac{\partial \dot{q}_k}{\partial p_i}$$

$$= \dot{q}_i + \sum_{k=1}^{n} \left(p_k - \frac{\partial L}{\partial \dot{q}_k} \right) \frac{\partial \dot{q}_k}{\partial p_i} = \dot{q}_i,$$

where each term of the summation is 0 by definition of p_k. Furthermore, note that Lagrange's equation reduces to $\partial L / \partial q_i = \dot{p}_i$. Thus, taking derivatives with respect to q_i, we get

$$\frac{\partial H}{\partial q_i} = \sum_{k=1}^{n} p_k \frac{\partial \dot{q}_k}{\partial q_i} - \left(\frac{\partial L}{\partial q_i} + \sum_{k=1}^{n} \frac{\partial L}{\partial \dot{q}_k} \frac{\partial \dot{q}_k}{\partial q_i} \right)$$

$$= -\frac{\partial L}{\partial q_i} + \sum_{k=1}^{n} \left(p_k - \frac{\partial L}{\partial \dot{q}_k} \right) \frac{\partial \dot{q}_k}{\partial q_i}$$

$$= -\frac{\partial L}{\partial q_i} = -\dot{p}_i.$$

Therefore, given the definition in Equation (7.6), the Euler-Lagrange Equation (7.5) is equivalent to

$$\dot{q}_k = \frac{\partial H}{\partial p_k},$$
$$\dot{p}_k = -\frac{\partial H}{\partial q_k}. \tag{7.7}$$

This system of equations is called *Hamilton's equations of motion*. They consist of $2n$ first-order, ordinary, differential equations in n unknown functions, each involving $2n$ variables, whereas Lagrange's equations of motion consisted of n second-order, ordinary, differential equations in n unknown functions.

For simple dynamic systems, the kinetic energy T is a homogeneous quadratic function in the variables $\dot{q}_k$. If this is the case, then it is not hard to show that

$$\sum_{k=1}^{n} \dot{q}_k p_k = 2T, \tag{7.8}$$

where T is the kinetic energy. If in addition, the forces acting on the system are conservative, then

$$H = 2T - (T - V) = T + V,$$

which is the total energy of the system.

Example 7.1.2 (The Spherical Pendulum). As a longer example that compares the Lagrange and the Hamilton equations of motion, consider the spherical pendulum

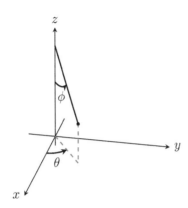

Figure 7.4: Spherical pendulum.

as shown in Figure 7.4. This classical problem consists of a point mass that is hanging from a string and is free to move not just in a vertical plane but in both its natural degrees of freedom. We label the mass of the object at the end of the string as m and the length of the string as l. For simplicity, we assume that the mass of the string is negligible and that there is no friction where the string attaches at a fixed point. This scenario is called the spherical pendulum problem because the same equations govern the motion of an object moving in a spherical bowl under the action of gravity and with no (negligible) friction.

We use a Cartesian frame of reference, in which the origin is the fixed point to which the string is attached and the z-axis lines up with the vertical axis that the string makes when at rest and hanging straight down. Furthermore, we orient the z-axis downward. With this setup, the degrees of freedom are the usual angles θ and φ from spherical coordinates. To obtain the Lagrange equations of motion, we need to first identify the kinetic energy T and potential energy V.

The velocity vector for the particle moving at the end of the string is

$$\vec{v} = l(\dot{\varphi}\cos\varphi\cos\theta - \dot{\theta}\sin\varphi\sin\theta, \dot{\varphi}\cos\varphi\sin\theta + \dot{\theta}\sin\varphi\cos\theta, -\dot{\varphi}\sin\varphi),$$

so after simplifications, the kinetic energy is

$$T = \frac{1}{2}ml^2(\dot{\varphi}^2 + \dot{\theta}^2\sin^2\varphi).$$

The potential energy is $V = mgl(1 - \cos\varphi)$, so the Lagrangian is

$$L = \frac{1}{2}ml^2(\dot{\varphi}^2 + \dot{\theta}^2\sin^2\varphi) - mgl(1 - \cos\varphi).$$

The Lagrange equations of motion are

$$\frac{d}{dt}\left(\frac{\partial L}{\partial \dot{\theta}}\right) = \frac{\partial L}{\partial \theta} \quad \text{and} \quad \frac{d}{dt}\left(\frac{\partial L}{\partial \dot{\varphi}}\right) = \frac{\partial L}{\partial \varphi},$$

which in this specific example give the system of differential equations

$$\begin{cases} \dfrac{d}{dt}\left(ml^2\dot{\theta}\sin^2\varphi\right) = 0 \\ \dfrac{d}{dt}\left(ml^2\dot{\varphi}\right) = ml^2\dot{\theta}^2\sin\varphi\cos\varphi - mgl\sin\varphi. \end{cases}$$

As we try to extract a more convenient set of equations that govern this system, we could take the derivative on the left hand side of the first equation. However, it is more useful to see that $p_\theta = \partial L/\partial\dot{\theta} = ml^2\dot{\theta}\sin^2\varphi$ is a constant. This is the θ-momentum. Then we can write the second equation as

$$\ddot{\varphi} = \sin\varphi(\dot{\theta}^2\cos\varphi - \frac{g}{l}).$$

Since p_θ is a constant, we can solve for $\dot{\theta}$ in terms of p_θ and write the second equation only in terms of φ to get the following system of equations:

$$\begin{cases} p_\theta = ml^2\dot{\theta}\sin^2\varphi, \\ \ddot{\varphi} = \dfrac{p_\theta^2}{m^2l^4}\dfrac{\cos\varphi}{\sin^3\varphi} - \dfrac{g}{l}\sin\varphi. \end{cases} \qquad (7.9)$$

These are still essentially the Lagrange equations of motion, with the understanding the p_θ is constant.

It might appear that the $\sin^3\varphi$ in the denominator in the second equation in (7.9) could be a cause for concern at $\varphi = 0$ but it is not, as we now explain. Recall that p_θ is constant. If $p_\theta = 0$, then the second equation in (7.9) does not possess a singularity at $\varphi = 0$. On the other hand, if $p_\theta \neq 0$, then $\sin\varphi \neq 0$ so φ is never 0.

In order to solve the equations of motion, we first solve the equation that involves only $\varphi(t)$. Once we know $\varphi(t)$, we find $\theta(t)$ by integrating

$$\dot{\theta} = \frac{p_\theta}{ml^2\sin^2(\varphi(t))},$$

using the fact that p_θ is a constant.

To establish Hamilton's equations of motion, we first find the generalized momenta of the coordinates as $p_\theta = ml^2\dot{\theta}\sin^2\varphi$ and $p_\varphi = ml^2\dot{\varphi}$. To get the Hamiltonian of this system, we first point out that the momenta give us values for $\dot{\theta}$ and $\dot{\varphi}$. Hence

$$\begin{aligned} H &= p_\theta\dot{\theta} + p_\varphi\dot{\varphi} - L \\ &= \frac{p_\theta^2}{ml^2\sin^2\varphi} + \frac{p_\varphi^2}{ml^2} - \frac{1}{2}ml^2\left(\left(\frac{p_\varphi}{ml^2}\right)^2 + \left(\frac{p_\theta}{ml^2\sin^2\varphi}\right)^2\sin^2\varphi\right) + mgl(1-\cos\varphi) \\ &= \frac{p_\theta^2}{2ml^2\sin^2\varphi} + \frac{p_\varphi^2}{2ml^2} + mgl(1-\cos\varphi). \end{aligned}$$

In this example, we see that the Hamiltonian is indeed the total energy $T + V$. Then Hamilton's equations (7.7) for this system are

$$
\begin{cases}
\dot{p}_\theta = 0 \\
\dot{p}_\varphi = -mgl \sin \varphi + \dfrac{p_\theta^2 \cos \varphi}{ml^2 \sin^3 \varphi} \\
\dot{\theta} = \dfrac{p_\theta}{ml^2 \sin^2 \varphi} \\
\dot{\varphi} = \dfrac{p_\varphi}{ml^2}.
\end{cases}
$$

7.1.2 Symplectic Manifolds

We now introduce the notion of a symplectic manifold and show how Hamilton's equations of motion arise naturally in this context. The theory of symplectic geometry is a branch of geometry in and of itself so we do not pretend to cover it extensively here. Instead, we refer the reader to [8] or [1] for a more thorough introduction. In this section, we simply illustrate how the theory of manifolds, equipped with some additional structure, is ideally suited for this area of mathematical physics.

Definition 7.1.3. Let W be a vector space over a field K. A *symplectic form* is a bilinear form

$$\omega : V \times V \to K$$

that is:

1. antisymmetric: $\omega(v, v) = 0$ for all $v \in W$;

2. nondegenerate: if $\omega(v, w) = 0$ for all $w \in W$, then $v = 0$.

The pair (V, ω) is called a *symplectic vector space*.

Proposition 7.1.4. *Let (V, ω) be a finite-dimensional, symplectic vector space. There exists a basis $\mathcal{B}$ of V relative to which the matrix of ω is*

$$
[\omega]_\mathcal{B} = \begin{pmatrix} 0 & I_n \\ -I_n & 0 \end{pmatrix}.
$$

where I_n is the $n \times n$ identity matrix. In addition, V has even dimension.

Proof. (Left as an exercise for the reader. See Problem 7.1.4.) $\qquad\square$

Since the form ω is antisymmetric and bilinear, then $\omega \in \bigwedge^2 V$. Suppose that V has a basis $\mathcal{B} = \{e_1, \ldots, e_{2n}\}$, and let $\mathcal{B}^* = \{e_1^*, \ldots, e_{2n}^*\}$ be the associated dual basis (see Section 4.1). Then in coordinates, we can write ω as

$$\omega = \sum_{1 \le i < j \le n} \omega_{ij} e_i^* \wedge e_j^*.$$

However, from Proposition 7.1.4 follows immediately a nice corollary.

Corollary 7.1.5. *Let (V, ω) be a symplectic vector space of dimension $2n$. Then there exists a basis $\mathcal{B} = \{e_1, \ldots, e_{2n}\}$ such that ω can be written as*

$$\omega = \sum_{i=1}^{n} e_i^* \wedge e_{n+i}^*.$$

The expression in Corollary 7.1.5 is called the *canonical form* of the symplectic form ω.

Definition 7.1.6. A *symplectic manifold* (M, ω) is a smooth manifold M equipped with a 2-form ω that is closed ($d\omega = 0$) and nondegenerate. In other words, M is a smooth manifold such that for each $p \in M$, T_pM is a symplectic vector space with symplectic form ω_p and ω_p varies smoothly with p.

By Proposition 7.1.4, one sees that a symplectic manifold has even dimension.

Definition 7.1.7. If (M, ω) and $(\tilde{M}, \tilde{\omega})$ are two symplectic manifolds, then a smooth map $F : M \to \tilde{M}$ is called *symplectic* if

$$F^* \tilde{\omega} = \omega.$$

We say that F preserves the symplectic structure. If in addition, F^{-1} is also a smooth symplectic map, then F is called a *symplectomorphism*.

Darboux's Theorem, a fundamental result in the theory of symplectic manifolds, establishes that given any two symplectic forms ω and $\tilde{\omega}$ such that $\omega_P = \tilde{\omega}_P$ at some point $P \in M$, there exists a neighborhood U of P and a diffeomorphism $F : U \to F(U) \subset M$ such that $F(P) = P$ and $F^* \tilde{\omega} = \omega$. (We refer the reader to [8, Section 2.2] for a proof.) Darboux's Theorem is equivalent to the following formulation.

Theorem 7.1.8. *Let (M, ω) be a symplectic manifold. For each point $P \in M$, there exists an open neighborhood U of P and a symplectomorphism F of U onto $F(U) \subset \mathbb{R}^{2n}$ such that $(F^{-1})^* \omega$ takes the canonical form in $\mathbb{R}^{2n}$.*

As a consequence of this theorem, at every point $P \in M$, there exists a coordinate neighborhood U of P with coordinates x in which

$$\omega = \sum_{i=1}^{n} dx_i \wedge dx_{n+i}.$$

The formalism of symplectic manifolds applies to Hamiltonian mechanics in the following way. Consider the configuration space Q for a physical system. Suppose that Q is a manifold of dimension n. The cotangent bundle $M = TQ^*$ is a manifold in itself of dimension $2n$. If U is a coordinate neighborhood of Q with coordinates $(q_1, \ldots, q_n)$, then $\tilde{U} = \pi^{-1}(U)$ is a coordinate neighborhood for the manifold TQ^*, where $\pi : TQ^* \to Q$ is the bundle projection map. The quantities $(q_1, \ldots, q_n, p_1, \ldots, p_n)$ of position coordinates and corresponding generalized momenta form a coordinate system on $\tilde{U}$. By the identification of $T_p\mathbb{R}^n \cong \mathbb{R}^n$, it is not hard to show that $TM = T(TQ^*) \cong TQ \oplus TQ^*$.

Proposition 7.1.9. *The 2-form defined over a particular coordinate patch $\pi^{-1}(U)$ by*

$$\omega = \sum_{i=1}^{n} dq_i \wedge dp_i \tag{7.10}$$

extends to a 2-form $\omega \in \Omega^2(TQ^*)$ over the whole phase space TQ^*. Furthermore, it is defined in exactly the same way as in Equation (7.10) over every coordinate patch on TQ^* obtained as $\pi^{-1}(\bar{U})$, where $\bar{U}$ is any other coordinate patch of Q. Consequently, the form ω endows TQ^* with the structure of a symplectic manifold.

Proof. Let $F : U \cap \bar{U} \to U \cap \bar{U}$ be a coordinate transformation from (q_i) to $(\bar{q}_i)$ coordinates, and let $G : \pi^{-1}(U \cap \bar{U}) \to \pi^{-1}(U \cap \bar{U})$ be the corresponding coordinate transformation from (q_i, p_i) to $(\bar{q}_i, \bar{p}_i)$ on TQ^*. Since p_i are coordinates in the cotangent space, the differential of G has coordinate functions

$$[dG] = \begin{pmatrix} \dfrac{\partial \bar{q}_i}{\partial q_j} & 0 \\ 0 & \dfrac{\partial q_k}{\partial \bar{q}_l} \end{pmatrix} = \begin{pmatrix} [dF] & 0 \\ 0 & [dF]^{-1} \end{pmatrix}.$$

In particular, we deduce that

$$d\bar{q}_i = \frac{\partial \bar{q}_i}{\partial q_j} dq_j \qquad \text{and} \qquad d\bar{p}_i = \frac{\partial q_k}{\partial \bar{q}_i} dp_k.$$

Thus,

$$\sum_{i=1}^{n} d\bar{q}_i \wedge d\bar{p}_i = \sum_{i=1}^{n} \left(\frac{\partial \bar{q}_i}{\partial q_j} dq_j \right) \wedge \left(\frac{\partial q_k}{\partial \bar{q}_i} dp_k \right) = \sum_{j=1}^{n} \sum_{k=1}^{n} \left(\sum_{i=1}^{n} \frac{\partial \bar{q}_i}{\partial q_j} \frac{\partial q_k}{\partial \bar{q}_i} \right) dq_j \wedge dp_k$$

$$= \sum_{j=1}^{n} \sum_{k=1}^{n} \delta_{jk} dq_j \wedge dp_k = \sum_{j=1}^{n} dq_j \wedge dp_j. \qquad \square$$

The Hamiltonian function H is a smooth function $TQ^* \to \mathbb{R}$. We define the *Hamiltonian vector field* X_H as the unique vector field that satisfies

$$i_{X_H}\omega = dH, \tag{7.11}$$

where i_{X_H} is the contraction operator i_{X_H}, which on forms is equivalent to the interior product (see Problem 5.4.16). Specifically in this case, $i_{X_H}\omega$ is the 1-form

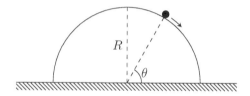

Figure 7.5: A point mass sliding off a hemisphere.

defined by $i_{X_H}\omega(Y) = \omega(X_H, Y)$ at all $P \in M$ and for all $Y \in \mathfrak{X}(M)$. It is not too hard to show that in coordinates of $T(TQ^*)$, the vector field X_H is

$$X_H = \sum_{i=1}^{n} \frac{\partial H}{\partial p_i} \frac{\partial}{\partial q_i} - \sum_{i=1}^{n} \frac{\partial H}{\partial q_i} \frac{\partial}{\partial p_i}. \tag{7.12}$$

Proposition 7.1.10. *A curve γ on the phase space TQ^* is an integral curve of the vector field X_H if and only if in each coordinate system the components $\gamma(t) = (q_k(t), p_k(t))$ satisfy Hamilton's equations of motion from Equation (7.7).*

Proof. As a vector field on the curve γ, the derivative $\dot{\gamma}(t)$ is written in coordinates as

$$\dot{\gamma}(t) = \sum_{i=1}^{n} \dot{q}_i \frac{\partial}{\partial q_i} + \sum_{i=1}^{n} \dot{p}_i \frac{\partial}{\partial p_i}. \tag{7.13}$$

By Equation (7.12), the Hamiltonian vector field X_H at points along the curve is expressed in coordinates as

$$(X_H)_{\gamma(t)} = \sum_{i=1}^{n} \frac{\partial H}{\partial p_i}\Big|_{\gamma(t)} \frac{\partial}{\partial q_i} - \sum_{i=1}^{n} \frac{\partial H}{\partial q_i}\Big|_{\gamma(t)} \frac{\partial}{\partial p_i}. \tag{7.14}$$

The proposition follows by identification of Equations (7.13) and (7.14). $\qquad\square$

In other words, Proposition 7.1.10 states that a solution to Hamilton's equations of motion corresponds to a curve $\gamma(t)$ in the phase space TQ^* such that

$$\dot{\gamma}(t) = (X_H)_{\gamma(t)}.$$

Because of the importance of this formulation, it has its own terminology. If (M, ω) is a symplectic manifold and $H \in C^\infty(M)$, then with X_H defined by Equation (7.11), the triple (M, ω, X_H) is called a *Hamiltonian system*.

PROBLEMS

7.1.1. The special orthogonal group in $\mathbb{R}^3$, denoted SO(3), consists of all 3×3 matrices that are orthogonal and have a determinant of 1. Explain why the configuration space of the position and orientation of a general solid in Euclidean three-space is $Q = \mathbb{R}^3 \times \text{SO}(3)$. Explain why SO(3) is diffeomorphic to $\mathbb{RP}^3$.

7.1.2. Determine the Lagrangian, Lagrange's equations of motion, and Hamilton's equation of motion for a point mass m sliding off a hemisphere of radius R. (See Figure 7.5.)

7.1.3. Determine the Lagrangian, Lagrange's equations of motion, and Hamilton's equations of motion for an elastic pendulum: a particle of mass m attached to a (massless) elastic string of elasticity constant k and unstretched length ℓ.

7.1.4. Determine the Lagrangian, Lagrange's equations of motion, and Hamilton's equation of motion for the coupled harmonic oscillations depicted below:

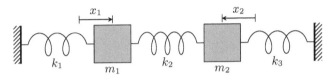

Use x_1 and x_2 as the displacement from where the masses labeled m_1 and m_2 are in equilibrium. Assume that there is no friction on the ground. For simplicity, also assume that when the masses are in equilibrium, all three springs are relaxed.

7.1.5. Consider the motion of the earth around the sun. Placing the sun at the origin, use polar coordinates (r, θ) to locate the center of the earth with respect to the sun. The force of gravity of the sun acting on the earth has a potential energy function of $V(r) = -GM_S M_E/r$, where G is Newton's universal constant of gravity, M_S is the mass of the sun and M_E is the mass of the earth. Take into account the fact that the earth rotates on its own axis. Use the additional angle ψ to orient the earth around its axis. Write down the Hamiltonian function for this system, taking into account earth's rotation. Show that, despite the fact that the rotation of the earth affects the Hamiltonian, the rotation does not affect the motion of the earth around the sun.

7.1.6. Suppose that Q is the configuration space for a physical system involving a particle of mass m, and suppose that Q is a Riemannian manifold with metric $g = \langle\,,\,\rangle$. Then the kinetic energy of a particle traveling along a curve $\gamma(t)$ is

$$T = \frac{1}{2}m\langle\dot{\gamma}(t), \dot{\gamma}(t)\rangle.$$

(a) Consider the sphere $\mathbb{S}^2$ of radius R, and use the coordinates (θ, ϕ). Write down the Lagrangian, the Hamiltonian, and Hamilton's equations of motion of a particle of mass m affected by a potential $V = f(\theta, \phi)$.

(b) Let Q be any Riemannian manifold with metric g and with the associated Levi-Civita connection. Show that if the potential V is constant, then a solution to Hamilton's equations of motion defines a geodesic on Q.

7.1.7. Friction is a non-conservative force. Suppose that an object of mass m with motion in one space variable $x(t)$ is affected by conservative forces with a combined potential energy function $V(x, t)$ and the force of friction of $F = -\gamma\dot{x}$, where γ is a positive constant. Prove that

$$L = e^{t\gamma/m}\left(\frac{1}{2}m\dot{x}^2 - V\right)$$

is such that the Euler-Lagrange equation (7.5) leads to the correct equation of motion. Calculate the Hamiltonian associated with this Lagrangian, and write down Hamilton's equations of motion.

7.1.8. *Classical electromagnetism.* Consider a charged particle of mass m and charge e under the influence of a static electric field $\vec{E}$ and magnetic field $\vec{B}$. The non-relativistic theory of electromagnetism [46] states that the force applied to the particle is

$$\vec{F} = e(\vec{E} + \frac{1}{c}\vec{v} \times \vec{B}),$$

where $\vec{v} = d\vec{x}/dt$ is the velocity vector of the particle and c is the speed of light. (The presence of c is a mere scaling factor due to the choice of units.) The electric field is induced from an electric potential ϕ so that $\vec{E} = -\vec{\nabla}\phi$. The magnetic force, however, is not a conservative force. Show that the Lagrangian

$$L = \frac{1}{2}mv^2 + e\phi + \frac{e}{c}\vec{v} \cdot \vec{A}$$

yields Newton's equation of motion from Equation (7.2), where $\vec{A}$ is the vector potential satisfying $\vec{B} = \vec{\nabla} \times \vec{A}$. Show that the Hamiltonian of this system given in coordinates (x_i, p_i) is

$$H(\vec{x}, \vec{p}) = \frac{1}{2m}(p_1^2 + p_2^2 + p_3^2) - e\phi(\vec{x}) - \frac{e}{mc}(p_1 A_1 + p_2 A_2 + p_3 A_3).$$

7.1.9. Prove Proposition 7.1.4.

7.1.10. Let V be a vector space of dimension $2n$, and let ω be any bilinear form on V. Show that ω is nondegenerate if and only if $\omega^n = \omega \wedge \cdots \wedge \omega$ is nonzero.

7.1.11. Let (V, ω) be a real symplectic vector space. Let $\mathcal{B} = \{e_1, \cdots, e_{2n}\}$ be a basis of V that gives ω a canonical form.

   (a) Show that if a linear transformation $T : V \to V$ leaves the form invariant, i.e.,

$$\omega(T(\vec{v}), T(\vec{w})) = \omega(\vec{v}, \vec{w}) \qquad \text{for all } \vec{v}, \vec{w} \in V,$$

   then the matrix A of T with respect to the basis $\mathcal{B}$ satisfies

$$A^T J A = J, \quad \text{where} \quad J = \begin{pmatrix} 0 & I_n \\ -I_n & 0 \end{pmatrix}.$$

   (b) Suppose that T leaves ω invariant. Show that if λ is an eigenvalue of T with multiplicity k, then 1λ, $\bar{\lambda}$, and $1/\bar{\lambda}$ are also eigenvalues of T with multiplicity k.

7.1.12. An alternative way to define the Hamiltonian vector field X_H involves using the process of raising indices as defined in Equation (6.11) in Section 6.1. Show that $X_H = dH^\sharp$, relative to the canonical form ω on TQ^*.

7.1.13. Prove Equation (7.12). [Hint: Use the embedding of $\wedge^2 TQ^*$ in $TQ^* \otimes TQ^*$ given by $dq_i \wedge dp_i = dq_i \otimes dp_i - dp_i \otimes dq_i$.]

7.1.14. Let Q be a configuration space and let M be the associated phase space $M = TQ^*$. Let $\pi : TQ^* \to Q$ be the canonical projection. Define the *Liouville form* $\vartheta \in \Omega^1(M)$ by

$$\vartheta_m(X) \overset{\text{def}}{=} \lambda_q\big(d\pi_m(X)\big)$$

for any point $m = (q, \lambda_q)$ of the phase space M and for any vector $X \in T_m M$.

(a) Using the standard coordinates on $\pi^{-1}(U)$ in TQ^*, where U is a coordinate patch of Q, show that the Liouville form has the expression

$$\vartheta = \sum_{i=1}^{n} p_i \, dq_i.$$

(b) Conclude that the canonical symplectic form on TQ^* satisfies $\omega = -d\vartheta$.

7.1.15. *Poisson Bracket.* Consider the phase space $M = TQ^*$ for a configuration space Q. Define the Poisson bracket $\{\,,\,\}$ on the function space $C^\infty(TQ^*)$ by

$$\{f, g\} = \sum_{i=1}^{n} \left(\frac{\partial f}{\partial q_i} \frac{\partial g}{\partial p_i} - \frac{\partial f}{\partial p_i} \frac{\partial g}{\partial q_i} \right).$$

(a) Show that $\{\,,\,\}$ is a differential in each entry, i.e., $\{f_1 f_2, g\} = \{f_1, g\} f_2 + f_1 \{f_2, g\}$ and similarly for the second entry.

(b) Prove that $\{\,,\,\}$ gives $C^\infty(TQ^*)$ the structure of a Lie algebra, i.e., $\{\,,\,\}$ satisfies the first three items of Proposition 5.3.4.

(c) Show that Hamilton's equations of motion from Equation (7.7) are equivalent to

$$\dot{q}_k = \{q_k, H\} \quad \text{and} \quad \dot{p}_k = \{p_k, H\} \qquad \text{for } k = 1, \dots, n.$$

7.2 Special Relativity; Pseudo-Riemannian Manifolds

7.2.1 Concepts from Special Relativity

In Chapter 2 we discussed why, in classical mechanics, it is not proper to assume the existence or the possibility of finding an absolutely fixed frame. However, one of the foundation principles of classical physics, namely the principle of inertia, also known as Newton's First Law of Motion, affirms that a body with no net force acting on it moves with constant velocity (or stays at rest, which corresponds to zero velocity). However, if an observer is in a moving frame, by virtue of that movement, the observer may see a particle with no net force have an acceleration. Some authors refer to this effect as inertial forces, which are not true forces but only exist because of the motion of the observer. This leads to the concept of an inertial frame as one in which a particle with no net force acting on it appears to move in a straight line.

The "principle of relativity" in classical mechanics states that the laws of dynamics are the same in all inertial frames. We saw in Section 2.2 that if a frame

$\mathcal{F} = (O, \vec{e}_1, \vec{e}_2, \vec{e}_3)$ is inertial, then another frame $\mathcal{F}' = (O', \vec{f}_1, \vec{f}_2, \vec{f}_3)$ is also inertial if the origin of $\mathcal{F}'$ travels along a line $\vec{b} + \vec{v}t$ in reference to $\mathcal{F}$ and where the orthonormal set of vectors of $\mathcal{F}$ is a fixed rotation from the orthonormal set of vectors in $\mathcal{F}$. It is a standard result in geometry that a *direct isometry* $f : \mathbb{R}^3 \to \mathbb{R}^3$ has the form $f(\vec{x}) = A\vec{x} + \vec{b}$, where $A \in \mathrm{SO}(3)$ and $\vec{b}$ is any fixed vector in $\mathbb{R}^3$. This direct isometry corresponds to $\overrightarrow{OO'} = \vec{b}$ and $\vec{f}_i = A\vec{e}_i$ for $i = 1, 2, 3$. Consequently, the frame $\mathcal{F}'$ differs from $\mathcal{F}$ by a fixed direct isometry composed with a translation by $\vec{v}t$, which corresponds to movement along a fixed velocity vector, which could be $\vec{0}$ if $\mathcal{F}'$ is stationary.

Of particular interest, the change of coordinates

$$x' = x - vt, \quad y' = y, \quad z' = z, \tag{7.15}$$

where v is a constant velocity, preserves the inertial property of frames. This is called the Galilean transformation. It corresponds to an observer in the frame $\mathcal{F}'$ moving at a constant speed v along the x-axis. with all other basis vectors between frames staying the same. The laws of mechanics expressed in one system of coordinates will be the same when expressed in the other. If P_1 and P_2 are two points in space with coordinates (x_1, y_1, z_1) and (x_2, y_2, z_2) in the frame $\mathcal{F}$ then the coordinates in the frame $\mathcal{F}'$ are (x'_1, y'_1, z'_1) and (x'_2, y'_2, z'_2), which could very well be different. The coordinates with respect to a frame are not a physical quantity in that no law of mechanics will depend on the specific value of the coordinates. However, if we denote $\Delta x = x_2 - x_1$ and likewise for the other coordinates, then the distance $P_1 P_2$ is preserved between inertial frames:

$$\Delta s \overset{\mathrm{def}}{=} \sqrt{(\Delta x)^2 + (\Delta y)^2 + (\Delta z)^2} = \sqrt{(\Delta x')^2 + (\Delta y')^2 + (\Delta z')^2} \tag{7.16}$$

So distance Δs between points is a physical quantity, independent of inertial reference frame.

Consider the situation of passengers on a plane. When the plane is sitting stationary on the tarmac the passengers will observe all the laws of physics to be the same as if they were not on the plane. At that point, a frame $\mathcal{F}'$ fixed to the plane is an inertial frame since we will assume that a frame $\mathcal{F}$ fixed to the Earth is inertial. When the plane is at altitude and cruising speed, and not effected by turbulence, except for the sound of the engines, there is no experiment that can be done internal to the plane that would allow a passenger to discern that it is moving. However, while the plane accelerates during take-off it is not an inertial frame: if a passenger drops an object, it will not fall straight toward the ground even though the only non-negligible force acting on it is gravity, which is vertical.

Classical mechanics implicitly treats the notion of time as absolute and independent of any frame. For centuries, no one could imagine anything different. To be more precise, in order to record events in different frames, we must use the space variables which come from the geometric frames but also a time variable. So we must imagine a clock attached to each frame. By calling time absolute, we mean that the only possible difference between clocks in classical frames is that they may have $t = 0$ corresponding to different points in time. In particular, suppose we use t and t' as the time variables in the frames $\mathcal{F}$ and $\mathcal{F}'$. If events P_1 and P_2 occurs as t_1 and t_2 in frame $\mathcal{F}$ and at t_1' and t_2' in frame $\mathcal{F}'$, then

$$\Delta t = \Delta t'.$$

So though the recorded point in time is not a physical quantity, an interval of time Δt is. In modern explanations of the Galilean transformation (7.15) it is common to add the equation $t = t'$, though physicists working before special relativity would never thought of needing to write this.

Through the 19th century, experiments on the nature of space, light, electro-magnetism, and ether (the hypothetical medium through which it was thought that light propagates like sound through air) produced unexpected results that began to call into question even these fundamental perspectives on the nature of space and time. Einstein's theory of special relativity resolved these observations by using modern developments in mathematics by reformulating the notion of spacetime according to two postulates:

Postulate 1 The laws of electrodynamics and optics are valid in all reference frames in which the laws of mechanics hold (inertial frames).

Postulate 2 Light is always propagated in empty space with a definite velocity c that is independent of the motion of the emitting body.

These principles bear out the surprising but experimentally observed fact that distance Δs as defined in (7.16) and Δt change between inertial frames and this change is particularly evident when v is large. Suppose that we locate an event in a frame $\mathcal{F}$ using (t, x, y, z) time and space variables and similarly for another frame $\mathcal{F}'$. Using work by Minkowski, Lorentz and others, Einstein showed that Postulate 2 implies that if $\mathcal{F}'$ is moving at a constant speed in the direction of the x-axis of $\mathcal{F}$ and if the unit basis vectors in each frame are the same, then the Galilean transformation should be replaced with

$$\begin{pmatrix} t' \\ x' \\ y' \\ z' \end{pmatrix} = \begin{pmatrix} \gamma & -\dfrac{\gamma v}{c^2} & 0 & 0 \\ -v\gamma & \gamma & 0 & 0 \\ 0 & 0 & 1 & 0 \\ 0 & 0 & 0 & 1 \end{pmatrix} \begin{pmatrix} t \\ x \\ y \\ z \end{pmatrix}, \quad \text{where } \gamma = \frac{1}{\sqrt{1 - \dfrac{v^2}{c^2}}}. \tag{7.17}$$

We sometimes write $\gamma(v)$ to indicate the dependence of γ on the magnitude of the velocity. In general, a *Lorentz transformation* is any change of coordinates from (t, x, y, z) to (t', x', y', z') that consists of compositions of transformations in (7.17) and rotations in the space variables.

Here are a few surprising consequences.

- Contraction of length. Suppose that $P_1 = (t_1, x_1, y_1, z_1)$ and $P_2 = (t_2, x_2, y_2, z_2)$ are two events in frame $\mathcal{F}$ with $t_1 = t_2$, $y_1 = y_2$, and $z_1 = z_2$. Then the length $P_1 P_2$ is Δx and $\overline{P_1 P_2}$ is a segment along the direction of travel. Then (7.17) implies that

$$\Delta x' = x'_2 - x'_1 = (\gamma x_2 - v\gamma t_2) - (\gamma x_1 - v\gamma t_1) = \gamma \Delta x.$$

Then $\Delta x'$ is the length between P_1 and P_2 as seen in the frame $\mathcal{F}'$. So while an observer in frame $\mathcal{F}'$ sees as a segment of length $L_0 = \Delta x'$, the observer in frame $\mathcal{F}$ will it see as having length

$$\Delta x = \frac{\Delta x'}{\gamma} = L_0 \sqrt{1 - \frac{v^2}{c^2}}.$$

- Loss of simultaneity. Consider the same two points as above. In frame $\mathcal{F}$ they are simultaneous because $t_1 = t_2$. However,

$$\Delta t' = t'_2 - t'_1 = \left(\gamma t_2 - \frac{v\gamma}{c^2} x_2\right) - \left(\gamma t_1 - \frac{v\gamma}{c^2} x_1\right) = -\frac{v\gamma}{c^2} \Delta x.$$

Thus, an observer in frame $\mathcal{F}'$ does not view P_1 and P_2 as occurring simultaneously.

- We do also note that if P_1 and P_2 were events with only a difference in their y coordinates, with the relative motion still along the x-axis, then $\Delta y' = \Delta y$. Hence, there is no observed contraction of length perpendicular to the direction of motion.

- Finally, the formulas for the Lorentz transformation imply that no particle can move faster than the speed of light c.

Though distances and time intervals are not preserved across inertial frames, the Minkowski line element given by

$$(\Delta s)^2 \stackrel{\text{def}}{=} -c^2 (\Delta t)^2 + (\Delta x)^2 + (\Delta y)^2 + (\Delta z)^2 \tag{7.18}$$

is preserved by any Lorentz transformation. Consequently, the postulates of special relativity require us to jettison the assumption that time and space coordinates are independent of each other. This perspective leads to the mental model of *spacetime*. A point in this spacetime is called an *event* and we use coordinates (t, x, y, z) with respect to some frame.

Scaling the right-hand side of (7.18) by any factor still gives us a quantity that is preserved by Lorentz transformations. The choice of signs reflects the fact that if $\Delta t = 0$ in some frame, then Δs is precisely the usual distance between points in that frame. The *proper time* interval $\Delta \tau$ between two events in spacetime is

$$(\Delta \tau)^2 \stackrel{\text{def}}{=} -(\Delta s)^2 = \frac{1}{c^2}(c^2 (\Delta t)^2 - (\Delta x)^2 - (\Delta y)^2 - (\Delta z)^2).$$

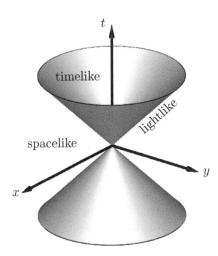

Figure 7.6: Light cone.

In this context, it is not as natural to talk about the "trajectory" of a particle, since this term usually assumes that the space variables are expressed as a function of time. In contrast, we can still model the motion of a particle by parametric equations $\vec{x}(\lambda) = (t(\lambda), x(\lambda), y(\lambda), z(\lambda))$ for some parameter λ.

Definition 7.2.1. If a particle has the property that for all λ in some interval $[\lambda_1, \lambda_2]$, the particle exists in space time at $\vec{x}(\lambda)$, then the image of this curve is called the *world line* of the particle.

If we wish λ to carry some sense of moving forward in time, we simply impose the assumption that $dt/d\lambda > 0$. Then the rate of change $d\tau/d\lambda$ of the particle's proper time with respect to λ satisfies

$$\left(\frac{d\tau}{d\lambda}\right)^2 = \left(\frac{dt}{d\lambda}\right)^2 - \frac{1}{c^2}\left(\frac{dx}{d\lambda}\right)^2 - \frac{1}{c^2}\left(\frac{dy}{d\lambda}\right)^2 - \frac{1}{c^2}\left(\frac{dz}{d\lambda}\right)^2.$$

Using chain rules so that $da/d\lambda = (da/dt)(dt/d\lambda)$ and simplifying by $dt/d\lambda$, we get

$$\left(\frac{d\tau}{dt}\right)^2 = 1 - \frac{1}{c^2}\left(\left(\frac{dx}{dt}\right)^2 + \left(\frac{dy}{dt}\right)^2 + \left(\frac{dz}{dt}\right)^2\right).$$

So if in a frame $\mathcal{F}$ a particle has a velocity vector function of $\vec{v}(t)$, then

$$d\tau = \sqrt{1 - \frac{v^2}{c^2}}\, dt \quad \text{and} \quad \frac{dt}{d\tau} = \gamma(v). \tag{7.19}$$

The quantity $d\tau$ is the *proper time differential* and the function

$$\tau = \int_0^\lambda d\tau \qquad (7.20)$$

is called the *proper time* of the particle traveling on its world line.

Proper time plays a central role in the theory of relativity since it is the same for all inertial observers, i.e., unchanged by any Lorentz transformation. Furthermore, this reminds us of the habit in elementary differential geometry to consider the parametrization of a curve by arclength: since (7.20) defines a function $\tau(\lambda)$ such that $d\tau/d\lambda > 0$ so an inverse $\lambda(\tau)$ exists; using this function we can reparametrize a particle's world line by proper time. The proper time function defined in (7.20) also emphasizes that the proper time between two events is the time ticked off by a clock which actually passes through both events.[50]

Suppose we have two events with a given Minkowski metric Δs^2 between them. They are called

- *timelike separated* if $\Delta s^2 < 0$. This means that $\Delta \tau^2 > 0$. Clearly, for two timelike separated events time must have elapsed. Also, since a particle cannot travel faster than the speed of light, any two events on the world line of a particle must be timelike separated. From another perspective, two events are called timelike separated if a particle can travel between them (without moving faster than the speed of light).

- *lightlike separated* if $\Delta s^2 = 0$. Only a particle traveling in a straight line at the speed of light can connect two lightlike separated events.

- *spacelike separated* if $\Delta s^2 > 0$. No particle can have a world line that connects two spacelike separated events. [60, Section 2.2]

The *light cone* based at an event P is the set of all events that are lightlike separated from P. Figure 7.6 shows the light cone for the origin, though we can only display the variables (t, x, y).

7.2.2 Minkowski Spacetime

Euclidean geometry takes place in the inner product space $(\mathbb{R}^n, \cdot)$, where the inner product is the dot product. In Chapter 4 we considered properties of vector spaces equipped with other bilinear forms. In particular, in Example 4.3.12 we already saw the following space.

Definition 7.2.2. We define $(n + 1)$-Minkowski spacetime as a real vector space with coordinates $(x^0, x^1, \ldots, x^n)$ equipped with the bilinear form η defined by

$$\eta = \eta_{ij} dx^i \otimes dx^j$$

with coefficients

$$\eta_{00} = -1, \quad \eta_{ii} = 1 \text{ if } i > 0, \quad \eta_{ij} = 0 \text{ if } i \neq j.$$

The bilinear form η is called the *Minkowski metric*. We denote the Minkowski spacetime by $\mathbb{R}^{n,1}$.

For simplicity of notation, we will write $\vec{a} \cdot \vec{b}$ instead of $\eta(\vec{a}, \vec{b})$.

This vector space is suited for special relativity because we can set

$$(x^0, x^1, x^2, x^3) = (ct, x, y, z). \tag{7.21}$$

Then under this coordinate change, the Minkowski line element in (7.18) corresponds to

$$\eta(\vec{a}, \vec{a}), \qquad \text{where } \vec{a} = \begin{pmatrix} c\Delta t \\ \Delta x \\ \Delta y \\ \Delta z \end{pmatrix}.$$

We can think of the difference between the (t, x, y, z) coordinates and the (x^0, x^1, x^2, x^3) coordinates as a change of units, so that in the (x^i) system the speed of light is 1.

On the other hand, because of the centrality of the speed of light, especially since we postulate that it is the same for every observer, we might as well use a system of units in which it is 1. Many people are familiar with the light-year: applied to time a light-year is a usual year and when applied to distance, it means the distance traveled by light in a year. Or we could use the unit of meter: when applied to time, 1 m of time refers to how long it takes for light to travel 1 meter. Since in the SI (international system) $c = 3 \times 10^8 \text{m/s}$, the conversion between a meter of time and a second of time is

$$1\,\text{m} = \frac{1}{3 \times 10^8}\,\text{s}.$$

This convention of units is common among specialists in general relativity but is not universal throughout other branches of physics. Consequently, this text refrains from using this convention of units. So when applying Minkowski space to special relativity, we continue to assume $x^0 = ct$. (In doing so, we hope that specialists will not be put off and that non-specialists will not be confused.)

In Example 4.3.12, we determined the automorphisms of the Minkowski metric. When $c = 1$, we had found precisely the format of Lorentz transformations as in (7.17), except that in (4.26) we had found a few possible differences of signs in $\varepsilon_1 = \pm 1$ and $\varepsilon_2 = \pm 1$. These signs do not appear in (7.17). Consequently, the allowed transformations between inertial frames in special relativity, namely Lorentz transformations, correspond to the restricted Lorentz transformation group discussed in Example 4.3.12. Over Minkowski space, this group of transformations plays a parallel role to the group of direct isometries in Euclidean geometry.

An inner product space, defined in Definition 4.2.11, generalizes the Euclidean space of $\mathbb{R}^n$ equipped with the dot product. All the geometry of angles, lengths, volumes that we can define in Euclidean space have identical definitions in any inner product space. Sylvester's Law of Inertia (Theorem 4.2.14) affirms that for any symmetric bilinear form, the signature is independent of the basis. Hence, we can generalize the notion of the Minkowski spacetime to the following.

Definition 7.2.3. A vector space V of dimension $n+1$ is called a *Minkowski space* if it is equipped with a bilinear form $\langle\,,\,\rangle$ that has signature $(1, n, 0)$ or $(n, 1, 0)$.

From the perspective of bilinear forms, the difference between an inner product space and a Minkowski space appears minor. However, as we already saw in the previous subsection, there are significant differences between the geometry of an inner product space and a Minkowski space. Most notably, for a vector $v \in V$, it is not always true that $\langle v, v\rangle \geq 0$ or that $\langle v, v\rangle = 0$ implies that $v = 0$. In an inner product space we define the length of a vector as $\sqrt{\langle v, v\rangle}$ but this notion of length does not exist in the same way. Instead, just as when we discussed the difference between timelike, lightlike and spacelike separated points above, in any Minkowski space, there are regions in which $\langle v, v\rangle$ is positive, 0, or negative. If we do define distances, we must do so differently in each of these regions.

For applications to special relativity, we usually use a Minkowski space with signature $(n, 1, 0)$. However, the difference between the geometry of vector spaces with bilinear forms of signature $(1, n, 0)$ versus $(n, 1, 0)$ is immaterial, mostly a matter of terms.

The concept of a light cone from special relativity inspires the following definition.

Definition 7.2.4. Let $(V, \langle\,,\,\rangle)$ be a Minkowski space. The *null cone* is the collection of points $\vec{x} = (x^0, x^1, \ldots, x^n)^\top$ such that $\langle \vec{x}, \vec{x}\rangle = 0$.

In an arbitrary Minkowski space, the null cone is a generalized cone with apex at the origin in that for all $\vec{x}$ in the null cone, $\lambda\vec{x}$ is also in the cone. The null cone separates the space into components in which $\langle \vec{x}, \vec{x}\rangle > 0$ or $\langle \vec{x}, \vec{x}\rangle < 0$. In special relativity, we called these regions spacelike and timelike separated.

7.2.3 Physical Quantities in Special Relativity

Let $\vec{x}(\lambda)$ be a parametric curve that traces out the world line of a particle. We define the *four-velocity* of the particle as the vector

$$\vec{U} = \frac{d\vec{x}}{d\tau} = \left(c\frac{dt}{d\tau}, \frac{dx}{d\tau}, \frac{dy}{d\tau}, \frac{dz}{d\tau}\right). \tag{7.22}$$

This vector is tangent to the world line.

In the geometry of curves in Euclidean space, the derivative $d\vec{x}/ds$, where s is the arclength parameter, is the unit tangent vector. This is geometrically significant since the arclength function is independent of any regular reparametrization and independent of the position and orientation of the curve in space. Similarly, the four-velocity is a vector whose identity is independent of the observer's frame. However, its components will change between frames by the corresponding Lorentz transformation.

We point out that

$$
\begin{aligned}
\vec{U} \cdot \vec{U} &= -c^2 \left(\frac{dt}{d\tau}\right)^2 + \left(\frac{dx}{d\tau}\right)^2 + \left(\frac{dy}{d\tau}\right)^2 + \left(\frac{dz}{d\tau}\right)^2 \\
&= \left(-c^2 + \left(\frac{dx}{dt}\right)^2 + \left(\frac{dy}{dt}\right)^2 + \left(\frac{dz}{dt}\right)^2\right) \left(\frac{dt}{d\tau}\right)^2 \\
&= (-c^2 + v^2)\gamma(v)^2 \\
&= -c^2.
\end{aligned}
$$

The four-momentum vector is defined as

$$
\vec{p} = m\vec{U}, \tag{7.23}
$$

where m is the rest mass of the particle. In a frame $\mathcal{F}$, the components of the four-momentum vector are

$$
[\vec{p}]_{\mathcal{F}} = \begin{pmatrix} p^0 \\ p^1 \\ p^2 \\ p^3 \end{pmatrix} = \begin{pmatrix} E/c \\ m\, dx/dt \\ m\, dy/dt \\ m\, dz/dt \end{pmatrix}, \tag{7.24}
$$

where $E = p^0 c$ is the energy of the particle in the frame $\mathcal{F}$ and (p^1, p^2, p^3) are the components of its spatial momentum.

The four-acceleration is

$$
\vec{a} = \frac{d\vec{U}}{d\tau}. \tag{7.25}
$$

Since $\vec{U} \cdot \vec{U}$ is constant, then $\vec{a} \cdot \vec{U} = 0$.

Special relativity requires careful study to develop an effective intuition. This text has not provided any of the historical developments or experimental results that support this theory. We refer the reader to [24, 21, 42], each offering a comprehensive treatment of the subject.

7.2.4 Pseudo-Riemannian Manifolds

The definition of a Riemannian manifold arose from assuming a smooth manifold came equipped with an inner product on every tangent space, that varied smoothly across the manifold. The usefulness of Minkowski space for special relativity illustrates that an inner product is not always what we might want for certain applications. This inspires the following definition.

Definition 7.2.5. A *pseudo-Riemannian metric* on a smooth manifold M is a symmetric tensor field g of type $(0, 2)$ on M that is nondegenerate at every point. The pair (M, g) is called a *pseudo-Reimannian manifold*.

In more detail, g is a global section of $\mathrm{Sym}^2 TM^*$ such that at each point $p \in M$, we can have $g_p(X, Y) = 0$ for all $Y \in T_pM$ if and only if $X = 0$ in T_pM. This definition is looser than that of a Riemannian manifold since it has removed the positive-definite condition of inner products. Some authors refer to g on a pseudo-Riemannian manifold as a metric, whereas other authors prefer the term pseudometric to emphasize that g is not positive definite.

From Sylvester's Law of Inertia (Theorem 4.2.14), the signature of a symmetric bilinear form on a vector space is independent of the basis. More can be said for pseudo-Riemannian manifolds.

Proposition 7.2.6. *Let (M, g) be a pseudo-Riemannian manifold. Then on every connected component of M, the signature of g is the same.*

Proof. Recall that if $\langle \, , \, \rangle$ is a symmetric bilinear form on a vector space V of dimension n, then $\langle \, , \, \rangle$ is nondegenerate if and only if the signature (s, t, r) has $r = 0$. This condition is also equivalent to the coefficient matrix of $\langle \, , \, \rangle$ with respect to some basis having a nonzero determinant.

For each $p \in M$, we consider the symmetric bilinear form g_p. By definition, $g : M \to \mathrm{Sym}^2 TM^*$ is a continuous function. The coefficients of the characteristic polynomial of a matrix are polynomials, and therefore continuous, in the entries of a matrix. By the Spectral Theorem, since the bilinear form g_p is symmetric, its matrix with respect to any basis is diagonalizable and all its eigenvalues are real. Consequently, we can order the eigenvalues of g_p as functions $\lambda_1(p) \geq \lambda_2(p) \geq \cdots \geq \lambda_n(p)$. It is a well known result that the zeros of a polynomial vary continuously with the coefficients, even over $\mathbb{C}$. [38, p.3]. Hence, the eigenvalue functions $\lambda_i : M \to \mathbb{R}$ are continuous.

However, $\det(g_p) = \lambda_1(p)\lambda_2(p) \cdots \lambda_n(p)$ is continuous and never 0. Hence, $\lambda_i(p) \neq 0$ for all $p \in M$. From the proof of (Theorem 4.2.14), in the signature (s, t, r), the value s represents the number of eigenvalues that are positive, while t represents the number of eigenvalues that are negative. Since the eigenvalue functions are continuous and never 0, the number of eigenvalues that are positive and the number that are negative stays constant over any connected component. $\square$

Definition 7.2.7. The *signature* of a pseudo-Riemannian manifold is the pair (s, t) of (M, g), where $(s, t, 0)$ is the signature of g_p for each $p \in M$.

As we will see in Section 7.5, the theory of general relativity requires a model of space that is not flat but nonetheless behaves locally like Minkowski spacetime. Gravitational effects will cause the Lorentz metric to vary through space. A pseudo-Riemannian manifold of signature $(3, 1)$ models this well.

A review of the proof of the Levi-Civita Theorem shows that the proof only used the symmetry and the nondegenerate (invertibility of the (g_{ij}) matrix) aspect of the metric. Consequently, the following holds.

Theorem 7.2.8. *The Levi-Civita Theorem, Theorem 6.2.11, also holds for pseudo-Riemannian manifolds. The coefficients for the Levi-Civita connection are also given by the Christoffel symbols defined in Proposition 6.2.13.*

Despite this theorem, a few relevant changes arise in the following contexts:

- We can no longer define the length of a tangent vector if $g_p(V, V) < 0$.

- We cannot define the arclength of a curve γ if $g_{\gamma(\sigma)}(\gamma'(\sigma), \gamma'(\sigma)) < 0$ for some $\sigma \in I$.

- We might not be able to define the volume of a region $\mathcal{R}$ of M.

Despite these possible obstructions, the equations for geodesics still satisfy the existence and uniqueness properties of Theorem 6.3.12. Furthermore, like Proposition 6.3.13, geodesics on pseudo-Riemannian manifolds have a constant $\langle \gamma'(\sigma), \gamma'(\sigma) \rangle$. Thus, geodesics come in three categories depending on the sign of $\langle \gamma'(\sigma), \gamma'(\sigma) \rangle = g_{ij}\dot\gamma^i(\sigma)\dot\gamma^j(\sigma)$.

In the context of a Minkowski spacetime $\mathbb{R}^{3,1}$, where the metric g has signature $(3, 1)$, we say that a geodesic is

- a *timelike geodesic* if $g(\gamma'(\sigma), \gamma'(\sigma)) < 0$;

- a *null geodesic* if $g(\gamma'(\sigma), \gamma'(\sigma)) = 0$;

- a *spacelike geodesic* if $g(\gamma'(\sigma), \gamma'(\sigma)) > 0$.

PROBLEMS

7.2.1. Use the interpretation of the four-momentum in (7.24) to recover the energy-momentum relation $E^2 = m^2 c^4 + p^2 c^2$.

7.2.2. *Action for a Relativistic Point Particle.* The action of a free (no external forces) non-relativistic particle traveling between $t = t_1$ and $t = t_2$ is simply

$$S = \int_{t_1}^{t_2} \frac{1}{2}mv^2 \, dt = \int_{t_1}^{t_2} \frac{1}{2}m \left\| \frac{d\vec{x}}{dt} \right\|^2 dt,$$

and thus the Lagrangian is $L = T = \frac{1}{2}mv^2$. To give a relativistic formulation for the action of a free particle, let us first assume we are in the context of a Minkowski space with coordinates described in Equation (7.21). We must describe the action in a way that is invariant under a Lorentz transformation. Therefore, we cannot directly use the particle velocity since the velocity is not a Lorentz invariant. This exercise seeks to justify the definition of the action of a relativistic point particle with rest mass of m_0 as

$$S = -mc^2 \int_{\mathcal{P}} d\tau, \tag{7.26}$$

where we integrate over a world line $\mathcal{P}$ of the particle. According to (7.19), the action in (7.26) has an associated Lagrangian of

$$L = -mc^2 \sqrt{1 - \frac{v^2}{c^2}}. \qquad (7.27)$$

(a) Calculate the 6th-order Taylor expansion of $\sqrt{1-x^2}$, and show that the quadratic approximation to L is

$$L \cong -mc^2 + \frac{1}{2}mv^2. \qquad (7.28)$$

(b) Using Equation (7.27), show that the generalized momentum vector $\vec{p}$ and the Hamiltonian H satisfy

$$\vec{p} = \frac{m\vec{v}}{\sqrt{1 - \frac{v^2}{c^2}}} \qquad \text{and} \qquad H = \frac{mc^2}{\sqrt{1 - \frac{v^2}{c^2}}}.$$

(This formula for H conforms with the formula [24, (1-16)] for the total energy of a free relativistic particle.)

7.2.3. Let (M, g) be a four-dimensional manifold with a Lorentzian metric that over a particular coordinate system (t, x, y, z) has the matrix

$$g_{ij} = \begin{pmatrix} k^2 - g^2 t^2 & 0 & 0 & gt \\ 0 & 1 & 0 & 0 \\ 0 & 0 & 1 & 0 \\ gt & 0 & 0 & 1 \end{pmatrix}.$$

Show that the geodesics that have the initial condition $(x, y, z, t) = (0, 0, 0, 0)$ when $s = 0$ satisfy

$$x = at, \qquad y = bt, \qquad \text{and} \quad z = -\frac{1}{2}gt^2 + ct.$$

Use this to give a physical interpretation of this metric.

7.2.4. Let M be a pseudo-Riemannian manifold of dimension 3 with the line element

$$ds^2 = -dt^2 + \frac{1}{1 - \lambda r^2}dr^2 + r^2 d\theta^2,$$

where we assume $r^2 < 1/\lambda$. Show that the null geodesics satisfy the relationship

$$\left(\frac{dr}{d\theta}\right)^2 = r^2(1 - \lambda r^2)(Cr^2 - 1),$$

where C is a constant. Use the substitution $u = 1/r^2$ to solve this differential equation, and show that the solutions are ellipses if we interpret r and θ as the usual polar coordinates.

7.2.5. Let $g > 0$ be a positive constant. Let M be a pseudo-Riemannian manifold of dimension 4 that has a line element of

$$-ds^2 = (1 - 2gx)dt^2 - \frac{1}{1 - 2gx}dx^2 - dy^2 - dz^2,$$

Show that the curve defined by $(1 - 2gx)\cosh^2(gt) = 1$, $y = z = 0$ is a geodesic passing through the origin.

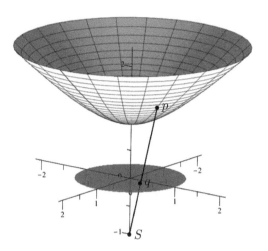

Figure 7.7: Stereographic projection from a hyperboloid.

7.2.6. Determine the geodesics in a pseudo-Riemannian manifold that has the line element metric

$$ds^2 = -x dt^2 + \frac{1}{x} dx^2 + dy^2 + dz^2.$$

7.2.7. Consider a Lorentzian metric given by $ds^2 = -dt^2 + f(t)^2 dx^2$, where $f(t)$ is any smooth function of t. Show that the Einstein tensor is identically 0.

7.2.8. Let H_R^n be the upper half of the two-sheeted hyperboloid in $\mathbb{R}^{n+1}$, defined by

$$(x^1)^2 + \cdots + (x^n)^2 - (x^{n+1})^2 = R^2 \qquad \text{and} \quad x^{n+1} > 0.$$

Equip H_R^n with the metric $g = i^*\eta$, where $i : H_R^n \to \mathbb{R}^{n+1}$ is the inclusion map and η is the Minkowski metric expressed as

$$\eta = (dx^1)^2 + \cdots + (dx^n)^2 - (dx^{n+1})^2.$$

Define the manifold $\mathbb{B}_R^n$ as the n-dimensional open ball in the $x^{n+1} = 0$ hyperplane of $\mathbb{R}^{n+1}$ with center at the origin and radius R. Equip $\mathbb{B}_R^n$ with the metric $\tilde{g}$ defined in Problem 6.1.12, namely

$$\tilde{g} = \frac{4R^4}{(R^2 - \|x\|^2)^2} \left((dx^1)^2 + \cdots + (dx^n)^2 \right).$$

We define the stereographic projection $\pi : H_R^n \to \mathbb{B}_R^n$ such that $\pi(p) = q$ is the unique point in $\mathbb{B}_R^n$ on the line segment $\overline{Sp}$, where $S = (0, 0, \ldots, 0, -R)$. Figure 7.7 depicts this projection for $n = 2$.

(a) Prove that $\pi(x^1, \ldots, x^n, x^{n+1}) = \dfrac{R}{R + x^{n+1}} (x^1, \ldots, x^n).$

(b) For $u \in \mathbb{B}_R^n \subset \mathbb{R}^n$, show that $\pi^{-1}(u) = \left(\dfrac{2R^2 u}{R^2 - \|u\|^2}, \dfrac{R(R^2 + \|u\|^2)}{R^2 - \|u\|^2} \right).$

(c) Show that $(\pi^{-1})^* g = \tilde{g}$.

(d) Deduce that (H^n_R, g) is a Riemannian manifold (even though the metric is a pull-back from a pseudo-Riemannian metric) and that π is an isometry between Riemannian manifolds.

7.3 Electromagnetism

7.3.1 Maxwell's Equations

The goal of this section is to summarize the dynamics of a charged particle moving under the influence of an electric field $\vec{E}$ and a magnetic field $\vec{B}$, both of which are time and space dependent. In no way does this brief section attempt to encapsulate all of the theory of electromagnetism. Rather we show how to pass from a classical formulation of a few of the basic laws of electromagnetism to a modern formulation that uses Minkowski metrics and the language of forms. (Note: all formulas in this section use CGS units, i.e., centimeters-grams-seconds units. In this system, force is measured in dyne, energy in erg, electric charge in esu, electric potential in statvolt, and the magnetic field strength in gauss.)

The mathematical theory relies on the model (based on experiment) that point charges exist, i.e., particles of negligible size with charge. For example, the electron and the proton fit this bill. In contrast, magnetic monopoles – point-like particles with a magnetic charge – do not (appear to) exist. The observation of a single magnetic monopole would change the rest of the theory (by adding an extra magnetic charge density and magnetic current) but even this "would not alter the fact that in matter as we know it, the only sources of the magnetic field are electric currents." [46, p. 405]

Coulomb's law of electrostatic force states that the force between two point charges is inversely proportional to the square of the distance between them, namely,

$$\vec{F} = \frac{q_1 q_2}{r^2} \hat{r}, \tag{7.29}$$

where q_1 and q_2 are the respective charges of the particles, r is the distance between them, and $\hat{r}$ is the unit vector pointing from the location of the point charge 1 to the point charge 2. One then considers systems of charges, modeled by a charge density $\rho(x, y, z)$, acting on a point particle with charge q. The electric field of a charged system is the vector field $\vec{E}(x, y, z) = \frac{1}{q}\vec{F}$ where $\vec{F}$ is the force the system would exert on a particle of charge q at position (x, y, z). It is calculated by

$$\vec{E} = \iiint_{\mathbb{R}^3} \rho(x', y', z') \frac{(x - x', y - y', z - z')}{((x - x')^2 + (y - y')^2 + (z - z')^2)^{3/2}} \, dx' \, dy' \, dz', \tag{7.30}$$

If the charge density ρ depends on time t as well, the $\vec{E}$ is a time dependent vector field $\vec{E}(x, y, z, t)$. An application of Gauss's Theorem from vector calculus gives

Gauss's Law for electrostatics, i.e.,

$$\operatorname{div} \vec{E} = 4\pi\rho, \tag{7.31}$$

where the divergence is only taken in the space variables.

Consider the following function defined in terms of the charge density ρ:

$$\varphi(x, y, z, t) = \iiint_{\mathbb{R}^3} \frac{\rho(x', y', z', t)}{\sqrt{(x - x')^2 + (y - y')^2 + (z - z')^2}} \, dx' \, dy' \, dz'. \tag{7.32}$$

By taking the gradient with respect to the space variables (x, y, z) passing under the integral, we see that

$$\vec{E} = -\vec{\nabla}\varphi. \tag{7.33}$$

This shows that $\vec{E}$ is a conservative vector field. The function φ is called the *electric potential*. The potential energy of the electric force field acting on a particle with charge q is $V = q\varphi$. If the system of electrical charges is moving, then $\vec{E}$, φ, and ρ are also functions of time, but (7.30) and (7.33) still hold with the caveat that the integration and the gradient only involve the space variables.

A system of time-dependent current density also induces what are called *electrical currents*. The *current density* is the vector field $\vec{J}$ that at each point (x, y, z) measures the direction of the current and how much current is passing per area and per time. A direct application of Gauss's Theorem from vector calculus gives

$$\operatorname{div} \vec{J} = -\frac{\partial\rho}{\partial t}. \tag{7.34}$$

At the heart of electromagnetism lies an interdependence between magnetic fields and electric fields. A charged particle that is moving in the presence of a current experiences a force perpendicular to its velocity. That force acting on the particle is called the *magnetic force*. The magnetic field of a system of charges is the field $\vec{B}$ defined implicitly by

$$\vec{F} = q(\vec{E} + \frac{1}{c}\vec{v} \times \vec{B}). \tag{7.35}$$

This overall effect on a particle with charge q is called the *electromagnetic force*. It is no longer conservative due to the presence of $\vec{v}$. Nonetheless, we define the magnetic vector potential $\vec{A}$ by

$$\vec{A}(x, y, z) = \frac{1}{c} \iiint_{\mathbb{R}^3} \frac{\vec{J}(x', y', z', t)}{\sqrt{(x - x')^2 + (y - y')^2 + (z - z')^2}} \, dx' \, dy' \, dz'. \tag{7.36}$$

Furthermore, Faraday discovered that not only does a time-dependent distribution of charge induce a magnetic field, a variable magnetic field similarly affects the electric field. The relationship between the electric and magnetic fields can be summarized by two separate sets of equations: Faraday's law for potential, i.e.,

$$\vec{E} = -\vec{\nabla}\varphi - \frac{1}{c}\frac{\partial \vec{A}}{\partial t} \qquad \vec{B} = \vec{\nabla} \times \vec{A}, \qquad (7.37)$$

and the celebrated Maxwell's equations, i.e.,

$$\vec{\nabla} \cdot \vec{E} = 4\pi\rho, \qquad \vec{\nabla} \cdot \vec{B} = 0,$$
$$\vec{\nabla} \times \vec{E} = -\frac{1}{c}\frac{\partial \vec{B}}{\partial t}, \qquad \vec{\nabla} \times \vec{B} = \frac{1}{c}\frac{\partial \vec{E}}{\partial t} + \frac{4\pi}{c}\vec{J}. \qquad (7.38)$$

Maxwell's equations stand as a crowning achievement in electromagnetism. They encapsulate the interdependent phenomena of induction and the static source of the various fields. Furthermore, solving the equations for empty space (i.e., $\rho = 0$ and $\vec{J} = \vec{0}$) leads to an interpretation of light as an electromagnetic wave.

Hidden in Maxwell's equations lie relativistic effects. If a charged particle travels fast (a non-trivial fraction of the speed of light), then due to relativistic effects, its electric field appears distorted to a stationary observer. Lorentz transformations in (7.17) describe how the electric and magnetic fields look different in different moving frames of reference.

7.3.2   Covariant Formula of Electromagnetism

Having developed considerable analytical machinery in the previous chapters, we are in a position to reformulate the theory of electromagnetism in a more concise way. We work in a four-dimensional Minkowski spacetime, which means we use the pseudometric $g = \eta$, as defined in Section 7.2.2. As before, we label the coordinates as $x^0 = ct$, $x^1 = x$, $x^2 = y$, and $x^3 = z$.

Define the 4-vector potential $\mathbf{A}$ as the covector (1-form) with components

$$A_i = (-\varphi, A_1, A_2, A_3). \qquad (7.39)$$

We call the electromagnetic tensor $\mathbf{F}$ the 2-form

$$F = -\sum_{i=1}^{3} E_i dx^0 \wedge dx^i + \sum_{i=1}^{3} B_i(\tilde{\star}dx^i)$$
$$= -E_1 dx^0 \wedge dx^1 - E_2 dx^0 \wedge dx^2 - E_3 dx^0 \wedge dx^3$$
$$+ B_1 dx^2 \wedge dx^3 - B_2 dx^1 \wedge dx^3 + B_3 dx^1 \wedge dx^2,$$

where by $\tilde{\star}$ we mean the Hodge star operator acting only on the space variables. If we exhibit the components of $\mathbf{F}$ in an antisymmetric matrix, we write

$$F_{\mu\nu} = \begin{pmatrix} 0 & -E_1 & -E_2 & -E_3 \\ E_1 & 0 & B_3 & -B_2 \\ E_2 & -B_3 & 0 & B_1 \\ E_3 & B_2 & -B_1 & 0 \end{pmatrix}. \qquad (7.40)$$

(As always, we use the convention that in $F_{\mu\nu}$, the index μ corresponds to the row and ν corresponds to the column of the representing matrix.) In Problem 5.4.7, we showed that Faraday's law for potential from Equation (7.37) can be expressed as

$$F_{\alpha\beta} = \partial_\alpha A_\beta - \partial_\beta A_\alpha. \tag{7.41}$$

Example 4.5.9 showed that a collection of component functions defined this way in terms of a covariant field A does define a tensor field of type $(0, 2)$.

We also define the 4-current vector by $\mathbf{J} = (c\rho, J^1, J^2, J^3)$, where ρ is the charge density and $(J^1, J^2, J^3) = \vec{J}$ is the classic current density vector. Using the Minkowski metric η, recall that by $F^{\alpha\beta}$ we mean the raising-indices operation $F^{\alpha\beta} = \eta^{\alpha\mu}\eta^{\beta\nu}F_{\mu\nu}$ and similarly for the lowering operation $J_\alpha = \eta_{\alpha\beta}J^\beta$. Recall that we write $\mathbf{J}^\flat$ for the covector associated to $\mathbf{J}$. In coordinates, we have

$$F^{\alpha\beta} = \begin{pmatrix} 0 & E_1 & E_2 & E_3 \\ -E_1 & 0 & B_3 & -B_2 \\ -E_2 & -B_3 & 0 & B_1 \\ -E_3 & B_2 & -B_1 & 0 \end{pmatrix} \quad \text{and} \quad J_\alpha = (-c\rho, J^1, J^2, J^3).$$

With this setup, it is not hard to show that Maxwell's equations can be written in tensor form as

$$\partial_\alpha F^{\beta\alpha} = \frac{4\pi}{c} J^\beta, \quad \text{and} \quad \varepsilon^{\alpha\beta\gamma\delta}(\partial_\gamma F_{\alpha\beta}) = 0, \tag{7.42}$$

where the last equation holds for all $\delta = 0, 1, 2, 3$. The second equation in (7.42) can be written equivalently as

$$\varepsilon^{\alpha\beta\gamma\delta}(\partial_\gamma F_{\alpha\beta}) = \partial_\gamma F_{\alpha\beta} + \partial_\alpha F_{\beta\gamma} + \partial_\beta F_{\gamma\alpha} = 0. \tag{7.43}$$

Using 4-vectors, one can describe the potential between the current 4-vector and the potential 4-vector. First, we define the *D'Alembertian* operator as

$$\Box = \frac{\partial^2}{\partial x^2} + \frac{\partial^2}{\partial y^2} + \frac{\partial^2}{\partial z^2} - \frac{1}{c^2}\frac{\partial^2}{\partial t^2}. \tag{7.44}$$

We point out that the D'Alembertian is equivalent to the Laplacian in the (x^0, x^1, x^2, x^3) with the Minkowski metric. Since $x^0 = ct$, we have

$$\frac{\partial}{\partial x^0} = \frac{1}{c}\frac{\partial}{\partial t}.$$

So $\nabla = (1/c\partial_t, \partial_x, \partial_y, \partial_z)$. The Laplacian is $\nabla^2 = \nabla \cdot \nabla$ so

$$\nabla^2 = -\frac{\partial^2}{\partial(x^0)^2} + \frac{\partial^2}{\partial(x^1)^2} + \frac{\partial^2}{\partial(x^2)^2} + \frac{\partial^2}{\partial(x^3)^2}$$

Since $\partial/\partial x^0 = \frac{1}{c}\partial/\partial t$, we see that the D'Alembertian operator is the same as the Laplacian for this context.

Applying Equation (7.34) to Equations (7.36) and (7.32), one can show that $\vec{\nabla} \cdot \vec{A} = -\frac{1}{c}\partial\varphi/\partial t$. Thus, taking the divergence of $\vec{E}$ expressed in Equation (7.37), we obtain

$$\Box\varphi = -4\pi\rho. \tag{7.45}$$

Using similar calculations, we can also show that

$$\Box A_i = -\frac{4\pi}{c}J_i \qquad \text{for } i = 1, 2, 3. \tag{7.46}$$

7.3.3 Electromagnetism Expressed in Differential Forms

Much of the reformulation of Faraday's and Maxwell's equations in the previous paragraphs can be expressed in even simpler terms using differential forms. We still work under the assumption that we work in Minkowski space $\mathbb{R}^{3,1}$. Faraday's law, expressed classically as (7.37) and in covariant components in 7.41, simply means

$$d\mathbf{A} = \mathbf{F}. \tag{7.47}$$

Interestingly enough, this formula does not refer to any metric but simply claims that the electromagnetic tensor $\mathbf{F}$ is an exact 2-form.

Since $\mathbf{F}$ is exact, it is also closed with $d\mathbf{F} = 0$. This property again has nothing to do with a metric. It is easy to check that it corresponds to the second and third Maxwell equations in (7.38). In Section 6.2.5, we mentioned the divergence operator on any tensor field over a Riemannian manifold. In order to take the divergence on a covariant index, we first need to raise that index. If we take the divergence operator of $\mathbf{F}$ in the first index, by (6.25) in components it is

$$(g^{ij}F_{j\beta})_{;i} = g^{ij}F_{j\beta;i},$$

where we are using the covariant derivative associated to the Levi-Civita connection. It is straightforward to prove that Maxwell's first and fourth equations are equivalent to $\text{div } F = \frac{4\pi}{c}\mathbf{J}^\flat$. Hence, we can write Maxwell's equations as

$$\text{div } \mathbf{F} = \frac{4\pi}{c}\mathbf{J}^\flat \quad \text{and} \quad d\mathbf{F} = 0. \tag{7.48}$$

This formulation of Faraday's equation and Maxwell's equations lends itself to generalization from Minkowski space to a pseudo-Riemannian manifold of signature $(3, 1)$. In fact, $d\mathbf{F} = 0$ follows immediately from $\mathbf{F} = d\mathbf{A}$ but the divergence operator in the first equation in the above pair depends on the metric of the manifold.

PROBLEMS

7.3.1. Suppose we are in $\mathbb{R}^{3+1}$, and let $\mathcal{F}$ be the standard reference frame. Suppose that another frame $\mathcal{F}'$ keeps the x-, y-, and z-axes in the same orientation but has an origin O' that travels at velocity v along the x-axis of $\mathcal{F}$. Let $\vec{E}$ and $\vec{B}$ be joint electric and magnetic force fields with coordinates (E_1, E_2, E_3) and (B_1, B_2, B_3) as observed in $\mathcal{F}$. Use the electromagnetic tensor from Equation (7.40) and the coordinate transformation described in (7.17) to show that in $\mathcal{F}'$ the components of the same vector fields are observed as having the components

$$
\begin{aligned}
E_1' &= E_1, & E_2' &= \gamma(E_2 - \beta B_3), & E_3' &= \gamma(E_3 + \beta B_2), \\
B_1' &= B_1, & B_2' &= \gamma(B_2 + \beta E_3), & B_3' &= \gamma(B_3 - \beta E_2).
\end{aligned}
\tag{7.49}
$$

(This result conforms to standard results of special relativistic effects in electromagnetism. [46, (58) Chap. 6].)

7.3.2. Show that (7.42) is equivalent to (7.38).

7.3.3. Let f be a smooth function defined over the Minkowski space $\mathbb{R}^{3,1}$. As always, set $x^0 = ct$, $x^1 = x$, $x^2 = y$, and $x^3 = z$. Prove that

$$
d(\star(df)) = c \Box f\, dt \wedge dx \wedge dy \wedge dz.
$$

7.3.4. Suppose we are in Minkowski spacetime.

(a) Prove that $\frac{1}{2} F^{\alpha\beta} F_{\alpha\beta} = \|\vec{E}\|^2 - \|\vec{B}\|^2$. Conclude that $\|\vec{E}\|^2 - \|\vec{B}\|^2$ is preserved under any Lorentz transformation.

(b) Prove that $-\frac{1}{4} \eta_{ij}(\star F)^{jk} \eta^{il} F_{lk} = \vec{E} \cdot \vec{B}$. Conclude that $\vec{E} \cdot \vec{B}$ is preserved under any Lorentz transformation.

7.3.5. Recall $\star$ as the Hodge star operator. Show that in Minkowski spacetime with the metric η, the operator $\star d\star$ is the same as the divergence operator div over the first index. Conclude that Maxwell's equations equations can be expressed as

$$
d\mathbf{F} = 0
$$
$$
\star d \star F = \frac{4\pi}{c} J^\flat.
\tag{7.50}
$$

7.3.6. Let M be any pseudo-Riemannian manifold. Consider the operation that consists of the compositions $\star d \star d$.

(a) Show that $\star d \star d$ is an $\mathbb{R}$-linear operator $\Omega^k(M) \to \Omega^k(M)$ for $k < \dim M$.

(b) Let $M\mathbb{R}^n$ be a standard Euclidean space. Recalling that $\Omega^0(M) = C^\infty(M)$, show that for any smooth function f,

$$
\star d \star df = \nabla^2 f,
$$

where ∇^2 is the usual Laplacian $\nabla^2 = \frac{\partial^2}{\partial (x^1)^2} + \cdots + \frac{\partial^2}{\partial (x^n)^2}$.

(c) Suppose we are in Minkowski space. Show that $\Box = \star d \star d$, and conclude that Equations (7.45) and (7.46) can be summarized by

$$
\star d \star d\mathbf{A} = \frac{4\pi}{c} \mathbf{J}^\flat.
$$

7.3.7. In [60, (5.38)], the author states that "the full action for the electrically charged point particle is"

$$S = -m_0 c \int_{\mathcal{P}} d\tau + \frac{q}{c} \int_{\mathcal{P}} A_\mu \, dx^\mu, \qquad (7.51)$$

where $d\tau$ is given by (7.19) and A_μ are the components of the potential covector given in Equation (7.39). Suppose a charged particle travels along a path $(x^1(t), x^2(t), x^3(t))$.

(a) Write the action in Equation (7.51) as an integral of time t alone.

(b) Determine the Lagrangian for this system, and write down Lagrange's equations of motion.

(c) Write down Hamilton's equations of motion.

7.4 Geometric Concepts in String Theory

What is generically understood in physics as string theory is a collection of theories called superstring theories. The name of these models derives from the fact that in many of the first proposed theories, elementary particles were viewed as strings. Since then, theories have been formulated in terms of points or surfaces. The string can be either open on the ends or be a closed loop. For theoretical reasons, the length of the strings should be on the order of the Planck length, $\ell_P = 1.6162 \times 10^{-35}$ m. This size is so small as to render it impossible to directly observe the string structure with present technology or, so it would seem, with technology that will be available in the near future. In this model, observed properties of the particle, such as mass or electric charge, arise as specific properties of the vibration of the string.

A string in common day occurence is made of some material like thread or wire. One could ask what these strings are made of, i.e., what is the nature of the "thread." This type of question is, however, vacuous because the string is not made up of any constituent parts. One should rather think of the particle-wave duality that drew considerable debate during the inception of quantum mechanics. In this duality, under different circumstances, a particle would exhibit behavior like a billiard ball while in other circumstances it would display a wave-like behavior. While some physicists discussed the fundamental nature of particles, many simply emphasized the fact that growing experimental evidence supported the probability wave function model, without worrying about the ontology.

As a refinement to the Standard Model of quantum mechanics, string theory bears a similar duality in that one thinks of the particle as having a string nature as well as a probability wave nature. The space of the "state" functions (i.e., functions that describe the state of the particle) is the same, but there are more operators than in the point-particle theory. In practice, instead of debating the nature of the string, the theories work out mathematical consequences of this formulation in the hope that the resulting theory agrees with experimental observations and unifies without irreparable inconsistencies with previously established theories.

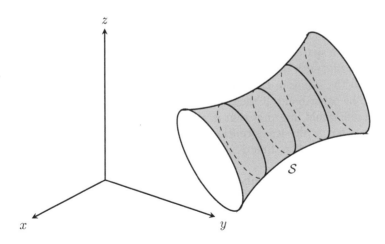

Figure 7.8: The world sheet of a (nonrelativistic) closed string.

Our goal in this section is to introduce a few of the geometric notions that underlie the relativistic dynamics of a string. Issues of quantization of these dynamics exceed the scope of this book.

We first consider the nonrelativistic dynamics of a string of length L in Euclidean $\mathbb{R}^n$. If the string is open, we can pick an end of the string and use the arclength parameter to locate a point on the string. If it is closed, we pick a specific point on the string and locate other points on the string using the same arclength parameter. The position of the string in space at time t is described by a smooth function $X : [0, L] \times \mathbb{R} \to \mathbb{R}^n$, where $X(s, t)$ is the location of the point of position s on the string at time t. Therefore, while the trajectory of a classic particle is described by a curve in $\mathbb{R}^n$, the "trajectory" of a string is a surface (see Figure 7.8). In keeping with the terminology of "world line" for a relativistic point particle, the surface $\mathcal{S}$ is called the *world sheet* of the string.

To study the dynamics of a relativistic string, we must work in the context of a Lorentzian spacetime. (This can be curved or flat and can have any number of space dimensions but only one time dimension. In other words, the pseudometric on the space has index 1.) As always, the coordinates in the spacetime are $x^\mu = (x^0, x^1, \cdots, x^d)$, with d being the number of space dimensions and $x^0 = ct$.

One can no longer parametrize the world sheet $\mathcal{S}$ with the time parameter t since $x^0 = ct$ is one of the coordinates in the target space. Nonetheless, the world sheet requires two parameters, say ξ^1 and ξ^2. Furthermore, we can no longer give the same definition of the domain of X as in the nonrelativistic description of moving strings. One refers to the domain of X as the parameter space for the world sheet.

Now we encounter something new in Lorentzian spacetime that we never encountered in the study of Riemannian manifolds. The world sheet $\mathcal{S}$ must be such that at each point there exists at least one spacelike tangent vector and at least one timelike tangent vector (recall Section 7.2.4 for the definitions). It is not hard to see the need for a spacelike tangent vector. Any point in time corresponds to a slice of x^0. Intersecting such a slice with $\mathcal{S}$ gives the locus of the string at a given time. Any point on the string in this slice will have a tangent vector that is a spacelike vector. On the other hand, if there did not exist a timelike tangent vector at some point on $\mathcal{S}$, one would interpret that as that point not having any evolution through time. This is not a physical situation. Hence, at each point P of $\mathcal{S}$, the tangent space has both a timelike direction and a spacelike direction. This is the criterion for motion of the string.

If g is the pseudo-Riemannian metric of the spacetime target space, then the induced metric $\tilde{g}$ on the tangent bundle $T\mathcal{S}$ is defined by

$$\tilde{g}_{ij} = g\left(\frac{\partial X}{\partial \xi^i}, \frac{\partial X}{\partial \xi^j}\right).$$

The criterion of motion for the string is equivalent to $\tilde{g}$ having index 1 at all points of $\mathcal{S}$.

Proposition 7.4.1. *Let $X(\xi^1, \xi^2)$ be a parametrization for a surface $\mathcal{S}$ in Lorentzian space with metric g such that at each point of $\mathcal{S}$, X has at least one nontrivial spacelike tangent vector and at least one nontrivial timelike vector. Then*

$$\det(\tilde{g}_{ij}) = \det\left(g\left(\frac{\partial X}{\partial \xi^i}, \frac{\partial X}{\partial \xi^j}\right)\right) < 0 \qquad (7.52)$$

at all points of $\mathcal{S}$.

Proof. Let p be a point on $\mathcal{S}$, and let $V(\alpha)$ be the vector in $T_p\mathcal{S}$ defined by

$$V(\alpha) = \cos\alpha \frac{\partial X}{\partial \xi^1} + \sin\alpha \frac{\partial X}{\partial \xi^2}$$

for $\alpha \in [0, 2\pi]$. Then

$$\|V(\alpha)\|^2 = \cos^2\alpha\, g\left(\frac{\partial X}{\partial \xi^1}, \frac{\partial X}{\partial \xi^1}\right) + 2\sin\alpha\cos\alpha\, g\left(\frac{\partial X}{\partial \xi^1}, \frac{\partial X}{\partial \xi^2}\right) + \sin^2\alpha\, g\left(\frac{\partial X}{\partial \xi^2}, \frac{\partial X}{\partial \xi^2}\right)$$

$$= \cos^2\alpha\, \tilde{g}_{11} + 2\sin\alpha\cos\alpha\, \tilde{g}_{12} + \sin^2\alpha\, \tilde{g}_{22}. \qquad (7.53)$$

The property of tangent vectors of being timelike or spacelike is independent of the length or sign of the vector. Thus, for some α_1, there exists a vector $V(\alpha_1)$ such that $\|V(\alpha_1)\|^2 < 0$, and for some α_2 there exists a $V(\alpha_2)$ such that $\|V(\alpha_2)\|^2 > 0$. Furthermore, $\|V(\alpha_i + \pi)\|^2 = \|V(\alpha_i)\|^2$. Hence, since $\|V(\alpha)\|^2$ changes sign twice over $\alpha \in [0, \pi]$, it must have at least two distinct roots. Therefore, Equation (7.53) leads to quadratic equations in $\tan\alpha$ or $\cot\alpha$. Either way, according to the quadratic formula, the equation for $\tan\alpha$ or for $\cot\alpha$ has two distinct roots if and only if

$$\tilde{g}_{12}^2 - \tilde{g}_{11}\tilde{g}_{22} > 0.$$

The proposition follows immediately. $\qquad\qquad\square$

It is customary to parametrize S with two variables labeled σ and τ defined in such a way that for all points in the parameter space, $\partial X/\partial\sigma$ is a spacelike tangent vector and $\partial X/\partial\tau$ is a timelike tangent vector. The parameters σ and τ no longer directly represent position along the string or time, respectively. One could say that σ and τ approximately represent position and time along the world sheet. In fact, with the sole exception of the endpoints, when considering the motion of open strings, one cannot know the movement of individual points on the string. In general, σ ranges over a finite interval $[0, \sigma_1]$, while τ ranges over all of $\mathbb{R}$.

The derivatives $\partial X/\partial\sigma$ and $\partial X/\partial\tau$ occur often enough that it is common to use the symbols X' and $\dot{X}$ for them, respectively. In components, we write

$$X'^\mu = \frac{\partial X^\mu}{\partial\sigma} \quad \text{and} \quad \dot{X}^\mu = \frac{\partial X^\mu}{\partial\tau}.$$

The area element of the world sheet in this Lorentzian space is defined as

$$dA = \sqrt{-\det\tilde{g}} = \sqrt{g(\dot{X}, X')^2 - g(\dot{X}, \dot{X})g(X', X')}. \tag{7.54}$$

We finish this section by briefly discussing the Nambu-Goto action for a free relativistic string and the resulting equations of motion for the string.

Definition 7.4.2. Let $g = \langle\,,\,\rangle$ be a pseudometric of signature $(1, 1)$. The Nambu-Goto action of the string is defined as

$$S \stackrel{\text{def}}{=} -\frac{T_0}{c}\iint_S dA = -\frac{T_0}{c}\int_{\tau_1}^{\tau_2}\int_0^{\sigma_1}\sqrt{-\det\tilde{g}_{\alpha\beta}}\,d\sigma\,d\tau$$

$$= -\frac{T_0}{c}\int_{\tau_1}^{\tau_2}\int_0^{\sigma_1}\sqrt{\langle\dot{X}, X'\rangle^2 - \|\dot{X}\|^2\,\|X'\|^2}\,d\sigma\,d\tau, \tag{7.55}$$

where T_0 is called the string tension and c is the speed of light.

Before proceeding, we must give some justification for this definition. First of all, it mimics the action for a free relativistic particle given in (7.26). The difference is that instead of defining the action as a multiple of the length of the path in the ambient Minkowski space, we define it as a multiple of the area of the world-sheet. Furthermore, this action is obviously invariant under reparametrization since the area is a geometrical quantity.

As a more convincing argument, we consider a classical vibrating string of length ℓ. Using Figure 7.9 as a guide, we model the motion of the string by a function $y(x, t)$ that measures the deviation of the string from rest at horizontal position x and at time t. If the string has constant density μ_0 and tension T_0, then the differential equation of motion for a string with small deviations is

$$\mu_0\frac{\partial^2 y}{\partial t^2} = T_0\frac{\partial^2 y}{\partial x^2}.$$

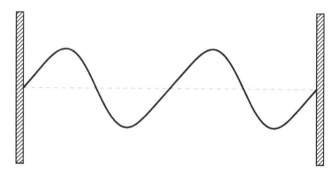

Figure 7.9: A vibrating string.

The fraction μ_0/T_0 has the units of time^2/length^2 and is in fact equal to $1/c^2$, where c is the speed of propagation of the wave. It is not hard to reason that at any point in time t, the total kinetic energy of the string is

$$T = \int_0^\ell \frac{1}{2}\mu_0\left(\frac{\partial y}{\partial t}\right)^2 dx,$$

and that the total potential energy is

$$V = \int_0^\ell \frac{1}{2}T_0\left(\frac{\partial y}{\partial x}\right)^2 dx.$$

Thus the Lagrangian of the system is

$$L = \int_0^\ell \frac{1}{2}\mu_0\left(\frac{\partial y}{\partial t}\right)^2 - \frac{1}{2}T_0\left(\frac{\partial y}{\partial x}\right)^2 dx. \tag{7.56}$$

The integrand of Equation (7.56) is called the *Lagrangian density* and is denoted by $\mathcal{L}$. It is explicitly a function of $\partial y/\partial t$ and $\partial y/\partial x$. The action of the system for $t \in [t_1, t_2]$ is

$$S = \int_{t_1}^{t_2} \int_0^\ell \frac{1}{2}\mu_0\left(\frac{\partial y}{\partial t}\right)^2 - \frac{1}{2}T_0\left(\frac{\partial y}{\partial x}\right)^2 dx\,dt. \tag{7.57}$$

Now assume that we are in a Minkowski space $\mathbb{R}^{1,2}$ with the flat pseudometric $-ds^2 = -(dx^0)^2 + (dx^1)^2 + (dx^2)^2$, where $x^0 = ct$, $x^1 = x$, and $x^2 = y$. After some manipulation, we can rewrite that string action as

$$S = -\frac{T_0}{c} \int_{ct_1}^{ct_2} \int_0^\ell \frac{1}{2}\left(-\left(\frac{\partial y}{\partial x^0}\right)^2 + \left(\frac{\partial y}{\partial x^1}\right)^2\right) dx^1\,dx^0. \tag{7.58}$$

The motion of the string can be parametrized in $\mathbb{R}^{1,2}$ by $\vec{f}(x^0, x^1) = (x^0, x^1, y(x^0, x^1))$.

With the inner product induced by this metric, the area element becomes

$$\sqrt{-\left(\left\langle \frac{\partial \vec{f}}{\partial x^0}, \frac{\partial \vec{f}}{\partial x^0}\right\rangle\left\langle \frac{\partial \vec{f}}{\partial x^1}, \frac{\partial \vec{f}}{\partial x^1}\right\rangle - \left\langle \frac{\partial \vec{f}}{\partial x^0}, \frac{\partial \vec{f}}{\partial x^1}\right\rangle^2\right)} = \sqrt{1 - \left(\frac{\partial y}{\partial x^0}\right)^2 + \left(\frac{\partial y}{\partial x^1}\right)^2}$$

$$\cong 1 + \frac{1}{2}\left(-\left(\frac{\partial y}{\partial x^0}\right)^2 + \left(\frac{\partial y}{\partial x^1}\right)^2\right).$$

Adjusting for $x^0 = ct$, the Lagrangian associated to Equation (7.58) differs from the linear approximation to the Nambu-Goto action

$$-T_0 \int_0^\ell 1 + \frac{1}{2}\left(-\left(\frac{\partial y}{\partial x^0}\right)^2 + \left(\frac{\partial y}{\partial x^1}\right)^2\right) dx^1$$

by

$$-T_0 \int_0^\ell 1\, dx^1 = -T_0 \ell = -\mu_0 \ell c^2 = -mc^2.$$

Similar to the linear approximation to the Lagrangian for the free relativistic particle in Equation (7.28), this difference is precisely the negative of the rest energy mc^2 of the string. Since this is a constant, it leaves the Euler-Lagrange equations unchanged. This shows how the classic Lagrangian of a wave is a linear approximation for the Lagrangian associated to the Nambu-Goto action.

We now wish to obtain the equations of motion associated to the Nambu-Goto action. The Lagrangian density in Equation (7.55) is

$$\mathcal{L}(\dot{X}^\mu, X'^\mu) = -\frac{T_0}{c}\sqrt{\langle \dot{X}, X'\rangle^2 - \|\dot{X}\|^2 \|X'\|^2}. \tag{7.59}$$

This is an explicit function of the eight variables X'^μ and $\dot{X}^\mu$ for $\mu = 0, 1, 2, 3$. Hamilton's principle states that the system will evolve in such a way as to minimize the action. According to a generalization of the Euler-Lagrange Theorem in the calculus of variations (see Problem 7.4.2), the Nambu-Goto action is minimized if and only if the $X^\mu(s,t)$ satisfy

$$\frac{d}{d\sigma}\left(\frac{\partial \mathcal{L}}{\partial X'^\mu}\right) + \frac{d}{d\tau}\left(\frac{\partial \mathcal{L}}{\partial \dot{X}^\mu}\right) = 0$$

for all μ. These are the equations of motion for a relativistic string, whether open or closed. More explicitly, the equations of motion read

$$\frac{\partial}{\partial \sigma}\left(\frac{\langle \dot{X}, X'\rangle g_{\mu\nu}\dot{X}^\nu - \|\dot{X}\|^2 g_{\mu\nu}X'^\nu}{\sqrt{\langle \dot{X}, X'\rangle^2 - \|\dot{X}\|^2 \|X'\|^2}}\right) + \frac{\partial}{\partial \tau}\left(\frac{\langle \dot{X}, X'\rangle g_{\mu\nu}X'^\nu - \|X'\|^2 g_{\mu\nu}\dot{X}^\nu}{\sqrt{\langle \dot{X}, X'\rangle^2 - \|\dot{X}\|^2 \|X'\|^2}}\right) = 0$$

$$\tag{7.60}$$

for $\mu = 0, 1, 2, 3$. At first glance, these equations are incredibly complicated. They involve a system of four second-order partial differential equations of four functions each in two variables. A remarkable fact among the basic results of string theory is that it is possible to solve Equation (7.60) once one makes a suitable choice of σ and τ.

Using the notion of generalized momenta defined in Equation (7.6), we define two momenta densities $\mathcal{P}^\sigma$ and $\mathcal{P}^\tau$ that are cotangent vectors on the world sheet with components

$$\mathcal{P}_\mu^\sigma \stackrel{\text{def}}{=} -\frac{T_0}{c} \frac{\langle \dot{X}, X' \rangle g_{\mu\nu} \dot{X}^\nu - \|\dot{X}\|^2 g_{\mu\nu} X'^\nu}{\sqrt{\langle \dot{X}, X' \rangle^2 - \|\dot{X}\|^2 \|X'\|^2}},$$

$$\mathcal{P}_\mu^\tau \stackrel{\text{def}}{=} -\frac{T_0}{c} \frac{\langle \dot{X}, X' \rangle g_{\mu\nu} X'^\nu - \|X'\|^2 g_{\mu\nu} \dot{X}^\nu}{\sqrt{\langle \dot{X}, X' \rangle^2 - \|\dot{X}\|^2 \|X'\|^2}}.$$

(7.61)

(One should note that in this case the superscript σ and τ in $\mathcal{P}_\mu^\sigma$ and $\mathcal{P}_\mu^\tau$ are not indices but are parameter indicators.) Then the equations of motion read

$$\frac{\partial \mathcal{P}_\mu^\sigma}{\partial \sigma} + \frac{\partial \mathcal{P}_\mu^\tau}{\partial \tau} = 0. \tag{7.62}$$

This is all we will say about the underlying geometry in string theory. String theory extends well beyond the scope of this book, and we encourage the reader to consult [60] for an artful and accessible introduction to the subject.

PROBLEMS

7.4.1. Show that at some point on the world-sheet of a string, if the point moves at the speed of light, there is no timelike direction.

7.4.2. Use the methods of calculus of variations provided for the proof of Theorem B.3.1 to prove the following result. Let $x^1(s,t), \ldots, x^n(s,t)$ be n twice-differentiable functions in two variables. Denote derivatives by $x'^i = dx^i/ds$ and $\dot{x}^i = dx^i/dt$. Suppose that a function f is given explicitly in terms of $x^i, x'^i, \dot{x}^i, s$, and t. Show that the integral

$$\int_{t_1}^{t_2} \int_{s_1}^{s_2} f(x^1, \ldots, x^n, x'^1, \ldots, x'^1, \dot{x}^1, \ldots, \dot{x}^n, s, t) \, ds \, dt$$

is optimized when

$$\frac{\partial f}{\partial x^i} - \frac{d}{ds}\left(\frac{\partial f}{\partial x'^i}\right) - \frac{d}{dt}\left(\frac{\partial f}{\partial \dot{x}^i}\right) = 0 \qquad \text{for all } i = 1, \ldots, n.$$

7.4.3. Consider a free relativistic string with σ-length σ_1. The Hamiltonian for the system is

$$H = \int_0^{\sigma_1} \mathcal{P}_\mu^\tau \dot{X}^\mu - \mathcal{L} \, d\sigma.$$

(a) Recover the equations of motion as in Equation (7.62) from Hamilton's equations of motion.

(b) Show that H vanishes identically for all τ.

(c) Let $\mathcal{L}$ be as in Equation (7.59). Consider the matrix with entries $\dfrac{\partial^2 \mathcal{L}}{\partial \dot{X}^\mu \partial \dot{X}^\nu}$. Show that this matrix has two 0 eigenvalues, with eigenvectors $\dot{X}$ and X'. Deduce the following conditions on the momentum $\mathcal{P}^\tau$:

$$i_{X'}(\mathcal{P}^\tau) = \mathcal{P}^\tau_\mu X'^\mu = 0,$$

$$\|\mathcal{P}^\tau\|^2 + \frac{T_0^2}{c^2}\|X'\|^2 = g^{\mu\nu}\mathcal{P}^\tau_\mu \mathcal{P}^\tau_\nu + \frac{T_0^2}{c^2}g_{\mu\nu}X'^\mu X'^\nu = 0.$$

7.4.4. Show that according to the relativistic string equations of motion, the endpoints of an open string move with the speed of light.

7.4.5. Consider a relativistic string in Minkowski space $\mathbb{R}^{d,1}$ but only consider the history of the string in the real space. We parametrize this history as $\vec{X}(\sigma, \tau)$. (We use the vector superscript to indicate vectors in the Euclidean $\mathbb{R}$ part of the spacetime.) Define $s(\sigma)$ to be the length of the string along $[0, \sigma]$, so that $s(0) = 0$ and $s(\sigma_1)$ is the length of the string. Also set $t = \tau$.

(a) Prove that $\dfrac{\partial \vec{X}}{\partial s}$ is a unit vector.

(b) Define the vector $\vec{v}_\perp$ as the component of the velocity vector $\dfrac{\partial \vec{X}}{\partial t}$ that is perpendicular to the string. Thus,

$$\vec{v}_\perp = \frac{\partial \vec{X}}{\partial t} - \left(\frac{\partial \vec{X}}{\partial t} \cdot \frac{\partial \vec{X}}{\partial s}\right)\frac{\partial \vec{X}}{\partial s},$$

where we use the usual dot product. Prove that one can write the Nambu-Goto string action as

$$S = -T_0 \int_{t_1}^{t_2} \int_0^{\sigma_1} \frac{ds}{d\sigma}\sqrt{1 - \frac{v_\perp^2}{c^2}}\, d\sigma\, dt$$

7.4.6. (*) *The Nambu-Goto Bubble Action.* Suppose that instead of considering particles as strings, we model them as bubbles. Then a world sheet $\mathcal{S}$ is given by a function $X(\sigma, \bar{\sigma}, \tau)$ into a pseudo-Riemannian manifold M with signature $(2, 1)$.

(a) Explain why it still makes sense to define the action of the free motion of the relativistic bubble for $\tau_1 \le \tau \le \tau_2$ by

$$S = -\frac{T_0}{c}\iiint_{\mathcal{S}}\sqrt{-\det \tilde{g}_{\alpha\beta}}\, d\sigma\, d\bar{\sigma}\, d\tau,$$

where T_0 is now a surface tension and $\tilde{g}$ is the metric induced from M on $\mathcal{S}$.

(b) Write down the equations of motion associated to this action.

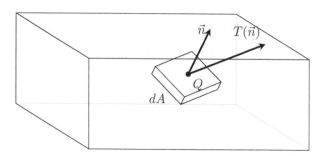

Figure 7.10: Stress tensor: an area element in a continuous medium.

7.5   Brief Introduction to General Relativity

As with the previous sections, the reader might consider it outlandish that we only allow one section to discuss general relativity. General relativity is a vast subject with contributions from a host of scientists and mathematicians, and it stands alongside quantum mechanics as one of the most revolutionary ideas in physics of the 20th century.

On the other hand, most textbooks on general relativity take a considerable amount of time to develop the techniques of analysis on manifolds, in particular, pseudo-Riemannian manifolds. However, these are precisely the mathematical methods we have developed in the previous chapters, so we are in a position to introduce some differential geometric concepts in general relativity as applications.

From the perspective of mathematical structures, general relativity builds on special relativity. The postulates of special relativity brought us to the notion of spacetime, which is a Minkowski space $\mathbb{R}^{3,1}$. In general relativity, we will want to consider our space as *locally Minkowski*, meaning that each tangent space is a Minkowski space. Hence, we model the universe as a pseudo-Riemannian manifold with signature $(3,1)$. More importantly, the Einstein field equations propose a relationship between the presence of energy and the curvature of this spacetime manifold.

7.5.1   Stress-Energy Tensor

In the mechanics of elastic media, one encounters the concept of a stress tensor, which is a tensor-valued function defined at each point within the body or medium. Suppose the body is in equilibrium but subject to external forces and/or body forces (i.e., forces that act through the whole body). Then there must exist internal forces. Let Q be a point, $\vec{n}$ a vector based at Q, and consider the area element ΔA that is in the plane perpendicular to $\vec{n}$ and has area equal to $\|\vec{n}\|$ (see Figure 7.10). Let $\Delta\vec{F}$ be the overall internal forces distributed over the area element ΔA. The stress vector through the area element ΔA is the vector

$$\vec{T}(\vec{n}) = \lim_{\Delta A \to 0} \frac{\Delta \vec{F}}{\Delta A}. \tag{7.63}$$

It is not hard to show that the function $\vec{T}(\vec{n})$ is a linear function.[56, Section 10.6] Thus the stress tensor with respect to an orthogonal basis $\mathcal{B}$ based at the point Q is the matrix σ such that

$$\vec{T}(\vec{n}) = \sigma [\vec{n}]_{\mathcal{B}}.$$

Consider now a small rectangular parallelepiped with sides parallel to the coordinate planes. The stress acts on each face as depicted in Figure 7.11. Then the columns of σ are given by

$$\sigma \vec{e}_i = T(\vec{e}_i) = \begin{pmatrix} \sigma_{i1} \\ \sigma_{i2} \\ \sigma_{i3} \end{pmatrix}.$$

Minimal assumptions that are logical for physics (angular momentum in a medium cannot grow to be infinite at a point) imply that the stress tensor σ is symmetric.

As a simple example, in an ideal fluid, the stress on any small area element is composed only of pressure, and there is no shearing force. Consequently, the stress tensor is $\sigma = P\mathbf{I}$, where P is the pressure and $\mathbf{I}$ is the 3×3 identity matrix. (This restates the claim given in calculus texts on the applications of integration to hydrostatics when one says that "at any point in a liquid the pressure is the same in all directions."[55, p. 576]) The stress tensor arises also in the dynamics of viscous fluids where it is no longer necessarily diagonal. The stress tensor at a point "may be a function of the density and temperature, of the relative positions and velocities of elements near [the point], and perhaps also the previous history of the medium."[56, p. 434] This characterization describes the stress tensor as a function of many ambient quantities, but the reference to "relative positions" indicates that the stress tensor need not be diagonal.

Einstein's equation in general relativity involves the so-called *stress-energy tensor*. This tensor is different from the stress tensor but is based on the same concept. We assume that we are in a Minkowski space with metric $\mathbf{g}$ of signature $(3, 1)$.

As in special relativity, the four-velocity of a particle on a world line $\mathcal{P}$ parametrized by $\vec{x}$ is the tangent vector along $\mathcal{P}$ given by

$$\vec{U} = \frac{d\vec{x}}{d\tau}, \tag{7.64}$$

where, the proper time τ of an object along its world line is given in (7.20). From the theory of special relativity, by (7.22), the four-velocity is

$$\vec{U} = (U^0, U^1, U^2, U^3) = (\gamma c, \gamma v_x, \gamma v_y, \gamma v_z), \tag{7.65}$$

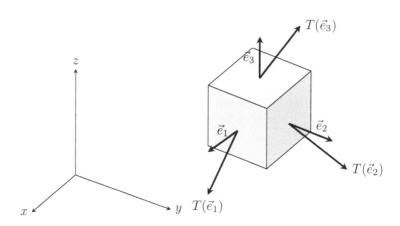

Figure 7.11: Action of stress on an infinitesimal coordinate cube.

where $\gamma = (1 - v^2/c^2)^{-1/2}$ and $\vec{v} = (dx/dt, dy/dt, dz/dt)$. In (7.23), we defined the momentum 4-vector of a particle of rest mass m by $\vec{p} = m\vec{u}$. Equally useful is the four-momentum covector given by

$$\mathbf{p} = \vec{p}^{\flat}. \tag{7.66}$$

In special relativity where the metric $\mathbf{g} = \eta$ is the standard Minkowski metric, the components of the momentum covector are

$$(p_0, p_1, p_2, p_3) = \left(-\frac{E}{c}, p_x, p_y, p_z\right), \tag{7.67}$$

where $(p_x, p_y, p_z) = m_0\gamma(v^1, v^2, v^3)$ is the relativistic 3-vector momentum.

Underlying the assumptions that define the stress-energy tensor, we assume that "spacetime contains a flowing river of 4-momentum"[41, p.130]. Any mass that is moving or anything with energy contributes to the 4-momentum. We could think of an individual particle, in which case the 4-momentum would only be defined on the particle's world line, or we could consider a system of many particles carrying this 4-momentum. In the latter case, we should think of the 4-momentum as a covector field on the spacetime manifold M, that is, as a 1-form.

Let $\mathbf{n}$ be any 1-form on M. Then at each point $Q \in M$, $\mathbf{n}_Q$ is perpendicular (using $\mathbf{g} = \langle\,,\rangle$ at Q) to a three-dimensional subspace of $T_Q M$. This subspace can be spanned by vectors A_Q, B_Q, and C_Q such that

$$\mathbf{n}_Q(u) = -\text{Vol}_{\mathbf{g}}(u, A_Q, B_Q, C_Q), \tag{7.68}$$

where on the right-hand side we mean the 4-volume (with respect to $\mathbf{g}$) of the 4-parallelepiped spanned by $u, A_Q, B_Q, C_Q \in T_Q M$. Then the 3-volume of the 3-

parallelepiped A_Q, B_Q, C_Q is the length $\|\mathbf{n}\|$. In this way, every 1-form represents a volume element in $\mathbb{R}^4$.

We define the *stress-energy tensor* $\mathbf{T}$ of type $(2,0)$ by $\mathbf{T}(\mathbf{p}, \mathbf{n})$ being the flux of the momentum covector $\mathbf{p}$ across a volume element represented by $\mathbf{n}$. At each point $Q \in M$, we have

$$\mathbf{T}(\mathbf{p}, \mathbf{n}) = \langle \mathbf{p}, \mathbf{n} \rangle_{\mathbf{g}}. \tag{7.69}$$

Over any coordinate chart (x^i), the components of $\mathbf{T}$ are

$$T^{\alpha\beta} = \mathbf{T}(dx^\alpha, dx^\beta)$$

to represent that flux of unit four-momentum in direction dx^α across a volume-element of constant β.

The stress-energy tensor $\mathbf{T}$ is also called the *energy-momentum tensor* because it contains information pertaining to the momentum flowing through space and the presence of static or moving energy in space. The name "stress-energy tensor" is commonly used since it is modeled off the stress tensor in mechanics of elastic media.

The following gives a summary of the information included in the stress-energy tensor. Assuming $j, k > 0$,

$$T^{00} = \text{density of energy (including mass)}, \tag{7.70}$$

$$T^{j0} = j\text{th component of the momentum density}, \tag{7.71}$$

$$T^{0k} = k\text{th component of the energy flux}, \tag{7.72}$$

$$T^{jk} = (j, k)\text{th component of the momentum stress} \tag{7.73}$$

$$= k\text{th component of the flux of the } j\text{th component of momentum}.$$

The notion of *flux* in this context refers to a similar limit as in Equation (7.63) but in the situation where one is concerned with the movement of something (energy, fluid momentum, heat,...) through the infinitessimal area element $d\vec{A}$. In fact, with this particular concept of flux, one can define the stress-energy tensor in short by saying that T^{kj} is the kth component of the flux of the jth component of the 4-momentum.

We now state two facts about the stress-energy tensor that we do not fully justify here.

Proposition 7.5.1. *The (contravariant) stress-energy tensor* $\mathbf{T}$ *is symmetric.*

The symmetry in the components $1 \le i, j \le 3$ follows from the same physical reasoning for why the stress tensor in fluid dynamics is symmetric.

Proposition 7.5.2 (Einstein's Conservation Law)**.** *The conservation of energy is equivalent to the identity*

$$\text{div } \mathbf{T} = T^{\alpha\beta}{}_{;\beta} = 0.$$

Proof Sketch. Suppose that energy is conserved in a certain region of M. In other words, though energy and mass may move around, no energy or mass is created or annihilated in M. Then given any four-dimensional submanifold $\mathcal{V}$ with boundary $\partial\mathcal{V}$, the total flux of 4-momentum passing through $\partial\mathcal{V}$ must be 0. We can restate this as

$$\iiint_{\partial\mathcal{V}} \mathbf{T} \cdot d\mathbf{V}^{(3)} = 0,$$

where $d\mathbf{V}^{(3)}$ is the 3-volume element with direction along the outward-pointing normal vector to $\partial\mathcal{V}$. (We can view this as a volume 1-form.) By the product $\cdot$ we mean the contraction of $\mathbf{T}$ with the volume 1-form element $d\mathbf{V}^{(3)}$. Stokes' Theorem applied to pseudo-Riemannian manifolds gives

$$\iiiint_{\mathcal{V}} \operatorname{div} \mathbf{T} dV^{(4)} = \iiint_{\partial\mathcal{V}} \mathbf{T} \cdot d\mathbf{V}^{(3)} = 0.$$

Since this is true for all $\mathcal{V}$ as described above, using a limiting argument similar to that used to show Gauss' Law that $\operatorname{div} \vec{E} = 0$ for an electric field, a limiting argument establishes $\operatorname{div} \mathbf{T} = 0$ everywhere. $\qquad\square$

Example 7.5.3 (Perfect Fluid Stress-Energy Tensor). A perfect fluid is a fluid in which the pressure p is the same in any direction. The fluid must be free of heat conduction and viscosity and any process that can cause internal sheers. Using the interpretation of $\mathbf{T}$ from Equations (7.70)–(7.73), we see that $T^{jk} = 0$ if $j \neq k$ and $0 < j, k$. Furthermore, since the pressure is the same in all directions, $T^{jj} = p$ for $j = 1, 2, 3$. For components involving $j = 0$ or $k = 0$, we first have $T^{00} = \rho$, the energy density. This quantity includes mass density but also other types of energy such as compression energy. For the remaining off diagonal terms $T^{0j} = T^{j0}$, these are 0 because of the assumption that there is no heat conduction in the perfect fluid. Thus, the stress-energy tensor has components

$$T^{\alpha\beta} = \begin{pmatrix} \rho & 0 & 0 & 0 \\ 0 & p & 0 & 0 \\ 0 & 0 & p & 0 \\ 0 & 0 & 0 & p \end{pmatrix}. \tag{7.74}$$

If we suppose that an observer is in the Lorentz frame that is at rest with respect to the movement of the fluid, then the velocity has components $u^{\alpha} = (1, 0, 0, 0)$. With respect to the Minkowski metric η, we can write (7.74) as

$$T^{\alpha\beta} = (\rho + p)u^{\alpha}u^{\beta} + p\eta^{\alpha\beta}.$$

We can rewrite this in a coordinate-free way in any metric as

$$\mathbf{T} = p\mathbf{g}^{-1} + (p + \rho)\mathbf{u} \otimes \mathbf{u}, \tag{7.75}$$

where we have written $\mathbf{g}^{-1}$ for the contravariant tensor of type $(2, 0)$ associated to the metric tensor $\mathbf{g}$.

Example 7.5.4 (Electromagnetic Stress-Energy Tensor). Directly using the interpretation of **T** given in Equations (7.70)–(7.73) and results from electromagnetism, which we do not recreate here, one can determine the components of the stress-energy tensor for the electromagnetic field in free space. If $F^{\mu\nu}$ are the components of the electromagnetic field tensor, then

$$T^{\alpha\beta} = \frac{1}{\mu_0}\left(F^{\alpha\mu}g_{\mu\nu}F^{\beta\nu} - \frac{1}{4}g^{\alpha\beta}F^{\mu\nu}F_{\mu\nu}\right) \qquad \text{in SI units} \qquad (7.76)$$

$$= \frac{1}{4\pi}\left(F^{\alpha\mu}g_{\mu\nu}F^{\beta\nu} - \frac{1}{4}g^{\alpha\beta}F^{\mu\nu}F_{\mu\nu}\right) \qquad \text{in CGS units} \qquad (7.77)$$

where $\mu_0 = 4\pi \times 10^{-7}\,\mathrm{N/A}^{-2}$ is a constant sometimes called the vacuum permeability.

As the context requires, we may need the stress-energy tensor to be of type $(2,0)$ as defined, of type $(1,1)$ or of type $(0,2)$. To pass between any of these, we raise or lower the indices as needed using the metric.

7.5.2 Einstein Field Equations

The Einstein field equations (EFE) are the heart of general relativity. They stem from the juxtaposition of the two following principles:

1. Every aspect of gravity is a description of the spacetime geometry.

2. Mass (energy) is the source of gravity.

The metric tensor **g** encapsulates all the information about the geometry of the spacetime. The metric **g** has the associated Levi-Civita connection ∇, the Riemann curvature tensor **R** of type $(1,3)$, the Ricci curvature tensor **Rc**, and the scalar curvature function R, defined in Chapter 6. (Note: In math texts on Riemannian geometry, one often denotes by S the scalar curvature while texts on general relativity invariably denote it by R. Using the bold font **R** to indicate the curvature tensor alleviates any confusion between the scalar and tensor curvature.)

On the other hand, the stress-energy tensor describes the spacetime content of mass-energy. In fact, any observer with 4-velocity $\vec{U}$ measures the density of mass-energy as

$$\rho = \mathbf{u} \cdot \mathbf{T} \cdot \mathbf{u} = T_{\alpha\beta}u^\alpha u^\beta.$$

In order to put together the two above principles, we should be able to write the tensor **T** exclusively in terms of the components of the metric tensor **g**. The conservation of energy states that div $\mathbf{T} = 0$. Also, if **T** is to serve as a measure of the curvature of spacetime, we propose that it should explicitly involve only components of **R** and of **g** (no derivatives of any of these terms) and it should be linear in the components of **R**. It turns out that under these restrictions, there are only a few options for a geometric description of $\mathbf{T}$. In Problem 7.5.2, we show that for purely mathematical reasons, these constraints impose that

$$T_{\alpha\beta} = C\left(R_{\alpha\beta} - \frac{1}{2}Rg_{\alpha\beta} + \Lambda g_{\alpha\beta}\right), \qquad (7.78)$$

where $R_{\alpha\beta}$ are the components of the Ricci curvature tensor, R is the scalar curvature, and Λ and C are real constants. This leads to Einstein's field equations.

Let $\mathbf{G}$ be the Einstein curvature tensor described in Definition 6.5.4. General relativity is summarized in this following equation. The presence of mass-energy deforms spacetime according to

$$\mathbf{G} + \Lambda\mathbf{g} = \frac{8\pi G}{c^4}\mathbf{T} \qquad \text{in SI units}, \qquad (7.79)$$

where $G = 6.67 \times 10^{-11}\,\mathrm{m}^3\mathrm{s}^{-2}\mathrm{kg}^{-1}$ is the gravity constant and Λ is the *cosmological constant*. If we assume that empty (devoid of energy) spacetime is flat, then

$$\mathbf{G} = \frac{8\pi G}{c^4}\mathbf{T}. \qquad (7.80)$$

Equation (7.80) is called collectively the *Einstein field equations* (EFE), and the formulas in (7.79) are the Einstein field equations with cosmological constant. These equations are as important in astrophysics as Newton's second law of motion is in classic mechanics. Though we do not give the calculation here, the constant $8\pi G/c^4$, called *Einstein's constant* is chosen so that (1) when the gravitational field is weak and (2) velocities are small compare to the speed of light, the theory reduces to the Newtonian theory of gravitation in approximation.

In Equation (2) of Einstein's original paper on general relativity [19], Einstein made the assumption that $\mathbf{G}$ vanishes when spacetime is empty of mass-energy. This corresponds to the mathematical assumption that $\Lambda = 0$. However, Equation (7.80) predicts a dynamic universe. This result did not appeal to Einstein and, at the time, there existed no astronomical evidence to support this. In 1917, he introduced the constant Λ because it allows for a static universe. Physically, $\Lambda \neq 0$ would imply the presence of an otherwise unexplained force that counteracts gravity or a sort of negative pressure.

When Hubble discovered that the universe is expanding, the cosmological constant no longer appeared to be necessary and many physicists did away with it. In fact, in his autobiography, George Gamow relays that Einstein told Gamow that he considered the introduction of the cosmological constant as "the biggest blunder of my life." [26] However, the possibility of a small nonzero Λ has resurfaced and regularly enters into the conversations around the current most vexing problems in physics, namely, the nature of dark energy and the effort to unify gravity and quantum mechanics.

We should note the Einstein field equations (EFE) are very complicated. Finding a solution to the EFE means finding the metric tensor $\mathbf{g}$ that satisfies (7.80), which consists of 10 second-order, nonlinear, partial differential equations of 10 functions $g_{ij}(x^0, x^1, x^2, x^3)$, with $0 \leq i \leq j \leq 3$. Then, determining the trajectory of a particle or of radiation amounts to determining the geodesics in this metric. Surprisingly, under some circumstances, especially scenarios that involve a high level of symmetry, it is possible to provide an exact solution.

Whole books have been written about consequences of solutions to (7.80) or (7.79) that deviate from Newtonian mechanics. After the introduction of general relativity, some scientists balked at such mathematical complexity. Yet experiment has repeatedly confirmed predictions in favor of general relativity over Newtonian mechanics. The approach taken here to justify the Einstein field equations cited mathematical esthetics, the condition of being divergence-free, and linear in the components of the curvature tensor $\mathbf{R}$. During the 20th century, scientists arrived at (7.79) from other, more physical principles. Of note, in [36], Lovelock showed that the Einstein field equations arise as the unique second-order equations that can follow from the Euler-Lagrange equations of a Lagrange density involving g_{ij} and its derivatives up to second order. Furthermore, there exist natural generalizations to Einstein's theory of general relativity. However, any theory that can unify gravity and quantum mechanics should be able to derive (7.79) as an approximation.

7.5.3 Schwarzschild Metric

We finish this section with one of the earliest proven consequences of general relativity, i.e., the Schwarzschild metric which is an exact solution to Einstein's field equations. Instead of simply showing that the Schwarzschild metric satisfies EFE, we show how the metric was discovered.

One of the main contexts in which one can expect to see the effects of general relativity against Newtonian mechanics is in the context of astronomy. The simplest dynamical problem in astronomy involves calculating the orbit of a single planet around the sun. One can hope that the EFE for the effect of the sun on the space around it will become simple under the following two assumptions (approximations):

1. The sun is a spherically symmetric distribution of mass-energy density.

2. Outside of the sun, the stress-energy tensor should vanish.

The spherical symmetry implies that the components of the metric tensor should be given as functions of x^0 and r alone, where $r^2 = (x^1)^2 + (x^2)^2 + (x^3)^2$. Since we are looking only for solutions outside the sun, we are looking for solutions in a vacuum. Thus $\mathbf{T} = 0$, from which we deduce that $\mathbf{G} = 0$. Thus, $\mathrm{Tr}_{\mathbf{g}}\,\mathbf{G} = 0$. However,

$$\mathrm{Tr}_{\mathbf{g}}\,\mathbf{G} = \mathrm{Tr}_{\mathbf{g}}\left(\mathbf{Rc} - \frac{1}{2}R\mathbf{g}\right) = R - \frac{1}{2}R \cdot 4 = -R,$$

where this follows from (6.56) and the fact that $\mathrm{Tr}_{\mathbf{g}}\,\mathbf{g} = \dim M = 4$. Thus, $R = 0$ and the fact that $\mathbf{G} = 0$ implies that we are looking for spherically symmetric solutions to the equation

$$R_{\alpha\beta} = 0. \tag{7.81}$$

Since we are looking for solutions in a vacuum, it seems as though we have lost information, but, as we shall see, that is not the case.

The following derivation follows the treatment in [54]. We leave some of the details as exercises for the reader.

A judicious choice of coordinates and a few coordinate transformations will simplify the problem. We first start with the coordinates

$$(\bar{x}^0, x^1, x^2, x^3) = (\bar{x}^0, \bar{r}, \theta, \varphi),$$

where $\bar{x}^0 = ct$ for some timelike variable t, $\bar{r}^2 = (x^1)^2 + (x^2)^2 + (x^3)^2$ and where θ and φ are given in the physics style of defining spherical coordinates, i.e., so that φ is the longitudinal angle and θ is the latitude angle measured down from a "positive" vertical direction. We know that the standard line element in spherical coordinates is

$$ds^2 = d\bar{r}^2 + \bar{r}^2 d\theta^2 + \bar{r}^2 \sin^2\theta d\varphi^2.$$

Though we are not working with the Euclidean metric, spherical symmetry does imply that the metric tensor in the space coordinates is orthogonal and that no perpendicular direction to the radial direction is singled out. Thus, the metric tensor has the form

$$g_{\alpha\beta} = \begin{pmatrix} g_{00}(\bar{x}^0,\bar{r}) & g_{01}(\bar{x}^0,\bar{r}) & g_{02}(\bar{x}^0,\bar{r}) & g_{03}(\bar{x}^0,\bar{r}) \\ g_{10}(\bar{x}^0,\bar{r}) & g_{11}(\bar{x}^0,\bar{r}) & 0 & 0 \\ g_{20}(\bar{x}^0,\bar{r}) & 0 & f(\bar{x}^0,\bar{r})^2 & 0 \\ g_{30}(\bar{x}^0,\bar{r}) & 0 & 0 & f(\bar{x}^0,\bar{r})^2 \sin^2\theta \end{pmatrix}, \tag{7.82}$$

where f is any smooth function. We actually have some choice on θ and φ because they are usually given in reference to some preferred x-axis and z-axis. We choose θ and φ (which may change over time with respect to some fixed Cartesian frame) so that $g_{20} = g_{30} = 0$, and then the metric looks like

$$g_{\alpha\beta} = \begin{pmatrix} g_{00}(\bar{x}^0,\bar{r}) & g_{01}(\bar{x}^0,\bar{r}) & 0 & 0 \\ g_{10}(\bar{x}^0,\bar{r}) & g_{11}(\bar{x}^0,\bar{r}) & 0 & 0 \\ 0 & 0 & f(\bar{x}^0,\bar{r})^2 & 0 \\ 0 & 0 & 0 & f(\bar{x}^0,\bar{r})^2 \sin^2\theta \end{pmatrix}. \tag{7.83}$$

We make the coordinate transformation $r = f(\bar{x}^0,\bar{r})$ and all the other coordinates remain the same. In this coordinate system, the metric looks like

$$g_{\alpha\beta} = \begin{pmatrix} g_{00}(\bar{x}^0,r) & g_{01}(\bar{x}^0,r) & 0 & 0 \\ g_{10}(\bar{x}^0,r) & g_{11}(\bar{x}^0,r) & 0 & 0 \\ 0 & 0 & r^2 & 0 \\ 0 & 0 & 0 & r^2 \sin^2\theta \end{pmatrix}. \tag{7.84}$$

Finally, we can orthogonalize the metric tensor by a suitable coordinate transformation of $x^0 = ct = h(\bar{x}^0, r)$, with r staying fixed (see Problem 7.5.4). Since we know that the metric has signature $(3, 1)$, we can write the metric in the coordinate system $(x^0, r, \theta, \varphi)$ as

$$
g_{\alpha\beta} = \begin{pmatrix} -e^{\nu(x^0, r)} & 0 & 0 & 0 \\ 0 & e^{\lambda(x^0, r)} & 0 & 0 \\ 0 & 0 & r^2 & 0 \\ 0 & 0 & 0 & r^2 \sin^2\theta \end{pmatrix}, \tag{7.85}
$$

where λ and ν are smooth functions. The metric in Equation (7.85) is an orthogonal metric that is spherically symmetric in the space variables.

Using the notation $\dot{u} = \partial u / \partial x^0$ and $u' = \partial u / \partial r$, we can show (see Problem 7.5.5) that the independent nonzero Christoffel symbols for the Levi-Civita connection are

$$
\Gamma^0_{00} = \frac{1}{2}\dot{\nu}, \quad \Gamma^0_{01} = \frac{1}{2}\nu', \quad \Gamma^0_{11} = \frac{1}{2}\dot{\lambda}e^{\lambda - \nu}, \quad \Gamma^1_{00} = \frac{1}{2}\nu' e^{\nu - \lambda}, \tag{7.86}
$$

$$
\Gamma^1_{01} = \frac{1}{2}\dot{\lambda}, \quad \Gamma^1_{11} = \frac{1}{2}\lambda', \quad \Gamma^1_{22} = -re^{-\lambda}, \quad \Gamma^1_{33} = -r\sin^2\theta e^{-\lambda}, \tag{7.87}
$$

$$
\Gamma^2_{12} = \frac{1}{r}, \quad \Gamma^2_{33} = -\sin\theta\cos\theta, \quad \Gamma^3_{13} = \frac{1}{r}, \quad \Gamma^3_{23} = \cot\theta. \tag{7.88}
$$

Though it is a little long to calculate (see Problem 7.5.6), we then determine that the only nonzero components of the Ricci tensor are

$$
R_{00} = e^{\nu - \lambda}\left(\frac{\nu''}{2} + \frac{(\nu')^2}{4} - \frac{\nu'\lambda'}{4} + \frac{\nu'}{r}\right) - \frac{\ddot{\lambda}}{2} - \frac{\dot{\lambda}^2}{4} + \frac{\dot{\lambda}\dot{\nu}}{4},
$$

$$
R_{01} = R_{10} = \frac{\dot{\lambda}}{r},
$$

$$
R_{11} = -\frac{\nu''}{2} - \frac{(\nu')^2}{4} + \frac{\nu'\lambda'}{4} + \frac{\lambda'}{r} + e^{\lambda - \nu}\left(\frac{\ddot{\lambda}}{2} + \frac{\dot{\lambda}^2}{4} - \frac{\dot{\lambda}\dot{\nu}}{4}\right), \tag{7.89}
$$

$$
R_{22} = -e^{-\lambda}\left(1 + \frac{r}{2}(\nu' - \lambda')\right) + 1,
$$

$$
R_{33} = \sin^2\theta\, R_{22}.
$$

Since we are trying to solve $R_{\alpha\beta} = 0$, we obtain conditions on the functions λ and ν. Since $R_{01} = 0$, we deduce immediately that $\dot{\lambda} = 0$, which means that λ is a function of r alone. Also, since $\partial R_{22} / \partial t = 0$, we find that $\partial\nu' / \partial t = 0$. Therefore, we can write the function ν as

$$
\nu = \nu(r) + f(t)
$$

for some function $f(t)$.

We now make one final coordinate change. In the metric line element, t appears only in the summand $e^\nu d(x^0)^2 = e^{\nu(r)} e^{f(t)} d(ct)^2$. So by choosing the variable $\bar{t}$ in such a way that

$$\frac{d\bar{t}}{dt} = e^{f(t)/2}$$

and then renaming $\bar{t}$ to just t, we obtain a metric which is independent of any timelike variable. (The variable $\bar{t}$, relabeled as t, is not necessarily time anymore so we cannot necessarily call a solution with $t = 0$ as a solution static.)

We can now assume there is no t dependence. Simplifying the expression $R_{00} + e^{\nu-\lambda}R_{11}$ leads to

$$\frac{1}{r}(\lambda' + \nu') = 0.$$

This implies that $\lambda(r) = -\nu(r)+C$ for some constant C. Without loss of generality, we can assume that $C = 0$ since we have not specified λ or ν. Thus, we set $\lambda(r) = -\nu(r)$. Then $R_{22} = 0$ in (7.89) implies that

$$e^{-\lambda}(1 - r\lambda') = 1.$$

Now setting $h(r) = e^{-\lambda(r)}$, this last equation becomes

$$h' + \frac{h}{r} = \frac{1}{r}.$$

This is a linear, first-order, ordinary, differential equation whose general solution is

$$h(r) = e^{-\lambda(r)} = 1 - \frac{2M}{r},$$

where M is a constant of integration. One can verify directly that $R_{11} = R_{00} = 0$ in Equation (7.89) are satisfied by this solution and therefore give no additional conditions. Therefore, the spherically symmetric vacuum solution to the EFE gives a metric with line element

$$ds^2 = -\left(1 - \frac{2M}{r}\right)c^2dt^2 + \left(1 - \frac{2M}{r}\right)^{-1} dr^2 + r^2d\theta^2 + r^2\sin^2\theta d\varphi^2. \quad (7.90)$$

This is called the Schwarzschild metric. This metric provided the first exact solution to the Einstein field equations. Though it is still complicated, this metric can be compared in fundamental importance to the solution in mechanics to the differential equations $d^2\vec{x}/dt^2 = -m\vec{g}$. Many of the verifiable predictions of general relativity arise from this metric.

In order to understand (7.90), we need to have some sense of the meaning of the constant M. Obviously, if $M = 0$, then the Schwarzschild metric is simply the flat Minkowski metric for spacetime.

To derive an interpretation for $M \neq 0$, we study some consequences of (7.90) for small velocities. If the velocity v is much smaller than the speed of light, i.e., $v \ll c$, then special relativity tells us that proper time is approximately coordinate time $\tau \cong t = x^0/c$. Furthermore, from (7.65) we can approximate the velocity of any particle as $\vec{U} = (c, 0, 0, 0)$. Plugging these into the geodesic equation

$$\frac{d^2 x^i}{d\tau^2} = -\Gamma^i_{jk} \frac{dx^j}{d\tau} \frac{dx^k}{d\tau},$$

we obtain the approximate relationship

$$\frac{d^2 x^i}{dt^2} = -\Gamma^i_{00} c^2 = \frac{c^2}{2} g^{il} \frac{\partial g_{00}}{\partial x^l}, \tag{7.91}$$

where the second equality follows from the formula for Christoffel symbols and the fact that the functions g_{ij} are not x^0 dependent. However, since $\mathbf{g}$ is diagonal and g_{00} depends only on r, we find that the only nonzero derivative is

$$\frac{d^2 r}{dt^2} = -\frac{c^2}{2} \left(1 - \frac{2M}{r}\right) \frac{\partial}{\partial r} \left(1 - \frac{2M}{r}\right) = -\frac{Mc^2}{r^2} \left(1 - \frac{2M}{r}\right).$$

We must compare this to the formula for gravitational attraction in Newtonian mechanics, namely,

$$\frac{d^2 r}{dt^2} = -\frac{GM_S}{r^2}$$

where M_S is the mass of the attracting body (and G is the gravitational constant). Thus, we find as a first approximation that the constant of integration M is

$$M \cong \frac{GM_S}{c^2}. \tag{7.92}$$

Hence, M is a constant multiple of the mass of the attracting body. The constant $2M$ has the dimensions of length, and one calls $r_G = 2M$ the *Schwarzschild radius*. The formula for it is

$$r_G = \frac{2G}{c^2} \cdot M_S = (1.48 \times 10^{-27} \text{ m/kg}) M_S.$$

The Schwarzschild radius is 2.95 km for the sun and 8.8 mm for the Earth. Evidently, for spherically symmetric objects that one encounters in common experience, the Schwarzschild radius is much smaller than the object's actual radius. In fact, if a spherically symmetric object has radius R_S and mass M_S, then

$$r_G < R_S \iff \frac{2G}{c^2} M_S < R_S.$$

A sphere with $r_G > R_S$ would need to have an enormous density. Furthermore, this situation would seem to be physically impossible for the following reason. It is understood that the Schwarzschild metric holds only in the vacuum outside of the body (planet or star). However, if $r_G > R_S$, then the Schwarzschild radius would correspond to a sphere outside of the spherical body where the Schwarzschild metric has a singularity $g_{11} = 1/0$. For this reason, some physicists initially claimed this to be a result of the successive approximations or simply a physically impossible situation.

The history of science has occasionally shown that singularities in the equations do not immediately imply that the scenario is impossible. The possibility of traveling at the speed of sound was thought to be impossible because of the consequences for the Doppler effect equation. Now, military jets regularly fly faster than the speed of sound. Similarly, in recent decades, physicists regularly study objects considered to be so dense that $2GM_S/c^2 > R_S$. Such objects are called *black holes*. For a time, the existence of black holes remained in the realm of hypothesis, but now astronomers are convinced they have observed many such objects, and astrophysicists have worked out many of their dynamic properties.

PROBLEMS

7.5.1. Show that we can rephrase the explanation for Equation (7.68) by saying that for any three vectors A, B, and C in $T_P M$,

$$(\star \mathbf{n}_P)(A, B, C) = \text{Vol}_{\mathbf{g}}(\mathbf{n}_P^\sharp, A, B, C),$$

where $\text{Vol}_{\mathbf{g}}$ is the volume form with respect to the metric $\mathbf{g}$.

7.5.2. Let $\mathbf{L}$ be a symmetric tensor of type $(0, 2)$ consisting of components that are constructible from those of $\mathbf{R}$ and Tg and are linear in the components of the Riemann curvature tensor $\mathbf{R}$.

(a) Show that $\mathbf{L}$ can only have the form

$$L_{\alpha\beta} = a R_{\alpha\beta} + b R g_{\alpha\beta} + \lambda g_{\alpha\beta},$$

where $R_{\alpha\beta}$ are the components of the Ricci curvature tensor, R is the scalar curvature, and a, b, and λ are real constants. [Hint: Consider Bianchi identities.]

(b) Show that $\text{div}\, \mathbf{L} = 0$ if and only if $b = -\frac{1}{2}a$.

(c) If $\mathbf{g} = \eta$, the standard Minkowski metric, show that $\mathbf{L} = 0$ if and only if $\lambda = 0$.

7.5.3. Calculate the curvature tensor and the Ricci curvature tensor for the Schwarzschild metric.

7.5.4. Find the "suitable" coordinate transformation h that allows one to pass from Equation (7.84) to Equation (7.85).

7.5.5. Prove that Equation (7.87) is correct.

7.5.6. Prove Equation (7.89).

7.5.7. *Light Propagation in the Schwarzschild Metric.* In the Schwarzschild metric, light travels along the null-geodesics, i.e., where $ds^2 = 0$.

(a) Explain why setting $\theta = 0$ does not lose any generality to finding the null-geodesics.

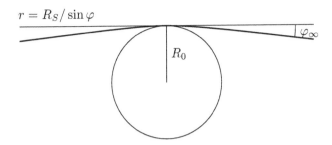

$r = R_S / \sin \varphi$

R_0

φ_∞

Figure 7.12: Deviation of light near a massive body.

(b) Prove that if one sets $u = 1/r$, then $ds^2 = 0$ implies that

$$\frac{d^2 u}{d\varphi^2} + u = 3Mu^2. \tag{7.93}$$

(c) Deduce that in the vicinity of a black hole, light travels in a circle precisely at the radius $r = \frac{3}{2} r_G$.

(d) Solve Equation (7.93) for $M = 0$. This corresponds to empty space (no mass present). Call this solution $u_0(\varphi)$.

(e) Now look for general solutions u to Equation (7.93) by setting $u = u_1 + u_0$. Then $u_1(\varphi)$ must satisfy

$$\frac{d^2 u_1}{d\varphi^2} + u_1 = \frac{3M}{R_0} \sin^2(\varphi - \varphi_0).$$

Solve this differential equation explicitly, and find the complete solution to Equation (7.93).

(f) In the complete solution, show that $M = 0$ (empty space) corresponds to traveling along a straight line $u = \frac{1}{R_0} \sin(\varphi - \varphi_0)$, where R_0 is the distance from the line to the origin.

(g) Show that the general solution to Equation (7.93) is asymptotically a line.

(h) We now consider Eddington's famous experiment to measure the deviation of light by the sun. Consider a geodesic $\mathcal{G}$ in the Schwarzschild metric that passes right alongside the sun, i.e., passes through the point $r = R_S$ and $\varphi = 0$. Define φ_∞ as the limiting angle of deviation between the line $r = R_S / \sin \varphi$ and the geodesic $\mathcal{G}$ (see Figure 7.12). The sun bends the light away from the straight line by a total of $2\varphi_\infty$. Using (at a judicious point) the approximation that $\sin \varphi \cong \varphi$, prove that the total deviation of light is

$$2\varphi_\infty = \frac{4GM_S}{R_S c^2},$$

where M_S is the mass of the sun and R_S is the sun's radius.

7.5.8. Consider the metric with line element $ds^2 = -e^{2ax} dt^2 + dx^2 + dy^2 + dz^2$.

(a) From the geodesic equation associated to this metric, show that at every point in this spacetime, a free particle experiences an acceleration of $d^2x/d\tau^2 = -a$.

(b) Show that the only nonzero Christoffel symbols for this metric are $\Gamma^0_{10} = \Gamma^0_{01} = a$ and $\Gamma^1_{00} = ae^{2ax}$.

(c) Show that the only nonzero components of the $(0, 4)$ curvature tensor are R_{0101} and its permutations. Calculate R_{0101}.

(d) Find the Ricci curvature tensor and notice that it is diagonal.

APPENDIX A

Point Set Topology

Though mathematicians, when developing a new area of mathematics, may define and study any object as they choose, the "natural" notion of a surface in $\mathbb{R}^3$ requires a rather intricate definition (Definition 3.1.1). Though at first somewhat unwieldy, this definition and also the definition for a differentiable manifold are necessary to appropriately generalize calculus and geometry to non-Euclidean spaces.

On the other hand, numerous concepts from geometry and calculus can be generalized not by formulating more constrained definitions but by expanding the context in which we define these concepts. The first wider context presented in this appendix is that of a metric space, a set equipped with some notion of distance. Many concepts from Euclidean geometry, including continuity, have natural generalizations to metric spaces. As it turns out, many useful concepts for analysis, like continuity, limit of sequences, or connectedness, arise in the yet more general context of topological spaces, where instead of a distance function, we have a looser notion of "nearness."

Though topology is a vast branch of mathematics, this appendix presents just the basic notions that support this book's presentation of differential geometry. A reader might encounter many of these concepts in a typical analysis course. We refer the reader to [27] for a gentle but thorough introduction to point set topology and to [43] and [2] for an introduction to topology that includes homology, the fundamental group, algebraic topology, and the classification of surfaces.

A.1 Metric Spaces

A.1.1 Metric Spaces: Definition

A metric space is a set that comes with the notion of "distance" between two points, where this distance function involves a few numerical conditions that mimic geometry in Euclidean spaces.

Definition A.1.1. Let X be any set. A *metric* on X is a function $D : X \times X \to \mathbb{R}^{\geq 0}$ such that

1. equality: $D(x, y) = 0$ if and only if $x = y$;
2. symmetry: $D(x, y) = D(y, x)$ for all $x, y \in X$;
3. triangle inequality: $D(x, y) + D(y, z) \geq D(x, z)$ for all $x, y, z \in X$.

A pair (X, D) where X is a set with a metric D is called a *metric space*.

Example A.1.2 (Euclidean Spaces). The Euclidean space $\mathbb{R}^n$ is a metric space where D is the usual Euclidean distance formula between two points, namely, if $P = (p_1, p_2, \ldots, p_n)$ and $Q = (q_1, q_2, \ldots, q_n)$, then

$$D(P, Q) = \sqrt{\sum_{i=1}^{n} (q_i - p_i)^2}.$$

Many notions in usual geometry (circles, parallelism, midpoint...) depend vitally on this particular distance formula. Furthermore, if $n = 1$, this formula simplifies to the usual distance formula on the real line $\mathbb{R}$, namely,

$$d(x, y) = |y - x|.$$

To prove that $(\mathbb{R}^n, D)$ is indeed a metric space, we must verify the three axioms in Definition A.1.1. The first holds because

$$D(P, Q) = 0 \iff \sum_{i=1}^{n} (q_i - p_i)^2 = 0,$$

which is equivalent to $(q_i - p_i)^2 = 0$ for all $1 \leq i \leq n$, and hence $q_i = p_i$ for all $1 \leq i \leq n$. The second obviously holds, and we prove the third axiom as follows. The Cauchy-Schwarz inequality on the vectors $\overrightarrow{PQ}$ and $\overrightarrow{QR}$ gives

$$\overrightarrow{PQ} \cdot \overrightarrow{QR} \leq |\overrightarrow{PQ} \cdot \overrightarrow{QR}| \leq \|\overrightarrow{PQ}\| \|\overrightarrow{QR}\|$$

so

$$2 \sum_{i=1}^{n} (q_i - p_i)(r_i - q_i) \leq 2 \sqrt{\sum_{i=1}^{n} (q_i - p_i)^2} \sqrt{\sum_{i=1}^{n} (r_i - q_i)^2}.$$

Using the property that $2ab = (a + b)^2 - a^2 - b^2$ with $a = q_i - p_i$ and $b = r_i - q_i$, we get

$$\sum_{i=1}^{n} (r_i - p_i)^2 \leq 2 \sqrt{\sum_{i=1}^{n} (q_i - p_i)^2} \sqrt{\sum_{i=1}^{n} (r_i - q_i)^2} + \sum_{i=1}^{n} (q_i - p_i)^2 + \sum_{i=1}^{n} (r_i - q_i)^2.$$

Hence,

$$\sum_{i=1}^{n} (r_i - p_i)^2 \leq \left(\sqrt{\sum_{i=1}^{n} (q_i - p_i)^2} + \sqrt{\sum_{i=1}^{n} (r_i - q_i)^2} \right)^2$$

from which it follows that $D(P, Q) + D(Q, R) \geq D(P, R)$.

Of course, the triangle inequality is used in Definition A.1.1 precisely because it is one of the fundamental properties of the Euclidean distance function. However, we needed to verify the triangle inequality based on the formula given for the Euclidean metric, and the example illustrates what is required in order to establish the three axioms.

Example A.1.3. There exists a variety of other metrics on Euclidean space, and we illustrate a few of these alternate metrics for $\mathbb{R}^2$. Let $P = (x_1, y_1)$ and $Q = (x_2, y_2)$. We leave to the reader the proofs that the following functions are metrics on $\mathbb{R}^2$:

$$D_1(P, Q) = |x_2 - x_1| + |y_2 - y_1|,$$
$$D_3(P, Q) = \sqrt[3]{|x_2 - x_1|^3 + |y_2 - y_1|^3},$$
$$D_\infty(P, Q) = \max\{|x_2 - x_1|, |y_2 - y_1|\}.$$

Example A.1.4 (Six Degrees of Kevin Bacon). A humorous example of a metric space is the set of syndicated actors A equipped with the function D defined as follows. Consider the graph whose set of vertices is A and has an edge between two actors a_1 and a_2 if they acted in a movie together. Define $D(a_1, a_2)$ as 0 if $a_1 = a_2$ and, otherwise, as the minimum number of edges it takes to create a path connecting a_1 and a_2. The pair (A, D) is a metric space.

The party game called "Six Degrees of Kevin Bacon" asks players to find $D(a_1, a_2)$ given any pair (a_1, a_2).

Having a notion of distance in a set, we may want to consider the subset of all points that are within a certain distance of a fixed point.

Definition A.1.5. Let (X, D) be a metric space, and let $p \in X$ be a point. We define the *open ball* of radius r around p as the set

$$B_r(p) = \{y \in X \mid D(p, y) < r\}.$$

The reader who is new to topology should note that the terminology "open ball" might be initially misleading since the set $B_r(p)$ only takes the shape of an actual ball (disk, sphere, etc.) in the case of the Euclidean metric on $\mathbb{R}^n$.

Example A.1.6. Consider the metric D_1 from Example A.1.3 above, and let $O = (0, 0)$. The ball of radius 1 around the origin O using the metric D_1 is the set

$$B_1(O) = \{(x, y) \in \mathbb{R}^2 \mid |x| + |y| < 1\}.$$

Notice that the equation $|x| + |y| = 1$ has a locus that is symmetric about the x-axis and about the y-axis. So to determine its locus we only need to see what happens in the first quadrant. In the first quadrant, the equation $|x| + |y| = 1$ becomes $x + y = 1$, which is a line segment from $(1, 0)$ to $(0, 1)$. Thus, the open ball $B_1(O)$ is the open square with corners at $\{(1, 0), (0, 1), (-1, 0), (0, -1)\}$.

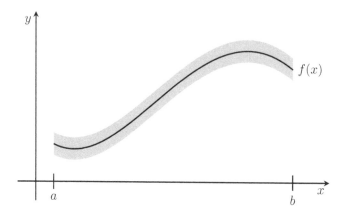

Figure A.1: Example A.1.7: a collar around $f(x)$.

Example A.1.7. Metric spaces can encompass a much wider range than the above examples have illustrated so far. Let $X = C^0([a, b])$ be the set of continuous real functions defined on the closed interval $[a, b]$, or let $X = F_{\text{bounded}}([a, b])$ be the set of all bounded functions on interval $[a, b]$. (A theorem of calculus tells us that any function f continuous over $[a, b]$ is bounded so $C^0([a, b]) \subset F_{\text{bounded}}([a, b])$.) Define the function $D : X \times X \to \mathbb{R}^{\geq 0}$ as

$$D(f, g) = \text{lub}\{|g(x) - f(x)| \, : \, x \in [a, b]\},$$

where lub refers to the least upper bound of a subset of reals.

The open ball of radius r around a function f is the set of all the functions $g \in X$ such that $|f(x) - g(x)| < r$ for all $x \in [a, b]$, or in other words,

$$f(x) - r < g(x) < f(x) + r \qquad \text{for all } x \in [a, b].$$

In this context, we call the region $f(x) - r < y < f(x) + r$ with $a \leq x \leq b$ the *r-collar* of $f(x)$. See Figure A.1.

Above, we introduced the notion of an open ball in any metric space. Though this appendix introduces notions that primarily support an overview of point-set topology, the notion of distance between two points allows us to generalize concepts from geometry to any metric space (X, D). We list here below a few of these concepts, which exist in any metric space but do not generalize further to topological spaces.

- A point $C \in X$ is said to be *between* two points A and B if $D(A, B) = D(A, C) + D(C, B)$. When D is the Euclidean metric on $\mathbb{R}^n$, this equality occurs for a degenerate triangle and only in the case when C lies on the segment $\overline{AB}$. It is in this sense that this definition directly generalizes the notion of betweenness from Euclidean geometry.

- The *bisector* of two points A and B is the set of points

$$\{P \in X \mid D(P, A) = D(P, B)\}.$$

This is the usual definition for the segment bisector in Euclidean geometry but in other metric spaces this set may look quite different.

- If $A, B \in X$ and $c \in \mathbb{R}$ with $2c > D(A, B)$, then the set of points

$$\{M \in X \mid D(A, M) + D(B, M) = 2c\}$$

is the ellipse with foci A and B and with half axis c.

- Let $S \subset X$ be a subset. We define the *diameter* of S to be

$$\operatorname{diam} S = \operatorname{lub}\{D(x, y) \mid x, y \in S\}.$$

- A subset S of X is called *bounded* if $\operatorname{diam} S < \infty$.

- Let S_1 and S_2 be two subsets of the metric space X. Then the distance between S_1 and S_2 is

$$D(S_1, S_2) = \operatorname{glb}\{D(x, y) \mid x \in S_1, \, y \in S_2\},$$

where glb is the greatest lower bound. The distance between a point $x \in X$ and a subset $A \subset X$ is $D(\{x\}, A)$. We observe that this definition of distance between subsets does not establish a metric on $\mathcal{P}(X)$, the set of subsets of X. Indeed, for any two subsets S_1 and S_2 in X such that $S_1 \neq S_2$ and $S_1 \cap S_2 \neq 0$, the distance between them is $D(S_1, S_2) = 0$, and hence, even the first axiom for metric spaces fails. However, in geometry, the notion of distance between sets, especially disjoint sets, is quite useful.

A.1.2 Open and Closed Sets

In the study of real functions, we often use the notions of open intervals and closed intervals. In this context, we simply say that a bounded interval is open if it does not include its endpoints and closed if it includes both of them; a similar definition is given for an unbounded interval. Then a subset of $\mathbb{R}$ is called open if it is a disjoint union of open intervals. In contrast, in $\mathbb{R}^n$ or in a metric space, given the wide range of possibilities for the shape of sets, we cannot legitimately talk about endpoints, though we could attempt to make sense of the concept of "including its boundary points." Regardless, a different definition for openness and closedness is required.

Definition A.1.8. Let (X, D) be a metric space. A subset $U \subseteq X$ is called *open* if for all $p \in U$ there exists $r > 0$ such that the open ball $B_r(p) \subset U$. A subset $F \subseteq X$ is called *closed* if the complement $F^c \stackrel{\text{def}}{=} X - F$ is open.

Intuitively, this definition states that a subset U of a metric space is called open if around every point there is an open ball, perhaps with a small radius, that is completely contained in U. Note that we may wish to consider more than one metric at the same time on the same set X. In this case, we will refer to a D-open set.

Proposition A.1.9. *Let (X, D) be a metric space. Then*

1. *X and $\emptyset$ are both open;*

2. *the intersection of any two open sets is open;*

3. *the union of any collection of open sets is open.*

Proof. For part 1, if $p \in X$, then any open ball satisfies $B_r(p) \subset X$. Also, since $\emptyset$ is empty, the criteria for openness holds trivially for $\emptyset$.

To prove part 2, let U_1 and U_2 be two open sets and let $p \in U_1 \cap U_2$. Since U_1 and U_2 are open, there exist r_1 and r_2 such that $B_{r_1}(p) \subset U_1$ and $B_{r_2}(p) \subset U_2$. Take $r = \min(r_1, r_2)$. Then $B_r(p) \subseteq B_{r_1}(p) \subset U_1$ and $B_r(p) \subseteq B_{r_2}(p) \subset U_2$ so $B_r(p) \subset U_1 \cap U_2$. Thus, $U_1 \cap U_2$ is open.

Finally, consider a collection of open sets U_α where α is an index taken from some indexing set I, which is not necessarily finite. Define

$$U = \bigcup_{\alpha \in I} U_\alpha .$$

For any $p \in U$, there exists some $\alpha_0 \in I$ such that $p \in U_{\alpha_0}$. Since U_{α_0} is open, there exists r such that $B_r(p) \subset U_{\alpha_0}$, and thus, $B_r(p) \subset U$. Consequently, U is open. $\square$

Using Proposition A.1.9(2), it is easy to show that any intersection of a finite number of open sets is again open. In contrast, part 3 states that the union of any collection of open subsets of X is again open, regardless of whether this collection is finite or not. This difference between unions and intersections of open sets is not an insufficiency of this proposition but rather a fundamental aspect of open sets in a metric space. In fact, as the following simple example shows, the infinite intersection of open sets need not be open.

Example A.1.10. For each integer $n \geq 1$, consider the open intervals $I_n = \left(0, 1 + \frac{1}{n}\right)$, and define

$$S = \bigcap_{n=1}^{\infty} I_n .$$

Obviously, $I_{n+1} \subsetneq I_n$, and so the intervals form a decreasing, nested chain. Since $\lim_{n \to \infty} \frac{1}{n} = 0$, we expect S to contain $(0, 1)$, but we must determine whether it

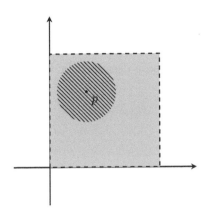

Figure A.2: Example A.1.11.

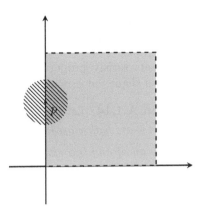

Figure A.3: Example A.1.12.

contains anything more. If $r > 1$, then if n is large enough so that $\frac{1}{n} < r - 1$, we have $r \notin I_n$. On the other hand, for all $n \in \mathbb{Z}^{\geq 1}$, $\frac{1}{n} > 0$, so $1 < 1 + \frac{1}{n}$. Hence, $1 \in I_n$ for all $n \in \mathbb{Z}^{\geq 1}$, and thus, $1 \in S$. Thus, we conclude that $S = (0, 1]$. This shows that the infinite intersection of open sets need not be open.

Example A.1.11. As a more down-to-earth example, we wish to show that according to this definition, the set $S = \{(x, y) \in \mathbb{R}^2 \mid 0 < x < 1 \text{ and } 0 < y < 1\}$ is open in $\mathbb{R}^2$ equipped with the Euclidean metric. Let $p = (x_0, y_0)$ be a point in S. Since $p \in S$, we see that $x_0 > 0$, $1 - x_0 > 0$, $y_0 > 0$, and $1 - y_0 > 0$. Since the closed distance from a point p to any line L is along a perpendicular to L, then the closest distance between p and any of the lines $x = 0$, $x = 1$, $y = 0$, and $y = 1$ is $\min\{x_0, 1 - x_0, y_0, 1 - y_0\}$. Consequently, if r is any positive real number such that

$$r \leq \min\{x_0, 1 - x_0, y_0, 1 - y_0\},$$

then $B_r(p) \subset S$. (See Figure A.2.)

Example A.1.12. In contrast to the previous example, consider the set

$$T = \{(x, y) \in \mathbb{R}^2 \mid 0 \leq x < 1 \text{ and } 0 < y < 1\},$$

where again we assume $\mathbb{R}^2$ is equipped with the Euclidean metric. The work in Example A.1.11 shows that for any point $p = (x_0, y_0)$, with $0 < x_0 < 1$ and $0 < y_0 < 1$, there exists a positive radius r such that $B_r(p) \subset S \subset T$. Thus, consider now points $p \in T$ with coordinates $(0, y_0)$. For all positive r, the open ball $B_r(p)$ contains the point $(-r/2, y_0)$, which is not in T. Hence, no open ball centered around points $(0, y_0)$ is contained in T, and hence T, is not open. (See Figure A.3.)

It is very common in proofs and definitions that rely on topology to refer to an open set that contains a particular point. Here is the common terminology.

Definition A.1.13. Let p be a point in a metric space (X, D). An *open neighborhood* (or simply *neighborhood*) of p is any open set of X that contains p.

Closed sets satisfy properties quite similar to those described in Proposition A.1.9, with a slight but crucial difference.

Proposition A.1.14. *Let (X, D) be a metric space. Then*

    1. X and $\emptyset$ are both closed;

    2. the union of any two closed sets is closed;

    3. the intersection of any collection of closed sets is closed.

Because a set is defined as closed if its complement is open and because of DeMorgan laws for sets, this proposition is actually a simple corollary of Proposition A.1.9. Therefore, we leave the details of the proof to the reader.

Note that in any metric space, the whole set X and the empty set $\emptyset$ are both open and closed. Depending on the particular metric space, these are not necessarily the only subsets of X that are both open and closed.

Proposition A.1.15. *Let (X, D) be a metric space, and let $x \in X$. The singleton set $\{x\}$ is a closed subset of X.*

Proof. To prove that $\{x\}$ is closed, we must prove that $X - \{x\}$ is open. Let y be a point in $X - \{x\}$. Since $x \neq y$, by the axioms of a metric space, $D(x, y) > 0$. Let $r = \frac{1}{2}D(x, y)$. The real number r is positive, and we consider the open ball $B_r(y)$. Since $D(x, y) > r$, then $x \notin B_r(y)$, and hence, $B_r(y) \subset X - \{x\}$. Hence, we have shown that $X - \{x\}$ is open and thus that $\{x\}$ is closed. $\qquad\square$

The notion of distance between sets provides an alternate characterization of closed sets in metric spaces. Recall that for any subset $A \subset X$, $x \in A$ implies that $D(x, A) = 0$. The following proposition shows that the converse holds precisely for closed sets.

Proposition A.1.16. *Let (X, D) be a metric space. A subset F is closed if and only if $D(x, F) = 0$ implies $x \in F$.*

Proof. Suppose first that F is closed. If $x \notin F$, then $x \in X - F$, which is open, so there exists an open ball $B_r(x)$ around x contained entirely in $X - F$. Hence, the distance between any point $a \in F$ and x is greater than the radius $r > 0$, thus, $D(x, F) > 0$, and in particular, $D(x, F) \neq 0$. Thus, $D(x, F) = 0$ implies that $x \in F$.

We now prove the converse. Suppose that F is a subset of X such that $D(x, F) = 0$ implies that $x \in F$. Then for all $x \in X - F$, we have $D(x, F) > 0$. Take the positive number $r = \frac{1}{2}D(x, F)$, and consider the open ball $B_r(x)$. Let p be any point in F and a any point in $B_r(x)$. Form the triangle inequality

$$D(p, x) \leq D(p, a) + D(a, x) \iff D(p, a) \geq D(p, x) - D(a, x).$$

The least possible value for $D(p, a)$ occurs when $D(p, x)$ is the least possible and when $D(a, x)$ is the greatest possible, that is when $D(p, x) = D(x, F)$ and $D(a, x) = r$. Thus, we find that $D(p, a) \geq r > 0$. Hence, for all $a \in B_r(x)$, we have $D(F, a) > 0$ and thus $B_r(x) \cap F = \emptyset$. Therefore, $B_r(x) \subset X - F$ so $X - F$ is open and F is closed. □

Proposition A.1.16 indicates that given any subset A of a metric space X, one can obtain a closed subset of X by adjoining all the points with 0 distance from A. This motivates the following definition.

Definition A.1.17. Let (X, D) be a metric space, and let $A \subset X$ be any subset. Define the *closure* of A as

$$\text{Cl}\, A = \{x \in X \mid D(x, A) = 0\}.$$

Proposition A.1.18. *Let (X, D) be a metric space and A any subset of X. $\text{Cl}\, A$ is the smallest closed set containing A. In other words,*

$$\text{Cl}\, A = \bigcap_{A \subset F,\ F\ closed} F.$$

Proof. (Left as an exercise for the reader. See Problem A.1.20.) □

A.1.3 Sequences

In standard calculus courses, one is introduced to the notion of a sequence of real numbers along with issues of convergence and limits. The definition given in such courses for when we say a sequence converges to a certain limit formalizes the idea of all terms in the sequence ultimately coming arbitrarily close to the limit point. Consequently, since limits formalize a concept about closeness and distance, the natural and most general context for convergence and limits is in a metric space.

Definition A.1.19. Let (X, D) be a metric space and let $\{x_n\}_{n \in \mathbb{N}}$ be a sequence in X. The sequence $\{x_n\}$ is said to converge to the limit $\ell \in X$ if for all $\varepsilon \in \mathbb{R}^{>0}$ there exists $N \in \mathbb{N}$ such that if $n > N$, then $D(x_n, \ell) < \varepsilon$ (i.e., $x_n \in B_\varepsilon(\ell)$). If $\{x_n\}$ converges to ℓ, then we write

$$\lim_{n \to \infty} x_n = \ell.$$

Note that we can restate Definition A.1.19 to say that $\{x_n\}$ converges to ℓ if for all positive $\varepsilon \in \mathbb{R}^{>0}$, only finitely many elements of the sequence $\{x_n\}$ are not in the open ball $B_\varepsilon(\ell)$.

Example A.1.20. Consider the sequence $\{x_n\}_{n \geq 1}$ in $\mathbb{R}^3$ given by $x_n = (3, \frac{1}{n+2}, \frac{2n}{n+1})$. We prove that $\{x_n\}$ converges to $(3, 0, 2)$. We know that as sequences of real numbers,

$$\lim_{n \to \infty} \frac{1}{n+2} = 0 \quad \text{and} \quad \lim_{n \to \infty} \frac{2n}{n+1} = 2.$$

Pick any positive ε. Choose N_1 such that $n > N_1$ implies that $\frac{1}{n+2} < \frac{\varepsilon}{\sqrt{2}}$, and choose N_2 such that $n > N_2$ implies that $\left| \frac{2n}{n+1} - 2 \right| < \frac{\varepsilon}{\sqrt{2}}$. Using the Euclidean distance

$$D(x_n, (3, 0, 2)) = \sqrt{(3-3)^2 + \left(\frac{1}{n+2} - 0\right)^2 + \left(\frac{2n}{n+1} - 2\right)^2},$$

one sees that if $n > N = \max(N_1, N_2)$, then

$$D(x_n, (3, 0, 2)) < \sqrt{\frac{\varepsilon}{2} + \frac{\varepsilon}{2}} = \varepsilon.$$

This proves that $\lim x_n = (3, 0, 2)$.

Note that we could have proved directly that $\lim x_n = (3, 0, 2)$ by considering the limit of $D(x_n, (3, 0, 2))$ as a sequence of real numbers and proving that this converges to 0.

Example A.1.21. Consider the set X of bounded, real-valued functions defined over the interval $[0, 1]$ equipped with the metric defined in Example A.1.7. For $n \geq 1$, consider the sequence of functions given by

$$f_n(x) = \begin{cases} 1 - nx, & \text{for } 0 \leq x \leq \frac{1}{n}, \\ 0, & \text{for } \frac{1}{n} \leq x \leq 1. \end{cases}$$

Figure A.4 shows the functions for $n = 1, 2, 3$. One might suspect that the limit of this sequence $f_n(x)$ would be the function

$$f(x) = \begin{cases} 1, & \text{if } x = 0, \\ 0, & \text{for } x > 0, \end{cases}$$

but this is not the case. Let $r = \frac{1}{4}$, and consider the r-collar around $f(x)$. There is no n such that $f_n(x)$ lies within the $\frac{1}{4}$-collar around $f(x)$. (See Figure A.5.) Consequently, $f_n(x)$ does not converge to $f(x)$ in the metric space (X, D). Note, however, that for all $x \in [0, 1]$, as sequences of real numbers $\lim_{n \to \infty} f_n(x) = f(x)$. We say that $f_n(x)$ converges *pointwise*.

Proposition A.1.22. *Let (X, D) be a metric space. Any sequence $\{x_n\}$ can converge to at most one limit point.*

Proof. Suppose that

$$\lim_{n \to \infty} x_n = \ell \qquad \text{and} \qquad \lim_{n \to \infty} x_n = \ell'.$$

Let ε be any positive real number. There exists N_1 such that $n > N_1$ implies that $D(x_n, \ell) < \frac{\varepsilon}{2}$, and there exists N_2 such that $n > N_s$ implies that $D(x_n, \ell') < \frac{\varepsilon}{2}$. Thus, taking some $n > \max(N_1, N_2)$, we deduce from the triangle inequality that

$$D(\ell, \ell') \leq D(x_n, \ell) + D(x_n, \ell') \leq \frac{\varepsilon}{2} + \frac{\varepsilon}{2} = \varepsilon.$$

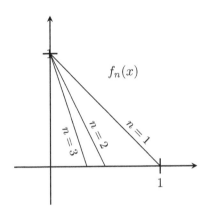

Figure A.4: Sequence of functions.

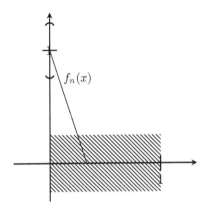

Figure A.5: $\frac{1}{4}$-collar around $f(x)$.

Thus, since $D(\ell, \ell')$ is less than any positive real number, we deduce that $D(\ell, \ell') = 0$ and hence that $\ell = \ell'$. $\square$

In any metric space, there are plenty of sequences that do not converge to any limit. For example, the sequence $\{a_n\}_{n\geq 1}$ of real numbers given by $a_n = (-1)^n + \frac{1}{n}$ does not converge toward anything but, in the long term, alternates between being very close to 1 and very close to -1. Referring to the restatement of Definition A.1.19, one can loosen the definition of limit to incorporate the behavior of such sequences as the one just mentioned.

Definition A.1.23. Let (X, D) be a metric space, and let $\{x_n\}$ be a sequence in X. A point $p \in X$ is called an *accumulation point* of $\{x_n\}$ if for all real $\varepsilon > 0$, an infinite number of elements x_n are in $B_\varepsilon(p)$. The accumulation set of $\{x_n\}$ is the set of all accumulation points.

Example A.1.24. Consider again the real sequence $a_n = (-1)^n + \frac{1}{n}$. Let ε be any positive real number. If $n > \frac{1}{\varepsilon}$ and n is even, then $a_n \in B_\varepsilon(1)$. If $n > \frac{1}{\varepsilon}$ and n is odd, then $a_n \in B_\varepsilon(-1)$. Hence, 1 and -1 are accumulation points. However, for any r different than 1 or -1, suppose we choose a ε such that $\varepsilon < \min(|r-1|, |r+1|)$. If n is large enough, then

$$\frac{1}{n} < \big| \min(|r - 1|, |r + 1|) - \varepsilon \big|,$$

and for such n, we have $a_n \notin B_\varepsilon(r)$. Thus, 1 and -1 are the only accumulation points of $\{a_n\}$. In the terminology of Definition, A.1.23, the accumulation set is $\{-1, 1\}$.

A.1.4 Continuity

For the same reason as for the convergence of sequences, the notion of continuity, first introduced in the context of real functions over an interval, generalizes naturally to the category of metric spaces. Here is the definition.

Definition A.1.25. Let (X, D) and (Y, D') be two metric spaces. A function $f : X \to Y$ is called *continuous at* $a \in X$ if for all $\varepsilon \in \mathbb{R}^{>0}$, there exists $\delta \in \mathbb{R}^{>0}$ such that $D(x, a) < \delta$ implies that $D(f(x), f(a)) < \varepsilon$. The function f is called *continuous* if it is continuous at all points $a \in X$.

Example A.1.26. As a first example of Definition A.1.25, consider the function $f : \mathbb{R}^2 \to \mathbb{R}$ given by $f(x, y) = x + y$, where we assume $\mathbb{R}^2$ and $\mathbb{R}$ are equipped with the usual Euclidean metrics. Consider some point $(a_1, a_2) \in \mathbb{R}^2$. Let $\varepsilon > 0$ be any positive real number. Choosing $\delta = \frac{\varepsilon}{2}$ will suffice, as we now show. First note that

$$D((x, y), (a_1, a_2)) = \sqrt{(x - a_1)^2 + (y - a_2)^2} < \frac{\varepsilon}{2}$$

implies that

$$|x - a_1| < \frac{\varepsilon}{2} \quad \text{and} \quad |y - a_2| < \frac{\varepsilon}{2}.$$

But if this is so, then

$$|f(x, y) - f(a_1, a_2)| = |x + y - (a_1 + a_2)| < |x - a_1| + |y - a_2| < \frac{\varepsilon}{2} + \frac{\varepsilon}{2} = \varepsilon.$$

Thus, f is continuous.

Example A.1.27. Definition A.1.25 allows one to study the continuity of functions in much more general contexts, as we show with this example. Let X be a proper subset of $\mathbb{R}^n$, and let $\vec{p}$ be a point in $\mathbb{R}^n - X$. We view X as a metric space by restricting the Euclidean metric to it. Let $\mathbb{S}^{n-1}$ be the unit sphere in $\mathbb{R}^n$ also with its metric coming from the Euclidean one in $\mathbb{R}^n$. Define a function $f : X \to \mathbb{S}^{n-1}$ by

$$f(\vec{x}) = \frac{\vec{x} - \vec{p}}{\|\vec{x} - \vec{p}\|}.$$

We will show that f is continuous.

Let $\vec{a} \in X$. The Euclidean metric is $D(\vec{x}, \vec{a}) = \|\vec{x} - \vec{a}\|$. Hence

$$D(f(\vec{a}), f(\vec{x})) = \left\| \frac{\vec{a} - \vec{p}}{\|\vec{a} - \vec{p}\|} - \frac{\vec{x} - \vec{p}}{\|\vec{x} - \vec{p}\|} \right\|$$

$$= \sqrt{\left(\frac{\vec{a} - \vec{p}}{\|\vec{a} - \vec{p}\|} - \frac{\vec{x} - \vec{p}}{\|\vec{x} - \vec{p}\|} \right) \cdot \left(\frac{\vec{a} - \vec{p}}{\|\vec{a} - \vec{p}\|} - \frac{\vec{x} - \vec{p}}{\|\vec{x} - \vec{p}\|} \right)}$$

$$= \sqrt{2 - 2 \frac{(\vec{a} - \vec{p}) \cdot (\vec{x} - \vec{p})}{\|\vec{a} - \vec{p}\| \, \|\vec{x} - \vec{p}\|}}$$

$$= \sqrt{2 - 2 \cos \alpha},$$

where α is the angle between the vectors $(\vec{a} - \vec{p})$ and $(\vec{x} - \vec{p})$. However, from the trigonometric identity $\sin^2 \theta = (1 - \cos 2\theta)/2$, we deduce that

$$D(f(\vec{a}), f(\vec{x})) = 2\sin\left(\frac{\alpha}{2}\right).$$

If $\vec{x}$ and $\vec{a}$ are close enough, then $(\vec{a} - \vec{p})$ and $(\vec{x} - \vec{p})$ form an acute angle, and hence, if d is the height from $\vec{a}$ to the segment between $\vec{p}$ and $\vec{x}$, we have

$$D(f(\vec{x}), f(\vec{a})) = 2\sin\left(\frac{\alpha}{2}\right) \le 2\sin\alpha = 2\frac{d}{\|\vec{p} - \vec{a}\|} \le 2\frac{\|\vec{x} - \vec{a}\|}{\|\vec{p} - \vec{a}\|}.$$

Therefore, choosing δ small enough so that the angle between $(\vec{a} - \vec{p})$ and $(\vec{x} - \vec{p})$ is acute and $\delta < \frac{1}{2}\|\vec{p} - \vec{a}\|\,\varepsilon$, we conclude that

$$\|\vec{x} - \vec{a}\| < \delta \implies D(f(\vec{x}), f(\vec{a})) < \varepsilon,$$

proving that f is continuous at all points $\vec{a} \in X$.

Proposition A.1.28. *Let (X, D), (Y, D'), and (Z, D'') be metric spaces. Let $f : X \to Y$ and $g : Y \to Z$ be functions such that f is continuous at a point $a \in X$ and g is continuous at $f(a) \in Y$. Then the composite function $g \circ f : X \to Z$ is continuous at a.*

Proof. Since f is continuous at a, for all $\varepsilon_1 \in \mathbb{R}^{>0}$, there exists $\delta_1 \in \mathbb{R}^{>0}$ such that $D(x, a) < \delta_1$ implies that $D(f(x), f(a)) < \varepsilon_1$. Since g is continuous at $f(a)$, for all $\varepsilon_2 \in \mathbb{R}^{>0}$, there exists $\delta_2 \in \mathbb{R}^{>0}$ such that $D(y, f(a)) < \delta_2$ implies that $D(g(y), g(f(a))) < \varepsilon_2$. Therefore, given any $\varepsilon > 0$, set $\varepsilon_2 = \varepsilon$ and choose ε_1 so that $\varepsilon_1 < \delta_2$. Then

$$D(x, a) < \delta_1 \Rightarrow D(f(x), f(a)) < \varepsilon_1 < \delta_2 \Rightarrow D(g(f(x)), g(f(a))) < \varepsilon,$$

showing that $g \circ f$ is continuous at a. $\qquad\square$

Using the concepts of open sets, we can give alternate formulations for when a function between metric spaces is continuous.

Proposition A.1.29. *Let (X, D) and (Y, D') be two metric spaces, and let $f : X \to Y$ be a function. The function f is continuous if and only if for all open subsets $U \subset Y$, the set*

$$f^{-1}(U) = \{x \in X \mid f(x) \in U\}$$

is an open subset of X.

Proof. First suppose that f is continuous. Let U be an open subset of Y, and let x be some point in $f^{-1}(U)$. Of course $f(x) \in U$. Since U is open, there exists a real

$\varepsilon > 0$ such that $B_\varepsilon(f(x)) \subset U$. Since f is continuous, there exists a $\delta > 0$ such that $y \in B_\delta(x)$ implies that $f(y) \in B_\varepsilon(f(x))$. Hence,

$$f(B_\delta(x)) \subset B_\varepsilon(f(x)) \subset U,$$

and thus, $B_\delta(x) \subset f^{-1}(U)$.

Conversely, suppose that $f^{-1}(U)$ is an open set in X for every open set U in Y. Let $f(x)$ be a point in U, and let ε be a positive real number. Then

$$f^{-1}\big(B_\varepsilon(f(x))\big)$$

is an open set in X. Since $x \in f^{-1}\big(B_\varepsilon(f(x))\big)$ is open, there exists some δ such that $B_\delta(x) \subset f^{-1}\big(B_\varepsilon(f(x))\big)$. Thus, $f(B_\delta(x)) \subset B_\varepsilon(f(x))$, and therefore, f is continuous. $\qquad\square$

The following proposition is an equivalent formulation to Proposition A.1.29 but often more convenient for proofs.

Proposition A.1.30. *Let (X, D) and (Y, D') be two metric spaces and let $f : X \to Y$ be a function. The function f is continuous if and only if for all open balls $B_r(p)$ in Y, the set $f^{-1}(B_r(p))$ is an open subset of X.*

Proof. (Left as an exercise for the reader. See Problem A.1.27.) $\qquad\square$

Much more could be included in an introduction to metric spaces. However, many properties of metric spaces and continuous functions between them hold simply because of the properties of open sets (Proposition A.1.9) and the characterization of continuous functions in terms of open sets (Proposition A.1.29). This fact motivates the definition of topological spaces.

PROBLEMS

A.1.1. Prove that D_1, D_3, and D_∞ from Example A.1.3 are in fact metrics on $\mathbb{R}^2$.

A.1.2. In the following functions on $\mathbb{R}^2 \times \mathbb{R}^2$, which axioms fail to make the function into a metric?

    (a) $D_1((x_1, y_1), (x_2, y_2)) = |x_1| + |x_2| + |y_1| + |y_2|$.

    (b) $D_2((x_1, y_1), (x_2, y_2)) = -\big((x_2 - x_1)^2 + (y_2 - y_1)^2\big)$.

    (c) $D_3((x_1, y_1), (x_2, y_2)) = |x_2 - x_1| \cdot |y_2 - y_1|$.

    (d) $D_4((x_1, y_1), (x_2, y_2)) = |x_2^2 - x_1^2| + |y_2 - y_1|$.

A.1.3. Let (X_1, D_1) and (X_2, D_2) be metric spaces. Consider the Cartesian product $X = X_1 \times X_2$. Prove that the following function is a metric on X:

$$D((p_1, p_2), (q_1, q_2)) = D(p_1, q_1) + D(p_2, q_2).$$

A.1.4. Let $X = \mathcal{P}_{\text{fin}}(\mathbb{Z})$ be the set of all finite subsets of the integers. Recall that the symmetric difference between two sets A and B is $A \triangle B = (A - B) \cup (B - A)$. Define the function $D : X \times X \to \mathbb{R}^{\geq 0}$ by

$$D(A, B) = |A \triangle B|,$$

the cardinality of $A \triangle B$. Prove that D is a metric on X.

A.1.5. In Euclidean geometry, the median line between two points p_1 and p_2 in $\mathbb{R}^2$ is defined as the set of points that are of equal distance from p_1 and p_2, i.e.,

$$M = \{q \in \mathbb{R}^2 \mid D(q, p_1) = D(q, p_2)\}.$$

What is the shape of the median lines in $\mathbb{R}^2$ for D_1, D_2, and D_∞ from Example A.1.3?

A.1.6. Prove that if (X, D) is any metric space, then $D(x, y)^n$ where n is any positive integer, is also metric on X.

A.1.7. Let (X, D) be a metric space, and let S be any subset of X. Prove that (S, D) is also a metric space. (The metric space (S, D) is referred to as the *restriction of D to S*.)

A.1.8. Let $\mathbb{S}^2$ be the unit sphere in $\mathbb{R}^3$, i.e.,

$$\mathbb{S}^2 = \{(x, y, z) \in \mathbb{R}^3 \mid x^2 + y^2 + z^2 = 1\}.$$

Sketch the open balls on $\mathbb{S}^2$ obtained by the restriction of the Euclidean metric to $\mathbb{S}^2$ (see Problem A.1.7). Setting the radius $r < 2$, for some point $p \in \mathbb{S}^2$, describe $B_r(p)$ algebraically by the equation $x^2 + y^2 + z^2 = 1$ and some linear inequality in x, y, and z.

A.1.9. Prove that Example A.1.7 is in fact a metric space.

A.1.10. Consider the metric space $(\mathbb{R}^2, D_1)$, where D_1 is defined in Example A.1.3.

    (a) Let A and B be two points in $\mathbb{R}^2$. Prove that the set of points between A and B is the rectangle with vertical or horizontal edges with A and B as opposite corners.

    (b) Determine the bisector of the points A and B in this metric.

A.1.11. Find the distance between the following pairs of sets in $\mathbb{R}^2$:

    (a) $A = \{(x, y) \mid x^2 + y^2 < 1\}$ and $B = \{(x, y) \mid (x - 3)^2 + (y - 2)^2 < 1\}$.

    (b) $A = \{(x, y) \mid xy = 1\}$ and $B = \{(x, y) \mid xy = 0\}$.

    (c) $A = \{(x, y) \mid xy = 2\}$ and $B = \{(x, y) \mid x^2 + y^2 < 1\}$.

A.1.12. Prove that a subset A of a metric space (X, D) is bounded if and only if $A \subset B_r(p)$ for some $r \in \mathbb{R}^{\geq 0}$ and $p \in X$.

A.1.13. *Infinite Intersections and Unions.* Let $A_n = [n, +\infty)$ and let $B_n = [\frac{1}{n}, \sin n]$. Find

$$\text{(a) } \bigcup_{n=0}^{\infty} A_n, \qquad \text{(b) } \bigcap_{n=0}^{\infty} A_n, \qquad \text{(c) } \bigcup_{n=2}^{\infty} B_n. \qquad\qquad \text{(A.1)}$$

A.1.14. Define the metric D on $\mathbb{R}^2$ as follows. For any $p = (p_x, p_y)$ and $q = (q_x, q_y)$, let $D(p, q) = \sqrt{4(p_x - q_x)^2 + (p_y - q_y)^2}$. Prove that this is in fact a metric. What is the shape of a unit ball $B_r\big((x, y)\big)$? What is the shape of the "median" between two points? [Hint: see Problem A.1.5.]

A.1.15. Are the following subsets of the plane (using the usual Euclidean metric) open, closed, or neither:

    (a) $\{(x, y) \in \mathbb{R}^2 : x^2 + y^2 < 1\}$.

    (b) $\{(x, y) \in \mathbb{R}^2 : x^2 + y^2 \geq 1\}$.

    (c) $\{(x, y) \in \mathbb{R}^2 : x + y = 0\}$.

    (d) $\{(x, y) \in \mathbb{R}^2 : x + y \neq 0\}$.

    (e) $\{(x, y) \in \mathbb{R}^2 : x^2 + y^2 \leq 1 \text{ or } x = 0\}$.

    (f) $\{(x, y) \in \mathbb{R}^2 : x^2 + y^2 < 1 \text{ or } x = 0\}$.

    (g) The complement A^c where $A = \{(x, y) \in \mathbb{R}^2 \mid x = 0 \text{ and } -1 \leq y \leq 1\}$.

A.1.16. Let L be a line in the plane $\mathbb{R}^2$. Prove that $\mathbb{R}^2 - L$ is open in the Euclidean metric and in the three metrics presented in Example A.1.3.

A.1.17. Let $X = \mathbb{R}^2$, and define D_2 as the Euclidean metric and D_1 as $D_1((x_1, y_1), (x_2, y_2)) = |x_2 - x_1| + |y_2 - y_1|$. Prove that any D_2-open ball contains a D_1-open ball and is also contained in a D_1-open ball. Conclude that a subset of $\mathbb{R}^2$ is D_2-open if and only if it is D_1-open.

A.1.18. Prove that the set $\{\frac{1}{n} \mid n \in \mathbb{Z}^{>0}\}$ is not closed in $\mathbb{R}$ whereas the set $\{\frac{1}{n} \mid n \in \mathbb{Z}^{>0}\} \cup \{0\}$ is.

A.1.19. Consider a metric space (X, D), and let x and y be two distinct points of X. Prove that there exists a neighborhood U of x and a neighborhood V of y such that $U \cap V = \emptyset$. (In general topology, this property is called the *Hausdorff property*, and this exercise shows that all metric spaces are Hausdorff.)

A.1.20. Prove Proposition A.1.18.

A.1.21. Let A be a subset of a metric space (X, D). Suppose that every sequence $\{x_n\}$ in A that converges in X converges to an element of A. Prove that A is closed.

A.1.22. Let $\{x_n\}_{n \in \mathbb{N}}$ be a sequence in a metric space (X, D). Prove that the closure of the set of elements $\{x_n\}$ is $\{x_n \mid n \in \mathbb{N}\}$ together with the accumulation set of the sequence $\{x_n\}$.

A.1.23. Using Definition A.1.25, prove that the real function $f(x) = 2x - 5$ is continuous over $\mathbb{R}$.

A.1.24. Using Definition A.1.25, prove that

    (a) the real function $f(x) = x^2$ is continuous over $\mathbb{R}$;

    (b) the real function of two variables $f(x, y) = 1/(x^2 + y^2 + 1)$ is continuous over all $\mathbb{R}^2$ (using the usual Euclidean metric).

A.1.25. Let (X, D) and (Y, D') be metric spaces, and let $f : X \to Y$ be a continuous function. Prove that if a sequence $\{x_n\}$ in X converges to a limit point ℓ, then the sequence $\{f(x_n)\}$ in Y converges to $f(\ell)$.

A.1.26. Let $f : X \to Y$ be a function from the metric space (X, D) to the metric space (Y, D'). Suppose that f is such that there exists a positive real number λ, with

$$\frac{D'(f(x_1), f(x_2))}{D(x_1, x_2)} \leq \lambda \qquad \text{for all } x_1 \neq x_2,$$

i.e., that the stretching ratio for f is bounded. Show that f is continuous.

A.1.27. Prove Proposition A.1.30.

A.1.28. Let $X = \mathbb{R}^m$ and $Y = \mathbb{R}^n$ equipped with the usual Euclidean metric. Prove that any linear transformation from X to Y is continuous.

A.2 Topological Spaces

A.2.1 Definitions and Examples

Definition A.2.1. A *topological space* is a pair (X, τ) where X is a set and where τ is a set of subsets of X satisfying the following:

1. X and $\emptyset$ are in τ.

2. For all U and V in τ, $U \cap V \in \tau$.

3. For any collection $\{U_\alpha\}_{\alpha \in I}$ of sets in τ, the union $\bigcup_{\alpha \in I} U_\alpha$ is in τ.

The elements in τ are called open subsets of X and a subset $F \subset X$ is called closed if $X - F$ is open, i.e., if $X - F \in \tau$.

As an alternate terminology we talk about τ satisfying the above three properties as a *topology* on X. In the introduction to this appendix, we promised that topology attempts to provide a mental model that generalizes the notion of nearness. The following concept is key to this way of thinking.

Definition A.2.2. Let (X, τ) be a topological space and let $x \in X$. Any $U \in \tau$ such that $x \in U$ is called a *neighborhood* of x.

If we work with more than one topology on the same underlying set X, we refer to τ-open and τ-closed subsets to avoid ambiguity.

As with the properties of open sets in metric spaces, in criterion (3) of Definition A.2.1, the indexing set I need not be countable, and hence, we should not assume that the collection $\{U_\alpha\}_{\alpha \in I}$ can be presented as a sequence of subsets.

Example A.2.3. According to Proposition A.1.9, if (X, D) is a metric space, it is also a topological space, where we use the topology τ to be the open sets as defined by Definition A.1.8. Historically, it was precisely Proposition A.1.9 along with the discovery by mathematicians that collections of sets with the properties described in this proposition arise naturally in numerous other contexts that led to the given definition of a topological space.

Example A.2.4 (Euclidean topology). Consider the metric space $(\mathbb{R}^2, D)$, where D is the Euclidean metric. The topology induced on $\mathbb{R}^n$ according to Example A.2.3 is called the Euclidean topology.

Example A.2.5 (Discrete topology). Let X be any set. Setting $\tau = \mathcal{P}(X)$ to be the set of all subsets of X is a topology on X called the *discrete topology* on X. In the discrete topology, all subsets of X are both closed and open.

Example A.2.6 (Trivial topology). On the opposite end of the spectrum from, setting $\tau = \{X, \emptyset\}$ also satisfies the axioms of a topology, and this is called the *trivial topology* on X. These two examples represent the largest and the smallest possible examples of topologies on a set X.

Example A.2.7. Let $X = \{a, b, c\}$ be a set with three elements. Consider the set of subsets $\tau = \{\emptyset, \{a\}, \{a, b\}, X\}$. A simple check shows that τ is a topology on X, namely that τ satisfies all the axioms for a topology. Notice that $\{a\}$ is open, $\{c\}$ is closed (since $\{a, b\}$ is open) and that $\{b\}$ is neither open nor closed. By Proposition A.1.15, there is no metric D on X such that the D-open sets of X are the open sets in the topology of τ. We say that (X, τ) is not metrizable.

It is not always easy to specify a subset of $\mathcal{P}(X)$ that satisfy the axioms for a topology on X. The concept of a basis makes this possible and Proposition A.2.9 gives a practical characterization of a basis.

Definition A.2.8. Let (X, τ) be a topological space. A collection of open sets $\mathcal{B} \subset \tau$ is called a *basis* of the topology if every open set is a union of elements in $\mathcal{B} \subset \tau$.

Proposition A.2.9. *Let (X, τ) be a topological space, and suppose that $\mathcal{B}$ is a basis. Then:*

1. *the elements of $\mathcal{B}$ cover X;*

2. *if $B_1, B_2 \in \mathcal{B}$, then for all $x \in B_1 \cap B_2$ there exists $B_3 \in \mathcal{B}$ such that $x \in B_3 \subset B_1 \cap B_2$.*

Conversely, if any collection $\mathcal{B}$ of open sets satisfies the above two properties, then there exists a unique topology on X for which $\mathcal{B}$ is a basis. (This topology is said to be generated by $\mathcal{B}$.)

Proof. (Left as an exercise for the reader.)   □

This characterization allows one to easily describe topologies by presenting a basis of open sets.

Example A.2.10. By Definition A.1.8, in a metric space, the topology associated to a metric has the set of open balls as a basis.

Example A.2.11. Let $X = \mathbb{R}^2$, and consider the collection $\mathcal{B}$ of sets of the form

$$U = \{(x, y) \in \mathbb{R}^2 \mid x > x_0 \text{ and } y > y_0\},$$

where x_0 and y_0 are constants. This collection $\mathcal{B}$ satisfies both of the criteria in Proposition A.2.9, hence there exists a unique topology τ on $\mathbb{R}^2$ with $\mathcal{B}$ as a basis. It is easy to see that τ is different from the usual Euclidean topology. Note that for any open set $U \in \tau$, if $(a, b) \in U$, then the half-infinite ray

$$\{(a + t, b + t) \mid t \geq 0\}$$

is a subset of U. This is not a property of the Euclidean topology on $\mathbb{R}^2$, so this gives a topology different from the Euclidean topology.

Proposition A.2.12. *Let (X, τ) be a topological space. Then the following are true about the τ-closed sets of X:*

1. *X and $\emptyset$ are closed.*

2. *The union of any two closed sets is closed.*

3. *The intersection of any collection of closed sets is closed.*

Proof. Part 1 is obviously true since both X and $\emptyset$ are open.

For part 2, let F_1 and F_2 be any two closed subsets of X. Then $F_1{}^c$ and $F_2{}^c$ are open sets. Thus $F_1{}^c \cap F_2{}^c$ is open. However, by the DeMorgan laws,

$$F_1{}^c \cap F_2{}^c = (F_1 \cup F_2)^c.$$

Thus, since $F_1 \cup F_2{}^c$ is open, $F_1 \cup F_2$ is closed.

For part 3, let $\{F_\alpha\}$, where α is in some indexing set I, be a collection of closed subsets. The collection $\{F_\alpha{}^c\}$ is a collection of open sets. Therefore, $\bigcup_\alpha F_\alpha{}^c$ is open. Thus,

$$\bigcap_{\alpha \in I} F_\alpha = \left(\bigcup_{\alpha \in I} F_\alpha{}^c \right)^c$$

is closed. $\qquad\qquad\square$

A converse to this proposition turns out to be useful for defining certain classes of topologies on sets.

Proposition A.2.13. *Let X be a set. Suppose that a collection $\mathcal{C}$ of subsets of X satisfies the following properties:*

1. *X and $\emptyset$ are in $\mathcal{C}$.*

2. *The union of any two sets in $\mathcal{C}$ is again in $\mathcal{C}$.*

3. *The intersection of any collection of sets in $\mathcal{C}$ is again in $\mathcal{C}$.*

Then the set of all complements of sets in $\mathcal{C}$ form a topology on X.

Proof. (Left as an exercise for the reader.)   □

Example A.2.14. Let X be any set. Consider the collection $\mathcal{C}$ of subsets of X that include X, $\emptyset$, and all finite subsets of X. The collection $\mathcal{C}$ satisfies all three criteria in Proposition A.2.13, so X, $\emptyset$, and the complements of finite subsets of X form a topology on X. This is often called the *finite complement* topology.

We now present a topology on $\mathbb{R}^n$ that has more open sets than the finite complement topology but fewer than the usual Euclidean topology.

Example A.2.15. Let $X = \mathbb{R}^n$. Let $\mathcal{C}$ be the collection of all finite unions of affine subspaces of $\mathbb{R}^n$, where by affine subspace we mean any set of points that is the solution set to a set of linear equations in $x_1, x_2, \ldots, x_n$ (i.e., points, lines, planes, etc.). Taking the empty set of linear equations or an inconsistent set of linear equations, we obtain X and $\emptyset$ as elements of $\mathcal{C}$. Since the union operation of sets is associative, a finite union of finite unions of affine spaces is again just a union of affine spaces.

To establish that the third criterion in A.2.13 holds for $\mathcal{C}$, we must prove that the intersection of any collection of finite unions of affine spaces is a finite union of affine spaces. Note first that if the intersection of two affine subspaces A_1 and A_2 is a strict subspace of both A_1 and A_2, then $\dim A_1 \cap A_2$ is strictly less than $\dim A_1$ and $\dim A_2$. Let $\{F_\alpha\}_{\alpha \in I}$ be a collection of sets in $\mathcal{C}$, and let $\{\alpha_i\}_{i \in \mathbb{N}}$ be a sequence of indices. Given the sequence $\{\alpha_i\}$, create a sequence of "intersection" trees according to the following recursive definition. The tree T_0 is the tree with a base node, and an edge for each $A_{\alpha_0, j}$ in

$$F_{\alpha_0} = \bigcup_j A_{\alpha_0, j},$$

with a corresponding leaf for each $A_{\alpha_0, j}$. For each tree T_i, construct T_{i+1} from T_i as follows. Writing

$$F_{\alpha_i} = \bigcup_j A_{\alpha_i, j},$$

for each leaf F of T_i and for each j such that $F \cap A_{\alpha_i, j} \neq F$, adjoin an edge labeled by $A_{\alpha_i, j}$ to F and label the resulting new leaf by $F \cap A_{\alpha_i, j}$.

As constructed, for each k, the leaves of the tree T_k are labeled by intersections of affine spaces so that

$$\bigcap_{i=0}^{k} F_{\alpha_i}$$

is the union of the leaves of T_k. Since only a finite number of edges gets added to every leaf, for all $i \geq 0$, there can be only a finite number of vertices at a fixed distance from the base node. Furthermore, since any nontrivial intersection $A \cap B$ of affine spaces has dimension

$$\dim(A \cap B) \leq \min(\dim A, \dim B) - 1$$

and since the ambient space is $\mathbb{R}^n$, each branch (descending path) in any tree T_i can have at most $n + 1$ edges. In conclusion, for all T_i, there can only be a finite number of leaves. Thus,

$$\bigcap_{i=0}^{\infty} F_{\alpha_i}$$

is a finite union of intersections of affine spaces. Since this holds for all sequences $\{\alpha_i\}$ in I, this then proves that

$$\bigcap_{\alpha \in I} F_\alpha$$

is a finite union of intersections of affine spaces.

This rather lengthy proof shows that the collection $\mathcal{C}$ of finite unions of affine spaces in $\mathbb{R}^n$ satisfies the criteria of Proposition A.2.13. Consequently, the set τ of subsets of $\mathbb{R}^n$ that are complements of a finite union of affine spaces forms a topology on $\mathbb{R}^n$.

Example A.2.15, along with the result of Problem A.2.3, shows that given a set X, it is possible to have two topologies τ and τ' on X such that $\tau \subsetneq \tau'$, or in other words, that every τ-open set in X is τ'-open but not vice versa. This leads to a useful notion.

Definition A.2.16. Let X be a set and let τ and τ' be two topologies on X. If $\tau \subseteq \tau'$, then we say that τ' is *finer* than τ and that τ is *coarser* than τ'. If in addition $\tau \neq \tau'$, we say that τ' is *strictly finer* than τ and that τ is *strictly coarser* than τ'.

As a few examples, consider $X = \mathbb{R}^n$. Then the finite complement topology (Example A.2.14) is coarser than the topology defined in Example A.2.15, which is in turn coarser than the Euclidean topology.

In Section A.1.2, Proposition A.1.18 proved that the closure of a subset A (as defined by Definition A.1.17) of a metric space (X, D) is the intersection of all closed subsets of X containing A. This formulation of the closure of a set does not rely explicitly on the the metric D and carries over without changes to topological spaces. The closure of a subset $A \subset X$ is an example of a *topological operator*, three of which we mention here.

Definition A.2.17. Let (X, τ) be a topological space and A a subset of X. We define

1. the *closure* of A, written $\operatorname{Cl} A$, as the intersection of all closed sets in X containing A;

2. the *interior* of A, written A°, as the union of all open sets of X contained in A;

3. the *frontier* of A, written $\operatorname{Fr} A$, as the set of all $x \in X$ such that for every neighborhood U of x intersects A and A^c nontrivially, i.e., $U \cap A \neq \emptyset$ and $U \cap A^c \neq \emptyset$.

We leave some of the basic properties of the above topological operators to the exercises but present one common characterization of closed sets.

Proposition A.2.18. *Let (X, τ) be a topological space, and let A be any subset of X. The set A is closed if and only if $A = \operatorname{Cl} A$.*

Proof. If A is closed, then A is among the closed sets $F \subset X$ that contain A, and there is no smaller closed subset containing A. Hence $A = \operatorname{Cl} A$. Conversely, if $A = \operatorname{Cl} A$, then since the intersection of any collection of closed sets is closed, $\operatorname{Cl} A$ is closed, so A is closed. $\square$

Another useful characterization of closed sets relies not on a topological operator but on properties of its limit points.

Definition A.2.19. Let A be any subset of a topological space (X, τ). A *limit point* of A is any point $p \in X$ such that $U \cap (A - \{p\}) \ne \emptyset$ for every open neighborhood U of p.

In the vocabulary of sequences in a metric space, if the subset A is a sequence $\{x_n\}$, then in Definition A.1.23, we would call the limit points of A the accumulation points of A. For this reason, some authors use the alternate terminology of accumulation point for limit points of a subset A in a topological space. The discrepancy in terminology is unfortunate, but in topology, the majority of authors use the vocabulary of Definition A.2.19.

Proposition A.2.20. *Let A be a subset of a topological space X. The set A is closed if and only if it contains all of its limit points.*

Proof. Assume first that A is closed. The complement $X - A$ is open and hence is a neighborhood of each of its points. Therefore, there is no point in $X - A$ that is a limit point of A. Hence, A contains all its limit points.

Assume now that A contains all of its limit points. Let $p \in X - A$. Since p is not a limit point of A, there exists an open neighborhood U of p such that $U \cap A = \emptyset$. Thus, $X - A$ is a neighborhood of each of its points, and hence, it is open. Thus, A is closed. $\square$

Finally, we mention one last term related to closures.

Definition A.2.21. Let (X, τ) be a topological space. A subset A of X is called *dense* in X if $\operatorname{Cl} A = X$.

By Proposition A.2.20 a set A is dense in X if every point of X is a limit point of A. We give a few common examples of dense subsets.

Example A.2.22. Let $I = [a, b]$ be a closed interval in $\mathbb{R}$, and equip I with the topology induced from the Euclidean metric on $\mathbb{R}$. The open subsets in this topology on I are of the form $U \cap I$, where U is an open subset of $\mathbb{R}$. Furthermore, the open interval (a, b) is dense in I.

Proposition A.2.23. *The set $\mathbb{Q}$ of rational numbers is dense in $\mathbb{R}$.*

Proof. A precise proof of this statement must rely on a definition of real numbers, as constructed from the rationals. One may find this definition in any introductory analysis text. However, using a high school understanding of real numbers as numbers with an infinite decimal expansion that is not periodic, we can supply a simple proof of this fact.

Let $x_0 \in \mathbb{R}$ be any real number, and let U be an open neighborhood of x_0. By definition, there exists a positive real $\varepsilon > 0$ such that $(x_0 - \varepsilon, x_0 + \varepsilon) \in U$. Let $N = 1 - \lfloor \log_{10} \varepsilon \rfloor$. Consider q the fraction that represents the decimal approximation of x_0 that stops at N digits after the decimal period. Then, $q \in \mathbb{Q}$ and $q \in U$. Hence, x_0 is a limit point of $\mathbb{Q}$. $\qquad\square$

Knowing that $\mathbb{Q}$ is a countable and dense subset of $\mathbb{R}$, we can introduce a property of the Euclidean topology that is a key part of the definition of a topological manifold.

Definition A.2.24. A topological space (X, τ) is called *second countable* if there exists a basis of τ that is countable.

Example A.2.25. The Euclidean space $\mathbb{R}^n$ is second countable. Consider the collection $\mathcal{B}$ of open balls whose centers have rational coordinates and of rational radius. This collection is in bijection with $\mathbb{Q}^n \times \mathbb{Q}^{>0}$. Since the Cartesian product of countable sets is countable, $\mathbb{B}$ is countable. It is not hard to see that this collection satisfies the conditions of Proposition A.2.9, so this $\mathcal{B}$ is a countable basis of $\mathbb{R}^n$.

A.2.2 Continuity

When working with topological spaces (X, τ) and (Y, τ'), we are usually not interested in studying the properties of just any function between $f : X \to Y$ because a function without any special properties will not necessarily relate the topology on X to that on Y or vice versa. In Section A.1.4, Proposition A.1.29 provided a characterization of continuous functions between metric spaces only in terms of the open sets in the metric space topology. This motivates the definition of continuity for functions between topological spaces.

Definition A.2.26. Let (X, τ) and (Y, τ') be two topological spaces, and let $f : X \to Y$ be a function. We call f continuous (with respect to τ and τ') if for every open set $U \subset Y$, the set $f^{-1}(U)$ is open in X.

Example A.2.27. Proposition A.1.29 shows that any function called continuous between two metric spaces (X, D) and (Y, D') is continuous with respect to the topologies induced by the metrics on X and Y.

In previous contexts (e.g., functions from $\mathbb{R}^n$ to $\mathbb{R}^m$, and functions between metric spaces), we first defined the notion of continuity at a point and then expanded to continuous over a domain. In this case, as in all texts on topology, we first defined continuity over a whole domain. However, there is a natural definition for continuity at a point.

Definition A.2.28. Let $f : X \to Y$ be as in the previous definition. We call f continuous at $x \in X$, if for every neighborhood V of $f(x)$, there is a neighborhood U of x with $f(U) \subset V$.

The following proposition shows why this definition of continuity at a point is natural.

Proposition A.2.29. *Let (X, τ) and (Y, τ') be topological spaces, and let $f : X \to Y$ be a function. Then f is continuous if and only if f is continuous at all $x \in X$.*

Proof. First, suppose that f is continuous. Then for all $f(x) \in Y$ and for all open neighborhoods V of $f(x)$, $f^{-1}(V)$ is open in X. Furthermore, $x \in f^{-1}(V)$ so $f^{-1}(V)$ is an open neighborhood of x. Also, since $f(f^{-1}(V)) = V$ for any set, we see that setting $U = f^{-1}(V)$ proves one direction.

Second, assume that f is continuous at all $x \in X$. Let V be any open set in Y. If V contains no image $f(x)$, then $f^{-1}(V) = \emptyset$, which is open in X. Therefore, assume that V contains some image $f(x)$. Let $W = f^{-1}(V)$. According to the assumption, for all $x \in W$ there exists open neighborhoods U_x of x such that $f(U_x) \subset V$. Since $f(U_x) \subset V$, then $U_x \subset f^{-1}(V) = W$. But then

$$W \subset \bigcup_{x \in W} U_x \subset W^\circ,$$

and since $W^\circ \subset W$ always, we conclude that $W = W^\circ$, which implies (see Problem A.2.7) that W is open. This shows that f is continuous. $\square$

Proposition A.2.30. *Let (X, τ), (Y, τ'), and (Z, τ'') be three topological spaces. If $f : X \to Y$ and $g : Y \to Z$ are continuous functions, then $g \circ f : X \to Z$ is also a continuous function.*

Proof. (Left as an exercise for the reader.) $\square$

For metric spaces, a continuous function is one that preserves "nearness" of points. Though topological spaces are not necessarily metric spaces, a continuous function $f : X \to Y$ between topological spaces preserves nearness in the sense that if two images $f(x_1)$ and $f(x_2)$ are in the same open set V, then there is an open set U that contains both x_1 and x_2, with $f(U) \subset V$.

In set theory, we view two sets as "the same" if there exists a bijection between them: they are identical except for how we label the specific elements. For topological spaces to be considered "the same," not only do the underlying sets need to be in bijection, but this bijection must preserve the topology. This is the concept of a homeomorphism.

Definition A.2.31. Let (X, τ) and (Y, τ') be two topological spaces, and let $f : X \to Y$ be a function. The function f is called a *homeomorphism* if

1. f is a bijection;

2. $f : X \to Y$ is continuous;

3. $f^{-1} : Y \to X$ is continuous.

If there exists a homeomorphism between two topological spaces, we call them *homeomorphic*.

Example A.2.32. Any two squares S_1 and S_2, as subsets of $\mathbb{R}^2$ with the Euclidean metric, are homeomorphic. For $i = 1, 2$, let t_i be the translation that brings the center of S_i to the origin, R_i a rotation that makes the edges of $t_i(S_i)$ parallel to the x- and y-axes, and let h_i be a scaling (homothetie) that changes $R_i \circ t_i(S_i)$ into the square with vertices $\{(-1, -1), (-1, 1), (1, 1), (1, -1)\}$. It is easy to see that translations are homeomorphisms of $\mathbb{R}^2$ to itself. Furthermore, by Problem A.1.28, we see that R_i and h_i are continuous, and since they are invertible linear transformations, their inverses are continuous as well. Thus, R_i and h_i are homeomorphisms. Thus, the function

$$f = t_2^{-1} \circ R_2^{-1} \circ h_2^{-1} \circ h_1 \circ R_1 \circ t_1$$

is a homeomorphism of $\mathbb{R}^2$ into itself that sends S_1 to S_2. Thus, S_1 and S_2 are homeomorphic.

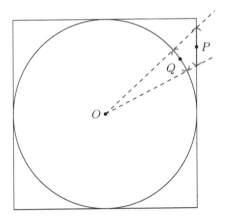

Figure A.6: A homeomorphism between a circle and a square from Example A.2.33.

Example A.2.33. Any circle and any square, as subsets of $\mathbb{R}^2$, are homeomorphic. We must present a homeomorphism between a circle and a square. By Example A.2.32 we see that any two squares are homeomorphic. By a similar argument, any two circles are homeomorphic. So without loss of generality, consider the consider the unit circle S and the unit square T, both with center $(0, 0)$. (See Figure A.6.)

Let $f : T \to S$ be the function defined by $f(x) = x/\|x\|$, where x is viewed as a point in $\mathbb{R}^2$. It is clear that this is a continuous function on its domain, $\mathbb{R}^2 - \{(0,0)\}$ so it is continuous on the subset T. For $\theta \in \mathbb{R}$, consider the function $g(\theta) = \max(|\sin\theta|, |\cos\theta|)$ and then the curve $\gamma : [0, 2\pi] \to \mathbb{R}^2$ parametrized by $\gamma(t) = (\cos(t)/g(t), \sin(t)/g(t))$. It is not hard to verify that this curve traces out the unit square T. Identifying a point on the circle with its angle, we can think of γ as a function $S \to T$. It is not hard to see that g is continuous and that γ is as well. Furthermore, $(f \circ \gamma)(p) = p$ for all $p \in S$ and also $(\gamma \circ f)(q) = q$ for all $q \in T$. Hence, f is a homeomorphism between the unit square and the unit circle.

Figure A.6 shows corresponding points P and Q as well as corresponding neighborhoods of these points.

Example A.2.34. The above two examples only begin to illustrate how different homeomorphic spaces may look. In this example, we prove that any closed, simple, parametrized curve γ in $\mathbb{R}^n$ is homeomorphic to the unit circle $\mathbb{S}^1$ (in $\mathbb{R}^2$). Let $\vec{x} : [0, l] \to \mathbb{R}^n$ be a parametrization by arclength for the curve γ such that $\vec{x}(t_1) = \vec{x}(t_2)$ implies that $t_1 = 0$ and $t_2 = l$, assuming $t_1 < t_2$. The function $f : \gamma \to \mathbb{S}^1$ defined by

$$f(P) = \left(\cos\left(\frac{2\pi}{l}\vec{x}^{-1}(P)\right), \sin\left(\frac{2\pi}{l}\vec{x}^{-1}(P)\right) \right) \tag{A.2}$$

produces the appropriate homeomorphism. Note that $\vec{x}^{-1}$ is not well defined only at the point $\vec{x}(0)$ because $\vec{x}^{-1}(\vec{x}(0)) = \{0, l\}$. However, using either 0 or l in Equation (A.2) is irrelevant.

Example A.2.35. In contrast to the previous example, consider the closed, regular curve γ parametrized by $\vec{x} : [0, 2\pi] \to \mathbb{R}^2$ with

$$\vec{x}(t) = (\cos t, \sin 2t).$$

The curve γ traces out a figure eight of sorts and is not simple because $\vec{x}(\pi/2) = (0, 0) = \vec{x}(3\pi/2)$ (and $\vec{x}'(\pi/2) \neq \vec{x}'(3\pi/2)$). We show that it is not homeomorphic to a circle $\mathbb{S}^1$. Call $P = (0, 0)$.

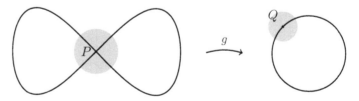

Figure A.7: Figure eight not homeomorphic to a circle.

There does exist a surjective continuous map f of γ onto the circle $\mathbb{S}^1$, namely, using the parametrization $\vec{x}(t)$,

$$f(\vec{x}(t)) = (\cos 2t, \sin 2t)$$

which amounts to folding the figure eight back onto itself so that the circle is covered twice. However, there exists no continuous bijection $g : \gamma \to \mathbb{S}^1$. To see this, call $Q = g(P)$, and let U be a small open neighborhood of Q. If g is a bijection, it has exactly one preimage for every element $x \in U$. Thus, $g^{-1}(U)$ is the image of a nonintersecting parametrized curve. However, every open neighborhood of P includes two segments of curves. Thus, the circle and γ are not homeomorphic.

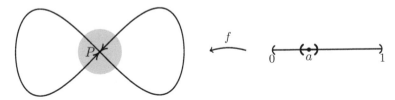

Figure A.8: Figure eight not homeomorphic to a segment.

Example A.2.36. Consider the same regular, closed parametrized curve as in the previous example. The function $f : (0,1) \to \gamma_2$ defined by $f(t) = \vec{x}\left(2\pi t - \frac{\pi}{2}\right)$ is a bijection. Furthermore, f is continuous since it is continuous as a function $(0,1) \to \mathbb{R}^2$. However, the inverse function is not continuous and in fact no continuous bijection $g : (0,1) \to \gamma_2$ can have a continuous inverse.

Let $a \in (0,1)$ be the real number such that $f(a) = P$, the point of self-intersection on γ_2. Take an open segment U around a. The image $f(U)$ is a portion of γ_2 through P (see the heavy lines in Figure A.8). However, this portion $f(U)$ is not an open subset of γ_2 since every open neighborhood of P includes $f((0,\varepsilon_1))$ and $f((1 - \varepsilon_2, 1))$. Thus there exists no homeomorphism between the open interval and the figure eight.

We conclude this section on continuity by mentioning one particular result, the proof of which exceeds the scope of this appendix.

Theorem A.2.37. *The Euclidean spaces $\mathbb{R}^n$ and $\mathbb{R}^m$ are homeomorphic if and only if $m = n$.*

This theorem states that Euclidean spaces can only be homeomorphic if they are of the same dimension. This might seem obvious to the casual reader but this fact hides a number of subtleties. First of all, the notion of dimension of a set in topology is not at all a simple one. Secondly, we must be careful to consider space-filling curves, such as the Peano curve, which is a continuous surjection of the closed interval $[0,1]$ onto the closed unit square $[0,1] \times [0,1]$. (See [28], Section 3-3, for a construction.) The construction for space-filling curves can be generalized to find continuous surjections of $\mathbb{R}^n$ onto $\mathbb{R}^m$ even if $n < m$. However, Theorem A.2.37 implies that no space-filling curve is bijective and has a continuous inverse. That Euclidean spaces of different dimensions are not homeomorphic means that they are distinct from the perspective of topology.

A.2.3   Derived Topological Spaces

Given any topological space (X, τ), there are a number of ways to create a new topological space. We present two common ways – subset topology and quotient spaces – which are used throughout this book.

Definition A.2.38. Let (X, τ) be a topological space, and let S be any subset of X. We define the *subset topology* τ' on S by calling a subset $A \subset S$ open if and only if there exists an open subset U of X such that $A = S \cap U$.

The subset topology is sometimes called the topology induced on S from X. There is an alternate way to characterize it.

Proposition A.2.39. *Let (X, τ) be a topological space, and let $S \subset X$. Let $i : S \to X$ be the inclusion function. The subset topology on S is the coarsest topology such that i is a continuous function.*

Proof. Given any subset A of X, we have $i^{-1}(A) = A \cap S$. If i is continuous, then for all open subsets $U \subset X$, the set $i^{-1}(U) = U \cap S$ is open. However, according to Definition A.2.38, the subset topology on S has no other open subsets and therefore is coarser than any other topology on S, making i continuous. □

Example A.2.40. Consider $\mathbb{R}$ equipped with the usual topology. Let $S = [a, b]$ be a closed interval. If $a < c < b$, in the subset topology on S, the interval $[a, c)$ is open. To see this, take any real $d < a$. Then

$$[a, c) = (d, c) \cap [a, b],$$

and (d, c) is open in $\mathbb{R}$.

A second and often rather useful way to create new topological spaces is to induce a topology on a quotient set.

Definition A.2.41. Let X be a set, and let R be an equivalence relation on X. The set of equivalence classes of R is denoted by X/R and is called the *quotient set* of X by R.

The concept of a quotient set arises in many areas of mathematics (congruence classes in number theory, quotient groups in group theory, quotient rings in ring theory, etc.) but also serves as a convenient way to define interesting objects in topology and geometry. We provide a few such examples before discussing topologies on quotient sets.

Example A.2.42 (Real Projective Space). The typical construction of the real projective space is given in Example 3.1.6. We present two other equivalent constructions.

Let X be the set of all lines in $\mathbb{R}^{n+1}$, and let R be the equivalence relation of parallelism on X. We can therefore discuss the set of equivalence classes X/R. Each line in X is uniquely parallel to a line that passes through the origin, and hence, X/R may be equated with the set of lines passing through the origin. This set is called the *real projective space* of dimension n and is usually denoted by $\mathbb{RP}^n$.

As an alternate characterization for $\mathbb{RP}^n$, consider the unit n-sphere $\mathbb{S}^n$ in $\mathbb{R}^{n+1}$ and centered at the origin. Each line through the origin intersects $\mathbb{S}^n$ at two antipodal points. Therefore, $\mathbb{RP}^n$ is the quotient set of $\mathbb{S}^n$ with respect to the equivalence relation, in which two points are called equivalent if they are antipodal (form the ends of a diameter through $\mathbb{S}^n$).

Example A.2.43 (Grassmannian). Let X_r be the set of all r dimensional vector subspaces in $\mathbb{R}^n$, and let R be the equivalence relation of parallelism between hyperplanes. The set of equivalence classes is called a Grassmannian and is denoted $\mathbb{G}(r,n)$. Again, since each r-dimensional hyperplane is uniquely parallel to one hyperplane through the origin, $\mathbb{G}(r,n)$ is the set of r-hyperplanes through the origin.

Of particular interest to geometry is the question of how to give a topology to X/R if X is equipped with a topology. Proposition A.2.39 illustrates how to make a reasonable definition.

Definition A.2.44. Let (X,τ) be a topological space and let R be an equivalence relation on X. Define $f : X \to X/R$ as the function that sends an element in X to its equivalence class; f is called the *quotient map* (or sometimes *identification map*). We call *quotient topology* (or *identification topology*) on X/R the finest topology that makes f continuous.

The above definition for the quotient topology does not make it too clear what the open sets of X/R should be. The following proposition provides a different characterization.

Proposition A.2.45. *Let (X,τ) be a topological space, let R be an equivalence relation on X and let f be the quotient map. The quotient topology on X/R is*

$$\tau' = \{U \in \mathcal{P}(X/R) \,|\, f^{-1}(U) \in \tau\}.$$

Proof. Let τ' be a topology on X/R such that $f : X \to X/R$ is continuous. Then for all open sets $U \in \tau'$, we must have $f^{-1}(U)$ be open in X. Note first that $f^{-1}(\emptyset) = \emptyset$ and $f^{-1}(X/R) = X$, which are both open in X.

From basic set theory, for any function $F : A \to B$ and any collection $\mathcal{C}$ of subsets of B, we have

$$F^{-1}\left(\bigcup_{S \in \mathcal{C}} S\right) = \bigcup_{S \in \mathcal{C}} F^{-1}(S) \qquad \text{and} \qquad F^{-1}\left(\bigcap_{S \in \mathcal{C}} S\right) = \bigcap_{S \in \mathcal{C}} F^{-1}(S).$$

(See Section 1.3 in [48].) Therefore, if U_1 and U_2 are such that $f^{-1}(U_1) \cap f^{-1}(U_2)$ is open, then $f^{-1}(U_1 \cap U_2)$ is open. Also, for any collection of sets $\{U_\alpha\}_{\alpha \in I}$ in X/R,

$$f^{-1}\left(\bigcup_{\alpha \in I} U_\alpha\right) = \bigcup_{\alpha \in I} f^{-1}(U_\alpha)$$

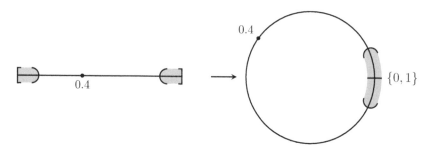

Figure A.9: Open set around $\{0, 1\}$.

so if the right-hand side is open, then so is the left-hand side. Consequently, the collection of subsets $\mathcal{B}$ of X/R such that $U \in \mathcal{B}$ if and only if $f^{-1}(U) \in \tau$ is a topology on X/R. However, any finer topology on X/R would include some subset $S \subset X/R$ such that $f^{-1}(S)$ would not be open in X and, hence, would make f not continuous. $\qquad\qquad\qquad\qquad\qquad\qquad\qquad\qquad\qquad\qquad\qquad\qquad\qquad\qquad$ $\square$

Example A.2.46 (Circles). Let I be the interval $[0, 1]$ equipped with the topology induced from $\mathbb{R}$. Consider the equivalence relation that identifies 0 with 1, and everything else inequivalent. The identification space I/R is homeomorphic to a circle. We may use the function $f : I/R \to \mathbb{S}^1$ defined by

$$f(t) = (\cos 2\pi t, \sin 2\pi t),$$

which is well-defined since $f(0) = f(1)$, so whether we take 0 or 1 for the equivalence class $\{0, 1\}$, we obtain the same image. This function f is clearly bijective.

To prove that f is continuous, let $P \in \mathbb{S}^1$, and let $x \in I/R$. If $P \neq (1, 0)$, then any open neighborhood U of P contains an open interval of angles $\theta_1 < \theta < \theta_2$, where $0 < \theta_1 < \theta_2 < 2\pi$ and the angle θ_0 corresponding to P satisfies $\theta_1 < \theta_0 < \theta_2$. Then

$$f\left(\frac{\theta_1}{2\pi}, \frac{\theta_2}{2\pi}\right)$$

contains P and is a subset of U. On the other hand, if $P = (1, 0)$, any open neighborhood U' of P in $\mathbb{S}^1$ contains an open arc of angles $\theta_1 < \theta < \theta_2$, with $\theta_1 < 0 < \theta_2$. Then if $g : I \to I/R$ is the quotient map,

$$f \circ g\left(\left[0, \frac{\theta_2}{2\pi}\right) \cup \left(1 - \frac{\theta_1}{2\pi}, 1\right]\right)$$

contains P and is contained in U'. Furthermore, by definition, $g([0, \theta_2/2\pi) \cup (1 - \theta_1/2\pi, 1])$ is open in I/R since $[0, \theta_2/2\pi) \cup (1 - \theta_1/2\pi, 1]$ is open in the subset topology of $[0, 1]$. By Proposition A.2.29, f is continuous.

Using a similar argument, it is not hard to show that f^{-1} is continuous, concluding that I/R is homeomorphic to $\mathbb{S}^1$.

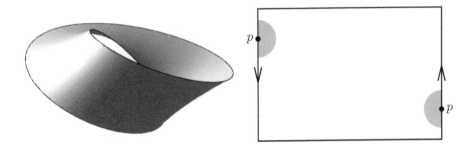

Figure A.10: Möbius strip.

Example A.2.47. Consider the real projective space $\mathbb{RP}^n$ given as the quotient space $\mathbb{S}^n / \sim$ where $\sim$ is the equivalence relation on $\mathbb{S}^n$ where $p_1 \sim p_2$ if and only if they are antipodal to each other, i.e., form a diameter of the sphere. The unit sphere $\mathbb{S}^n$ naturally inherits the subspace topology from $\mathbb{R}^{n+1}$. Definition A.2.44 provides the induced topology for $\mathbb{RP}^n$.

Example A.2.48 (Möbius Strip). Let $I = [0,1] \times [0,1]$ be the unit square with the topology induced from $\mathbb{R}^n$. Define the identification (equivalence relation $\sim$) between points by $(0, y) \sim (1, 1 - y)$, for all $y \in [0,1]$ and no other points are equivalent to any others. The topological space obtained is called the *Möbius strip*. In $\mathbb{R}^3$, the Möbius strip can be viewed as a strip of paper twisted once and with ends glued together. Figure A.10 shows an embedding of the Möbius strip in $\mathbb{R}^3$, as well as a diagrammatic representation of the Möbius strip. In the diagrammatic representation, the arrows indicate that the opposite edges are identified but in inverse direction. The shaded area shows a disk around a point p on the identified edge.

A.2.4   Compactness

In any calculus course, we encounter the Extreme Value Theorem, a result in analysis that forms an essential ingredient to Rolle's Theorem, hence the Mean Value Theorem, and therefore the Fundamental Theorem of Calculus.

Theorem A.2.49 (Extreme Value Theorem). *Let $f : [a, b] \to \mathbb{R}$ be a continuous real-valued function. Then f attains a maximum and a minimum over the interval $[a, b]$.*

As topology developed, a variety of attempts were made to generalize the idea contained in this fact. In the context of metric spaces, the key properties that would allow for a generalization are that the interval $[a, b]$ is closed and bounded. Though closed is a concept in topology, the notion of being bounded does not have an equivalent notion. It turns out that in topological spaces, the notion of *compactness* provides this desired generalization.

Definition A.2.50. Let (X, τ) be a topological space. Let $\mathcal{U} = \{U_i\}_{i \in I}$ be a collection of open sets in X. We call $\mathcal{U}$ an *open cover* of X if

$$X = \bigcup_{i \in I} U_i.$$

If $J \subset I$, the collection $\mathcal{V} = \{V_j\}_{j \in J}$ is called a *subcover* of $\mathcal{U}$ if $\mathcal{V}$ is itself an open cover of X.

Definition A.2.51. A topological space (X, τ) is called *compact* if every open cover of X has a finite subcover.

Remark A.2.52. A subset A of X is called compact if it is compact when equipped with the subspace topology induced from (X, τ). This is equivalent to the property that whenever there exists a collection of $\mathcal{U} = \{U_i\}_{i \in I}$ of open sets with $A \subset \bigcup_{i \in I} U_i$, there exists a finite subset $J \subset I$ with $A \subset \bigcup_{i \in J} U_j$.

Some examples of compact spaces are obvious, such as that any finite subset of a topological space is compact. However, the following theorem justifies, at least in part, the given definition of compactness.

Theorem A.2.53 (Heine-Borel). *A closed and bounded interval $[a, b]$ of $\mathbb{R}$ is compact.*

Proof. (A variety of proofs exist for the Heine-Borel Theorem. The following proof is called the "creeping along" proof.)
Let $\mathcal{U} = \{U_\alpha\}_{\alpha \in I}$ be an open cover of $[a, b]$. Define the subset of $[a, b]$ by

$$E = \{x \in [a, b] \mid [a, x] \text{ is contained in a finite subfamily of } \mathcal{U}\}.$$

Obviously, E is an interval, but a priori we do not know whether it is open or closed or even nonempty. We will show that $b \in E$ to establish the theorem.
Now $a \in E$ because there exists some U_{α_0} such that $a \in U_{\alpha_0}$. Let $c = \operatorname{lub} E$. Clearly $a \le c \le b$. Suppose that $c < b$. Since $\mathcal{U}$ covers $[a, b]$, there exists some index β such that $c \in U_\beta$, and therefore there exists ε such that $(c - \varepsilon, c + \varepsilon) \subset U_\beta$. Since c is the least upper bound of E, any $x \in (c - \varepsilon, c)$ is in E. Therefore, there exists a finite set $\{\alpha_1, \alpha_2, \ldots, \alpha_n\}$ such that

$$[a, x] \subset \bigcup_{i=1}^{n} U_{\alpha_i}.$$

But then the finite union of open sets

$$(c - \varepsilon, c + \varepsilon) \cup \bigcup_{i=1}^{n} U_{\alpha_i}$$

contains the point $c + \varepsilon/2$, contradicting the assumption that $c < b$ and $c = \operatorname{lub} E$. Thus $c = b$, and $b \in E$. $\qquad\square$

Compactness is an important property of topological spaces, and we refer the reader to Chapter 3 in [2], Section 6.6 in [15], or Chapter 7 in [27] for complete treatments. For the purposes of this book however, we are primarily interested in two results, namely Theorem A.2.57 and Theorem A.2.58, and a few necessary propositions to establish them. We will not prove Theorem A.2.57 completely but again refer the reader to the above sources.

Proposition A.2.54. *Let (X, τ) be a topological space, and let K be a compact subset of X. Every closed subset of K is compact.*

Proof. Let $F \subset K$ be a closed set. Then $X - F$ is open. If $\mathcal{U}$ is an open cover of F, then $\mathcal{U} \cup \{X - F\}$ is an open cover of K. Since K is compact, $\mathcal{U} \cup \{X - F\}$ must admit a finite subcover of K. This finite subcover of K is of the form $\mathcal{U}' \cup \{X - F\}$, where $\mathcal{U}'$ is a finite subcover of $\mathcal{U}$ of F. Thus, F is compact. $\square$

Problem A.1.19 established the fact that two distinct points p_1 and p_2 in a metric space possess, respectively, open neighborhoods U_1 and U_2 such that $U_1 \cap U_2 = \emptyset$. This type of property is called a *separation* property of a topological space because it gives some qualification for how much we can distinguish points in the topological space. There exists a variety of separation axioms, but we only present the one that is relevant for differential geometry.

Definition A.2.55. A topological space (X, τ) is called *Hausdorff* if given any two points p_1 and p_2 in X, there exist open neighborhoods U_1 of p_1 and U_2 of p_2 such that $U_1 \cap U_2 = \emptyset$.

Proposition A.2.56. *If (X, τ) is a Hausdorff topological space, then every compact subset K of X is closed.*

Proof. Let K be compact. Since X is Hausdorff, for every $x \in X - K$ and $y \in K$, there exist open sets U_{xy} and V_{xy}, with $x \in U_{xy}$ and $y \in V_{xy}$, such that $U_{xy} \cap V_{xy} = \emptyset$. For each $x \in X - K$,

$$\{V_{xy} \mid y \in K\}$$

is an open cover of K, so it must possess a finite subcover that we index with a finite number of points $y_1, y_2, \ldots, y_n$. But then for each x,

$$U_x = \bigcap_{i=1}^{n} U_{xy_i}$$

is open since it is a finite intersection of open sets. Since $K \subset \bigcup_{i=1}^{n} V_{xy_i}$, we conclude that $K \cap U_x = \emptyset$ for all $x \in X - K$. Thus, $X - K$ is a neighborhood of all of its points and hence it is open. Thus, K is closed. $\square$

Theorem A.2.57. *Let A be any subset of $\mathbb{R}^n$ (equipped with the Euclidean topology). The set A is compact if and only if it is closed and bounded.*

Proof. (We only prove $\Longrightarrow$.)

Suppose that A is compact. Since by Problem A.1.19 any metric space is Hausdorff, Proposition A.2.56 allows us to conclude that A is closed. Since every open set in a metric space is the union of open balls, any open cover of A can be viewed as an open cover of open balls. If A is compact, it is contained in only a finite number of such open balls $\{B_{r_i}(p_i)\}_{i=1}^m$. There exists an open ball $B_r(p)$ that contains all the $B_{r_i}(p_i)$. The radius r will be less than $(m-1)\max\{d(p_i, p_j)\} + 2\max\{r_i\}$. Then $K \subset B_r(P)$, and hence, K is bounded.

(The proof of the converse is more difficult and uses other techniques that we do not have the time to develop here.) $\qquad\square$

Theorem A.2.57 establishes that we might view closed and bounded subsets of $\mathbb{R}^n$ as the topological analog to $[a, b] \subset \mathbb{R}$ in Theorem A.2.49. We now complete the generalization to topological spaces.

Theorem A.2.58. *Let $f : X \to Y$ be a continuous function between topological spaces X and Y. If X is a compact space, then $f(X)$ is compact in Y.*

Proof. Let $\mathcal{U}$ be an open cover of $f(X)$. Since f is continuous, each $f^{-1}(U)$ is open and the collection $\{f^{-1}(U) \,|\, U \in \mathcal{U}\}$ is an open cover of X. Since X is compact, there exists a finite set $\{U_1, U_2, \ldots, U_n\} \subset \mathcal{U}$ such that

$$X = \bigcup_{i=1}^n f^{-1}(U_i).$$

Since for any functions $f(f^{-1}(A)) \subset A$ always and $f(\bigcup_\lambda A_\lambda) = \bigcup_\lambda f(A_\lambda)$, then

$$f(X) = f\left(\bigcup_{i=1}^n f^{-1}(U_i)\right) = \bigcup_{i=1}^n f(f^{-1}(U_i)) \subset \bigcup_{i=1}^n U_i.$$

Thus, $f(X)$ is compact. $\qquad\square$

Corollary A.2.59. *Let X be a compact topological space, and let $f : X \to \mathbb{R}$ be a real-valued function from X. Then f attains both a maximum and a minimum.*

Proof. By Theorem A.2.58, the image $f(X)$ is compact. By Theorem A.2.57, $f(X)$ is a closed and bounded subset of $\mathbb{R}$. Hence, lub $\{f(x) \,|\, x \in X\}$ and glb $\{f(x) \,|\, x \in X\}$ are both elements of $f(X)$ and hence are the maximum and minimum of f over X. $\qquad\square$

Corollary A.2.60. *Let (X, τ) and (Y, τ') be topological spaces, and let $f : X \to Y$ be a continuous function onto Y. If X is compact, then so is Y.*

Corollary A.2.61. *Let X be a compact topological space. If a space Y is homeomorphic to X, then Y is compact.*

A.2.5 Connectedness

We end this overview of point set topology with a discussion of connectedness. The concept of connectedness is rather natural. We simply mean that we cannot subdivide the topological space into two "parts, " i.e., nonempty disjoint open subsets.

Definition A.2.62. A nonempty topological space (X, τ) is called *connected* if whenever $X = U \cup V$, where U and V are open and disjoint, then either $U = \emptyset$ or $V = \emptyset$.

As we will see, when proving results concerning connectedness, it is often useful to assume the set is not connected and prove a contradiction. Consequently, we present a definition and terminology for the negation of connectedness.

Definition A.2.63. Let (X, τ) be a nonempty topological space. A *separation* of X is a pair (U, V) of nonempty open subsets such that $U \cap V = \emptyset$ and $U \cup V = X$. If (X, τ) has a separation then we say it is *disconnected*.

Proposition A.2.64. *Let (X, τ) be a topological space. A subset Y if connected if and only if there does not exist a pair of open subsets $U, V \in \tau$ such that $U \cap V \cap Y = \emptyset$ and $Y \subset U \cup V$.*

Proof. The subspace Y is connected if and only if there exists a pair (U', V') of open subsets in (the subspace topology of) Y such that $U' \cup V' = Y$ and $U' \cap V' = \emptyset$. By definition of the subspace topology, $U' = Y \cap U$ and $V' = Y \cap V$ for sets $U, V \in \tau$. Since

$$Y = U' \cup V' = (Y \cap U) \cup (Y \cap V) = Y \cap (U \cup V),$$

then $Y \subset U \cup V$. Similarly, since $U' \cap V' = \emptyset$, then $U \cap V \cap Y = \emptyset$. $\square$

Proposition A.2.65. *Let X be a topological space and let Y_1 and Y_2 be two connected subsets such that $Y_1 \cap Y_2 \neq \emptyset$. Then $Y_1 \cup Y_2$ is a connected subspace.*

Proof. Assume the contrary, that $Y_1 \cup Y_2$ is disconnected. Let (U, V) be a separation of $Y_1 \cup Y_2$. Then either $Y_1 \subset U$ or $Y_1 \subset V$; otherwise $(U \cap Y_1, v \cap Y_1)$ forms a separation of Y_1, which is a contradiction since Y_1 is connected. Similarly, $Y_2 \subset U$ or $Y_2 \subset V$. If $Y_1 \subset U$ and $Y_2 \subset V$ or if $Y_1 \subset V$ and $Y_2 \subset U$, then $Y_1 \cap Y_2 = \emptyset$, a contradiction. If $Y_1 \subset U$ and $Y_2 \subset U$ or if $Y_1 \subset V$ and $Y_2 \subset V$, then either $V = \emptyset$ or $U = \emptyset$, both contradictions. Hence, $Y_1 \cup Y_2$ is not disconnected. $\square$

The following proposition gives one of the key examples of connected topological spaces.

Theorem A.2.66. *Any interval I of $\mathbb{R}$ with the subspace topology is connected.*

Proof. Assume that there exists a separation (U, V) of I. Let $a \in U$ and $b \in V$, and without loss of generality suppose that $a < b$. By definition of an interval, we have $[a, b] \subset I$. Calling $A = U \cap [a, b]$ and $B = V \cap [a, b]$, we see that $[a, b] = A \cup B$ and that $A \cap B = \emptyset$.

Let $c = \sup A$. We show that c is in neither A nor B, which leads to a contradiction since, by construction, $c \leq b$ and $c \geq a$, and hence $c \in [a, b]$.

Assume that $c \in A$. Then $c \neq b$ so $c < b$. Since A is open in $[a, b]$, there exists an interval $[c, c+\varepsilon)$, with $c+\varepsilon < b$, contained in A. Then $c+\varepsilon/2 \in A$, contradicting the hypothesis that $c = \sup A$.

Assume that $c \in B$. Then $c \neq a$, so either $c = b$ or $c \in (a, b)$, the open interval. Since B is open, there is some interval $(c - \varepsilon, c]$ in B. If $c = b$, then $c - \varepsilon/2$ is greater than any element in A, contradicting $c = \sup A$. If $c < b$, then $(c.b] \subset B$ and thus $[c, b] \subset B$. Since B is open in I, then $(c - \varepsilon, b] \subset B$ so again $c - \varepsilon/2$ is greater than any element in A, contradicting $c = \sup A$.

The theorem holds by contradiction. $\qquad\square$

Definition A.2.67. Let (X, τ) be a topological space. A connected open subset U is called a *connected component* of X if $U = X$ or if $(U, X - U)$ is a separation of X.

Clearly, every topological space is a union of its connected components.

There is another common way to formulate a precise definition for the intuitive notion of being connected, namely, deciding where it is possible to "get there from here."

Definition A.2.68. A topological space (X, τ) is called *path-connected* if for any two points $p, q \in X$, there exists a continuous map $\gamma : [0, 1] \to X$ such that $\gamma(0) = p$ and $\gamma(1) = q$.

Proposition A.2.69. *If a topological space is path-connected, then it is connected.*

Proof. Let (X, τ) be path-connected and assume that X has a separation (U, V). Let $p, q \in X$ with $p \in U$ and $q \in V$, and let $\gamma : [0, 1] \to X$ be a continuous curve connecting p and q. Then

$$\gamma^{-1}(U) \cap \gamma^{-1}(V) = \gamma^{-1}(U \cap V) = \gamma^{-1}(\emptyset) = \emptyset.$$

On the other hand,

$$\gamma^{-1}(U) \cup \gamma^{-1}(V) = \gamma^{-1}(U \cup V) = \gamma^{-1})(X) = [0, 1].$$

By Proposition A.2.66, $[0, 1]$ is connected so this gives a contradiction because $\gamma^{-1}(U)$ and $\gamma^{-1}(V)$ would form a separation of $[0, 1]$. We deduce that X is connected. $\qquad\square$

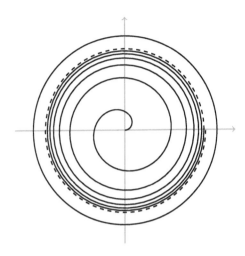

Figure A.11: Connected but not path-connected.

This proposition allows us to quickly conclude that $\mathbb{R}^n$ and open balls $B_r(p)$ in $\mathbb{R}^n$ are connected topological spaces.

Though path-connected implies connected, the reverse is not true. This means that connected and path-connected are not equivalent concepts. The following example describes a subset of $\mathbb{R}^2$ that is connected but not path-connected.

Example A.2.70. Let X be the subset of $\mathbb{R}^2$ that is the union of the unit circle and the image of the curve $\gamma : [1, \infty) \to \mathbb{R}^2$ with

$$\gamma(t) = \left(\left(1 - \frac{1}{t}\right) \cos(2\pi t), \left(1 - \frac{1}{t}\right) \cos(2\pi t) \right).$$

See Figure A.11. The subspace X is not path-connected since there is no path connecting a point $p \in \gamma([1, \infty))$ to a point on the unit circle: any path from p staying on $\gamma([1, \infty))$ moving toward the unit circle has infinite length, never reaching the unit circle.

On the other hand, we claim that X is connected. The set $\gamma([1, \infty))$ is clearly path-connected and hence, by Proposition A.2.69, connected, as is the unit circle. Assume there is a separation (U, V) of X, then $\gamma([1, \infty))$ is in either U or V. Without loss of generality, suppose that $\gamma([1, \infty)) \subset U$. Then the unit circle is in V. However, any open neighborhood of any point p on the unit circle intersects $\gamma([1, \infty))$. Hence, $U \cap V \neq \emptyset$. This contradicts the assumption that X has a separation and we conclude that X is connected.

PROBLEMS

A.2.1. Prove Proposition A.2.13.

A.2.2. Find all the topologies on the set $\{a, b, c\}$. How many different topologies exist on a set of four elements?

A.2.3. Consider the topology τ constructed in Example A.2.15. Prove that every open set in τ is also open in the topology τ' induced from the Euclidean metric. Give an example of an open set in τ' that is not τ-open. (When these two facts hold, one says that τ' is a strictly finer topology than τ.)

A.2.4. Let τ be the set of all subsets of $\mathbb{R}$ that are unions of intervals of the form $[a, b)$. Prove that τ is a topology on $\mathbb{R}$. Is τ the same topology as that induced by the absolute value (Euclidean) metric?

A.2.5. Prove that in a topological space (X, τ), a set A is open if and only if it is equal to its interior.

A.2.6. Prove Proposition A.2.9.

A.2.7. Let (X, τ) be a topological space, and let A and B be any subsets of X. Prove the following:

(a) $(A^\circ)^\circ = A^\circ$.

(b) $A^\circ \subset A$.

(c) $(A \cap B)^\circ = A^\circ \cap B^\circ$.

(d) A subset $U \subset X$ is open if and only $U = U^\circ$.

A.2.8. Find an example that shows that $(A \cup B)^\circ$ is not necessarily equal to $A^\circ \cup B^\circ$.

A.2.9. Let (X, τ) be a topological space, and let A and B be any subsets of X. Prove the following:

(a) $\mathrm{Cl}(\mathrm{Cl}\, A) = \mathrm{Cl}\, A$.

(b) $A \subset \mathrm{Cl}\, A$.

(c) $\mathrm{Cl}(A \cup B) = \mathrm{Cl}\, A \cup \mathrm{Cl}\, B$.

(d) A subset $F \subset X$ is closed if and only $\mathrm{Cl}\, F = F$.

A.2.10. Find an example that shows that $\mathrm{Cl}(A \cap B)$ is not necessarily equal to $\mathrm{Cl}\, A \cap \mathrm{Cl}\, B$.

A.2.11. Let (X, τ) be a topological space, and let A be any subset of X. Prove the following:

(a) $\mathrm{Cl}\, A = A^\circ \cup \mathrm{Fr}\, A$.

(b) $\mathrm{Cl}\, A - \mathrm{Fr}\, A = A^\circ$.

(c) $\mathrm{Fr}\, A = \mathrm{Fr}(X - A)$.

A.2.12. Show that every open subset of $\mathbb{R}$ is the union of disjoint open intervals.

A.2.13. Let (X, D) be any metric space, and let A be any subset of X. Consider the function $f : X \to \mathbb{R}^{\geq 0}$ defined by $D(x, A)$, where $\mathbb{R}^{\geq 0}$ is equipped with the usual topology. (Note that $[0, a)$ is open in this topology on $\mathbb{R}^{\geq 0}$.) Prove that f is continuous. (The set $f^{-1}([0, r))$ is sometimes called the open r-envelope of A.)

A.2.14. Let (X, D) be any metric space, and let A and B be closed subsets of X. Use the previous exercise to construct a continuous function $g : X \to \mathbb{R}$ such that $g(a) = 1$ for all $a \in A$ and $g(b) = -1$ for all $b \in B$.

A.2.15. Let $\mathbb{S}^2$ be the two-dimensional unit sphere in $\mathbb{R}^3$. Give $\mathbb{S}^2$ the topology of a metric induced on $\mathbb{S}^2$ as a subset of $\mathbb{R}^3$. Suppose we locate points on $\mathbb{S}^2$ using (θ, ϕ) in spherical coordinates. Let $f_\alpha : \mathbb{S}^2 \to \mathbb{S}^2$ be the rotation function such that

$$f(\theta, \phi) = (\theta + \alpha, \phi).$$

Prove that f is continuous.

A.2.16. Let X be a topological space, and let $f : X \to \mathbb{R}$ be a continuous function. Prove that the set of zeroes of f, namely $\{x \in X \mid f(x) = 0\}$, is closed.

A.2.17. Prove Proposition A.2.30.

A.2.18. Let (X, τ) and (Y, τ') be topological spaces, and let $f : X \to Y$ be a function. Prove that F is continuous if and only if for all closed sets $F \subset Y$, the set $f^{-1}(F)$ is closed in X.

A.2.19. Let $\mathbb{R}$ be the set of real numbers equipped with the absolute value topology. Prove the following:

(a) Any open interval (a, b) is homeomorphic to the open interval $(0, 1)$.

(b) Any infinite open interval $(a, +\infty)$ is homeomorphic to $(1, \infty)$.

(c) Any infinite open interval $(a, +\infty)$ is homeomorphic to $(-\infty, a)$.

(d) The open interval $(0, 1)$ is homeomorphic to the set of reals $\mathbb{R}$. [Hint: Use $f(x) = \tan x$.]

(e) The interval $(1, \infty)$ is homeomorphic to $(0, 1)$.

Conclude that all open intervals of $\mathbb{R}$ are homeomorphic.

A.2.20. Prove that a circle and a line segment are not homeomorphic.

A.2.21. Finish proving that the function f in Example A.2.34 is a homeomorphism.

A.2.22. Consider $\mathbb{Z}$ and $\mathbb{Q}$ as subsets of $\mathbb{R}$ equipped with the absolute value metric. Decide whether $\mathbb{Z}$ and $\mathbb{Q}$ are homeomorphic.

A.2.23. Let (X, τ) and (Y, τ') be topological spaces, and suppose that there exists a continuous surjective function $f : X \to Y$. Define the equivalence relation on X by

$$x \sim y \iff f(x) = f(y).$$

Prove that $X/\sim$ is homeomorphic to (Y, τ').

A.2.24. Find a quotient space of $\mathbb{R}^2$ homeomorphic to each of the following: (a) A straight line; (b) A sphere; (c) A (filled) rectangle; (d) A torus.

A.2.25. Describe each of the following spaces:

(a) A finite cylinder with each of its boundary circles identified to a point.

(b) The sphere $\mathbb{S}^2$ with an equator identified to a point.

(c) $\mathbb{R}^2$ with points identified according to $(x, y) \sim (-x, -y)$.

A.2.26. Find an open cover of the following sets that does not contain a finite subcover: (a) $\mathbb{R}$; (b) $[0, 1)$; (c) $(0, 1)$.

A.2.27. Let K be a subset of a metric space (X, D). Prove that K is compact if and only if every sequence in K has an accumulation point in K.

A.2.28. Let K be a compact subset of a metric space (X, D). Show that the diameter of K is equal to $D(x, y)$ for some pair $x, y \in K$. Prove that given any $x \in X$, $D(x, A)$ is equal to $D(x, y)$ for some $y \in K$.

A.2.29. Let X be a compact topological space, and let $\{K_n\}_{n=1}^{\infty}$ be a sequence of nonempty closed subsets of X, with $K_{n+1} \subset K_n$ for all n. Prove that $\bigcap_{n=1}^{\infty} K_n$ is nonempty.

A.2.30. Prove that the union of finitely many compact spaces is compact. Is the intersection of two compact sets necessarily compact?

A.2.31. Prove that the set $\mathbb{R}$ equipped with the finite complement topology (see Example A.2.14) is not Hausdorff.

A.2.32. Prove Corollary A.2.60.

A.2.33. Let $f : X \to Y$ be a continuous map between topological spaces. Show that if X is connected, then $f(X)$ is connected in Y.

A.2.34. Decide with proof if $A = \{(x, y) \in \mathbb{R}^2 \,|\, x > 0 \text{ and } (y = 0 \text{ or } y = 1/x)\}$, with the subset topology from $\mathbb{R}^2$, is connected or disconnected.

A.2.35. Show that the Cartesian product of two connected topological spaces is again connected.

A.2.36. Show that as subsets of $\mathbb{R}^2$, the union of two open balls $X = B_1(-1, 0) \cup B_1(1, 0)$ is disconnected but that $Y = X \cup \{(0, 0)\}$ is connected.

APPENDIX B

Calculus of Variations

B.1 Formulation of Several Problems

One of the greatest uses of calculus is the principle that extrema of a continuous function occur at critical points, i.e., at real values of the function, where the first derivative (partial derivatives when dealing with a multivariable function) is (are all) 0 or not defined. In practical applications, when we wish to optimize a certain quantity, we write down a function describing said quantity in terms of relevant independent variables, calculate the first partials, and solve the equations obtained by setting the derivatives equal to 0 or undefined.

Many other problems in math and physics, however, involve quantities that do not just depend on independent variables but on an independent function. Some classic examples are problems that ask us to find the shortest distance between two points, the shape with fixed perimeter enclosing the most area, and the curve of quickest descent between two points. Calculus of variations refers to a general method to deal with such problems.

Let $[x_1, x_2]$ be a fixed interval of real numbers. For any differentiable function $y : [x_1, x_2] \to \mathbb{R}$, the definite integral

$$I(y) = \int_{x_1}^{x_2} f(x, y, y') \, dx \tag{B.1}$$

is a well defined quantity that depends only on $y(x)$ when the integrand f is a function of the arguments x, y, and y'. We can view the above integral I as a function from $C^1([x_1, x_2], \mathbb{R})$, the set of all continuously differentiable functions from $[x_1, x_2]$, to $\mathbb{R}$. The problem is to find all functions $y(x)$ for which $I(y)$ attains a minimum or maximum value for all $y \in C^1([x_1, x_2], \mathbb{R})$. Unlike optimization problems in usual multivariable calculus that involve solving algebraic equations, this initial problem in the calculus of variations involves a second-order differential equation for which the constants of integration are fixed once we set $y(x_1) = y_1$ and $y(x_2) = y_2$.

Many generalizations to this first problem exist. For example, similar to optimization problems in multiple variables, we may impose certain conditions so that we consider only a subset of functions in $C^1([x_1, x_2], \mathbb{R})$ among those to optimize $I(y)$. In another direction, we may seek to optimize the double integral

$$I(w) = \iint_{\mathcal{D}} f\left(x, y, w, \frac{\partial w}{\partial x}, \frac{\partial w}{\partial y}\right) dA$$

where $\mathcal{D}$ is a region of $\mathbb{R}^2$ and w is a two-variable function. The solution would be a function $w \in C^1(\mathcal{D}, \mathbb{R})$ that produces the maximum or minimum value for the integral. Of course, we can consider situations where the unknown function w is a function of any number of variables. As a third type of generalization, we consider the integral

$$I(x, y) = \int_{t_1}^{t_2} f\left(t, x, \frac{dx}{dt}, y, \frac{dy}{dt}\right) dt,$$

where $I(x, y)$ involves two unknown functions of one independent variable t.

Finally, we may then consider any number of combinations to the above generalizations. For example, the isoperimetric problem – the problem of finding the shape with a fixed perimeter and maximum area – involves finding parametric equations $x(t)$ and $y(t)$ that produce a simple closed curve that maximizes area (a one-variable integration by Green's Theorem), subject to the condition that the perimeter is some fixed constant.

The following sections follow the excellent presentation given in [58].

B.2   Euler-Lagrange Equation

B.2.1   Main Theorem

Many problems in calculus of variations amount to solving a particular differential equation called the Euler-Lagrange equation and variants thereof. However, all the theorems that justify the use of the Euler-Lagrange equation hinge on one lemma and its subsequent generalizations.

Lemma B.2.1. *Let G be a continuous real-valued function on an interval $[x_1, x_2]$. If*

$$\int_{x_1}^{x_2} \eta(x)G(x)\,dx = 0 \tag{B.2}$$

for all continuously differentiable functions $\eta(x)$ that satisfy $\eta(x_1) = \eta(x_2) = 0$, then $G(x) = 0$ for all $x \in [x_1, x_2]$.

Proof. We prove the contrapositive, namely, if G is not identically 0 then there exists some function $\eta(x)$ on $[x_1, x_2]$ that does not satisfy Equation (B.2). If we assume that G is not identically 0, then there exists $c \in [x_1, x_2]$ such that $G(c) \neq 0$.

By continuity, there exist a, b such that $x_a \leq a < c < b \leq x_2$ and $G(x) \neq 0$ for all $x \in [a, b]$. Now consider the function

$$\eta(x) = \begin{cases} 0, & \text{for } x_1 \leq x \leq a, \\ G(c)(x-a)^2(x-b)^2, & \text{for } a \leq x \leq b, \\ 0, & \text{for } b \leq x \leq x_2. \end{cases}$$

The function $\eta(x)$ is continuously differentiable, and we have

$$\int_{x_1}^{x_2} \eta(x)G(x)\, dx = \int_a^b G(c)G(x)(x-a)^2(x-b)^2\, dx.$$

The integrand on the right is nonnegative since $G(x)$ has the same sign as $G(c)$ and, by construction, equal to 0 only at $x = a$ and $x = b$. Consequently, the integral on the right is positive. This proves the lemma. $\qquad \square$

Let us consider the first problem in the calculus of variations, in which we wish to optimize the integral in Equation (B.1), with the only condition that $y(x_1) = y_1$ and $y(x_2) = y_2$. The general tactic proceeds as follows. Assume $y(x)$ is a function that optimizes $I(y)$. Let $\eta(x)$ be an arbitrary continuously differentiable function on $[x_1, x_2]$, with $\eta(x_1) = \eta(x_2) = 0$. Define the one-parameter family of functions Y_ε by

$$Y_\varepsilon(x) = y(x) + \varepsilon\eta(x).$$

Obviously, for all ε, we have $Y_\varepsilon(x_1) = y(x_1) = y_1$ and $Y_\varepsilon(x_2) = y(x_2) = y_2$. For shorthand, we define

$$I(\varepsilon) = I(Y_\varepsilon) = \int_{x_1}^{x_2} f(x, Y_\varepsilon, Y_\varepsilon')\, dx.$$

With this notation, we see that $I(0) = I(y)$, and since $y(x)$ is an optimizing function, then

$$I'(0) = 0 \tag{B.3}$$

no matter the choice of arbitrary function $\eta(x)$.

To calculate the derivative in Equation (B.3), we obtain

$$I'(\varepsilon) = \int_{x_1}^{x_2} \left(\frac{\partial f}{\partial Y} \frac{\partial Y}{\partial \varepsilon} + \frac{\partial f}{\partial Y'} \frac{\partial Y'}{\partial \varepsilon} \right) dx = \int_{x_1}^{x_2} \left(\frac{\partial f}{\partial Y}\eta + \frac{\partial f}{\partial Y'}\eta' \right) dx,$$

where $\frac{\partial f}{\partial Y}$ means explicitly $\frac{\partial f}{\partial y}(x, Y_\varepsilon(x), Y_\varepsilon'(x))$ and similarly for $\frac{\partial f}{\partial Y'}$. Setting Equation (B.3) then becomes

$$I'(0) = \int_{x_1}^{x_2} \left(\frac{\partial f}{\partial y}\eta + \frac{\partial f}{\partial y'}\eta' \right) dx = 0.$$

Integrating the second term in this integral by parts, we obtain

$$I'(0) = \left[\frac{\partial f}{\partial y'}\eta(x)\right]_{x_1}^{x_2} + \int_{x_1}^{x_2} \frac{\partial f}{\partial y}\eta - \frac{d}{dx}\left(\frac{\partial f}{\partial y'}\right)\eta\, dx$$

$$= \int_{x_1}^{x_2} \left(\frac{\partial f}{\partial y} - \frac{d}{dx}\left(\frac{\partial f}{\partial y'}\right)\right)\eta\, dx = 0.$$

Applying Lemma B.2.1 to the above equation proves the following theorem.

Theorem B.2.2. *Let* $y : [x_1, x_2] \to \mathbb{R}$ *be a function that optimizes*

$$I(y) = \int_{x_1}^{x_2} f(x, y, y')\, dx.$$

Then y *satisfies the differential equation*

$$\frac{\partial f}{\partial y} - \frac{d}{dx}\left(\frac{\partial f}{\partial y'}\right) = 0, \tag{B.4}$$

which is called the Euler-Lagrange *equation.*

Just as a solution x_0 to $f'(x) = 0$ is not necessarily a maximum or minimum, a function that satisfies this equation is not necessarily an optimizing function. Consequently, we call a solution to Equation (B.4) an *extremizing function*. Understanding that $\frac{\partial f}{\partial y'}$ means $f_{y'}(x, y(x), y'(x))$, we notice that the Euler-Lagrange equation is a second-order differential equation of y in terms of x.

Since Equation (B.4), and in particular the left-hand side of this equation, occurs frequently, we define it as the *Lagrangian operator* $\mathcal{L}$ on a function $f(x, y, y')$, where y is a function of x, by

$$\mathcal{L}(f) = \frac{\partial f}{\partial y} - \frac{d}{dx}\left(\frac{\partial f}{\partial y'}\right).$$

This $\mathcal{L}(f)$ is a differential operator on functions y in $C^2([x_1, x_2], \mathbb{R})$ because for any given $y(x)$ function, $\mathcal{L}(f)(y)$ is a continuous function over $[x_1, x_2]$. Note that $\mathcal{L}$ is a linear transformation in f. On the other hand, whether the differential equation $\mathcal{L}(f) = 0$ is a linear operator in $y(x)$ depends on f.

B.2.2 Brachistochrone Problem

At the turn of the 18th century, Johann Bernoulli posed the problem of finding the path in space that a particle will take when travelling under the action of gravity between two fixed points but taking the shortest amount of time. To be precise, the problem assumes no friction, a simple constant force of gravity mg (where m is the mass of the particle and g the gravity constant), and an initial velocity v_1 that is not necessarily 0. This problem became known as the "brachistochrone" problem, the roots of which come from the Greek words *brachistos* (shortest) and *chronos* (time).

We suppose the two fixed points A and B lie in a vertical plane that we can label as the xy-plane, with the y-axis directed vertically upward and the x-axis oriented so that passing from A to B means an increase in x. Let $A = (x_1, y_1)$ and $B = (x_2, y_2)$ so that any curve $y(x)$ connecting A and B satisfies $y(x_1) = y_1$ and $y(x_2) = y_2$. Note that though the shape of a curve from A to B is a function $y(x)$, a particle moving along this curve under the action of gravity travels with nonconstant speed.

The speed along the curve is given by $v = \frac{ds}{dt}$, where the arclength function $s(x)$ satisfies

$$\frac{ds}{dx} = \sqrt{1 + (y'(x))^2}.$$

The total time T of descent along the path $y(x)$ is given by the integral

$$T = \int_{x=x_1}^{x=x_2} 1 , dt = \int_{x_1}^{x_2} \frac{ds}{v} = \int_{x_1}^{x_2} \frac{\sqrt{1 + (y')^2}}{v} , dx.$$

Since there is no friction and since gravity is a conservative force, the sum of the kinetic energy and potential energy remains constant, namely,

$$\frac{1}{2}mv_1^2 + mgy_1 = \frac{1}{2}mv^2 + mgy.$$

Solving for v we obtain

$$v = \sqrt{v_1^2 + 2gy_1 - 2gy} = \sqrt{2g}\sqrt{y_0 - y},$$

where $y_0 = y_1 + (v_1^2/2g)$ is the height from which the particle descended from rest to reach v_1 at height y_1. The time of travel is

$$T = \frac{1}{\sqrt{2g}} \int_{x_1}^{x_2} \frac{\sqrt{1 + (y')^2}}{\sqrt{y_0 - y}} , dx, \tag{B.5}$$

and finding the path with the shortest time of travel amounts to finding a function $y(x)$ that minimizes this integral.

Applying the Euler-Lagrange equation, we label the integrand in Equation (B.5) as

$$f(x, y, y') = \frac{\sqrt{1 + (y')^2}}{\sqrt{y_0 - y}}. \tag{B.6}$$

Notice that this problem has one simplification from the general Euler-Lagrange equation: f does not depend explicitly on x. This fact allows us to make a useful simplification. The chain rule gives

$$\frac{df}{dx} = \frac{\partial f}{\partial x} + y'\frac{\partial f}{\partial y} + y''\frac{\partial f}{\partial y'} = y'\frac{\partial f}{\partial y} + y''\frac{\partial f}{\partial y'}$$

since f does not depend directly on x. However,

$$\frac{d}{dx}\left(y'\frac{\partial f}{\partial y'}\right) = y''\frac{\partial f}{\partial y'} + y'\frac{d}{dx}\left(\frac{\partial f}{\partial y'}\right),$$

so

$$\frac{df}{dx} = \frac{d}{dx}\left(y'\frac{\partial f}{\partial y'}\right) + y'\left(\frac{\partial f}{\partial y} - \frac{d}{dx}\left(\frac{\partial f}{\partial y'}\right)\right) = \frac{d}{dx}\left(y'\frac{\partial f}{\partial y'}\right),$$

where the second term in the middle expression is identically 0 due to the Euler-Lagrange equation. Integrating both sides with respect to x we obtain

$$y'\frac{\partial f}{\partial y'} - f = C$$

for some constant C. Using the specific function in Equation (B.6), we obtain

$$\frac{(y')^2}{\sqrt{(y_0 - y)(1 + (y')^2)}} - \frac{\sqrt{1 + (y')^2}}{\sqrt{y_0 - y}} = C.$$

Solving for $y' = \frac{dy}{dx}$, we obtain

$$\frac{dy}{dx} = \frac{\sqrt{C^{-2} - (y_0 - y)}}{\sqrt{y_0 - y}},$$

which, upon taking the inverse and integrating with respect to y, becomes

$$x = \int \frac{\sqrt{y_0 - y}}{\sqrt{C^{-2} - (y_0 - y)}}\, dy. \tag{B.7}$$

Using the substitution

$$y_0 - y = \frac{1}{C^2}\sin^2\frac{\theta}{2}, \tag{B.8}$$

the integral in Equation (B.7) becomes

$$x = -\frac{1}{C^2}\int \sin^2\frac{\theta}{2}\, d\theta = -\frac{1}{2C^2}\int 1 - \cos\theta\, d\theta = \frac{1}{2C^2}(\sin\theta - \theta) + x_0, \tag{B.9}$$

where x_0 is some constant of integration. Rewriting Equation (B.8), setting $a = 1/(2C^2)$, and substituting $t = -\theta$, we obtain the equations

$$\begin{cases} x = x_0 + a(t - \sin t), \\ y = y_0 - a(1 - \cos t). \end{cases}$$

Obviously, these equations do not give y as an explicit function of x but do show that the path with most rapid descent is in the shape of an upside-down cycloid.

B.3 Several Dependent Variables

B.3.1 Main Theorem

A first generalization to the basic problem in the calculus of variations is to find n twice-differentiable functions $x_1(t), \ldots, x_n(t)$ defined over the interval $[t_1, t_2]$ that optimize the integral

$$I = \int_{t_1}^{t_2} f(x_1, \ldots, x_n, x_1', \ldots, x_n', t) \, dt. \tag{B.10}$$

We follow the same technique as in Section B.2. Label $x_1(t), \ldots, x_n(t)$ as the actual optimizing functions and define corresponding one-parameter families of functions by

$$X_i(t) = x_i(t) + \varepsilon \xi_i(t),$$

where $\xi_i(t)$ are any differentiable functions with

$$\xi_i(t_1) = \xi_i(t_2) = 0 \qquad \text{for } 1 \le i \le n.$$

With the one-parameter families X_i, we form the integral

$$I(\varepsilon) = \int_{t_1}^{t_2} f(X_1, \ldots, X_n, X_1', \ldots, X_n', t) \, dt.$$

Then $I(0) = I$, and since by assumption the functions $x_1, \ldots, x_n$ are the optimizing functions, we must also have $I'(0) = 0$.

Taking the derivative of $I(\varepsilon)$ and using the chain rule, we have

$$I'(\varepsilon) = \int_{t_1}^{t_2} \frac{\partial f}{\partial X_1} \xi_1 + \cdots + \frac{\partial f}{\partial X_n} \xi_n + \frac{\partial f}{\partial X_1'} \xi_1' + \cdots + \frac{\partial f}{\partial X_n'} \xi_n' \, dt,$$

where by $\partial f / \partial X_i$ we mean the partial derivative to f with respect to the variable that we evaluate to be the one parameter of functions X_i. Regardless of the arbitrary functions ξ_i, setting $\varepsilon = 0$ replaces the family of functions X_i with the function x_i. Using the same abuse of notation for $\partial f / \partial x_i$, we have

$$I'(0) = \int_{t_1}^{t_2} \frac{\partial f}{\partial x_1} \xi_1 + \cdots + \frac{\partial f}{\partial x_n} \xi_n + \frac{\partial f}{\partial x_1'} \xi_1' + \cdots + \frac{\partial f}{\partial x_n'} \xi_n' \, dt = 0.$$

Since this equation must hold for all choices of the functions x_i, we can in particular set $\xi_j = 0$ for all indices $j \ne i$. Then we deduce that

$$\int_{t_1}^{t_2} \frac{\partial f}{\partial x_i} \xi_i + \frac{\partial f}{\partial x_i'} \xi_i' \, dt = 0 \qquad \text{for all } i.$$

Integrating the second term in the above integral by parts and using the fact that $\xi_i(t_1) = \xi_i(t_2) = 0$, we obtain

$$\int_{t_1}^{t_2} \left(\frac{\partial f}{\partial x_i} - \frac{d}{dt}\left(\frac{\partial f}{\partial x_i'} \right) \right) \xi_i \, dt = 0.$$

Then using Lemma B.2.1, we deduce the following theorem.

Theorem B.3.1. *Consider the integral*

$$I = \int_{t_1}^{t_2} f(x_1, \ldots, x_n, x_1', \ldots, x_n', t) \, dt,$$

where f is a continuous function and each $x_i(t)$ is a twice-differentiable function defined over $[t_1, t_2]$. Then the functions $x_1, x_2, \ldots, x_n$ optimize the integral I if and only if

$$\frac{\partial f}{\partial x_i} - \frac{d}{dt}\left(\frac{\partial f}{\partial x_i'} \right) = 0 \qquad \text{for all } 1 \leq i \leq n.$$

Here again, if f is a function as defined in the above theorem, we define the Lagrangian operator $\mathcal{L}_i$ or $\mathcal{L}_{x_i}$ as

$$\mathcal{L}_i(f) = \frac{\partial f}{\partial x_i} - \frac{d}{dt}\left(\frac{\partial f}{\partial x_i'} \right).$$

B.4   Isoperimetric Problems and Lagrange Multipliers

B.4.1   Main Theorem

In this section, we approach a new class of problems in which we desire not only to optimize a certain integral but to do so considering only functions that satisfy an additional criterion besides the usual restriction of continuity. In all the problems we consider, the criteria consist of imposing a prescribed value on a certain integral related to our variable function. More precisely, we will wish to construct a function $x(t)$ defined over an interval $[t_1, t_2]$ that optimizes the integral

$$I = \int_{t_1}^{t_2} f(x, x', t) \, dt, \tag{B.11}$$

subject to the condition that

$$\int_{t_1}^{t_2} g(x, x', t) \, dt = J \tag{B.12}$$

for some fixed value of J. It is assumed that f and g are twice-differentiable functions in their variables. Such a problem is called an *isoperimetric problem*.

Following the same approach as in Section B.2, we label $x(t)$ as the actual optimizing function to the integral in Equation (B.11), which we assume also satisfies Equation (B.12), and we introduce a two-parameter family of functions

$$X(t) = x(t) + \varepsilon_1 \xi_1(t) + \varepsilon_2 \xi_2(t),$$

where $\xi_1(t)$ and $\xi_2(t)$ are any differentiable functions that satisfy

$$\xi_1(t_1) = \xi_2(t_1) = \xi_1(t_2) = \xi_2(t_2) = 0. \tag{B.13}$$

The condition in Equation (B.13) guarantees that $X(t_1) = x(t_1) = x_1$ and $X(t_2) = x(t_2) = x_2$ for all choices of the parameters ε_1 and ε_2. We use the family of functions $X(t)$ as a comparison to the optimizing function $x(t)$, but in contrast to Section B.2, we need a two-parameter family, as we shall see shortly.

We replace the function $x(t)$ with the family $X(t)$ in Equations (B.11) and (B.12) to obtain

$$I(\varepsilon_1, \varepsilon_2) = \int_{t_1}^{t_2} f(X, X', t) \, dt$$

and

$$J(\varepsilon_1, \varepsilon_2) = \int_{t_1}^{t_2} g(X, X', t) \, dt.$$

The parameters ε_1 and ε_2 cannot be independent if the family $X(t)$ is to always satisfy Equation (B.12). Indeed, since J is constant, ε_1 and ε_2 satisfy the equation

$$J(\varepsilon_1, \varepsilon_2) = J \qquad \text{(a constant).} \tag{B.14}$$

Since $x(t)$ is assumed to be the optimizing function, then $I(\varepsilon_1, \varepsilon_2)$ is optimized with respect to ε_1 and ε_2, subject to Equation (B.14) when $\varepsilon_1 = \varepsilon_2 = 0$, no matter the particular choice of $\xi_1(t)$ and $\xi_2(t)$.

Consequently, we can apply the method of Lagrange multipliers, usually presented in a multivariable calculus course. Following that method, $I(\varepsilon_1, \varepsilon_2)$ is optimized, subject to Equation (B.14), when

$$\begin{cases} \dfrac{\partial I}{\partial \varepsilon_i} = \lambda \dfrac{\partial J}{\partial \varepsilon_i}, & \text{for } i = 1, 2, \text{ and} \\ J(\varepsilon_1, \varepsilon_2) = J, \end{cases} \tag{B.15}$$

where λ is a free parameter called the *Lagrange multiplier*. In order to apply this to Euler-Lagrange methods of optimizing integrals, define the function

$$f^*(x, x', t) = f(x, x', t) - \lambda g(x, x', t).$$

Then the first two equations in Equation (B.15) are tantamount to solving

$$\frac{\partial f^*}{\partial \varepsilon_i} = 0.$$

Following a nearly identical approach as in Section B.2, the details of which we leave to the interested reader, we can prove the following theorem.

Theorem B.4.1. *Assume that f and g are twice-differentiable functions $\mathbb{R}^3 \to \mathbb{R}$. Let $x : [t_1, t_2] \to \mathbb{R}$ be a function that optimizes*

$$I = \int_{t_1}^{t_2} f(x, x', t) \, dt,$$

subject to the condition that

$$J = \int_{t_1}^{t_2} g(x, x', t) \, dt$$

remains constant. Then x satisfies the differential equation

$$\frac{\partial f^*}{\partial x} - \frac{d}{dt}\left(\frac{\partial f^*}{\partial x'}\right) = 0, \tag{B.16}$$

where $f^* = f - \lambda g$. Furthermore, the solution to Equation (B.16) produces an expression for $x(t)$ that depends on two constants of integration and the parameter λ and, if a solution to this isoperimetric problem exists, then these quantities are fixed by requiring that $x(t_1) = x_1$, $x(t_2) = x_2$, and J be a constant.

Many generalizations extend this theorem, but rather than presenting in great detail the variants thereof, we present an example that shows why we refer to the class of problems presented in this section as isoperimetric problems.

B.4.2 Problem of Maximum Enclosed Area

Though simple to phrase and yet surprisingly difficult to solve is the classic question, "What closed simple curve of fixed length encloses the most area?" Even Greek geometers "knew" that if we fix the length of a closed curve, the circle has the largest area, but no rigorous proof is possible without the techniques of calculus of variations.

To solve this problem, consider parametric curves $\vec{x} = (x(t), y(t))$ with $t \in [t_1, t_2]$. We assume the curve is closed so that $\vec{x}(t_1) = \vec{x}(t_2)$ and similarly for all derivatives of $\vec{x}$. The arclength formula for this curve is

$$S = \int_{t_1}^{t_2} \sqrt{(x')^2 + (y')^2} \, dt,$$

and by a corollary to Green's Theorem, the area of the enclosed region is

$$A = \int_{t_1}^{t_2} xy' \, dt.$$

Therefore, we wish to optimize the integral A, subject to the constraint that the integral S is fixed, say $S = p$.

Following Theorem B.4.1 but adapting it to the situation of more than one dependent variable, we define the function

$$f^*(x, x', y, y', t) = xy' - \lambda\sqrt{x'^2 + y'^2},$$

and conclude that the curve with the greatest area satisfies

$$\begin{cases} \mathcal{L}_x(f^*), \\ \mathcal{L}_y(f^*). \end{cases} \tag{B.17}$$

Taking appropriate derivatives, Equation (B.17) becomes

$$\begin{cases} y' + \lambda\dfrac{d}{dt}\left(\dfrac{x'}{\sqrt{x'^2 + y'^2}}\right) = 0, \\ \dfrac{d}{dt}\left(x - \lambda\dfrac{y'}{\sqrt{x'^2 + y'^2}}\right) = 0, \end{cases}$$

and integrating with respect to t, we obtain

$$\begin{cases} y + \lambda\dfrac{x'}{\sqrt{x'^2 + y'^2}} = C_1, \\ x - \lambda\dfrac{y'}{\sqrt{x'^2 + y'^2}} = C_2. \end{cases}$$

From this, we deduce the relation

$$(x - C_2)^2 + (y - C_1)^2 = \lambda^2, \tag{B.18}$$

which means that the curve with a given perimeter and with maximum area lies on a circle. Since the curve is closed and simple, the parametric curve $\vec{x}(t)$ is an injective function (except for $\vec{x}(t_1) = \vec{x}(t_2)$), and the image is in fact a circle, though there is no assumption that $\vec{x}$ travels around the circle at a uniform rate. That the Lagrange multiplier appears in Equation (B.18) is not an issue because, since we know that the perimeter is fixed at p, we know that $\lambda = p/2\pi$.

APPENDIX C

Further Topics in Multilinear Algebra

Chapter 4 offers only a brief introduction to multilinear algebra. This appendix supplies a few more short topics with close connections to k-volume formulas in $\mathbb{R}^n$.

C.1 Binet-Cauchy and k-Volume of Parallelepipeds

The article [30] develops the connection between the wedge product of vectors in $\mathbb{R}^n$ and analytic geometry. Most important for applications to differential geometry is a formula for the volume of a k-dimensional parallelepiped in $\mathbb{R}^n$. The authors of [30] give the following definition.

Definition C.1.1. The *dot product* of two pure antisymmetric tensors in $\bigwedge^k \mathbb{R}^n$ is

$$(\vec{a}_1 \wedge \vec{a}_2 \wedge \cdots \wedge \vec{a}_k) \cdot (\vec{b}_1 \wedge \vec{b}_2 \wedge \cdots \wedge \vec{b}_k) = \begin{vmatrix} \vec{a}_1 \cdot \vec{b}_1 & \vec{a}_1 \cdot \vec{b}_2 & \cdots & \vec{a}_1 \cdot \vec{b}_k \\ \vec{a}_2 \cdot \vec{b}_1 & \vec{a}_2 \cdot \vec{b}_2 & \cdots & \vec{a}_2 \cdot \vec{b}_k \\ \vdots & \vdots & \ddots & \vdots \\ \vec{a}_k \cdot \vec{b}_1 & \vec{a}_k \cdot \vec{b}_2 & \cdots & \vec{a}_k \cdot \vec{b}_k \end{vmatrix}.$$

It turns out that this definition is equivalent to the usual dot product on $\bigwedge^k \mathbb{R}^n$ with respect to its standard basis, namely,

$$\{\vec{e}_{i_1} \wedge \vec{e}_{i_2} \wedge \cdots \wedge \vec{e}_{i_k}\}, \qquad \text{with } 1 \le i_1 < i_2 < \cdots < i_k \le n.$$

The equivalence of these two definitions is a result of the following combinatorial proposition.

Proposition C.1.2 (Binet-Cauchy). *Let A and B be two $n \times m$ matrices, with $m \le n$. Call $\mathcal{I}(m, n)$ the set of subsets of $\{1, 2, \dots, n\}$ of size m and for any $S \in \mathcal{I}(m, n)$, denote A_S as the $m \times m$ submatrix consisting of the rows of A indexed by S (and similarly for B). Then*

$$\det(B^\top A) = \sum_{S \in \mathcal{I}(m,n)} \big(\det(B_S) \big) \big(\det(A_S) \big).$$

Proof. Let $A = (a_{ij})$ and $B = (b_{ij})$, with $1 \leq i \leq n$ and $1 \leq j \leq m$. The matrix $B^\top A$ is an $m \times m$-matrix with entries

$$\sum_{j=1}^{n} b_{ji} a_{jk},$$

indexed by $1 \leq i, k \leq m$. Therefore, the determinant of $B^T A$ is

$$\det(B^\top A) = \sum_{\sigma \in S_m} \text{sign}(\sigma) \left(\sum_{j_1=1}^{n} b_{j_1 1} a_{j_1 \sigma(1)} \right) \left(\sum_{j_2=1}^{n} b_{j_2 2} a_{j_2 \sigma(2)} \right) \cdots \left(\sum_{j_m=1}^{n} b_{j_m m} a_{j_m \sigma(m)} \right),$$

where S_m is the set of permutations on the set $\{1, 2, \ldots, m\}$. Then, after rearranging the order of summation, we have

$$\det(B^\top A) = \sum_{\sigma \in S_m} \sum_{j_1=1}^{n} \sum_{j_2=1}^{n} \cdots \sum_{j_m=1}^{n} \text{sign}(\sigma) b_{j_1 1} b_{j_2 2} \cdots b_{j_m m} a_{j_1 \sigma(1)} a_{j_2 \sigma(2)} \cdots a_{j_m \sigma(m)}$$

$$= \sum_{j_1=1}^{n} \sum_{j_2=1}^{n} \cdots \sum_{j_m=1}^{n} b_{j_1 1} b_{j_2 2} \cdots b_{j_m m} \left(\sum_{\sigma \in S_m} \text{sign}(\sigma) a_{j_1 \sigma(1)} a_{j_2 \sigma(2)} \cdots a_{j_m \sigma(m)} \right).$$

Because of the sign of the permutation, any term in the summation where not all the j_l are distinct is equal to 0. Therefore, we only need to consider the summation over sets of indices $\mathbf{j} = (j_1, j_2, \ldots, j_m) \in \{1, \ldots, m\}^n$, where all of the indices are distinct. We can parametrize this set in an alternative manner as follows. Let $\mathcal{I}(m, n)$ be the set of indices in increasing order, i.e.,

$$\mathcal{I}(m, n) = \left\{ (j_1, j_2, \ldots, j_m) \in \{1, \ldots, n\}^m \mid 1 \leq j_1 < j_2 < \cdots < j_m \leq n \right\}. \quad \text{(C.1)}$$

The set $\mathcal{I}(m, n) \times S_m$ is in bijection with the set of all m-tuples of indices that are distinct via

$$(\mathbf{j}, \sigma) \mapsto (j_{\sigma(1)}, \ldots, j_{\sigma(m)}).$$

We can now write

$$\det(B^\top A)$$

$$= \sum_{\mathbf{j} \in \mathcal{I}(m,n)} \sum_{\tau \in S_m} b_{j_{\tau(1)} 1} b_{j_{\tau(2)} 2} \cdots b_{j_{\tau(m)} m} \left(\sum_{\sigma \in S_m} \text{sign}(\sigma) a_{j_{\tau(1)} \sigma(1)} a_{j_{\tau(2)} \sigma(2)} \cdots a_{j_{\tau(m)} \sigma(m)} \right)$$

$$= \sum_{\mathbf{j} \in \mathcal{I}(m,n)} \sum_{\tau \in S_m} b_{j_{\tau(1)} 1} b_{j_{\tau(2)} 2} \cdots b_{j_{\tau(m)} m} \text{sign}(\tau) \left(\sum_{\sigma' \in S_m} \text{sign}(\sigma') a_{j_1 \sigma'(1)} a_{j_2 \sigma'(2)} \cdots a_{j_m \sigma'(m)} \right)$$

$$= \sum_{\mathbf{j} \in \mathcal{I}(m,n)} \sum_{\tau \in S_m} \text{sign}(\tau) b_{j_{\tau(1)} 1} b_{j_{\tau(2)} 2} \cdots b_{j_{\tau(m)} m} \det A_{\mathbf{j}},$$

where $A_{\mathbf{j}}$ is the $m \times m$ submatrix obtained from A by using only the rows given in the m-tuple index $\mathbf{j}$. Then we conclude that

$$\det(B^\top A) = \sum_{\mathbf{j} \in \mathcal{I}(m,n)} \sum_{\tau \in S_m} \operatorname{sign}(\tau) b_{j_1 \tau^{-1}(1)} b_{j_2 \tau^{-1}(2)} \cdots b_{j_m \tau^{-1}(m)} \det A_{\mathbf{j}}$$

$$= \sum_{\mathbf{j} \in \mathcal{I}(m,n)} \left(\det B_{\mathbf{j}}^\top \right) \left(\det A_{\mathbf{j}} \right),$$

and the proposition follows since $\det(C^\top) = \det(C)$ for any square matrix C. □

Corollary C.1.3. *Definition C.1.1 is equivalent to the dot product on $\bigwedge^k \mathbb{R}^n$ with respect to the standard basis.*

Proof. If $\vec{a}_1, \vec{a}_2, \ldots, \vec{a}_k$ is a k-tuple of vectors in $\mathbb{R}^n$, call A the $n \times k$-matrix that has the vector $\vec{a}_i$ as the ith column. Define $P(n,k)$ as in Proposition C.1.2. For any subset S of $\{1, 2, \ldots, n\}$ of cardinality k, define

$$\vec{e}_S = \vec{e}_{s_1} \wedge \vec{e}_{s_2} \wedge \cdots \wedge \vec{e}_{s_k},$$

where $S = \{s_1, s_2, \ldots, s_k\}$, with the elements listed in increasing order. It is not hard to check that

$$\vec{a}_1 \wedge \vec{a}_2 \wedge \cdots \wedge \vec{a}_k = \sum_{S \in P(n,k)} (\det(A_S)) \vec{e}_S. \tag{C.2}$$

The corollary follows immediately from Proposition C.1.2. □

As with a usual Euclidean vector space $\mathbb{R}^n$, we define the Euclidean norm in the following way.

Definition C.1.4. Let $a = \vec{a}_1 \wedge \vec{a}_2 \wedge \cdots \wedge \vec{a}_k \in \bigwedge^k \mathbb{R}^n$. The (Euclidean) norm of this vector is

$$\|\vec{a}_1 \wedge \vec{a}_2 \wedge \cdots \wedge \vec{a}_k\| = \sqrt{a \cdot a}.$$

Corollary C.1.5. *The k-dimensional volume of a parallelepiped in $\mathbb{R}^n$ spanned by k vectors $\vec{v}_1, \vec{v}_2, \ldots, \vec{v}_k$ is given by*

$$\|\vec{v}_1 \wedge \vec{v}_2 \wedge \cdots \wedge \vec{v}_k\|.$$

Proof. It is a standard fact in linear algebra (see [14, Fact 6.3.7]) that the k-volume of the described parallelepiped is $\sqrt{\det(A^\top A)}$, where A is the matrix that has the vector $\vec{v}_i$ as the ith column. The corollary follows from Definitions C.1.1 and C.1.4. □

PROBLEMS

C.1.1. Use the results of this section to calculate the surface of the parallelogram in $\mathbb{R}^3$ spanned by

$$\vec{v} = \begin{pmatrix} 1 \\ -3 \\ 7 \end{pmatrix} \quad \text{and} \quad \vec{w} = \begin{pmatrix} 4 \\ 5 \\ -2 \end{pmatrix}.$$

C.1.2. Calculate the 3-volume of the parallelepiped in $\mathbb{R}^4$ spanned by

$$\vec{a} = \begin{pmatrix} 0 \\ -2 \\ 2 \\ 1 \end{pmatrix}, \quad \vec{b} = \begin{pmatrix} 3 \\ 1 \\ -1 \\ 0 \end{pmatrix}, \quad \text{and} \quad \vec{c} = \begin{pmatrix} 5 \\ 1 \\ -2 \\ -3 \end{pmatrix}.$$

C.1.3. Using the same vectors $\vec{a}$, $\vec{b}$, and $\vec{c}$ in the previous exercise, determine all vectors $\vec{x}$ such that the four-dimensional parallelepiped spanned by $\vec{a}$, $\vec{b}$, $\vec{c}$, and $\vec{x}$ has dimension 0.

C.1.4. Verify the claim in Equation (C.2).

C.1.5. *A Higher Pythagorean Theorem.* Let $\vec{a}$, $\vec{b}$, and $\vec{c}$ be three vectors in $\mathbb{R}^n$ that are mutually perpendicular.

(a) Prove that

$$\|\vec{a} \wedge \vec{b} + \vec{a} \wedge \vec{c} + \vec{b} \wedge \vec{c}\|^2 = \|\vec{a} \wedge \vec{b}\|^2 + \|\vec{a} \wedge \vec{c}\|^2 + \|\vec{b} \wedge \vec{c}\|^2.$$

(b) Consider the tetrahedron spanned by $\vec{a}$, $\vec{b}$, and $\vec{c}$. Let S_C be the face spanned by $\vec{a}$ and $\vec{b}$, S_B be the face spanned by $\vec{a}$ and $\vec{c}$, S_A be the face spanned by $\vec{b}$ and $\vec{c}$, and let S_D be the fourth face of the tetrahedron. Deduce that

$$S_A^2 + S_B^2 + S_C^2 = S_D^2.$$

C.2   Volume Form Revisited

In Example 4.6.24, we introduced the volume form on $\mathbb{R}^n$ in reference to the standard basis. This is not quite satisfactory for our applications because the standard basis has internal properties, namely that it is orthonormal with respect to the dot product. The following proposition presents the volume form on a vector space in its most general context.

Proposition C.2.1. *Let V be an n-dimensional vector space with an inner product $\langle,\rangle$. Then there exists a unique form $\omega \in \bigwedge^n V^*$ such that $\omega(\vec{e}_1, \ldots, \vec{e}_n) = 1$ for all oriented bases $(\vec{e}_1, \ldots, \vec{e}_n)$ of V that are orthonormal with respect to $\langle,\rangle$. Furthermore, if $(\vec{u}_1, \ldots, \vec{u}_n)$ is any oriented basis of V, then*

$$\omega = \sqrt{\det A}\, \vec{u}^{*1} \wedge \cdots \wedge \vec{u}^{*n},$$

where A is the matrix with entries $A_{ij} = (\langle \vec{u}_i, \vec{u}_j \rangle)$.

Proof. Let $(\vec{u}_1, \ldots, \vec{u}_n)$ be any basis of V and let $\vec{v} = \sum_i a^i \vec{u}_i$ and $\vec{w} = \sum_i b^i \vec{u}_i$ be two vectors in V along with their coordinates with respect to $(\vec{u}_1, \ldots, \vec{u}_n)$. Then by the linearity of the form,

$$\langle \vec{v}, \vec{w} \rangle = \sum_{i,j=1}^{n} a^i b^j \langle \vec{u}_i, \vec{u}_j \rangle = \vec{v}^\top A \vec{w}.$$

We remark that $\det A \neq 0$ because otherwise there would exist some nonzero vector $\vec{v}$ such that $A\vec{v} = 0$ and then $\langle \vec{v}, \vec{v} \rangle = 0$, which would contradict the positive-definite property of the form.

The existence of an orthonormal basis with respect to $\langle , \rangle$ follows from the Gram-Schmidt orthonormalization process. If $(\vec{e}_1, \ldots, \vec{e}_n)$ is an orthonormal basis with respect to $\langle , \rangle$, then the associated matrix $(\langle \vec{e}_i, \vec{e}_j \rangle)$ is the identity matrix.

Given an orthonormal ordered basis $\mathcal{E} = (\vec{e}_1, \ldots, \vec{e}_n)$, let $(\vec{e}^{*1}, \ldots, \vec{e}^{*n})$ be the cobasis of V^*. Set $\omega = \vec{e}^{*1} \wedge \cdots \wedge \vec{e}^{*n}$. Obviously, $\omega(\vec{e}_1, \ldots, \vec{e}_n) = 1$. Now, if $\mathcal{B} = (\vec{u}_1, \ldots, \vec{u}_n)$ is any other orthonormal basis of V with the same orientation of $(\vec{e}_1, \ldots, \vec{e}_n)$, then $\det(M^\top M) = 1$, where M is the transition matrix from coordinates in $(\vec{e}_1, \ldots, \vec{e}_n)$ to coordinates in $(\vec{u}_1, \ldots, \vec{u}_n)$. Hence, $\det(M)^2 = 1$, and the assumption that $(\vec{u}_1, \ldots, \vec{u}_n)$ has the same orientation as $(\vec{e}_1, \ldots, \vec{e}_n)$ means that $\det(M)$ is positive. Thus, $\det M = 1$.

By Proposition 4.1.6, the transition matrix from coordinates in $(\vec{e}^{*1}, \ldots, \vec{e}^{*n})$ to coordinates in $(\vec{u}^{*1}, \ldots, \vec{u}^{*n})$ is M^{-1}. However, by Proposition 4.6.23, we then conclude that

$$\omega = \vec{e}^{*1} \wedge \cdots \wedge \vec{e}^{*n} = \det(M)^{-1} \vec{u}^{*1} \wedge \cdots \wedge \vec{u}^{*n} = \vec{u}^{*1} \wedge \cdots \wedge \vec{u}^{*n}.$$

Since $\vec{e}^{*1} \wedge \cdots \wedge \vec{e}^{*n}(e_1, \ldots, e_n) = 1$, then ω evaluates to 1 on all bases of V that are orthonormal and have the same orientation as $\{\vec{e}_1, \ldots, \vec{e}_n\}$.

Suppose now that $\{\vec{u}_1, \ldots, \vec{u}_n\}$ is any basis of V, not necessarily orthonormal. Again let M be the coordinate change matrix as above. By definition $[\vec{u}_i]_{\mathcal{E}} = M\vec{e}_i$. Then we can calculate the coefficients of A by

$$A_{ij} = \langle \vec{u}_i, \vec{u}_j \rangle = [\vec{u}_i]_{\mathcal{E}} \cdot [\vec{u}_j]_{\mathcal{E}} = (M\vec{e}_i)^\top (M\vec{e}_j) = \vec{e}_i^\top M^\top M \vec{e}_j.$$

Hence, we have shown that $A = M^\top M$. We conclude that

$$\omega = \det(M) \vec{u}^{*1} \wedge \cdots \wedge \vec{u}^{*n} = \sqrt{\det A} \, \vec{u}^{*1} \wedge \cdots \wedge \vec{u}^{*n}. \qquad \square$$

Definition C.2.2. Let $(V, \langle \cdot, \cdot \rangle)$ be an inner product space of dimension n. Then the element $\omega \in \bigwedge^n V^*$ defined in Proposition C.2.1 is called the volume form of V.

C.3   Hodge Star Operator

We conclude this section by introducing an operator on wedge product spaces $\bigwedge^k V$. In this subsection, we assume throughout that $(V, \langle \cdot, \cdot \rangle)$ is a finite dimensional inner product space.

Exercise 4.2.4 introduced the bijection $\psi : V \to V^*$ defined by $\psi(v) = \lambda_v$, where $\lambda_v(w) = \langle v, w \rangle$ for all $w \in V$. Exercise 4.2.5 gave steps to extend the inner product $\langle \cdot, \cdot \rangle$ to an inner product $\langle \cdot, \cdot \rangle^*$ on V^* by defining

$$\langle \lambda_v, \lambda_w \rangle^* = \langle w, v \rangle = \langle v, w \rangle,$$

since $\langle \cdot, \cdot \rangle$ is symmetric. In other words, for all $\eta, \tau \in V^*$,

$$\langle \eta, \tau \rangle^* = \langle \psi^{-1}(\eta), \psi^{-1}(\tau) \rangle. \tag{C.3}$$

From now on in this section, we drop the superscript $*$ on the inner product extended to the dual.

Proposition C.3.1. *Let* $\eta_1, \ldots, \eta_k, \tau_1, \ldots, \tau_k \in V^*$. *Setting*

$$\langle \eta_1 \wedge \ldots \wedge \eta_k, \tau_1 \wedge \ldots \wedge \tau_k \rangle = \det(\langle \eta_i, \tau_j \rangle)$$

defines an inner product on $\bigwedge^k V^*$.

Proof. (Left as an exercise for the reader. See Problem C.3.3.) $\square$

Definition C.3.2. Let $(V, \langle \cdot, \cdot \rangle)$ be an inner product space of dimension n, and let $\omega \in \bigwedge^n V^*$ be the volume form. The *Hodge star operator* is the operator $\star : \bigwedge^k V^* \to \bigwedge^{n-k} V^*$ that is uniquely determined by

$$\langle \star \eta, \tau \rangle \omega = \eta \wedge \tau$$

for all $\tau \in \bigwedge^{n-k} V^*$.

The Hodge star operator has the following nice properties, which we leave as exercises.

Proposition C.3.3. *Let* $(V, \langle \cdot, \cdot \rangle)$ *be an inner product space. Let* $\mathcal{B} = \{e_1, \ldots, e_n\}$ *be a basis that is orthonormal with respect to* $\langle \cdot, \cdot \rangle$, *and let* $\mathcal{B}$ *cobasis of* V^*. *Set* ω *as the volume form with respect to* $\langle \cdot, \cdot \rangle$.

1. *The Hodge star operator* $\star$ *is well defined and linear.*

2. *Viewing 1 as an element of* $\mathbb{R} = \bigwedge^0 V$, *then* $\star 1 = \omega$.

3. *For any* $k < n$, *then* $\star(e^{*1} \wedge \cdots \wedge e^{*k}) = e^{*(k+1)} \wedge \cdots \wedge e^{*n}$.

4. *For any* k-*tuple* $(i_1, \ldots, i_k)$ *of increasing indices,*

$$\star(e^{*i_1} \wedge \cdots \wedge e^{*i_k}) = (\operatorname{sign} \sigma) \, e^{*j_1} \wedge \cdots \wedge e^{*j_{n-k}},$$

where the j_l *indices are such that* $\{i_1, \ldots, i_k, j_1, \ldots, j_{n-k}\} = \{1, \ldots, n\}$ *and* σ *is the permutation that maps the* n-*tuple* $(i_1, \ldots, i_k, j_1, \ldots, j_{n-k})$ *to* $(1, 2, \ldots, n)$.

Example C.3.4. This Proposition allows us to easily calculate the Hodge star operator of any $(0, k)$-tensor over V. For example, suppose that $V = \mathbb{R}^4$ is equipped with the usual dot product and that (e_1, e_2, e_3, e_4) is the standard basis. Then using the above Proposition, we calcualte that

$$\star(e^{*1} + 2e^{*2} + 8e^{*4}) = e^{*2} \wedge e^{*3} \wedge e^{*4} - 2e^{*1} \wedge e^{*3} \wedge e^{*4} - 8e^{*1} \wedge e^{*2} \wedge e^{*3}.$$

The following proposition gives a formula for the coordinates of the $\star\eta$ in terms of the coordinates of η.

Proposition C.3.5. *Let $(V, \langle \cdot, \cdot \rangle)$ be an inner product space. Let $\mathcal{B} = \{u_1, \ldots, u_n\}$ be any basis of V, and denote by $\{u^{*1}, \ldots, u^{*n}\}$ its cobasis in V^*. Let A be the matrix with entries $a_{ij} = (\langle u_i, u_j \rangle)$, and label a^{ij} as the components of the inverse A^{-1}. If $\eta \in \bigwedge^k V^*$, with coordinates $\eta_{i_1 \cdots i_k}$, so that*

$$\eta = \sum_{1 \leq i_1 < \cdots < i_k \leq n} \eta_{i_1 \cdots i_k} \, u^{*i_1} \wedge \cdots \wedge u^{*i_k},$$

then the components of $\star\eta$ with respect to $\mathcal{B}^*$ are

$$(\star\eta)_{j_1 \cdots j_{n-k}} = \frac{\sqrt{\det A}}{k!} \varepsilon_{i_1 \cdots i_k j_1 \cdots j_{n-k}} a^{i_1 h_1} \cdots a^{i_k h_k} \eta_{h_1 \cdots h_k}, \tag{C.4}$$

where $\varepsilon_{h_1 \cdots h_n}$ is the permutation symbol defined in (4.35).

Proof. By a calculation similar to the one in the proof of Proposition C.2.1 and using the definition of the inner product on 1-forms given in Equation (C.3), we determine that

$$\langle u^{*i}, u^{*j} \rangle = a^{ij},$$

i.e., the (i, j)th entry of the the inverse A^{-1}.

As above, denote by ω the volume form on V associated to $\langle \cdot, \cdot \rangle$.

A few preliminary notations will render the rest of the proof shorter. Recall the set $\mathcal{I}(m, n)$ defined in Equation (C.1). For any sequence $\mathbf{i} = (i_1, \ldots, i_k) \in \mathcal{I}(k, n)$, we denote by $u^{*\mathbf{i}}$ the wedge product

$$u^{*\mathbf{i}} = u^{*i_1} \wedge \cdots \wedge u^{*i_k}.$$

Denote also by $\mathbf{i}'$ the increasing sequence of length $n - k$ such that $\{\mathbf{i}, \mathbf{i}'\} = \{1, 2, \ldots, n\}$. We call $\mathbf{i}'$ the complement of $\mathbf{i}$. We define the permutation $\sigma_{\mathbf{i}} \in S_n$ by the permutation that maps the sequence $(1, 2, \ldots, n)$ to the sequence $(\mathbf{i}, \mathbf{i}')$. Note that the sign of the permutation satisfies

$$\text{sign} \, \sigma_{\mathbf{i}} = \varepsilon_{i_1 \cdots i_k i'_1 \cdots i'_{n-k}}.$$

Consider the kth wedge product $u^*_{\mathbf{i}}$. According to Definition C.3.2,

$$\langle \star u^{*\mathbf{i}}, \tau \rangle \omega = u^{*\mathbf{i}} \wedge \tau. \tag{C.5}$$

We know that $\{u^{*\mathbf{j}}\}$ for $\mathbf{j} \in I(n-k, n)$ forms a basis of $\bigwedge^{n-k} V^*$. Thus, we can write

$$\star u^{*\mathbf{i}} = \sum_{\mathbf{j} \in \mathcal{I}(n-k,n)} c_{\mathbf{j}} u^{*\mathbf{j}}$$

for some constants $c_{\mathbf{j}}$. However, Equation (C.5) imposes that $\langle \star u^{*\mathbf{i}}, u^{*\mathbf{j}} \rangle = 0$ unless $\mathbf{j} = \mathbf{i}'$. We denote $K = \langle \star u^{*\mathbf{i}}, u^{*\mathbf{i}'} \rangle$.

By Definition C.3.2, $\langle \star u^{*\mathbf{i}}, u^{*\mathbf{i}'} \rangle \omega = u^{*\mathbf{i}} \wedge u^{*\mathbf{i}'}$ so by Proposition C.2.1,

$$\langle \star u^{*\mathbf{i}}, u^{*\mathbf{i}'} \rangle \sqrt{\det A} \, u^{*1} \wedge \cdots \wedge u^{*n} = (\mathrm{sign}\,\sigma_{\mathbf{i}}) \, u^{*1} \wedge \cdots \wedge u^{*n},$$

which implies that

$$\langle \star u^{*\mathbf{i}}, u^{*\mathbf{j}} \rangle = (\mathrm{sign}\,\sigma_{\mathbf{i}})/\sqrt{\det A} \, \delta_{\mathbf{j},\mathbf{i}'},$$

where $\delta_{\mathbf{j},\mathbf{i}'} = 1$ if $\mathbf{j} = \mathbf{i}'$ and equals 0 otherwise. On the other hand,

$$\langle \star u^{*\mathbf{i}}, u^{*\mathbf{j}} \rangle = \sum_{\mathbf{l} \in \mathcal{I}(n-k,n)} c_{\mathbf{l}} \langle u^{*\mathbf{l}}, u^{*\mathbf{j}} \rangle = \sum_{\mathbf{l} \in \mathcal{I}(n-k,n)} c_{\mathbf{l}} \det((A^{-1})_{\mathbf{l}\mathbf{j}}),$$

where by $A_{\mathbf{l}\mathbf{j}}$ we mean the minor of A consisting of the rows $\mathbf{l} = (l_1, \ldots, l_{n-k})$ and columns $\mathbf{j} = (j_1, \ldots, j_{n-k})$. So, we conclude that

$$\sum_{\mathbf{l} \in \mathcal{I}(n-k,n)} c_{\mathbf{l}} \det((A^{-1})_{\mathbf{l}\mathbf{j}}) = \frac{\mathrm{sign}\,\sigma_{\mathbf{i}}}{\sqrt{\det A}} \delta_{\mathbf{j},\mathbf{i}'}. \tag{C.6}$$

To find the values of $c_{\mathbf{j}}$ for a given $\mathbf{i}$, we need to invert the matrix product in Equation (C.6), or more precisely, find the inverse of the $\binom{n}{k} \times \binom{n}{k}$ matrix $\det((A^{-1})_{\mathbf{l}\mathbf{j}})$. Though a little tedious to show, the following formula generalizes the Laplace expansion formula for determinants. For any $n \times n$ matrix B, with notations as above,

$$\det B = \sum_{\mathbf{j} \in \mathcal{I}(k,n)} (\mathrm{sign}\,\sigma_{\mathbf{i}})(\mathrm{sign}\,\sigma_{\mathbf{j}}) \det B_{\mathbf{i}\mathbf{j}} \det B_{\mathbf{i}'\mathbf{j}'}. \tag{C.7}$$

A slightly stronger result gives

$$\sum_{\mathbf{j} \in \mathcal{I}(k,n)} \det B_{\mathbf{i}\mathbf{j}} (\mathrm{sign}\,\sigma_{\mathbf{h}})(\mathrm{sign}\,\sigma_{\mathbf{j}}) \det B_{\mathbf{h}'\mathbf{j}'} = \begin{cases} \det B & \text{if } \mathbf{h} = \mathbf{i}, \\ 0 & \text{otherwise,} \end{cases} \tag{C.8}$$

for all $\mathbf{h} \in \mathcal{I}(k, n)$. Now multiplying Equation (C.6) by

$$(\mathrm{sign}\,\sigma_{\mathbf{h}})(\mathrm{sign}\,\sigma_{\mathbf{j}}) \det(A^{-1})_{\mathbf{h}'\mathbf{j}'},$$

summing the result over $\mathbf{j} \in \mathcal{I}(n-k, n)$, and taking into account $\delta_{\mathbf{j},\mathbf{i}'}$, we obtain

$$\det(A^{-1}) c_{\mathbf{h}} = \frac{\mathrm{sign}\,\sigma_{\mathbf{i}}}{\sqrt{\det A}} (\mathrm{sign}\,\sigma_{\mathbf{h}})(\mathrm{sign}\,\sigma_{\mathbf{j}'}) \det(A^{-1})_{\mathbf{h}'\mathbf{i}},$$

which implies that
$$c_{\mathbf{h}} = (\operatorname{sign} \sigma_{\mathbf{h}}) \sqrt{\det A} \, \det(A^{-1})_{\mathbf{h'i}}.$$

From this we deduce that
$$\star(u^{*i_1} \wedge \cdots \wedge u^{*i_k}) = \sum_{j_1=1}^{n} \cdots \sum_{j_n=1}^{n} \frac{\sqrt{\det a}}{(n-k)!} \varepsilon_{j_1 \cdots j_n} a^{j_1 i_1} \cdots a^{j_k i_k} (u^{*j_{k+1}} \wedge \cdots \wedge u^{*j_n}),$$
(C.9)

and the proposition follows by linearity of the Hodge star operator. $\square$

We point out that one can loosen the conditions on the bilinear form $\langle \cdot, \cdot \rangle$ and still define the Hodge star operator and obtain many of the same results. If we only assume that $\langle \cdot, \cdot \rangle$ is symmetric and nondegenerate, then all the above propositions hold except that one must replace $\det A$ with $|\det A|$ in Proposition C.3.5.

PROBLEMS

C.3.1. Let $V = \mathbb{R}^4$ equipped with dot product and let (e_1, e_2, e_3, e_4) be the standard basis. Calculate

(a) $\star(2e^{*1} \wedge e^{*2} \wedge e^{*4} + 5e^{*2} \wedge e^{*3} \wedge e^{*4})$;

(b) $\star(17e^{*1} \wedge e^{*2} - 3e^{*1} \wedge e^{*4} + 4e^{*2} \wedge e^{*4})$.

C.3.2. Let $V = \mathbb{R}^3$ equipped with dot product and let (e_1, e_2, e_3) be the standard basis. Let (u_1, u_2, u_3) with coordinates with respect to the standard basis as

$$u_1 = \begin{pmatrix} 2 \\ 1 \\ 1 \end{pmatrix}, \quad u_2 = \begin{pmatrix} 1 \\ 2 \\ 1 \end{pmatrix}, \quad u_3 = \begin{pmatrix} 1 \\ 1 \\ 2 \end{pmatrix}.$$

Use Proposition C.3.5 to calculate:

(a) $\star(4u^{*1} - 7u^{*2} + 5u^{*3})$;

(b) $\star(2u^{*1} \wedge u^{*2} - 3u^{*1} \wedge u^{*3})$.

C.3.3. Prove Proposition C.3.1.

C.3.4. Prove Proposition C.3.3.

C.3.5. Let $(V, \langle \cdot, \cdot \rangle)$ be an inner product space. Prove that the composition $\star \circ \star : \bigwedge^k V^* \to \bigwedge^k V^*$ is tantamount to multiplication on $\bigwedge^k V^*$ by $(-1)^{k(n-k)}$. Suppose that $\langle \cdot, \cdot \rangle$ is a symmetric and nondegenerate bilinear form with signature $(p, q, 0)$. Prove that in this case $\star \circ \star : \bigwedge^k V^* \to \bigwedge^k V^*$ is tantamount to multiplication on V^* by $(-1)^{k(n-k)}(-1)^{p+q}$.

Bibliography

[1] Ralph Abraham and Jerrold E. Marsden. *Foundations of Mechanics*. Benjamin-Gummings, Reading, MA, 1978.

[2] M. A. Armstrong. *Basic Topology*. Undergraduate Texts in Mathematics. Springer-Verlag, New York, 1983.

[3] Vladimir I. Arnold. *Ordinary Differential Equations*. MIT Press, Cambridge, MA, 1973.

[4] Andreas Arvanitoyeorgos. *An Introduction to Lie Groups and the Geometry of Homogeneous Spaces*, volume 22 of *Student Mathematical Library*. American Mathematical Society, Providence, Rhode Island, 1999.

[5] Thomas F. Banchoff and Stephen T. Lovett. *Differential Geometry of Curves and Surfaces*. A. K. Peters, Inc., Wellesley, MA, 2010.

[6] Thomas F. Banchoff and Stephen T. Lovett. *Differential Geometry of Curves and Surfaces*. Taylor & Francis, Boca Raton, FL, 2nd edition, 2016.

[7] Victor Bangert. On the existence of closed geodesics on two-spheres. *International J. of Math.*, 4(1):1–10, 1993.

[8] Rolf Berndt. *An Introduction to Symplectic Geometry*, volume 26 of *Graduate Studies in Mathematics*. American Mathematical Society, Providence, RI, 2001.

[9] Arthur L. Besse. *Einstein Manifolds*. Springer-Verlag, New York, 2007.

[10] George David Birkhoff. Dynamical systems with two degrees of freedom. *Trans. Am. Math. Soc.*, 18(2):199–300, 1917.

[11] Roberto Bonola. *Non-Euclidean Geometry – A Critical and Historical Study of its Developments*. Dover Publications, New York, 1955.

[12] Glen E. Bredon. *Topology and Geometry*, volume 139 of *Graduate Texts in Mathematics*. Springer-Verlag, New York, 1993.

[13] David M. Bressoud. *A Radical Approach to Lebesgue's Theory of Integration*. Cambridge University Press, Cambridge, United Kingdom, 2008.

[14] Otto Bretscher. *Linear Algebra with Applications*. Pearson Prentice Hall, Upper Saddle River, New Jersey, 3rd edition, 2005.

[15] Andrew Browder. *Mathematical Analysis: An Introduction*. Undergraduate Texts in Mathematics. Springer-Verlag, New York, 1996.

[16] Shiing-Shen Chern. *Global Differential Geometry*, volume 27 of *MAA Studies*. Mathematical Association of America, Washington, D.C., 1989.

[17] Manfredo do Carmo. *Riemannian Geometry*. Birkhäuser, Boston, MA, 1992.

[18] C. Henry Edwards and David E. Penny. *Differential Equations: Computing and Modeling*. Pearson Prentice-Hall, Upper Saddle River, New Jersey, 3rd edition, 2004.

[19] Albert Einstein. Die Feldgleichungen der Gravitation. *Sitzungsberichte der Preussischen Akademie der Wissenschaften zu Berlin*, pages 844–847, 1915.

[20] Albert Einstein. *Relativity: The Special and the General Theory*. Crown Publishers, Inc., New York, 1961.

[21] Rafael Ferraro. *Einstein's Space-Time, An Introduction to Special and General Relativity*. Springer-Verlag, New York, 2007.

[22] Grant R. Fowles. *Analytical Mechanics*. Saunders College Publishing, Philadelphia, PA, 4th edition, 1986.

[23] John Franks. Geodesics on S^2 and periodic points of annulus diffeomorphisms. *Invent. Math.*, 108(2):403–418, 1992.

[24] Anthony P. French. *Special Relativity*. The M.I.T. Introductory Physics Series. W. W. Norton, New York, 1968.

[25] Joseph A. Gallian. *Contemporary Abstract Algebra*. Houghton Mifflin, New York, sixth edition, 2006.

[26] George Gamow. *My World Line: An Informal Autobiography*. Viking Press, New York, New York, 1970.

[27] Michael C. Gemignani. *Elementary Topology*. Dover Publications, New York, 2nd edition, 1972.

[28] John G. Hocking and Gail S. Young. *Topology*. Dover Publications, New York, 1961.

[29] Michel A. Kervaire and John W. Milnor. Groups of homotopy spheres: I. *Annals of Math.*, 77(3):504–537, 1963.

[30] Mehrdad Khosravi and Michael D. Taylor. The wedge product and analytic geometry. *American Mathematical Monthly*, 115(7):623–644, 2008.

[31] Serge Lang. *Algebra*. Addison-Wesley, Reading, MA, 3rd edition, 1993.

[32] John M. Lee. *Riemannian Manifolds: An Introduction to Curvature*, volume 176 of *Graduate Texts in Mathematics*. Springer-Verlag, New York, 1997.

[33] John M. Lee. *Introduction to Smooth Manifolds*, volume 218 of *Graduate Texts in Mathematics*. Springer-Verlag, New York, 2003.

[34] Martin M. Lipshutz. *Differential Geometry*. Schaum's Outline Series. McGraw-Hill, New York, 1969.

[35] Malcolm Longair. *Theoretical Concepts in Physics*. Cambridge University Press, Cambridge, U.K., 2nd edition, 2003.

[36] David Lovelock. The uniqueness of the Einstein field equations in a four-dimensional space. *Journal for Rational Mechanics and Analysis*, 33(1):54–70, 1969.

[37] L. Lusternik and L. Schnirelmann. Sur the problème de trois géodésiques fermées sur les surfaces de genre 0. *C. R. Acad. Sci. Sér. I Math*, 189:269–271, 1929.

[38] Morris Marden. *The geometry of zeros of a polynomial in a complex variable*, volume 3 of *Mathematical Surveys*. American Mathematical Society, New York, 1949.

[39] John McCleary. *Geometry from a Differentiable Viewpoint*. Cambridge University Press, Cambridge, United Kingdom, 1994.

[40] John W. Milnor and James D. Stasheff. *Characteristic Classes*, volume 76 of *Annals of Mathematics Studies*. Princeton University Press, Princeton, NJ, 1994.

[41] Charles W. Misner, Kip S. Thorne, and John Archibald Wheeler. *Gravitation*. W. H. Freeman and Company, San Francisco, CA, 1973.

[42] Thomas Moore. *Six Ideas that Shaped Physics - Unit R*. McGraw-Hill, New York, 2nd edition, 2002.

[43] James Munkres. *Topology*. Prentice Hall, Englewood Cliffs, NJ, 1975.

[44] Mikio Nakahara. *Geometry, Topology and Physics*. Taylor & Francis, Boca Raton, FL, 2nd edition, 2003.

[45] Barrett O'Neill. *Semi-Riemannian Geometry with Applications to Relativity*, volume 103 of *Pure and Applied Mathematics*. Academic Press, New York, 1983.

[46] Edward M. Purcell. *Electricity and Magnetism*, volume 2 of *Berkeley Physics Course*. McGraw-Hill, New York, 1985.

[47] G. F. Bernhard Riemann. On the hypotheses which lie at the foundation of geometry. In William Bragg Ewald, editor, *From Kant to Hilbert: A Source Book in the Foundations of Mathematics*, volume 2, pages 652–651. Oxford University Press, New York, 1996.

[48] Halsey L. Royden. *Real Analysis*. Prentice Hall, Englewood Cliffs, New Jersey, 1988.

[49] Walter Rudin. *Principles of Mathematical Analysis*. McGraw-Hill, New York, 1976.

[50] Bernard Schutz. *A First Course in General Relativity*. Cambridge University Press, New York, 2009.

[51] Melvin Schwartz. *Principles of Electromagnetism*. Dover Publications, New York, 1987.

[52] Michael Spivak. *A Comprehensive Guide to Differential Geometry, Volume One*. Publish or Perish Inc., Berkeley, CA, 1979.

[53] Norman Steenrod. *The Topology of Fibre Bundles*. Princeton University Press, Princeton, NJ, 1951.

[54] Hans Stephani. *General Relativity*. Cambridge University Press, Cambridge, United Kingdom, 1982.

[55] James Stewart. *Calculus*. Thomson Brooks / Cole, Belmont, CA, sixth edition, 2003.

[56] Keith R. Symon. *Mechanics*. Addison-Wesley Publishing, Reading, MA, 1971.

[57] E. R. van Kampen. On the argument functions of simple closed curves and simple arcs. *Compositio Mathematica*, 4:271–275, 1937.

[58] Robert Weinstock. *Calculus of Variations*. McGraw-Hill, New York, 1952.

[59] Hermann Weyl. *The Concept of a Riemannian Surface*. Addison-Wesley, Reading, MA, 3rd edition, 1955. translated from the German by Gerald R. Maclane.

[60] Barton Zwiebach. *A First Course in String Theory*. Cambridge University Press, Cambridge, U.K., 2005.

Index